Unit Conversions (Equivalents)

Length

1 in. = 2.54 cm (defined)
1 cm = 0.3937 in.
1 ft = 30.48 cm
1 m = 39.37 in. = 3.281 ft
1 mi = 5280 ft = 1.609 km
1 km = 0.6214 mi
1 nautical mile (U.S.) = 1.151 mi = 6076 ft = 1.852 km
1 fermi = 1 femtometer (fm) = 10^{-15} m
1 angstrom (Å) = 10^{-10} m = 0.1 nm
1 light-year (ly) = 9.461×10^{15} m
1 parsec = 3.26 ly = 3.09×10^{16} m

Volume

1 liter (L) = 1000 mL = 1000 cm^3 = 1.0×10^{-3} m^3 = 1.057 qt (U.S.) = 61.02 $in.^3$
1 gal (U.S.) = 4 qt (U.S.) = 231 $in.^3$ = 3.785 L = 0.8327 gal (British)
1 quart (U.S.) = 2 pints (U.S.) = 946 mL
1 pint (British) = 1.20 pints (U.S.) = 568 mL
1 m^3 = 35.31 ft^3

Speed

1 mi/h = 1.4667 ft/s = 1.6093 km/h = 0.4470 m/s
1 km/h = 0.2778 m/s = 0.6214 mi/h
1 ft/s = 0.3048 m/s (exact) = 0.6818 mi/h = 1.0973 km/h
1 m/s = 3.281 ft/s = 3.600 km/h = 2.237 mi/h
1 knot = 1.151 mi/h = 0.5144 m/s

Angle

1 radian (rad) = 57.30° = 57°18′
1° = 0.01745 rad
1 rev/min (rpm) = 0.1047 rad/s

Time

1 day = 8.640×10^4 s
1 year = 3.15581×10^7 s

Mass

1 atomic mass unit (u) = 1.6605×10^{-27} kg
1 kg = 0.06852 slug
[1 kg has a weight of 2.20 lb where $g = 9.80\ m/s^2$.]

Force

1 lb = 4.44822 N
1 N = 10^5 dyne = 0.2248 lb

Energy and Work

1 J = 10^7 ergs = 0.7376 ft·lb
1 ft·lb = 1.356 J = 1.29×10^{-3} Btu = 3.24×10^{-4} kcal
1 kcal = 4.19×10^3 J = 3.97 Btu
1 eV = 1.6022×10^{-19} J
1 kWh = 3.600×10^6 J = 860 kcal
1 Btu = 1.056×10^3 J

Power

1 W = 1 J/s = 0.7376 ft·lb/s = 3.41 Btu/h
1 hp = 550 ft·lb/s = 746 W

Pressure

1 atm = 1.01325 bar = $1.01325 \times 10^5\ N/m^2$ = 14.7 $lb/in.^2$ = 760 torr
1 $lb/in.^2$ = $6.895 \times 10^3\ N/m^2$
1 Pa = 1 N/m^2 = $1.450 \times 10^{-4}\ lb/in.^2$

SI Derived Units and Their Abbreviations

Quantity	Unit	Abbreviation	In Terms of Base Units†
Force	newton	N	$kg \cdot m/s^2$
Energy and work	joule	J	$kg \cdot m^2/s^2$
Power	watt	W	$kg \cdot m^2/s^3$
Pressure	pascal	Pa	$kg/(m \cdot s^2)$
Frequency	hertz	Hz	s^{-1}
Electric charge	coulomb	C	$A \cdot s$
Electric potential	volt	V	$kg \cdot m^2/(A \cdot s^3)$
Electric resistance	ohm	Ω	$kg \cdot m^2/(A^2 \cdot s^3)$
Capacitance	farad	F	$A^2 \cdot s^4/(kg \cdot m^2)$
Magnetic field	tesla	T	$kg/(A \cdot s^2)$
Magnetic flux	weber	Wb	$kg \cdot m^2/(A \cdot s^2)$
Inductance	henry	H	$kg \cdot m^2/(s^2 \cdot A^2)$

† kg = kilogram (mass), m = meter (length), s = second (time), A = ampere (electric current).

Metric (SI) Multipliers

Prefix	Abbreviation	Value
yotta	Y	10^{24}
zeta	Z	10^{21}
exa	E	10^{18}
peta	P	10^{15}
tera	T	10^{12}
giga	G	10^{9}
mega	M	10^{6}
kilo	k	10^{3}
hecto	h	10^{2}
deka	da	10^{1}
deci	d	10^{-1}
centi	c	10^{-2}
milli	m	10^{-3}
micro	μ	10^{-6}
nano	n	10^{-9}
pico	p	10^{-12}
femto	f	10^{-15}
atto	a	10^{-18}
zepto	z	10^{-21}
yocto	y	10^{-24}

FOURTH EDITION

Volume III

PHYSICS
for
SCIENTISTS & ENGINEERS
with Modern Physics

DOUGLAS C. GIANCOLI

Upper Saddle River, New Jersey 07458

Library of Congress Cataloging-in-Publication Data

Giancoli, Douglas C.
Physics for scientists and engineers with modern physics / Douglas C. Giancoli.—4th ed.
p. cm.
Includes bibliographical references and index.
ISBN 0-13-227400-0
1. Physics—Textbooks. I. Title.
QC21.3.G539 2008
530—dc22

2006039431

President, Science: Paul Corey
Sponsoring Editor: Christian Botting
Executive Development Editor: Karen Karlin
Production Editor: Clare Romeo
Senior Managing Editor: Scott Disanno
Art Director and Interior & Cover Designer: John Christiana
Manager, Art Production: Sean Hogan
Copy Editor: Jocelyn Phillips
Proofreaders: Karen Bosch, Gina Cheselka, Traci Douglas, Nancy Stevenson, and Susan Fisher
Senior Operations Specialist: Alan Fischer
Art Production Editor: Connie Long
Illustrators: Audrey Simonetti and Mark Landis
Photo Researchers: Mary Teresa Giancoli and Truitt & Marshall
Senior Administrative Coordinator: Trisha Tarricone
Composition: Emilcomp/Prepare Inc.;
Pearson Education/Lissette Quiñones, Clara Bartunek
Photo credits appear on page A-48 which constitutes a continuation of the copyright page.

PEARSON Prentice Hall
Published by Pearson Education, Inc.
Pearson Prentice Hall
Pearson Education, Inc.
Upper Saddle River, NJ 07458

Printed in the United States of America
3 2022

ISBN-13: 978-0-13-227400-5
ISBN-10: 0-13-227400-0

Pearson Education LTD., *London*
Pearson Education Australia PTY, Limited, *Sydney*
Pearson Education Singapore, Pte. Ltd.
Pearson Education North Asia Ltd., *Hong Kong*
Pearson Education Canada, Ltd., *Toronto*
Pearson Educación de Mexico, S.A. de C.V.
Pearson Education—Japan, *Tokyo*
Pearson Education Malaysia, Pte. Ltd.

Contents

Volume 1

Force
Displacement

12 Static Equilibrium; Elasticity and Fracture 311

13 Fluids 339

14 Oscillations 369

15 Wave Motion 395

16 Sound 424

Volume 2

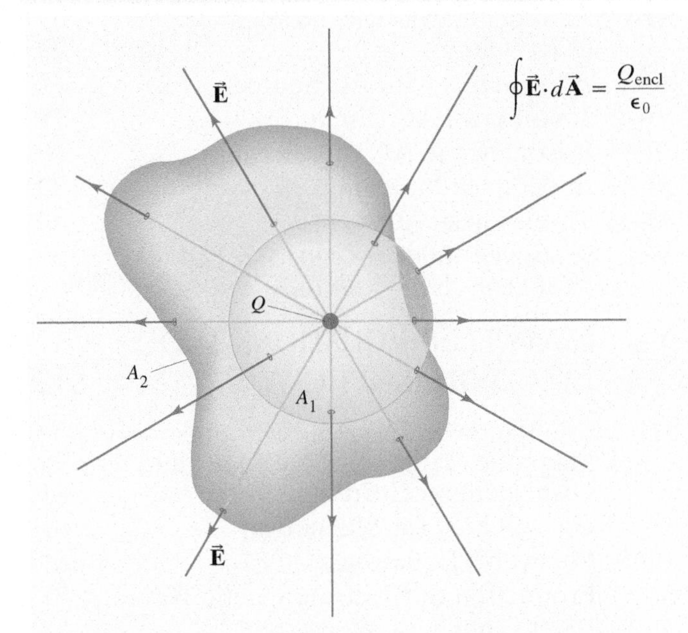

27 Magnetism 707

28 Sources of Magnetic Field 733

29 Electromagnetic Induction and Faraday's Law 758

30 Inductance, Electromagnetic Oscillations, and AC Circuits 785

31 Maxwell's Equations and Electromagnetic Waves 812

32 Light: Reflection and Refraction 837

33 Lenses and Optical Instruments 866

34 The Wave Nature of Light; Interference 900

35 Diffraction and Polarization 921

Volume 3

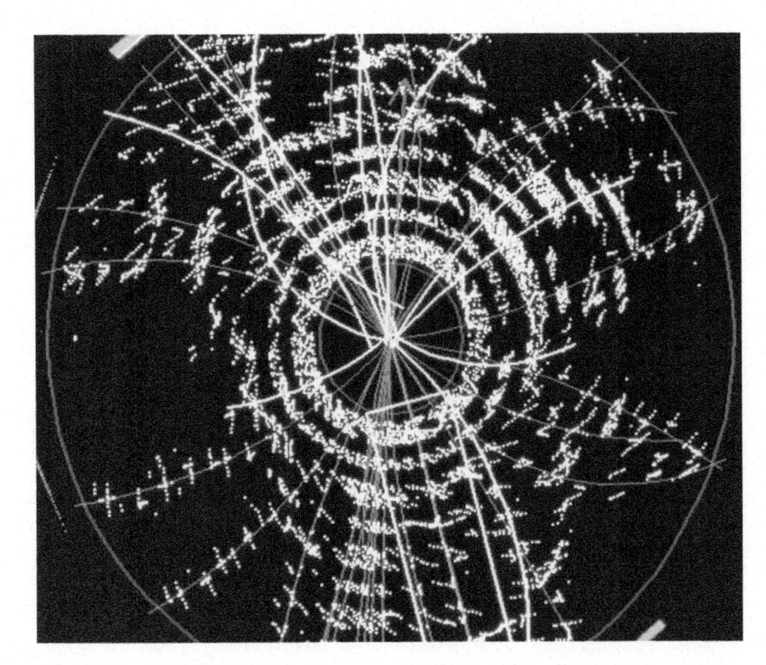

APPLICATIONS (SELECTED)

Preface

I was motivated from the beginning to write a textbook different from others that present physics as a sequence of facts, like a Sears catalog: "here are the facts and you better learn them." Instead of that approach in which topics are begun formally and dogmatically, I have sought to begin each topic with concrete observations and experiences students can relate to: start with specifics and only then go to the great generalizations and the more formal aspects of a topic, showing *why* we believe what we believe. This approach reflects how science is actually practiced.

Why a Fourth Edition?

Two recent trends in physics texbooks are disturbing: (1) their revision cycles have become short—they are being revised every 3 or 4 years; (2) the books are getting larger, some over 1500 pages. I don't see how either trend can be of benefit to students. My response: (1) It has been 8 years since the previous edition of this book. (2) This book makes use of physics education research, although it avoids the detail a Professor may need to say in class but in a book shuts down the reader. And this book still remains among the shortest.

This new edition introduces some important new pedagogic tools. It contains new physics (such as in cosmology) and many new appealing applications (list on previous page). Pages and page breaks have been carefully formatted to make the physics easier to follow: no turning a page in the middle of a derivation or Example. Great efforts were made to make the book attractive so students will want to *read* it.

Some of the new features are listed below.

What's New

Chapter-Opening Questions: Each Chapter begins with a multiple-choice question, whose responses include common misconceptions. Students are asked to answer before starting the Chapter, to get them involved in the material and to get any preconceived notions out on the table. The issues reappear later in the Chapter, usually as Exercises, after the material has been covered. The Chapter-Opening Questions also show students the power and usefulness of Physics.

APPROACH paragraph in worked-out numerical Examples: A short introductory paragraph before the Solution, outlining an approach and the steps we can take to get started. Brief NOTES after the Solution may remark on the Solution, may give an alternate approach, or mention an application.

Step-by-Step Examples: After many Problem Solving Strategies (more than 20 in the book), the next Example is done step-by-step following precisely the steps just seen.

Exercises within the text, after an Example or derivation, give students a chance to see if they have understood enough to answer a simple question or do a simple calculation. Many are multiple choice.

Greater clarity: No topic, no paragraph in this book was overlooked in the search to improve the clarity and conciseness of the presentation. Phrases and sentences that may slow down the principal argument have been eliminated: keep to the essentials at first, give the elaborations later.

$\vec{F}$, $\vec{v}$, $\vec{B}$

Vector notation, arrows: The symbols for vector quantities in the text and Figures now have a tiny arrow over them, so they are similar to what we write by hand.

Cosmological Revolution: With generous help from top experts in the field, readers have the latest results.

Page layout: more than in the previous edition, serious attention has been paid to how each page is formatted. Examples and all important derivations and arguments are on facing pages. Students then don't have to turn back and forth. Throughout, readers see, on two facing pages, an important slice of physics.

New Applications: LCDs, digital cameras and electronic sensors (CCD, CMOS), electric hazards, GFCIs, photocopiers, inkjet and laser printers, metal detectors, underwater vision, curve balls, airplane wings, DNA, how we actually *see* images. (Turn back a page to see a longer list.)

Examples modified: more math steps are spelled out, and many new Examples added. About 10% of all Examples are Estimation Examples.

This Book is Shorter than other complete full-service books at this level. Shorter explanations are easier to understand and more likely to be read.

Content and Organizational Changes

- **Rotational Motion**: Chapters 10 and 11 have been reorganized. All of angular momentum is now in Chapter 11.
- **First law of thermodynamics**, in Chapter 19, has been rewritten and extended. The full form is given: $\Delta K + \Delta U + \Delta E_{\text{int}} = Q - W$, where internal energy is E_{int}, and U is potential energy; the form $Q - W$ is kept so that $dW = P\,dV$.
- Kinematics and Dynamics of Circular Motion are now treated together in Chapter 5.
- Work and Energy, Chapters 7 and 8, have been carefully revised.
- Work done by friction is discussed now with energy conservation (energy terms due to friction).
- Chapters on Inductance and AC Circuits have been combined into one: Chapter 30.
- Graphical Analysis and Numerical Integration is a new optional Section 2–9. Problems requiring a computer or graphing calculator are found at the end of most Chapters.
- Length of an object is a script ℓ rather than normal l, which looks like 1 or I (moment of inertia, current), as in $F = I\ell B$. Capital L is for angular momentum, latent heat, inductance, dimensions of length $[L]$.
- Newton's law of gravitation remains in Chapter 6. Why? Because the $1/r^2$ law is too important to relegate to a late chapter that might not be covered at all late in the semester; furthermore, it is one of the basic forces in nature. In Chapter 8 we can treat real gravitational potential energy and have a fine instance of using $U = -\int \vec{\mathbf{F}} \cdot d\vec{\boldsymbol{\ell}}$.
- New Appendices include the differential form of Maxwell's equations and more on dimensional analysis.
- Problem Solving Strategies are found on pages 30, 58, 64, 96, 102, 125, 166, 198, 229, 261, 314, 504, 551, 571, 600, 685, 716, 740, 763, 849, 871, and 913.

Organization

Some instructors may find that this book contains more material than can be covered in their courses. The text offers great flexibility. Sections marked with a star * are considered optional. These contain slightly more advanced physics material, or material not usually covered in typical courses and/or interesting applications; they contain no material needed in later Chapters (except perhaps in later optional Sections). For a brief course, all optional material could be dropped as well as major parts of Chapters 1, 13, 16, 26, 30, and 35, and selected parts of Chapters 9, 12, 19, 20, 33, and the modern physics Chapters. Topics not covered in class can be a valuable resource for later study by students. Indeed, this text can serve as a useful reference for years because of its wide range of coverage.

Versions of this Book

Complete version: 44 Chapters including 9 Chapters of modern physics.

Classic version: 37 Chapters including one each on relativity and quantum theory.

3 Volume version: Available separately or packaged together (Vols. 1 & 2 or all 3 Volumes):

Volume 1: Chapters 1–20 on mechanics, including fluids, oscillations, waves, plus heat and thermodynamics.

Volume 2: Chapters 21–35 on electricity and magnetism, plus light and optics.

Volume 3: Chapters 36–44 on modern physics: relativity, quantum theory, atomic physics, condensed matter, nuclear physics, elementary particles, cosmology and astrophysics.

Thanks

Many physics professors provided input or direct feedback on every aspect of this textbook. They are listed below, and I owe each a debt of gratitude.

Mario Affatigato, Coe College
Lorraine Allen, United States Coast Guard Academy
Zaven Altounian, McGill University
Bruce Barnett, Johns Hopkins University
Michael Barnett, Lawrence Berkeley Lab
Anand Batra, Howard University
Cornelius Bennhold, George Washington University
Bruce Birkett, University of California Berkeley
Dr. Robert Boivin, Auburn University
Subir Bose, University of Central Florida
David Branning, Trinity College
Meade Brooks, Collin County Community College
Bruce Bunker, University of Notre Dame
Grant Bunker, Illinois Institute of Technology
Wayne Carr, Stevens Institute of Technology
Charles Chiu, University of Texas Austin
Robert Coakley, University of Southern Maine
David Curott, University of North Alabama
Biman Das, SUNY Potsdam
Bob Davis, Taylor University
Kaushik De, University of Texas Arlington
Michael Dennin, University of California Irvine
Kathy Dimiduk, University of New Mexico
John DiNardo, Drexel University
Scott Dudley, United States Air Force Academy
John Essick, Reed College
Cassandra Fesen, Dartmouth College
Alex Filippenko, University of California Berkeley
Richard Firestone, Lawrence Berkeley Lab
Mike Fortner, Northern Illinois University
Tom Furtak, Colorado School of Mines
Edward Gibson, California State University Sacramento
John Hardy, Texas A&M
J. Erik Hendrickson, University of Wisconsin Eau Claire
Laurent Hodges, Iowa State University
David Hogg, New York University
Mark Hollabaugh, Normandale Community College
Andy Hollerman, University of Louisiana at Lafayette
William Holzapfel, University of California Berkeley
Bob Jacobsen, University of California Berkeley
Teruki Kamon, Texas A&M
Daryao Khatri, University of the District of Columbia
Jay Kunze, Idaho State University
Jim LaBelle, Dartmouth College
M.A.K. Lodhi, Texas Tech
Bruce Mason, University of Oklahoma
Dan Mazilu, Virginia Tech
Linda McDonald, North Park College
Bill McNairy, Duke University
Raj Mohanty, Boston University
Giuseppe Molesini, Istituto Nazionale di Ottica Florence
Lisa K. Morris, Washington State University
Blaine Norum, University of Virginia
Alexandria Oakes, Eastern Michigan University
Michael Ottinger, Missouri Western State University
Lyman Page, Princeton and WMAP
Bruce Partridge, Haverford College
R. Daryl Pedigo, University of Washington
Robert Pelcovitz, Brown University
Vahe Peroomian, UCLA
James Rabchuk, Western Illinois University
Michele Rallis, Ohio State University
Paul Richards, University of California Berkeley
Peter Riley, University of Texas Austin
Larry Rowan, University of North Carolina Chapel Hill
Cindy Schwarz, Vassar College
Peter Sheldon, Randolph-Macon Woman's College
Natalia A. Sidorovskaia, University of Louisiana at Lafayette
James Siegrist, UC Berkeley, Director Physics Division LBNL
George Smoot, University of California Berkeley
Mark Sprague, East Carolina University
Michael Strauss, University of Oklahoma
Laszlo Takac, University of Maryland Baltimore Co.
Franklin D. Trumpy, Des Moines Area Community College
Ray Turner, Clemson University
Som Tyagi, Drexel University
John Vasut, Baylor University
Robert Webb, Texas A&M
Robert Weidman, Michigan Technological University
Edward A. Whittaker, Stevens Institute of Technology
John Wolbeck, Orange County Community College
Stanley George Wojcicki, Stanford University
Edward Wright, UCLA
Todd Young, Wayne State College
William Younger, College of the Albemarle
Hsiao-Ling Zhou, Georgia State University

I owe special thanks to Prof. Bob Davis for much valuable input, and especially for working out all the Problems and producing the Solutions Manual for all Problems, as well as for providing the answers to odd-numbered Problems at the end of this book. Many thanks also to J. Erik Hendrickson who collaborated with Bob Davis on the solutions, and to the team they managed (Profs. Anand Batra, Meade Brooks, David Currott, Blaine Norum, Michael Ottinger, Larry Rowan, Ray Turner, John Vasut, William Younger). I am grateful to Profs. John Essick, Bruce Barnett, Robert Coakley, Biman Das, Michael Dennin, Kathy Dimiduk, John DiNardo, Scott Dudley, David Hogg, Cindy Schwarz, Ray Turner, and Som Tyagi, who inspired many of the Examples, Questions, Problems, and significant clarifications.

Crucial for rooting out errors, as well as providing excellent suggestions, were Profs. Kathy Dimiduk, Ray Turner, and Lorraine Allen. A huge thank you to them and to Prof. Giuseppe Molesini for his suggestions and his exceptional photographs for optics.

For Chapters 43 and 44 on Particle Physics and Cosmology and Astrophysics, I was fortunate to receive generous input from some of the top experts in the field, to whom I owe a debt of gratitude: George Smoot, Paul Richards, Alex Filippenko, James Siegrist, and William Holzapfel (UC Berkeley), Lyman Page (Princeton and WMAP), Edward Wright (UCLA and WMAP), and Michael Strauss (University of Oklahoma).

I especially wish to thank Profs. Howard Shugart, Chair Frances Hellman, and many others at the University of California, Berkeley, Physics Department for helpful discussions, and for hospitality. Thanks also to Prof. Tito Arecchi and others at the Istituto Nazionale di Ottica, Florence, Italy.

Finally, I am grateful to the many people at Prentice Hall with whom I worked on this project, especially Paul Corey, Karen Karlin, Christian Botting, John Christiana, and Sean Hogan.

The final responsibility for all errors lies with me. I welcome comments, corrections, and suggestions as soon as possible to benefit students for the next reprint.

D.C.G.

email: Paul.Corey@Pearson.com

Post: Paul Corey
One Lake Street
Upper Saddle River, NJ 07458

About the Author

Douglas C. Giancoli obtained his BA in physics (summa cum laude) from the University of California, Berkeley, his MS in physics at the Massachusetts Institute of Technology, and his PhD in elementary particle physics at the University of California, Berkeley. He spent 2 years as a post-doctoral fellow at UC Berkeley's Virus lab developing skills in molecular biology and biophysics. His mentors include Nobel winners Emilio Segrè and Donald Glaser.

He has taught a wide range of undergraduate courses, traditional as well as innovative ones, and continues to update his textbooks meticulously, seeking ways to better provide an understanding of physics for students.

Doug's favorite spare-time activity is the outdoors, especially climbing peaks (here descending after a winter 2008 Sierra Nevada climb, California). He says climbing peaks is like learning physics: it takes effort and the rewards are great.

Online Supplements (partial list)

MasteringPhysics™ (www.masteringphysics.com)
is a sophisticated online tutoring and homework system developed specially for courses using calculus-based physics. Originally developed by David Pritchard and collaborators at MIT, MasteringPhysics provides **students** with individualized online tutoring by responding to their wrong answers and providing hints for solving multi-step problems when they get stuck. It gives them immediate and up-to-date assessment of their progress, and shows where they need to practice more. MasteringPhysics provides **instructors** with a fast and effective way to assign tried-and-tested online homework assignments that comprise a range of problem types. The powerful post-assignment diagnostics allow instructors to assess the progress of their class as a whole as well as individual students, and quickly identify areas of difficulty.

WebAssign (www.webassign.com)

CAPA and LON-CAPA (www.lon-capa.org)

Student Supplements (partial list)

Student Study Guide & Selected Solutions Manual (Volume I: 0-13-227324-1, Volumes II & III: 0-13-227325-X) by Frank Wolfs

Student Pocket Companion (0-13-227326-8) by Biman Das

Tutorials in Introductory Physics (0-13-097069-7)
by Lillian C. McDermott, Peter S. Schaffer, and the Physics Education Group at the University of Washington

Physlet® Physics (0-13-101969-4)
by Wolfgang Christian and Mario Belloni

Ranking Task Exercises in Physics, Student Edition (0-13-144851-X)
by Thomas L. O'Kuma, David P. Maloney, and Curtis J. Hieggelke

E&M TIPERs: Electricity & Magnetism Tasks Inspired by Physics Education Research (0-13-185499-2) by Curtis J. Hieggelke, David P. Maloney, Stephen E. Kanim, and Thomas L. O'Kuma

Mathematics for Physics with Calculus (0-13-191336-0)
by Biman Das

To Students

HOW TO STUDY

1. Read the Chapter. Learn new vocabulary and notation. Try to respond to questions and exercises as they occur.
2. Attend all class meetings. Listen. Take notes, especially about aspects you do not remember seeing in the book. Ask questions (everyone else wants to, but maybe you will have the courage). You will get more out of class if you read the Chapter first.
3. Read the Chapter again, paying attention to details. Follow derivations and worked-out Examples. Absorb their logic. Answer Exercises and as many of the end of Chapter Questions as you can.
4. Solve 10 to 20 end of Chapter Problems (or more), especially those assigned. In doing Problems you find out what you learned and what you didn't. Discuss them with other students. Problem solving is one of the great learning tools. Don't just look for a formula—it won't cut it.

NOTES ON THE FORMAT AND PROBLEM SOLVING

1. Sections marked with a star (*) are considered **optional**. They can be omitted without interrupting the main flow of topics. No later material depends on them except possibly later starred Sections. They may be fun to read, though.
2. The customary **conventions** are used: symbols for quantities (such as m for mass) are italicized, whereas units (such as m for meter) are not italicized. Symbols for vectors are shown in boldface with a small arrow above: $\vec{\mathbf{F}}$.
3. Few equations are valid in all situations. Where practical, the **limitations** of important equations are stated in square brackets next to the equation. The equations that represent the great laws of physics are displayed with a tan background, as are a few other indispensable equations.
4. At the end of each Chapter is a set of **Problems** which are ranked as Level I, II, or III, according to estimated difficulty. Level I Problems are easiest, Level II are standard Problems, and Level III are "challenge problems." These ranked Problems are arranged by Section, but Problems for a given Section may depend on earlier material too. There follows a group of General Problems, which are not arranged by Section nor ranked as to difficulty. Problems that relate to optional Sections are starred (*). Most Chapters have 1 or 2 Computer/Numerical Problems at the end, requiring a computer or graphing calculator. Answers to odd-numbered Problems are given at the end of the book.
5. Being able to solve **Problems** is a crucial part of learning physics, and provides a powerful means for understanding the concepts and principles. This book contains many aids to problem solving: (a) worked-out **Examples** and their solutions in the text, which should be studied as an integral part of the text; (b) some of the worked-out Examples are **Estimation Examples**, which show how rough or approximate results can be obtained even if the given data are sparse (see Section 1–6); (c) special **Problem Solving Strategies** placed throughout the text to suggest a step-by-step approach to problem solving for a particular topic—but remember that the basics remain the same; most of these "Strategies" are followed by an Example that is solved by explicitly following the suggested steps; (d) special problem-solving Sections; (e) "Problem Solving" marginal notes which refer to hints within the text for solving Problems; (f) **Exercises** within the text that you should work out immediately, and then check your response against the answer given at the bottom of the last page of that Chapter; (g) the Problems themselves at the end of each Chapter (point 4 above).
6. **Conceptual Examples** pose a question which hopefully starts you to think and come up with a response. Give yourself a little time to come up with your own response before reading the Response given.
7. **Math** review, plus some additional topics, are found in Appendices. Useful data, conversion factors, and math formulas are found inside the front and back covers.

USE OF COLOR

Vectors

A general vector
- resultant vector (sum) is slightly thicker
- components of any vector are dashed

Displacement ($\vec{\mathbf{D}}, \vec{\mathbf{r}}$)
Velocity ($\vec{\mathbf{v}}$)
Acceleration ($\vec{\mathbf{a}}$)
Force ($\vec{\mathbf{F}}$)
- Force on second or
- third object in same figure

Momentum ($\vec{\mathbf{p}}$ or $m\vec{\mathbf{v}}$)
Angular momentum ($\vec{\mathbf{L}}$)
Angular velocity ($\vec{\boldsymbol{\omega}}$)
Torque ($\vec{\boldsymbol{\tau}}$)
Electric field ($\vec{\mathbf{E}}$)
Magnetic field ($\vec{\mathbf{B}}$)

Electricity and magnetism

Electric field lines
Equipotential lines
Magnetic field lines
Electric charge (+) — + or • +
Electric charge (–) — – or • –

Electric circuit symbols

Wire, with switch S
Resistor
Capacitor
Inductor
Battery
Ground

Optics

Light rays
Object
Real image (dashed)
Virtual image (dashed and paler)

Other

Energy level (atom, etc.)
Measurement lines — |← 1.0 m →|
Path of a moving object
Direction of motion or current

An early science fantasy book (1940), called *Mr Tompkins in Wonderland* by physicist George Gamow, imagined a world in which the speed of light was only 10 m/s (20 mi/h). Mr Tompkins had studied relativity and when he began "speeding" on a bicycle, he "expected that he would be immediately shortened, and was very happy about it as his increasing figure had lately caused him some anxiety. To his great surprise, however, nothing happened to him or to his cycle. On the other hand, the picture around him completely changed. The streets grew shorter, the windows of the shops began to look like narrow slits, and the policeman on the corner became the thinnest man he had ever seen. 'By Jove!' exclaimed Mr Tompkins excitedly, 'I see the trick now. This is where the word *relativity* comes in.' "

Relativity does indeed predict that objects moving relative to us at high speed, close to the speed of light c, are shortened in length. We don't notice it as Mr Tompkins did, because $c = 3 \times 10^8$ m/s is incredibly fast. We will study length contraction, time dilation, simultaneity non-agreement, and how energy and mass are equivalent ($E = mc^2$).

The Special Theory of Relativity

CHAPTER-OPENING QUESTION—Guess now!

A rocket is headed away from Earth at a speed of $0.80c$. The rocket fires a missile at a speed of $0.70c$ (the missile is aimed away from Earth and leaves the rocket at $0.70c$ relative to the rocket). How fast is the missile moving relative to Earth?

(a) $1.50c$;
(b) a little less than $1.50c$;
(c) a little over c;
(d) a little under c;
(e) $0.75c$.

Physics at the end of the nineteenth century looked back on a period of great progress. The theories developed over the preceding three centuries had been very successful in explaining a wide range of natural phenomena. Newtonian mechanics beautifully explained the motion of objects on Earth and in the heavens. Furthermore, it formed the basis for successful treatments of fluids, wave motion, and sound. Kinetic theory explained the behavior of gases and other materials. Maxwell's theory of electromagnetism not only brought together and explained electric and magnetic phenomena, but it predicted the existence of electromagnetic waves that would behave in every way just like light—so light came to be thought of as an electromagnetic wave. Indeed, it seemed that the natural world, as seen through the eyes of physicists, was very well explained. A few puzzles remained, but it was felt that these would soon be explained using already known principles.

CONTENTS

FIGURE 36–1 Albert Einstein (1879–1955), one of the great minds of the twentieth century, was the creator of the special and general theories of relativity.

It did not turn out so simply. Instead, these puzzles were to be solved only by the introduction, in the early part of the twentieth century, of two revolutionary new theories that changed our whole conception of nature: the *theory of relativity* and *quantum theory*.

Physics as it was known at the end of the nineteenth century (what we've covered up to now in this book) is referred to as **classical physics**. The new physics that grew out of the great revolution at the turn of the twentieth century is now called **modern physics**. In this Chapter, we present the special theory of relativity, which was first proposed by Albert Einstein (1879–1955; Fig. 36–1) in 1905. In Chapter 37, we introduce the equally momentous quantum theory.

36–1 Galilean–Newtonian Relativity

Einstein's special theory of relativity deals with how we observe events, particularly how objects and events are observed from different frames of reference. This subject had, of course, already been explored by Galileo and Newton.

The special theory of relativity deals with events that are observed and measured from so-called **inertial reference frames** (Sections 4–2 and 11–8), which are reference frames in which Newton's first law is valid: if an object experiences no net force, the object either remains at rest or continues in motion with constant speed in a straight line. It is usually easiest to analyze events when they are observed and measured by observers at rest in an inertial frame. The Earth, though not quite an inertial frame (it rotates), is close enough that for most purposes we can consider it an inertial frame. Rotating or otherwise accelerating frames of reference are noninertial frames,† and won't concern us in this Chapter (they are dealt with in Einstein's general theory of relativity).

A reference frame that moves with constant velocity with respect to an inertial frame is itself also an inertial frame, since Newton's laws hold in it as well. When we say that we observe or make measurements from a certain reference frame, it means that we are at rest in that reference frame.

Both Galileo and Newton were aware of what we now call the **relativity principle** applied to mechanics: that *the basic laws of physics are the same in all inertial reference frames*. You may have recognized its validity in everyday life. For example, objects move in the same way in a smoothly moving (constant-velocity) train or airplane as they do on Earth. (This assumes no vibrations or rocking which would make the reference frame noninertial.) When you walk, drink a cup of soup, play pool, or drop a pencil on the floor while traveling in a train, airplane, or ship moving at constant velocity, the objects move just as they do when you are at rest on Earth. Suppose you are in a car traveling rapidly at constant velocity. If you drop a coin from above your head inside the car, how will it fall? It falls straight downward with respect to the car, and hits the floor directly below the point of release, Fig. 36–2a.

†On a rotating platform (say a merry-go-round), for example, an object at rest starts moving outward even though no object exerts a force on it. This is therefore not an inertial frame. See Section 11–8.

FIGURE 36–2 A coin is dropped by a person in a moving car. The upper views show the moment of the coin's release, the lower views are a short time later. (a) In the reference frame of the car, the coin falls straight down (and the tree moves to the left). (b) In a reference frame fixed on the Earth, the coin has an initial velocity (= to car's) and follows a curved (parabolic) path.

(a)
Reference frame = car

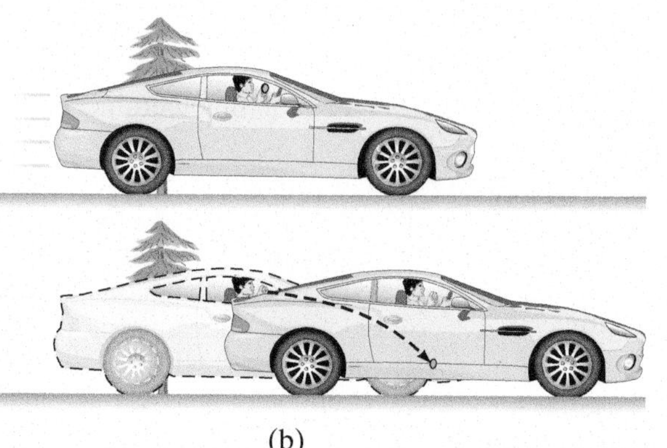

(b)
Reference frame = Earth

This is just how objects fall on the Earth—straight down—and thus our experiment in the moving car is in accord with the relativity principle. (If you drop the coin out the car's window, this won't happen because the moving air drags the coin backward relative to the car.)

Note in this example, however, that to an observer on the Earth, the coin follows a curved path, Fig. 36–2b. The actual path followed by the coin is different as viewed from different frames of reference. This does not violate the relativity principle because this principle states that the *laws* of physics are the same in all inertial frames. The same law of gravity, and the same laws of motion, apply in both reference frames. The acceleration of the coin is the same in both reference frames. The difference in Figs. 36–2a and b is that in the Earth's frame of reference, the coin has an initial velocity (equal to that of the car). The laws of physics therefore predict it will follow a parabolic path like any projectile (Chapter 3). In the car's reference frame, there is no initial velocity, and the laws of physics predict that the coin will fall straight down. The laws are the same in both reference frames, although the specific paths are different.

CAUTION
Laws are the same, but paths may be different in different reference frames

Galilean–Newtonian relativity involves certain unprovable assumptions that make sense from everyday experience. It is assumed that the lengths of objects are the same in one reference frame as in another, and that time passes at the same rate in different reference frames. In classical mechanics, then, space and time intervals are considered to be **absolute**: their measurement does not change from one reference frame to another. The mass of an object, as well as all forces, are assumed to be unchanged by a change in inertial reference frame.

The position of an object, however, is different when specified in different reference frames, and so is velocity. For example, a person may walk inside a bus toward the front with a speed of $2\,\mathrm{m/s}$. But if the bus moves $10\,\mathrm{m/s}$ with respect to the Earth, the person is then moving with a speed of $12\,\mathrm{m/s}$ with respect to the Earth. The acceleration of an object, however, is the same in any inertial reference frame according to classical mechanics. This is because the change in velocity, and the time interval, will be the same. For example, the person in the bus may accelerate from 0 to $2\,\mathrm{m/s}$ in 1.0 seconds, so $a = 2\,\mathrm{m/s^2}$ in the reference frame of the bus. With respect to the Earth, the acceleration is $(12\,\mathrm{m/s} - 10\,\mathrm{m/s})/(1.0\,\mathrm{s}) = 2\,\mathrm{m/s^2}$, which is the same.

CAUTION
Position and velocity are different in different reference frames, but length is the same (classical)

Since neither F, m, nor a changes from one inertial frame to another, then Newton's second law, $F = ma$, does not change. Thus Newton's second law satisfies the relativity principle. It is easily shown that the other laws of mechanics also satisfy the relativity principle.

That the laws of mechanics are the same in all inertial reference frames implies that no one inertial frame is special in any sense. We express this important conclusion by saying that **all inertial reference frames are equivalent** for the description of mechanical phenomena. No one inertial reference frame is any better than another. A reference frame fixed to a car or an aircraft traveling at constant velocity is as good as one fixed on the Earth. When you travel smoothly at constant velocity in a car or airplane, it is just as valid to say you are at rest and the Earth is moving as it is to say the reverse.[†] There is no experiment you can do to tell which frame is "really" at rest and which is moving. Thus, there is no way to single out one particular reference frame as being at absolute rest.

A complication arose, however, in the last half of the nineteenth century. Maxwell's comprehensive and successful theory of electromagnetism (Chapter 31) predicted that light is an electromagnetic wave. Maxwell's equations gave the velocity of light c as $3.00 \times 10^8\,\mathrm{m/s}$; and this is just what is measured. The question then arose: in what reference frame does light have precisely the value predicted by Maxwell's theory? It was assumed that light would have a different speed in different frames of reference. For example, if observers were traveling on a rocket ship at a speed of $1.0 \times 10^8\,\mathrm{m/s}$ away from a source of light, we might expect them to measure the speed of the light reaching them to be $(3.0 \times 10^8\,\mathrm{m/s}) - (1.0 \times 10^8\,\mathrm{m/s}) = 2.0 \times 10^8\,\mathrm{m/s}$. But Maxwell's equations have no provision for relative velocity. They predicted the speed of light to be $c = 3.0 \times 10^8\,\mathrm{m/s}$, which seemed to imply that there must be some preferred reference frame where c would have this value.

[†]We are ignoring the rotation and curvature of the Earth.

We discussed in Chapters 15 and 16 that waves can travel on water and along ropes or strings, and sound waves travel in air and other materials. Nineteenth-century physicists viewed the material world in terms of the laws of mechanics, so it was natural for them to assume that light too must travel in some *medium*. They called this transparent medium the **ether** and assumed it permeated all space.[†] It was therefore assumed that the velocity of light given by Maxwell's equations must be with respect to the ether.

At first it appeared that Maxwell's equations did *not* satisfy the relativity principle. They were simplest in the frame where $c = 3.00 \times 10^8\,\text{m/s}$; that is, in a reference frame at rest in the ether. In any other reference frame, extra terms would have to be added to take into account the relative velocity. Thus, although most of the laws of physics obeyed the relativity principle, the laws of electricity and magnetism apparently did not. Einstein's second postulate (Section 36–3) resolved this problem: Maxwell's equations do satisfy relativity.

Scientists soon set out to determine the speed of the Earth relative to this absolute frame, whatever it might be. A number of clever experiments were designed. The most direct were performed by A. A. Michelson and E. W. Morley in the 1880s. They measured the difference in the speed of light in different directions using Michelson's interferometer (Section 34–6). They expected to find a difference depending on the orientation of their apparatus with respect to the ether. For just as a boat has different speeds relative to the land when it moves upstream, downstream, or across the stream, so too light would be expected to have different speeds depending on the velocity of the ether past the Earth.

Strange as it may seem, they detected no difference at all. This was a great puzzle. A number of explanations were put forth over a period of years, but they led to contradictions or were otherwise not generally accepted. This **null result** was one of the great puzzles at the end of the nineteenth century.

Then in 1905, Albert Einstein proposed a radical new theory that reconciled these many problems in a simple way. But at the same time, as we shall see, it completely changed our ideas of space and time.

*36–2 The Michelson–Morley Experiment

The Michelson–Morley experiment was designed to measure the speed of the *ether*—the medium in which light was assumed to travel—with respect to the Earth. The experimenters thus hoped to find an absolute reference frame, one that could be considered to be at rest.

One of the possibilities nineteenth-century scientists considered was that the ether is fixed relative to the Sun, for even Newton had taken the Sun as the center of the universe. If this were the case (there was no guarantee, of course), the Earth's speed of about $3 \times 10^4\,\text{m/s}$ in its orbit around the Sun could produce a change of 1 part in 10^4 in the speed of light $(3.0 \times 10^8\,\text{m/s})$. Direct measurement of the speed of light to this precision was not possible. But A. A. Michelson, later with the help of E. W. Morley, was able to use his interferometer (Section 34–6) to measure the difference in the speed of light in different directions to this precision.

This famous experiment is based on the principle shown in Fig. 36–3. Part (a) is a diagram of the Michelson interferometer, and it is assumed that the "ether wind" is moving with speed v to the right. (Alternatively, the Earth is assumed to move to the left with respect to the ether at speed v.) The light from a source is split into two beams by a half-silvered mirror M_S. One beam travels to mirror M_1 and the other to mirror M_2. The beams are reflected by M_1 and M_2 and are joined again after passing through M_S. The now superposed beams interfere with each other and the resultant is viewed by the observer's eye as an interference pattern (discussed in Section 34–6).

Whether constructive or destructive interference occurs at the center of the interference pattern depends on the relative phases of the two beams after they have traveled their separate paths. Let us consider an analogy of a boat traveling up and

[†]The medium for light waves could not be air, since light travels from the Sun to Earth through nearly empty space. Therefore, another medium was postulated, the ether. The ether was not only transparent but, because of difficulty in detecting it, was assumed to have zero density.

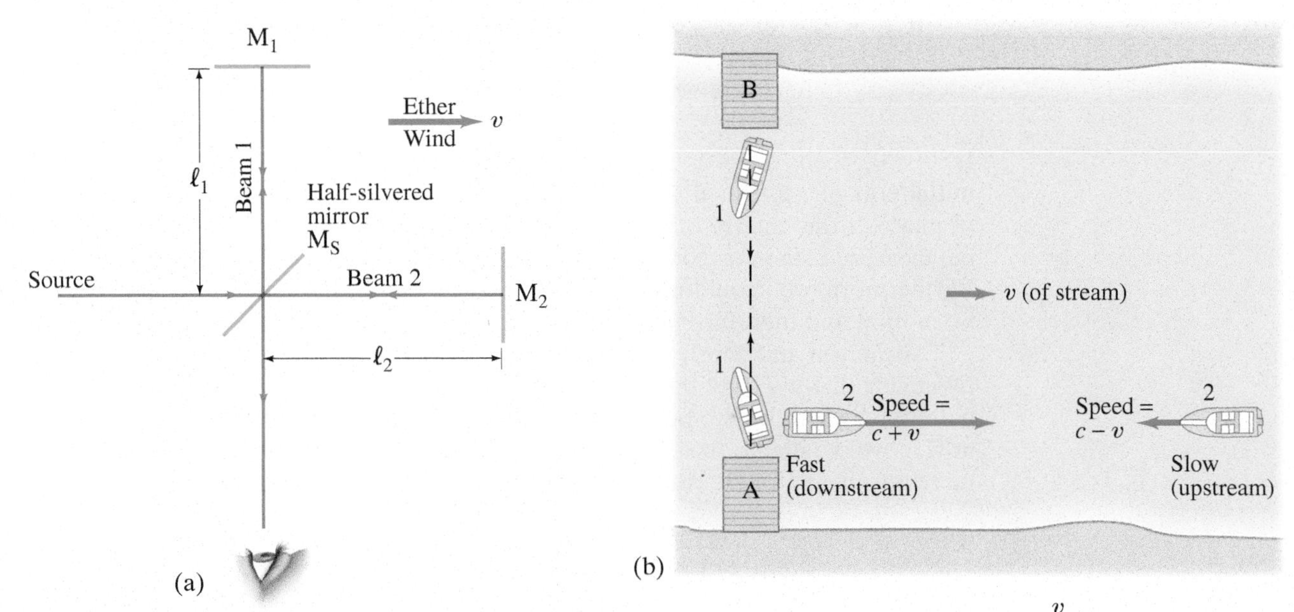

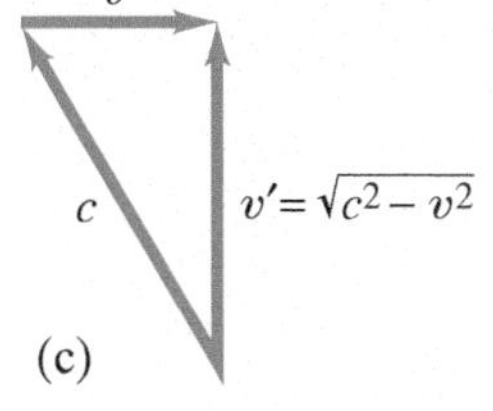

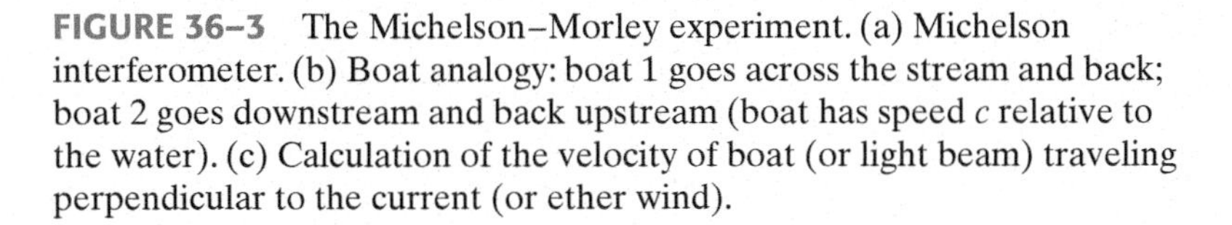

FIGURE 36–3 The Michelson–Morley experiment. (a) Michelson interferometer. (b) Boat analogy: boat 1 goes across the stream and back; boat 2 goes downstream and back upstream (boat has speed c relative to the water). (c) Calculation of the velocity of boat (or light beam) traveling perpendicular to the current (or ether wind).

down, and across, a river whose current moves with speed v, as shown in Fig. 36–3b. In still water, the boat can travel with speed c (not the speed of light in this case).

First we consider beam 2 in Fig. 36–3a, which travels parallel to the "ether wind." In its journey from M_S to M_2, the light would travel with speed $c + v$, according to classical physics, just as for a boat traveling downstream (see Fig. 36–3b) we add the speed of the river water to the boat's own speed (relative to the water) to get the boat's speed relative to the shore. Since the beam travels a distance ℓ_2, the time it takes to go from M_S to M_2 would be $t = \ell_2/(c + v)$. To make the return trip from M_2 to M_S, the light moves against the ether wind (like the boat going upstream), so its relative speed is expected to be $c - v$. The time for the return trip would be $\ell_2/(c - v)$. The total time for beam 2 to travel from M_S to M_2 and back to M_S is

$$t_2 = \frac{\ell_2}{c + v} + \frac{\ell_2}{c - v} = \frac{2\ell_2}{c\left(1 - v^2/c^2\right)}.$$

Now let us consider beam 1, which travels crosswise to the ether wind. Here the boat analogy (Fig. 36–3b) is especially helpful. The boat is to go from wharf A to wharf B directly across the stream. If it heads directly across, the stream's current will drag it downstream. To reach wharf B, the boat must head at an angle upstream. The precise angle depends on the magnitudes of c and v, but is of no interest to us in itself. Part (c) of Fig. 36–3 shows how to calculate the velocity v' of the boat relative to Earth as it crosses the stream. Since c, v, and v' form a right triangle, we have that $v' = \sqrt{c^2 - v^2}$. The boat has the same speed when it returns. If we now apply these principles to light beam 1 in Fig. 36–3a, we expect the beam to travel with speed $\sqrt{c^2 - v^2}$ in going from M_S to M_1 and back again. The total distance traveled is $2\ell_1$, so the time required for beam 1 to make the round trip would be $2\ell_1/\sqrt{c^2 - v^2}$, or

$$t_1 = \frac{2\ell_1}{c\sqrt{1 - v^2/c^2}}.$$

Notice that the denominator in this equation for t_1 involves a square root, whereas that for t_2 does not.

If $\ell_1 = \ell_2 = \ell$, we see that beam 2 will lag behind beam 1 by an amount

$$\Delta t = t_2 - t_1 = \frac{2\ell}{c}\left(\frac{1}{1 - v^2/c^2} - \frac{1}{\sqrt{1 - v^2/c^2}}\right).$$

If $v = 0$, then $\Delta t = 0$, and the two beams will return in phase since they were initially in phase. But if $v \neq 0$, then $\Delta t \neq 0$, and the two beams will return out of phase. If this change of phase from the condition $v = 0$ to that for $v \neq 0$ could be measured, then v could be determined. But the Earth cannot be stopped. Furthermore, we should not be too quick to assume that lengths are not affected by motion and therefore to assume $\ell_1 = \ell_2$.

Michelson and Morley realized that they could detect the difference in phase (assuming that $v \neq 0$) if they rotated their apparatus by 90°, for then the interference pattern between the two beams should change. In the rotated position, beam 1 would now move parallel to the ether and beam 2 perpendicular to it. Thus the roles could be reversed, and in the rotated position the times (designated by primes) would be

$$t_1' = \frac{2\ell_1}{c\,(1 - v^2/c^2)} \quad \text{and} \quad t_2' = \frac{2\ell_2}{c\sqrt{1 - v^2/c^2}}.$$

The time lag between the two beams in the nonrotated position (unprimed) would be

$$\Delta t = t_2 - t_1 = \frac{2\ell_2}{c\,(1 - v^2/c^2)} - \frac{2\ell_1}{c\sqrt{1 - v^2/c^2}}.$$

In the rotated position, the time difference would be

$$\Delta t' = t_2' - t_1' = \frac{2\ell_2}{c\sqrt{1 - v^2/c^2}} - \frac{2\ell_1}{c\,(1 - v^2/c^2)}.$$

When the rotation is made, the fringes of the interference pattern (Section 34–6) will shift an amount determined by the difference:

$$\Delta t - \Delta t' = \frac{2}{c}(\ell_1 + \ell_2)\left(\frac{1}{1 - v^2/c^2} - \frac{1}{\sqrt{1 - v^2/c^2}}\right).$$

This expression can be considerably simplified if we assume that $v/c \ll 1$. In this case we can use the binomial expansion (Appendix A), so

$$\frac{1}{1 - v^2/c^2} \approx 1 + \frac{v^2}{c^2} \quad \text{and} \quad \frac{1}{\sqrt{1 - v^2/c^2}} \approx 1 + \frac{1}{2}\frac{v^2}{c^2}.$$

Then

$$\Delta t - \Delta t' \approx \frac{2}{c}(\ell_1 + \ell_2)\left(1 + \frac{v^2}{c^2} - 1 - \frac{1}{2}\frac{v^2}{c^2}\right)$$
$$\approx (\ell_1 + \ell_2)\frac{v^2}{c^3}.$$

Now we assume $v = 3.0 \times 10^4\,\mathrm{m/s}$, the speed of the Earth in its orbit around the Sun. In Michelson and Morley's experiments, the arms ℓ_1 and ℓ_2 were about 11 m long. The time difference would then be about

$$\frac{(22\,\mathrm{m})(3.0 \times 10^4\,\mathrm{m/s})^2}{(3.0 \times 10^8\,\mathrm{m/s})^3} \approx 7.3 \times 10^{-16}\,\mathrm{s}.$$

For visible light of wavelength $\lambda = 5.5 \times 10^{-7}\,\mathrm{m}$, say, the frequency would be $f = c/\lambda = (3.0 \times 10^8\,\mathrm{m/s})/(5.5 \times 10^{-7}\,\mathrm{m}) = 5.5 \times 10^{14}\,\mathrm{Hz}$, which means that wave crests pass by a point every $1/(5.5 \times 10^{14}\,\mathrm{Hz}) = 1.8 \times 10^{-15}\,\mathrm{s}$. Thus, with a time difference of $7.3 \times 10^{-16}\,\mathrm{s}$, Michelson and Morley should have noted a movement in the interference pattern of $(7.3 \times 10^{-16}\,\mathrm{s})/(1.8 \times 10^{-15}\,\mathrm{s}) = 0.4$ fringe. They could easily have detected this, since their apparatus was capable of observing a fringe shift as small as 0.01 fringe.

But they found *no significant fringe shift whatever*! They set their apparatus at various orientations. They made observations day and night so that they would be at various orientations with respect to the Sun (due to the Earth's rotation).

They tried at different seasons of the year (the Earth at different locations due to its orbit around the Sun). Never did they observe a significant fringe shift.

This **null result** was one of the great puzzles of physics at the end of the nineteenth century. To explain it was a difficult challenge. One possibility to explain the null result was put forth independently by G. F. Fitzgerald and H. A. Lorentz (in the 1890s) in which they proposed that any length (including the arm of an interferometer) contracts by a factor $\sqrt{1 - v^2/c^2}$ in the direction of motion through the ether. According to Lorentz, this could be due to the ether affecting the forces between the molecules of a substance, which were assumed to be electrical in nature. This theory was eventually replaced by the far more comprehensive theory proposed by Albert Einstein in 1905—the special theory of relativity.

36–3 Postulates of the Special Theory of Relativity

The problems that existed at the start of the twentieth century with regard to electromagnetic theory and Newtonian mechanics were beautifully resolved by Einstein's introduction of the theory of relativity in 1905. Unaware of the Michelson–Morley null result, Einstein was motivated by certain questions regarding electromagnetic theory and light waves. For example, he asked himself: "What would I see if I rode a light beam?" The answer was that instead of a traveling electromagnetic wave, he would see alternating electric and magnetic fields at rest whose magnitude changed in space, but did not change in time. Such fields, he realized, had never been detected and indeed were not consistent with Maxwell's electromagnetic theory. He argued, therefore, that it was unreasonable to think that the speed of light relative to any observer could be reduced to zero, or in fact reduced at all. This idea became the second postulate of his theory of relativity.

In his famous 1905 paper, Einstein proposed doing away completely with the idea of the ether and the accompanying assumption of a preferred or absolute reference frame at rest. This proposal was embodied in two postulates. The first postulate was an extension of the Galilean–Newtonian relativity principle to include not only the laws of mechanics but also those of the rest of physics, including electricity and magnetism:

***First postulate (the relativity principle)*: The laws of physics have the same form in all inertial reference frames.**

The first postulate can also be stated as: *There is no experiment you can do in an inertial reference frame to tell if you are at rest or moving uniformly at constant velocity.*

The second postulate is consistent with the first:

***Second postulate (constancy of the speed of light)*: Light propagates through empty space with a definite speed *c* independent of the speed of the source or observer.**

These two postulates form the foundation of Einstein's **special theory of relativity**. It is called "special" to distinguish it from his later "general theory of relativity," which deals with noninertial (accelerating) reference frames (Chapter 44). The special theory, which is what we discuss here, deals only with inertial frames.

The second postulate may seem hard to accept, for it seems to violate common sense. First of all, we have to think of light traveling through empty space. Giving up the ether is not too hard, however, since it had never been detected. But the second postulate also tells us that the speed of light in vacuum is always the same, 3.00×10^8 m/s, no matter what the speed of the observer or the source. Thus, a person traveling toward or away from a source of light will measure the same speed for that light as someone at rest with respect to the source. This conflicts with our everyday experience: we would expect to have to add in the velocity of the observer. On the other hand, perhaps we can't expect our everyday experience to be helpful when dealing with the high velocity of light. Furthermore, the null result of the Michelson–Morley experiment is fully consistent with the second postulate.[†]

[†]The Michelson–Morley experiment can also be considered as evidence for the first postulate, since it was intended to measure the motion of the Earth relative to an absolute reference frame. Its failure to do so implies the absence of any such preferred frame.

Einstein's proposal has a certain beauty. By doing away with the idea of an absolute reference frame, it was possible to reconcile classical mechanics with Maxwell's electromagnetic theory. The speed of light predicted by Maxwell's equations *is* the speed of light in vacuum in *any* reference frame.

Einstein's theory required us to give up common sense notions of space and time, and in the following Sections we will examine some strange but interesting consequences of special relativity. Our arguments for the most part will be simple ones. We will use a technique that Einstein himself did: we will imagine very simple experimental situations in which little mathematics is needed. In this way, we can see many of the consequences of relativity theory without getting involved in detailed calculations. Einstein called these "thought" experiments.

36–4 Simultaneity

An important consequence of the theory of relativity is that we can no longer regard time as an absolute quantity. No one doubts that time flows onward and never turns back. But the time interval between two events, and even whether or not two events are simultaneous, depends on the observer's reference frame. By an **event**, which we use a lot here, we mean something that happens at a particular place and at a particular time.

Two events are said to occur simultaneously if they occur at exactly the same time. But how do we know if two events occur precisely at the same time? If they occur at the same point in space—such as two apples falling on your head at the same time—it is easy. But if the two events occur at widely separated places, it is more difficult to know whether the events are simultaneous since we have to take into account the time it takes for the light from them to reach us. Because light travels at finite speed, a person who sees two events must calculate back to find out when they actually occurred. For example, if two events are *observed* to occur at the same time, but one actually took place farther from the observer than the other, then the more distant one must have occurred earlier, and the two events were not simultaneous.

We now imagine a simple thought experiment. Assume an observer, called O, is located exactly halfway between points A and B where two events occur, Fig. 36–4. Suppose the two events are lightning that strikes the points A and B, as shown. For brief events like lightning, only short pulses of light (blue in Fig. 36–4) will travel outward from A and B and reach O. Observer O "sees" the events when the pulses of light reach point O. If the two pulses reach O at the same time, then the two events had to be simultaneous. This is because the two light pulses travel at the same speed (postulate 2), and since the distance OA equals OB, the time for the light to travel from A to O and B to O must be the same. Observer O can then definitely state that the two events occurred simultaneously. On the other hand, if O sees the light from one event before that from the other, then the former event occurred first.

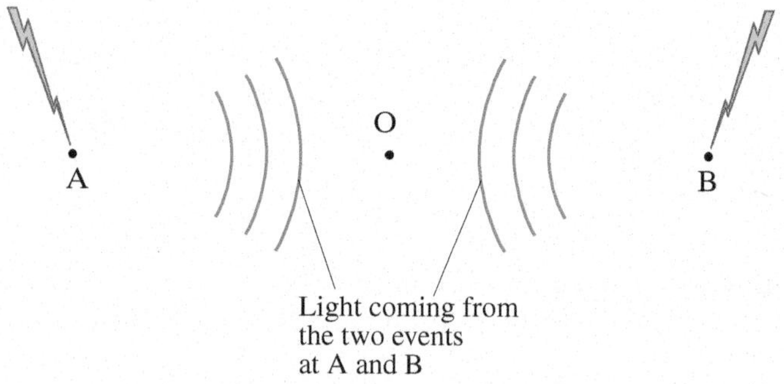

FIGURE 36–4 A moment after lightning strikes at points A and B, the pulses of light (shown as blue waves) are traveling toward the observer O, but O "sees" the lightning only when the light reaches O.

The question we really want to examine is this: if two events are simultaneous to an observer in one reference frame, are they also simultaneous to another observer moving with respect to the first? Let us call the observers O_1 and O_2 and assume they are fixed in reference frames 1 and 2 that move with speed v relative to one another. These two reference frames can be thought of as two rockets or two trains (Fig. 36–5). O_2 says that O_1 is moving to the right with speed v, as in Fig. 36–5a; and O_1 says O_2 is moving to the left with speed v, as in Fig. 36–5b. Both viewpoints are legitimate according to the relativity principle. [There is no third point of view which will tell us which one is "really" moving.]

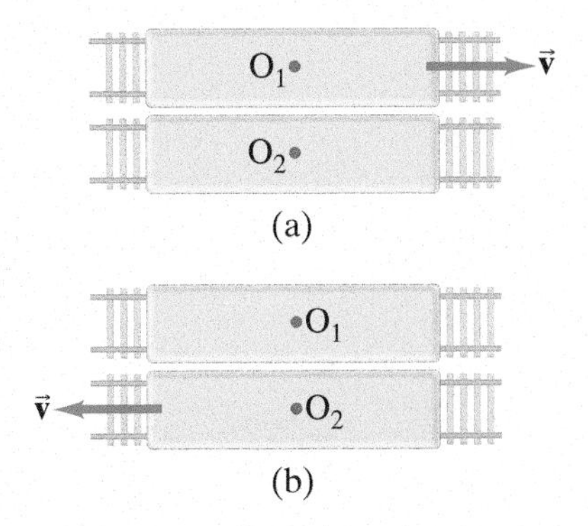

FIGURE 36–5 Observers O_1 and O_2, on two different trains (two different reference frames), are moving with relative speed v. O_2 says that O_1 is moving to the right (a); O_1 says that O_2 is moving to the left (b). Both viewpoints are legitimate: it all depends on your reference frame.

Now suppose that observers O_1 and O_2 observe and measure two lightning strikes. The lightning bolts mark both trains where they strike: at A_1 and B_1 on O_1's train, and at A_2 and B_2 on O_2's train, Fig. 36–6a. For simplicity, we assume that O_1 is exactly halfway between A_1 and B_1, and that O_2 is halfway between A_2 and B_2. Let us first put ourselves in O_2's reference frame, so we observe O_1 moving to the right with speed v. Let us also assume that the two events occur *simultaneously* in O_2's frame, and just at the instant when O_1 and O_2 are opposite each other, Fig. 36–6a. A short time later, Fig. 36–6b, the light from A_2 and B_2 reaches O_2 at the same time (we assumed this). Since O_2 knows (or measures) the distances O_2A_2 and O_2B_2 as equal, O_2 knows the two events are simultaneous in the O_2 reference frame.

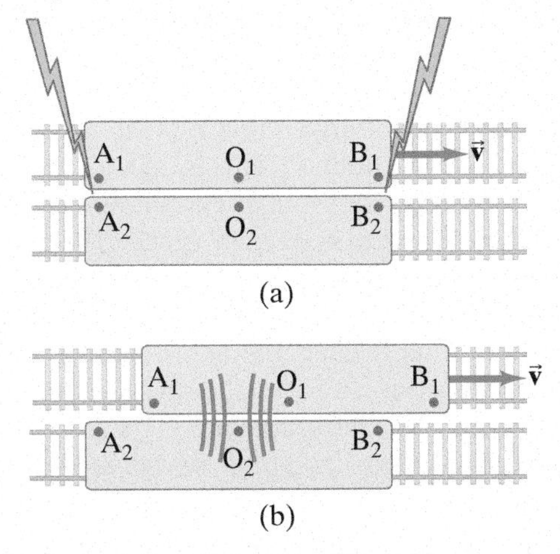

FIGURE 36–6 Thought experiment on simultaneity. In both (a) and (b) we are in the reference frame of observer O_2, who sees the reference frame of O_1 moving to the right. In (a), one lightning bolt strikes the two reference frames at A_1 and A_2, and a second lightning bolt strikes at B_1 and B_2. (b) A moment later, the light (shown in blue) from the two events reaches O_2 at the same time. So according to observer O_2, the two bolts of lightning struck simultaneously. But in O_1's reference frame, the light from B_1 has already reached O_1, whereas the light from A_1 has not yet reached O_1. So in O_1's reference frame, the event at B_1 must have preceded the event at A_1. Simultaneity in time is not absolute.

But what does observer O_1 observe and measure? From our (O_2) reference frame, we can predict what O_1 will observe. We see that O_1 moves to the right during the time the light is traveling to O_1 from A_1 and B_1. As shown in Fig. 36–6b, we can see from our O_2 reference frame that the light from B_1 has already passed O_1, whereas the light from A_1 has not yet reached O_1. That is, O_1 observes the light coming from B_1 before observing the light coming from A_1. Given (1) that light travels at the same speed c in any direction and in any reference frame, and (2) that the distance O_1A_1 equals O_1B_1, then observer O_1 can only conclude that the event at B_1 occurred before the event at A_1. The two events are *not* simultaneous for O_1, even though they are for O_2.

We thus find that two events which take place at different locations and are simultaneous to one observer, are actually not simultaneous to a second observer who moves relative to the first.

It may be tempting to ask: "Which observer is right, O_1 or O_2?" The answer, according to relativity, is that they are *both* right. There is no "best" reference frame we can choose to determine which observer is right. Both frames are equally good. We can only conclude that *simultaneity is not an absolute concept*, but is relative. We are not aware of this lack of agreement on simultaneity in everyday life because the effect is noticeable only when the relative speed of the two reference frames is very large (near c), or the distances involved are very large.

EXERCISE A Examine the experiment of Fig. 36–6 from O_1's reference frame. In this case, O_1 will be at rest and will see event B_1 occur before A_1. Will O_1 recognize that O_2, who is moving with speed v to the left, will see the two events as simultaneous? [*Hint*: Draw a diagram equivalent to Fig. 36–6.]

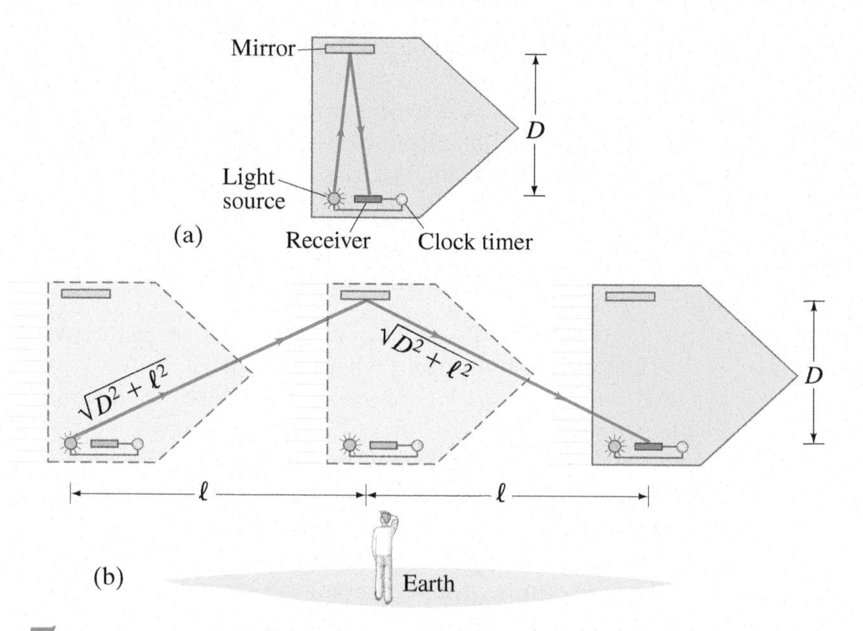

FIGURE 36–7 Time dilation can be shown by a thought experiment: the time it takes for light to travel across a spaceship and back is longer for the observer on Earth (b) than for the observer on the spaceship (a).

36–5 Time Dilation and the Twin Paradox

The fact that two events simultaneous to one observer may not be simultaneous to a second observer suggests that time itself is not absolute. Could it be that time passes differently in one reference frame than in another? This is, indeed, just what Einstein's theory of relativity predicts, as the following thought experiment shows.

Figure 36–7 shows a spaceship traveling past Earth at high speed. The point of view of an observer on the spaceship is shown in part (a), and that of an observer on Earth in part (b). Both observers have accurate clocks. The person on the spaceship (Fig. 36–7a) flashes a light and measures the time it takes the light to travel directly across the spaceship and return after reflecting from a mirror (the rays are drawn at a slight angle for clarity). In the reference frame of the spaceship, the light travels a distance $2D$ at speed c; so the time required to go across and back, which we call Δt_0, is

$$\Delta t_0 = 2D/c.$$

The observer on Earth, Fig. 36–7b, observes the same process. But to this observer, the spaceship is moving. So the light travels the diagonal path shown going across the spaceship, reflecting off the mirror, and returning to the sender. Although the light travels at the same speed to this observer (the second postulate), it travels a greater distance. Hence the time required, as measured by the observer on Earth, will be *greater* than that measured by the observer on the spaceship.

Let us determine the time interval Δt measured by the observer on Earth between sending and receiving the light. In time Δt, the spaceship travels a distance $2\ell = v\,\Delta t$ where v is the speed of the spaceship (Fig. 36–7b). The light travels a total distance on its diagonal path (Pythagorean theorem) of $2\sqrt{D^2 + \ell^2}$, where $\ell = v\,\Delta t/2$. Therefore

$$c = \frac{2\sqrt{D^2 + \ell^2}}{\Delta t} = \frac{2\sqrt{D^2 + v^2(\Delta t)^2/4}}{\Delta t}.$$

We square both sides,

$$c^2 = \frac{4D^2}{(\Delta t)^2} + v^2,$$

and solve for Δt, to find

$$\Delta t = \frac{2D}{c\sqrt{1 - v^2/c^2}}.$$

We combine this equation for Δt with the formula above, $\Delta t_0 = 2D/c$, and find

TIME DILATION

$$\Delta t = \frac{\Delta t_0}{\sqrt{1 - v^2/c^2}}. \qquad \textbf{(36–1a)}$$

Since $\sqrt{1 - v^2/c^2}$ is always less than 1, we see that $\Delta t > \Delta t_0$. That is, the time interval between the two events (the sending of the light, and its reception on the spaceship) is *greater* for the observer on Earth than for the observer on the spaceship. This is a general result of the theory of relativity, and is known as **time dilation**. Stated simply, the time dilation effect says that

clocks moving relative to an observer are measured to run more slowly (as compared to clocks at rest relative to that observer).

However, we should not think that the clocks are somehow at fault. Time is actually measured to pass more slowly in any moving reference frame as compared to your own. This remarkable result is an inevitable outcome of the two postulates of the theory of relativity.

The factor $1/\sqrt{1 - v^2/c^2}$ occurs so often in relativity that we often give it the shorthand symbol γ (the Greek letter "gamma"), and write Eq. 36–1a as

$$\Delta t = \gamma\, \Delta t_0 \qquad \textbf{(36–1b)}$$

where

$$\gamma = \frac{1}{\sqrt{1 - v^2/c^2}}. \qquad \textbf{(36–2)}$$

Note that γ is never less than one, and has no units. At normal speeds, $\gamma = 1$ to a few decimal places; in general, $\gamma \geq 1$.

The concept of time dilation may be hard to accept, for it contradicts our experience. We can see from Eqs. 36–1 that the time dilation effect is indeed negligible unless v is reasonably close to c. If v is much less than c, then the term v^2/c^2 is much smaller than the 1 in the denominator of Eq. 36–1a, and then $\Delta t \approx \Delta t_0$ (see Example 36–2). The speeds we experience in everyday life are much smaller than c, so it is little wonder we don't ordinarily notice time dilation. Experiments have tested the time dilation effect, and have confirmed Einstein's predictions. In 1971, for example, extremely precise atomic clocks were flown around the Earth in jet planes. The speed of the planes $(10^3\ \text{km/h})$ was much less than c, so the clocks had to be accurate to nanoseconds $(10^{-9}\ \text{s})$ in order to detect any time dilation. They were this accurate, and they confirmed Eqs. 36–1 to within experimental error. Time dilation had been confirmed decades earlier, however, by observations on "elementary particles" which have very small masses (typically 10^{-30} to 10^{-27} kg) and so require little energy to be accelerated to speeds close to the speed of light, c. Many of these elementary particles are not stable and decay after a time into lighter particles. One example is the muon, whose mean lifetime is 2.2 μs when at rest. Careful experiments showed that when a muon is traveling at high speeds, its lifetime is measured to be longer than when it is at rest, just as predicted by the time dilation formula.

EXAMPLE 36–1 Lifetime of a moving muon. (*a*) What will be the mean lifetime of a muon as measured in the laboratory if it is traveling at $v = 0.60c = 1.80 \times 10^8\ \text{m/s}$ with respect to the laboratory? Its mean lifetime at rest is $2.20\ \mu\text{s} = 2.20 \times 10^{-6}\ \text{s}$. (*b*) How far does a muon travel in the laboratory, on average, before decaying?

APPROACH If an observer were to move along with the muon (the muon would be at rest to this observer), the muon would have a mean life of 2.20×10^{-6} s. To an observer in the lab, the muon lives longer because of time dilation. We find the mean lifetime using Eq. 36–1a and the average distance using $d = v\,\Delta t$.

SOLUTION (*a*) From Eq. 36–1a with $v = 0.60c$, we have

$$\Delta t = \frac{\Delta t_0}{\sqrt{1 - v^2/c^2}} = \frac{2.20 \times 10^{-6}\ \text{s}}{\sqrt{1 - 0.36c^2/c^2}} = \frac{2.20 \times 10^{-6}\ \text{s}}{\sqrt{0.64}} = 2.8 \times 10^{-6}\ \text{s}.$$

(*b*) Relativity predicts that a muon with speed 1.80×10^8 m/s would travel an average distance $d = v\,\Delta t = (1.80 \times 10^8\ \text{m/s})(2.8 \times 10^{-6}\ \text{s}) = 500\ \text{m}$, and this is the distance that is measured experimentally in the laboratory.

NOTE At a speed of 1.8×10^8 m/s, classical physics would tell us that with a mean life of 2.2 μs, an average muon would travel $d = vt = (1.8 \times 10^8\ \text{m/s})(2.2 \times 10^{-6}\ \text{s}) = 400\ \text{m}$. This is shorter than the distance measured.

EXERCISE B What is the muon's mean lifetime (Example 36–1) if it is traveling at $v = 0.90c$? (*a*) 0.42 μs; (*b*) 2.3 μs; (*c*) 5.0 μs; (*d*) 5.3 μs; (*e*) 12.0 μs.

CAUTION

Proper time Δt_0 is for 2 events at the same point in space.

We need to clarify how to use Eqs. 36–1, and the meaning of Δt and Δt_0. The equation is true only when Δt_0 represents the time interval between the two events in a reference frame where the two events occur at *the same point in space* (as in Fig. 36–7a where the two events are the light flash being sent and being received). This time interval, Δt_0, is called the **proper time**. Then Δt in Eqs. 36–1 represents the time interval between the two events as measured in a reference frame moving with speed v with respect to the first. In Example 36–1 above, Δt_0 (and not Δt) was set equal to 2.2×10^{-6} s because it is only in the rest frame of the muon that the two events ("birth" and "decay") occur at the same point in space. The proper time Δt_0 is the shortest time between the events any observer can measure. In any other moving reference frame, the time Δt is greater.

EXAMPLE 36–2 **Time dilation at 100 km/h.** Let us check time dilation for everyday speeds. A car traveling 100 km/h covers a certain distance in 10.00 s according to the driver's watch. What does an observer at rest on Earth measure for the time interval?

APPROACH The car's speed relative to Earth is $100\,\mathrm{km/h} = (1.00 \times 10^5\,\mathrm{m})/(3600\,\mathrm{s}) = 27.8\,\mathrm{m/s}$. The driver is at rest in the reference frame of the car, so we set $\Delta t_0 = 10.00$ s in the time dilation formula.

SOLUTION We use Eq. 36–1a:

$$\Delta t = \frac{\Delta t_0}{\sqrt{1 - \dfrac{v^2}{c^2}}} = \frac{10.00\,\mathrm{s}}{\sqrt{1 - \left(\dfrac{27.8\,\mathrm{m/s}}{3.00 \times 10^8\,\mathrm{m/s}}\right)^2}} = \frac{10.00\,\mathrm{s}}{\sqrt{1 - (8.59 \times 10^{-15})}}.$$

If you put these numbers into a calculator, you will obtain $\Delta t = 10.00$ s, since the denominator differs from 1 by such a tiny amount. Indeed, the time measured by an observer on Earth would show no difference from that measured by the driver, even with the best instruments. A computer that could calculate to a large number of decimal places would reveal a difference between Δt and Δt_0. We can estimate the difference using the binomial expansion (Appendix A),

PROBLEM SOLVING

Use of the binomial expansion

$$(1 \pm x)^n \approx 1 \pm nx. \qquad [\text{for } x \ll 1]$$

In our time dilation formula, we have the factor $\gamma = (1 - v^2/c^2)^{-\frac{1}{2}}$. Thus

$$\Delta t = \gamma\,\Delta t_0 = \Delta t_0\left(1 - \frac{v^2}{c^2}\right)^{-\frac{1}{2}} \approx \Delta t_0\left(1 + \frac{1}{2}\frac{v^2}{c^2}\right)$$

$$\approx 10.00\,\mathrm{s}\left[1 + \frac{1}{2}\left(\frac{27.8\,\mathrm{m/s}}{3.00 \times 10^8\,\mathrm{m/s}}\right)^2\right] \approx 10.00\,\mathrm{s} + 4 \times 10^{-14}\,\mathrm{s}.$$

So the difference between Δt and Δt_0 is predicted to be 4×10^{-14} s, an extremely small amount.

EXERCISE C A certain atomic clock keeps perfect time on Earth. If the clock is taken on a spaceship traveling at a speed $v = 0.60c$, does this clock now run slow according to the people (*a*) on the spaceship, (*b*) on Earth?

EXAMPLE 36–3 **Reading a magazine on a spaceship.** A passenger on a high-speed spaceship traveling between Earth and Jupiter at a steady speed of $0.75c$ reads a magazine which takes 10.0 min according to her watch. (*a*) How long does this take as measured by Earth-based clocks? (*b*) How much farther is the spaceship from Earth at the end of reading the article than it was at the beginning?

APPROACH (*a*) The time interval in one reference frame is related to the time interval in the other by Eq. 36–1a or b. (*b*) At constant speed, distance is speed × time. Since there are two times (a Δt and a Δt_0) we will get two distances, one for each reference frame. [This surprising result is explored in the next Section (36–6).]

SOLUTION (*a*) The given 10.0-min time interval is the proper time—starting and finishing the magazine happen at the same place on the spaceship. Earth clocks measure

$$\Delta t = \frac{\Delta t_0}{\sqrt{1 - \frac{v^2}{c^2}}} = \frac{10.00 \text{ min}}{\sqrt{1 - (0.75)^2}} = 15.1 \text{ min}.$$

(*b*) In the Earth frame, the rocket travels a distance $D = v\,\Delta t = (0.75c)(15.1 \text{ min}) = (0.75)(3.0 \times 10^8 \text{ m/s})(15.1 \text{ min} \times 60 \text{ s/min}) = 2.04 \times 10^{11}$ m. In the spaceship's frame, the Earth is moving away from the spaceship at $0.75c$, but the time is only 10.0 min, so the distance is measured to be $D_0 = v\,\Delta t_0 = (2.25 \times 10^8 \text{ m/s})(600\text{s}) = 1.35 \times 10^{11}$ m.

Values for $\gamma = 1/\sqrt{1 - v^2/c^2}$ at a few speeds v are given in Table 36–1.

TABLE 36–1 Values of γ

v	γ
0	1.000
$0.01c$	1.000
$0.10c$	1.005
$0.50c$	1.15
$0.90c$	2.3
$0.99c$	7.1

Space Travel?

Time dilation has aroused interesting speculation about space travel. According to classical (Newtonian) physics, to reach a star 100 light-years away would not be possible for ordinary mortals (1 light-year is the distance light can travel in 1 year $= 3.0 \times 10^8 \text{ m/s} \times 3.16 \times 10^7 \text{ s} = 9.5 \times 10^{15}$ m). Even if a spaceship could travel at close to the speed of light, it would take over 100 years to reach such a star. But time dilation tells us that the time involved could be less. In a spaceship traveling at $v = 0.999c$, the time for such a trip would be only about $\Delta t_0 = \Delta t\sqrt{1 - v^2/c^2} = (100 \text{ yr})\sqrt{1 - (0.999)^2} = 4.5$ yr. Thus time dilation allows such a trip, but the enormous practical problems of achieving such speeds may not be possible to overcome, certainly not in the near future.

In this example, 100 years would pass on Earth, whereas only 4.5 years would pass for the astronaut on the trip. Is it just the clocks that would slow down for the astronaut? No. All processes, including aging and other life processes, run more slowly for the astronaut according to the Earth observer. But to the astronaut, time would pass in a normal way. The astronaut would experience 4.5 years of normal sleeping, eating, reading, and so on. And people on Earth would experience 100 years of ordinary activity.

Twin Paradox

Not long after Einstein proposed the special theory of relativity, an apparent paradox was pointed out. According to this **twin paradox**, suppose one of a pair of 20-year-old twins takes off in a spaceship traveling at very high speed to a distant star and back again, while the other twin remains on Earth. According to the Earth twin, the astronaut twin will age less. Whereas 20 years might pass for the Earth twin, perhaps only 1 year (depending on the spacecraft's speed) would pass for the traveler. Thus, when the traveler returns, the earthbound twin could expect to be 40 years old whereas the traveling twin would be only 21.

This is the viewpoint of the twin on the Earth. But what about the traveling twin? If all inertial reference frames are equally good, won't the traveling twin make all the claims the Earth twin does, only in reverse? Can't the astronaut twin claim that since the Earth is moving away at high speed, time passes more slowly on Earth and the twin on Earth will age less? This is the opposite of what the Earth twin predicts. They cannot both be right, for after all the spacecraft returns to Earth and a direct comparison of ages and clocks can be made.

There is, however, no contradiction here. One of the viewpoints is indeed incorrect. The consequences of the special theory of relativity—in this case, time dilation—can be applied only by observers in an inertial reference frame. The Earth is such a frame (or nearly so), whereas the spacecraft is not. The spacecraft accelerates at the start and end of its trip and when it turns around at the far point of its journey. During the acceleration, the twin on the spacecraft is not in an inertial frame. In between, the astronaut twin may be in an inertial frame (and is justified in saying the Earth twin's clocks run slow), but it is not always the same frame. So she cannot use special relativity to predict their relative ages when she returns to Earth. The Earth twin stays in the same inertial frame, and we can thus trust her predictions based on special relativity. Thus, there is no paradox. The prediction of the Earth twin that the traveling twin ages less is the proper one.

*Global Positioning System (GPS)

Airplanes, cars, boats, and hikers use **global positioning system (GPS)** receivers to tell them quite accurately where they are, at a given moment. The 24 global positioning system satellites send out precise time signals using atomic clocks. Your receiver compares the times received from at least four satellites, all of whose times are carefully synchronized to within 1 part in 10^{13}. By comparing the time differences with the known satellite positions and the fixed speed of light, the receiver can determine how far it is from each satellite and thus where it is on the Earth. It can do this to a typical accuracy of 15 m, if it has been constructed to make corrections such as the one below due to special relativity.

CONCEPTUAL EXAMPLE 36–4 **A relativity correction to GPS.** GPS satellites move at about $4\,\text{km/s} = 4000\,\text{m/s}$. Show that a good GPS receiver needs to correct for time dilation if it is to produce results consistent with atomic clocks accurate to 1 part in 10^{13}.

RESPONSE Let us calculate the magnitude of the time dilation effect by inserting $v = 4000\,\text{m/s}$ into Eq. 36–1a:

$$\Delta t = \frac{1}{\sqrt{1 - \frac{v^2}{c^2}}}\,\Delta t_0$$

$$= \frac{1}{\sqrt{1 - \left(\frac{4 \times 10^3\,\text{m/s}}{3 \times 10^8\,\text{m/s}}\right)^2}}\,\Delta t_0$$

$$= \frac{1}{\sqrt{1 - 1.8 \times 10^{-10}}}\,\Delta t_0 .$$

We use the binomial expansion: $(1 \pm x)^n \approx 1 \pm nx$ for $x \ll 1$ (see Appendix A) which here is $(1 - x)^{-\frac{1}{2}} \approx 1 + \frac{1}{2}x$. That is

$$\Delta t = \left(1 + \tfrac{1}{2}(1.8 \times 10^{-10})\right)\Delta t_0 = \left(1 + 9 \times 10^{-11}\right)\Delta t_0 .$$

The time "error" divided by the time interval is

$$\frac{(\Delta t - \Delta t_0)}{\Delta t_0} = 1 + 9 \times 10^{-11} - 1 = 9 \times 10^{-11} \approx 1 \times 10^{-10}.$$

Time dilation, if not accounted for, would introduce an error of about 1 part in 10^{10}, which is 1000 times greater than the precision of the atomic clocks. Not correcting for time dilation means a receiver could give much poorer position accuracy.

NOTE GPS devices must make other corrections as well, including effects associated with general relativity.

36–6 Length Contraction

Time intervals are not the only things different in different reference frames. Space intervals—lengths and distances—are different as well, according to the special theory of relativity, and we illustrate this with a thought experiment.

Observers on Earth watch a spacecraft traveling at speed v from Earth to, say, Neptune, Fig. 36–8a. The distance between the planets, as measured by the Earth observers, is ℓ_0. The time required for the trip, measured from Earth, is

$$\Delta t = \frac{\ell_0}{v}. \qquad \text{[Earth observer]}$$

In Fig. 36–8b we see the point of view of observers on the spacecraft. In this frame of reference, the spaceship is at rest; Earth and Neptune move† with speed v. The time between departure of Earth and arrival of Neptune (observed from the spacecraft) is the "proper time," since the two events occur at the same point in space (i.e., on the spacecraft). Therefore the time interval is less for the spacecraft

†We assume v is much greater than the relative speed of Neptune and Earth, so the latter can be ignored.

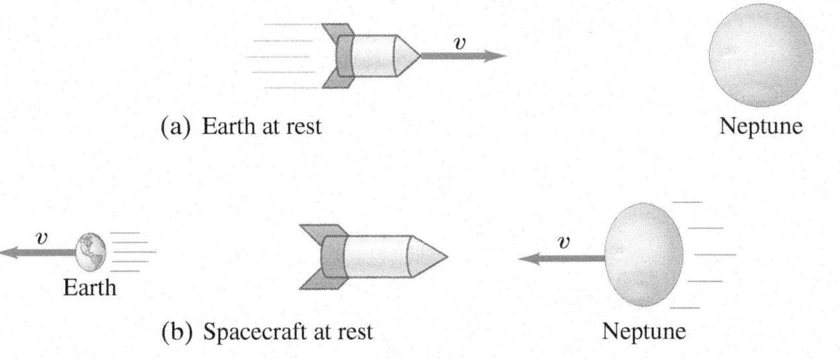

FIGURE 36–8 (a) A spaceship traveling at very high speed from Earth to the planet Neptune, as seen from Earth's frame of reference. (b) According to an observer on the spaceship, Earth and Neptune are moving at the very high speed v: Earth leaves the spaceship, and a time Δt_0 later Neptune arrives at the spaceship.

observers than for the Earth observers. That is, because of time dilation (Eq. 36–1a), the time for the trip as viewed by the spacecraft is

$$\Delta t_0 = \Delta t\sqrt{1 - v^2/c^2} = \Delta t/\gamma. \qquad \text{[spacecraft observer]}$$

Because the spacecraft observers measure the same speed but less time between these two events, they also measure the distance as less. If we let ℓ be the distance between the planets as viewed by the spacecraft observers, then $\ell = v\,\Delta t_0$, which we can rewrite as $\ell = v\,\Delta t_0 = v\,\Delta t\sqrt{1 - v^2/c^2} = \ell_0\sqrt{1 - v^2/c^2}$. Thus we have the important result that

$$\ell = \ell_0\sqrt{1 - v^2/c^2} \qquad \textbf{(36–3a)}$$

LENGTH CONTRACTION

or, using γ (Eq. 36–2),

$$\ell = \frac{\ell_0}{\gamma}. \qquad \textbf{(36–3b)}$$

This is a general result of the special theory of relativity and applies to lengths of objects as well as to distance between objects. The result can be stated most simply in words as:

the length of an object moving relative to an observer is measured to be shorter along its direction of motion than when it is at rest.

This is called **length contraction**. The length ℓ_0 in Eqs. 36–3 is called the **proper length.** It is the length of the object (or distance between two points whose positions are measured at the same time) as determined by *observers at rest* with respect to the object. Equations 36–3 give the length ℓ that will be measured by observers when the object travels past them at speed v.

CAUTION

Proper length is measured in reference frame where the two positions are at rest

It is important to note that length contraction occurs *only along the direction of motion*. For example, the moving spaceship in Fig. 36–8a is shortened in length, but its height is the same as when it is at rest.

Length contraction, like time dilation, is not noticeable in everyday life because the factor $\sqrt{1 - v^2/c^2}$ in Eq. 36–3a differs from 1.00 significantly only when v is very large.

FIGURE 36–9 Example 36–5.

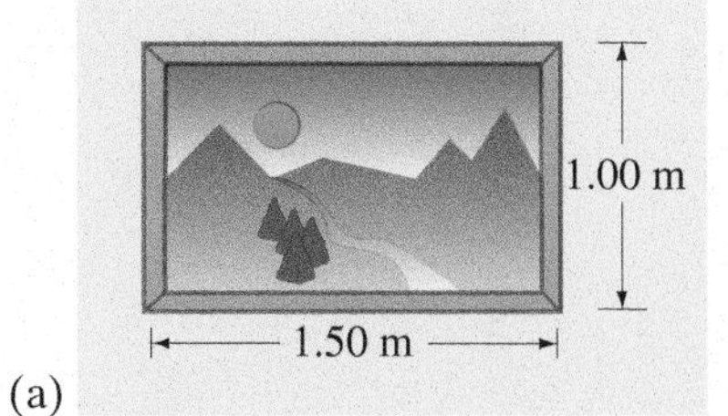

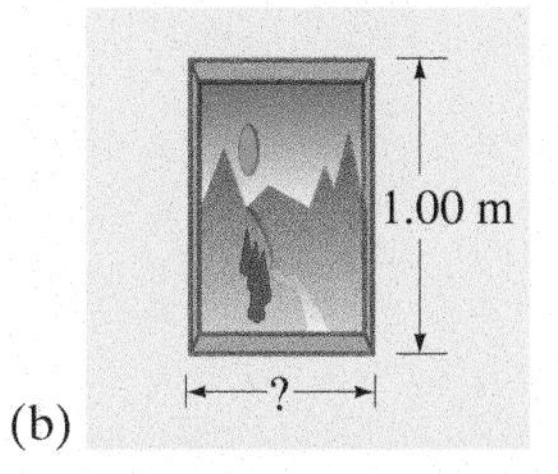

EXAMPLE 36–5 Painting's contraction. A rectangular painting measures 1.00 m tall and 1.50 m wide. It is hung on the side wall of a spaceship which is moving past the Earth at a speed of $0.90c$. See Fig. 36–9a. (*a*) What are the dimensions of the picture according to the captain of the spaceship? (*b*) What are the dimensions as seen by an observer on the Earth?

APPROACH We apply the length contraction formula, Eq. 36–3a, to the dimension parallel to the motion; v is the speed of the painting relative to the observer.

SOLUTION (*a*) The painting is at rest $(v = 0)$ on the spaceship so it (as well as everything else in the spaceship) looks perfectly normal to everyone on the spaceship. The captain sees a 1.00-m by 1.50-m painting.

(*b*) Only the dimension in the direction of motion is shortened, so the height is unchanged at 1.00 m, Fig. 36–9b. The length, however, is contracted to

$$\ell = \ell_0\sqrt{1 - \frac{v^2}{c^2}}$$
$$= (1.50\,\text{m})\sqrt{1 - (0.90)^2} = 0.65\,\text{m}.$$

So the picture has dimensions $1.00\,\text{m} \times 0.65\,\text{m}$.

EXAMPLE 36–6 **A fantasy supertrain.** A very fast train with a proper length of 500 m is passing through a 200-m-long tunnel. Let us imagine the train's speed to be so great that the train fits completely within the tunnel as seen by an observer at rest on the Earth. That is, the engine is just about to emerge from one end of the tunnel at the time the last car disappears into the other end. What is the train's speed?

APPROACH Since the train just fits inside the tunnel, its length measured by the person on the ground is 200 m. The length contraction formula, Eq. 36–3a or b, can thus be used to solve for v.

SOLUTION Substituting $\ell = 200\text{ m}$ and $\ell_0 = 500\text{ m}$ into Eq. 36–3a gives

$$200\text{ m} = 500\text{ m}\sqrt{1 - \frac{v^2}{c^2}};$$

dividing both sides by 500 m and squaring, we get

$$(0.40)^2 = 1 - \frac{v^2}{c^2}$$

or

$$\frac{v}{c} = \sqrt{1 - (0.40)^2}$$

and

$$v = 0.92c.$$

NOTE No real train could go this fast. But it is fun to think about.

NOTE An observer on the *train* would *not* see the two ends of the train inside the tunnel at the same time. Recall that observers moving relative to each other do not agree about simultaneity.

EXERCISE D What is the length of the tunnel as measured by observers on the train in Example 36–6?

CONCEPTUAL EXAMPLE 36–7 **Resolving the train and tunnel length.** Observers at rest on the Earth see a very fast 200-m-long train pass through a 200-m-long tunnel (as in Example 36–6) so that the train momentarily disappears from view inside the tunnel. Observers on the train measure the train's length to be 500 m and the tunnel's length to be only 80 m (Exercise D, using Eq. 36–3a). Clearly a 500-m-long train cannot fit inside an 80-m-long tunnel. How is this apparent inconsistency explained?

RESPONSE Events simultaneous in one reference frame may not be simultaneous in another. Let the engine emerging from one end of the tunnel be "event A," and the last car disappearing into the other end of the tunnel "event B." To observers in the Earth frame, events A and B are simultaneous. To observers on the train, however, the events are not simultaneous. In the train's frame, event A occurs before event B. As the engine emerges from the tunnel, observers on the train observe the last car as still $500\text{ m} - 80\text{ m} = 420\text{ m}$ from the entrance to the tunnel.

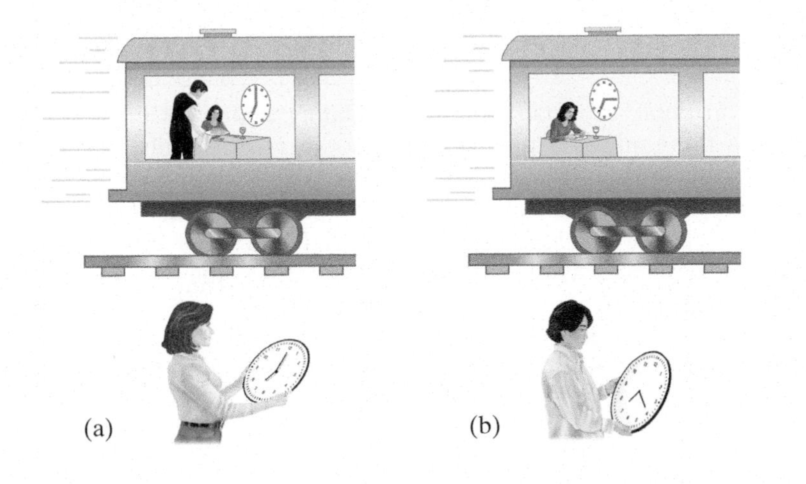

FIGURE 36–10 According to an accurate clock on a fast-moving train, a person (a) begins dinner at 7:00 and (b) finishes at 7:15. At the beginning of the meal, two observers on Earth set their watches to correspond with the clock on the train. These observers measure the eating time as 20 minutes.

36–7 Four-Dimensional Space–Time

Let us imagine a person is on a train moving at a very high speed, say $0.65c$, Fig. 36–10. This person begins a meal at 7:00 and finishes at 7:15, according to a clock on the train. The two events, beginning and ending the meal, take place at the same point on the train. So the proper time between these two events is 15 min. To observers on Earth, the meal will take longer—20 min according to Eqs. 36–1. Let us assume that the meal was served on a 20-cm-diameter plate. To observers on the Earth, the plate is only 15 cm wide (length contraction). Thus, to observers on the Earth, the meal looks smaller but lasts longer.

In a sense the two effects, time dilation and length contraction, balance each other. When viewed from the Earth, what an object seems to lose in size it gains in length of time it lasts. Space, or length, is exchanged for time.

Considerations like this led to the idea of **four-dimensional space–time**: space takes up three dimensions and time is a fourth dimension. Space and time are intimately connected. Just as when we squeeze a balloon we make one dimension larger and another smaller, so when we examine objects and events from different reference frames, a certain amount of space is exchanged for time, or vice versa.

Although the idea of four dimensions may seem strange, it refers to the idea that any object or event is specified by four quantities—three to describe where in space, and one to describe when in time. The really unusual aspect of four-dimensional space–time is that space and time can intermix: a little of one can be exchanged for a little of the other when the reference frame is changed.

It is difficult for most of us to understand the idea of four-dimensional space–time. Somehow we feel, just as physicists did before the advent of relativity, that space and time are completely separate entities. Yet we have found in our thought experiments that they are not completely separate. And think about Galileo and Newton. Before Galileo, the vertical direction, that in which objects fall, was considered to be distinctly different from the two horizontal dimensions. Galileo showed that the vertical dimension differs only in that it happens to be the direction in which gravity acts. Otherwise, all three dimensions are equivalent, a viewpoint we all accept today. Now we are asked to accept one more dimension, time, which we had previously thought of as being somehow different. This is not to say that there is no distinction between space and time. What relativity has shown is that space and time determinations are not independent of one another.

In Galilean–Newtonian relativity, the time interval between two events, Δt, and the distance between two events or points, Δx, are invariant quantities no matter what inertial reference frame they are viewed from. Neither of these quantities is invariant according to Einstein's relativity. But there is an invariant quantity in four-dimensional space–time, called the **space–time interval**, which is $(\Delta s)^2 = (c\,\Delta t)^2 - (\Delta x)^2$. We leave it as a Problem (97) to show that this quantity is indeed invariant under a Lorentz transformation (Section 36–8).

36–8 Galilean and Lorentz Transformations

We now examine in detail the mathematics of relating quantities in one inertial reference frame to the equivalent quantities in another. In particular, we will see how positions and velocities *transform* (that is, change) from one frame to the other.

We begin with the classical or Galilean viewpoint. Consider two inertial reference frames S and S′ which are each characterized by a set of coordinate axes, Fig. 36–11. The axes x and y (z is not shown) refer to S and x' and y' to S′. The x' and x axes overlap one another, and we assume that frame S′ moves to the right in the x direction at constant speed v with respect to S. For simplicity let us assume the origins 0 and 0′ of the two reference frames are superimposed at time $t = 0$.

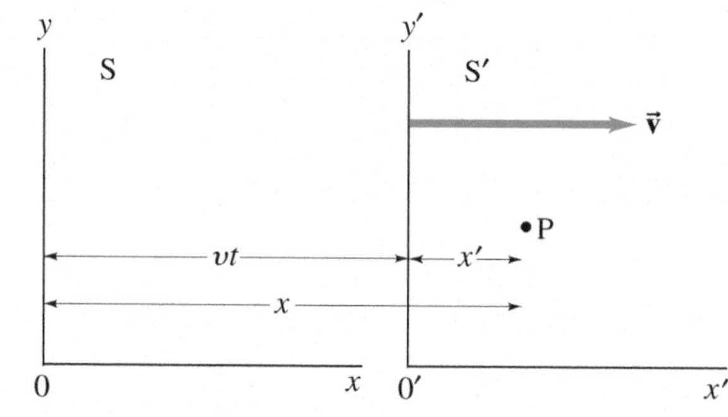

FIGURE 36–11 Inertial reference frame S′ moves to the right at constant speed v with respect to frame S.

Now consider an event that occurs at some point P (Fig. 36–11) represented by the coordinates x', y', z' in reference frame S′ at the time t'. What will be the coordinates of P in S? Since S and S′ initially overlap precisely, after a time t', S′ will have moved a distance vt'. Therefore, at time t', $x = x' + vt'$. The y and z coordinates, on the other hand, are not altered by motion along the x axis; thus $y = y'$ and $z = z'$. Finally, since time is assumed to be absolute in Galilean–Newtonian physics, clocks in the two frames will agree with each other; so $t = t'$. We summarize these in the following **Galilean transformation equations**:

$$\begin{aligned} x &= x' + vt' \\ y &= y' \\ z &= z' \\ t &= t'. \end{aligned} \qquad \text{[Galilean]} \quad \textbf{(36–4)}$$

These equations give the coordinates of an event in the S frame when those in the S′ frame are known. If those in the S frame are known, then the S′ coordinates are obtained from

$$x' = x - vt, \qquad y' = y, \qquad z' = z, \qquad t' = t. \qquad \text{[Galilean]}$$

These four equations are the "inverse" transformation and are very easily obtained from Eqs. 36–4. Notice that the effect is merely to exchange primed and unprimed quantities and replace v by $-v$. This makes sense because from the S′ frame, S moves to the left (negative x direction) with speed v.

Now suppose the point P in Fig. 36–11 represents a particle that is moving. Let the components of its velocity vector in S′ be u'_x, u'_y, u'_z. (We use u to distinguish it from the relative velocity of the two frames, v.) Now $u'_x = dx'/dt'$, $u'_y = dy'/dt'$ and $u'_z = dz'/dt'$. The velocity of P as seen from S will have components u_x, u_y, and u_z. We can show how these are related to the velocity components in S′ by differentiating Eqs. 36–4. For u_x we get

$$u_x = \frac{dx}{dt} = \frac{d(x' + vt')}{dt'} = u'_x + v$$

since v is assumed constant. For the other components, $u'_y = u_y$ and $u'_z = u_z$, so we have

$$\begin{aligned} u_x &= u'_x + v \\ u_y &= u'_y \\ u_z &= u'_z . \end{aligned} \qquad \text{[Galilean]} \quad \textbf{(36–5)}$$

These are known as the **Galilean velocity transformation equations**. We see that the y and z components of velocity are unchanged, but the x components differ by v: $u_x = u'_x + v$. This is just what we have used before (see Chapter 3, Section 3–9) when dealing with relative velocity.

The Galilean transformations, Eqs. 36–4 and 36–5, are valid only when the velocities involved are much less than c. We can see, for example, that the first of Eqs. 36–5 will not work for the speed of light: light traveling in S′ with speed $u'_x = c$ would have speed $c + v$ in S, whereas the theory of relativity insists it must be c in S. Clearly, then, a new set of transformation equations is needed to deal with relativistic velocities.

We derive the required equation, looking again at Fig. 36–11. We will try the simple assumption that the transformation is linear and of the form

$$x = \gamma(x' + vt'), \qquad y = y', \qquad z = z'. \qquad \textbf{(i)}$$

That is, we modify the first of Eqs. 36–4 by multiplying by a constant γ which is yet to be determined† ($\gamma = 1$ non-relativistically). But we assume the y and z equations are unchanged since there is no length contraction in these directions. We will not assume a form for t, but will derive it. The inverse equations must have the same form with v replaced by $-v$. (The principle of relativity demands it, since S′ moving to the right with respect to S is equivalent to S moving to the left with respect to S′.) Therefore

$$x' = \gamma(x - vt). \qquad \textbf{(ii)}$$

Now if a light pulse leaves the common origin of S and S′ at time $t = t' = 0$, after a time t it will have traveled a distance $x = ct$ or $x' = ct'$ along the x axis. Therefore, from Eqs. (i) and (ii) above,

$$ct = \gamma(ct' + vt') = \gamma(c + v)\,t', \qquad \textbf{(iii)}$$

$$ct' = \gamma(ct - vt) = \gamma(c - v)t. \qquad \textbf{(iv)}$$

We substitute t' from Eq. (iv) into Eq. (iii) and find $ct = \gamma(c + v)\gamma(c - v)(t/c) = \gamma^2(c^2 - v^2)t/c$. We cancel out the t on each side and solve for γ to find

$$\gamma = \frac{1}{\sqrt{1 - v^2/c^2}}.$$

The constant γ here has the same value as the γ we used before, Eq. 36–2. Now that we have found γ, we need only find the relation between t and t'. To do so, we combine $x' = \gamma(x - vt)$ with $x = \gamma(x' + vt')$:

$$x' = \gamma(x - vt) = \gamma(\gamma[x' + vt'] - vt).$$

We solve for t and find $t = \gamma(t' + vx'/c^2)$. In summary,

$$\begin{aligned} x &= \gamma(x' + vt') \\ y &= y' \\ z &= z' \\ t &= \gamma\left(t' + \frac{vx'}{c^2}\right) \end{aligned} \qquad \textbf{(36–6)}$$

LORENTZ TRANSFORMATIONS

These are called the **Lorentz transformation equations**. They were first proposed, in a slightly different form, by Lorentz in 1904 to explain the null result of the Michelson–Morley experiment and to make Maxwell's equations take the same form in all inertial reference frames. A year later Einstein derived them independently based on his theory of relativity. Notice that not only is the x equation modified as compared to the Galilean transformation, but so is the t equation; indeed, we see directly in this last equation how the space and time coordinates mix.

† γ here is not assumed to be given by Eq. 36–2.

Deriving Length Contraction

We now derive the length contraction formula, Eq. 36–3, from the Lorentz transformation equations. We consider two reference frames S and S′ as in Fig. 36–11.

Let an object of length ℓ_0 be at rest on the x axis in S. The coordinates of its two end points are x_1 and x_2, so that $x_2 - x_1 = \ell_0$. At any instant in S′, the end points will be at x_1' and x_2' as given by the Lorentz transformation equations. The length measured in S′ is $\ell = x_2' - x_1'$. An observer in S′ measures this length by measuring x_2' and x_1' at the same time (in the S′ reference frame), so $t_2' = t_1'$. Then, from the first of Eqs. 36–6,

$$\ell_0 = x_2 - x_1 = \frac{1}{\sqrt{1 - v^2/c^2}}\left(x_2' + vt_2' - x_1' - vt_1'\right).$$

Since $t_2' = t_1'$, we have

$$\ell_0 = \frac{1}{\sqrt{1 - v^2/c^2}}\left(x_2' - x_1'\right) = \frac{\ell}{\sqrt{1 - v^2/c^2}},$$

or

$$\ell = \ell_0\sqrt{1 - v^2/c^2},$$

which is Eq. 36–3.

Deriving Time Dilation

We now derive the time-dilation formula, Eq. 36–1a, using the Lorentz transformation equations.

The time Δt_0 between two events that occur at the same place $(x_2' = x_1')$ in S′ is measured to be $\Delta t_0 = t_2' - t_1'$. Since $x_2' = x_1'$, then from the last of Eqs. 36–6, the time Δt between the events as measured in S is

$$\Delta t = t_2 - t_1 = \frac{1}{\sqrt{1 - v^2/c^2}}\left(t_2' + \frac{vx_2'}{c^2} - t_1' - \frac{vx_1'}{c^2}\right)$$

$$= \frac{1}{\sqrt{1 - v^2/c^2}}\left(t_2' - t_1'\right) = \frac{\Delta t_0}{\sqrt{1 - v^2/c^2}},$$

which is Eq. 36–1a. Note that we chose S′ to be the frame in which the two events occur at the same place, so that $x_1' = x_2'$ and the terms containing x_1' and x_2' cancel out.

Relativistic Addition of Velocities

The relativistically correct velocity equations are readily obtained by differentiating Eqs. 36–6 with respect to time. For example (using $\gamma = 1/\sqrt{1 - v^2/c^2}$ and the chain rule for derivatives):

$$u_x = \frac{dx}{dt} = \frac{d}{dt}\left[\gamma(x' + vt')\right]$$

$$= \frac{d}{dt'}\left[\gamma(x' + vt')\right]\frac{dt'}{dt} = \gamma\left[\frac{dx'}{dt'} + v\right]\frac{dt'}{dt}.$$

But $dx'/dt' = u_x'$ and $dt'/dt = 1/(dt/dt') = 1/\left[\gamma(1 + vu_x'/c^2)\right]$ where we have differentiated the last of Eqs. 36–6 with respect to time. Therefore

$$u_x = \frac{\left[\gamma(u_x' + v)\right]}{\left[\gamma(1 + vu_x'/c^2)\right]} = \frac{u_x' + v}{1 + vu_x'/c^2}.$$

The others are obtained in the same way and we collect them here:

LORENTZ VELOCITY TRANSFORMATIONS

$$u_x = \frac{u_x' + v}{1 + vu_x'/c^2} \qquad \textbf{(36–7a)}$$

$$u_y = \frac{u_y'\sqrt{1 - v^2/c^2}}{1 + vu_x'/c^2} \qquad \textbf{(36–7b)}$$

$$u_z = \frac{u_z'\sqrt{1 - v^2/c^2}}{1 + vu_x'/c^2}. \qquad \textbf{(36–7c)}$$

Note that even though the relative velocity $\vec{\mathbf{v}}$ is in the x direction, if the object has y or z components of velocity, they too are affected by v and the x component of the object's velocity. This was not true for the Galilean transformation, Eqs. 36–5.

EXAMPLE 36–8 Adding velocities. Calculate the speed of rocket 2 in Fig. 36–12 with respect to Earth.

APPROACH Consider Earth as reference frame S, and rocket 1 as reference frame S′. Rocket 2 moves with speed $u' = 0.60c$ with respect to rocket 1. Rocket 1 has speed $v = 0.60c$ with respect to Earth. The velocities are along the same straight line which we take to be the x (and x') axis. We need use only the first of Eqs. 36–7.

SOLUTION The speed of rocket 2 with respect to Earth is

$$u = \frac{u' + v}{1 + \frac{vu'}{c^2}} = \frac{0.60c + 0.60c}{1 + \frac{(0.60c)(0.60c)}{c^2}} = \frac{1.20c}{1.36} = 0.88c.$$

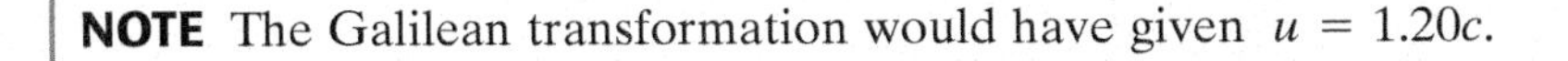

NOTE The Galilean transformation would have given $u = 1.20c$.

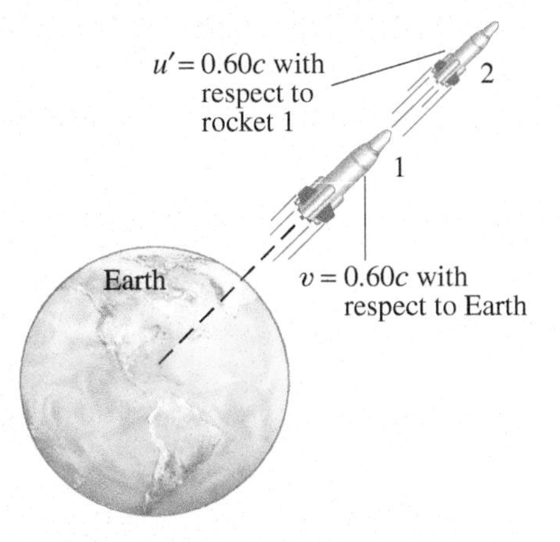

FIGURE 36–12 Rocket 1 moves away at speed $v = 0.60c$. Rocket 2 is fired from rocket 1 with speed $u' = 0.60c$. What is the speed of rocket 2 with respect to the Earth?

EXERCISE E Use Eqs. 36–7 to calculate the speed of rocket 2 in Fig. 36–12 relative to Earth if it was shot from rocket 1 at a speed $u' = 3000\text{ km/s} = 0.010c$. Assume rocket 1 had a speed $v = 6000\text{ km/s} = 0.020c$.

EXERCISE F Return to the Chapter-Opening Question, page 951, and answer it again now. Try to explain why you may have answered differently the first time.

Notice that Eqs. 36–7 reduce to the classical (Galilean) forms for velocities small compared to the speed of light, since $1 + vu'/c^2 \approx 1$ for v and $u' \ll c$. At the other extreme, let rocket 1 in Fig. 36–12 send out a beam of light, so that $u' = c$. Then Eq. 36–7a tells us the speed of light relative to Earth is

$$u = \frac{0.60c + c}{1 + \frac{(0.60c)(c)}{c^2}} = c,$$

which is consistent with the second postulate of relativity.

36–9 Relativistic Momentum

So far in this Chapter, we have seen that two basic mechanical quantities, length and time intervals, need modification because they are relative—their value depends on the reference frame from which they are measured. We might expect that other physical quantities might need some modification according to the theory of relativity, such as momentum, energy, and mass.

The analysis of collisions between two particles shows that if we want to preserve the law of conservation of momentum in relativity, we must redefine momentum as

$$p = \frac{mv}{\sqrt{1 - v^2/c^2}} = \gamma mv. \qquad \textbf{(36–8)}$$

Here γ is shorthand for $1/\sqrt{1 - v^2/c^2}$ as before (Eq. 36–2). For speeds much less than the speed of light, Eq. 36–8 gives the classical momentum, $p = mv$.

Relativistic momentum has been tested many times on tiny elementary particles (such as muons), and it has been found to behave in accord with Eq. 36–8. We derive Eq. 36–8 in the optional subsection on the next page.

EXAMPLE 36–9 **Momentum of moving electron.** Compare the momentum of an electron when it has a speed of (*a*) 4.00×10^7 m/s in the CRT of a television set, and (*b*) $0.98c$ in an accelerator used for cancer therapy.

APPROACH We use Eq. 36–8 for the momentum of a moving electron.

SOLUTION (*a*) At $v = 4.00 \times 10^7\,\mathrm{m/s}$, the electron's momentum is

$$p = \frac{mv}{\sqrt{1 - \frac{v^2}{c^2}}} = \frac{mv}{\sqrt{1 - \frac{(4.00 \times 10^7\,\mathrm{m/s})^2}{(3.00 \times 10^8\,\mathrm{m/s})^2}}} = 1.01mv.$$

The factor $\gamma = 1/\sqrt{1 - v^2/c^2} \approx 1.01$, so the momentum is only about 1% greater than the classical value. (If we put in the mass of an electron, $m = 9.11 \times 10^{-31}\,\mathrm{kg}$, the momentum is $p = 1.01mv = 3.68 \times 10^{-23}\,\mathrm{kg \cdot m/s}$.)

(*b*) With $v = 0.98c$, the momentum is

$$p = \frac{mv}{\sqrt{1 - \frac{v^2}{c^2}}} = \frac{mv}{\sqrt{1 - \frac{(0.98c)^2}{c^2}}} = \frac{mv}{\sqrt{1 - (0.98)^2}} = 5.0mv.$$

An electron traveling at 98% the speed of light has $\gamma = 5.0$ and a momentum 5.0 times its classical value.

Newton's second law, stated in its most general form, is

$$\vec{\mathbf{F}} = \frac{d\vec{\mathbf{p}}}{dt} = \frac{d}{dt}(\gamma m\vec{\mathbf{v}}) = \frac{d}{dt}\left(\frac{m\vec{\mathbf{v}}}{\sqrt{1 - v^2/c^2}}\right) \tag{36–9}$$

and is valid relativistically.

*Derivation of Relativistic Momentum

Classically, momentum is a conserved quantity. We hope to find a formula for momentum that will also be valid relativistically. To do so, let us assume it has the general form given by $p = fmv$ where f is some function of v: $f(v)$. We consider a hypothetical collision between two objects—a thought experiment—and see what form $f(v)$ must take if momentum is to be conserved.

Our thought experiment involves the elastic collision of two identical balls, A and B. We consider two inertial reference frames, S and S′, moving along the x axis with a speed v with respect to each other, Fig. 36–13. That is, frame S′ moves to the right with velocity $\vec{\mathbf{v}}$ as seen by observers on frame S; and frame S moves to the left with $-\vec{\mathbf{v}}$ as seen by observers on S′. In reference frame S, ball A is thrown with speed u in the $+y$ direction. In reference frame S′, ball B is thrown with speed u in the negative y' direction. The two balls are thrown at just the right time so that they collide. We assume that they rebound elastically and, from *symmetry*, that each moves with the same speed u back in the opposite direction in its thrower's reference frame. Figure 36–13a shows the collision as seen by an observer in reference frame S; and Fig. 36–13b shows the collision as seen from reference frame S′. In reference frame S, ball A has $v_x = 0$ both before and after the collision; it has $v_y = +u$ before the collision and $-u$ after the collision. In frame S′, ball A has x component of velocity $u'_x = -v$ both before and after the collision, and a y' component (Eq. 36–7b with $u'_x = -v$) of magnitude

$$u'_y = u\sqrt{1 - v^2/c^2}.$$

The same holds true for ball B, except in reverse. The velocity components are indicated in Fig. 36–13.

We now apply the law of conservation of momentum, which we hope remains valid in relativity, even if momentum has to be redefined. That is, we assume that the total momentum before the collision is equal to the total momentum after the

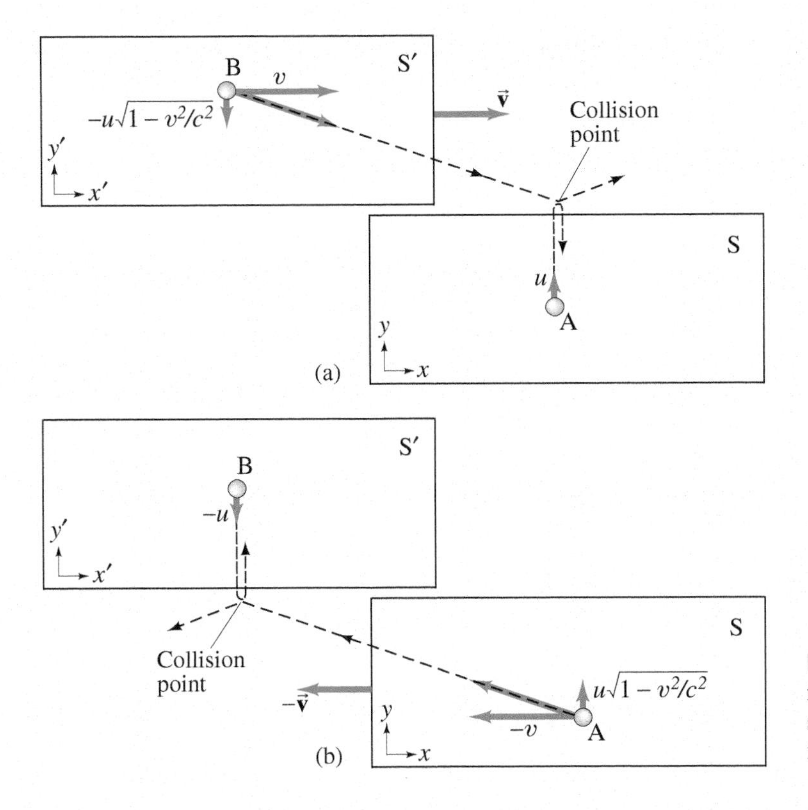

FIGURE 36–13 Deriving the momentum formula. Collision as seen by observers (a) in reference frame S, (b) in reference frame S′.

collision. We apply conservation of momentum to the y component of momentum in reference frame S (Fig. 36–13a). To make our task easier, let us assume $u \ll v$ so that the speed of ball B as seen in reference frame S is essentially v. Then B's y component of momentum in S before collision is $-f(v)mu\sqrt{1 - v^2/c^2}$ and after the collision is $+f(v)mu\sqrt{1 - v^2/c^2}$. Ball A in S has y component $f(u)mu$ before and $-f(u)mu$ after the collision. (We use $f(u)$ for A because its speed in S is only u.) Conservation of momentum in S for the y component is

$$(p_{\mathrm{A}} + p_{\mathrm{B}})_{\mathrm{before}} = (p_{\mathrm{A}} + p_{\mathrm{B}})_{\mathrm{after}}$$

$$f(u)mu - f(v)mu\sqrt{1 - v^2/c^2} = -f(u)mu + f(v)mu\sqrt{1 - v^2/c^2}.$$

We solve this for $f(v)$ and obtain

$$f(v) = \frac{f(u)}{\sqrt{1 - v^2/c^2}}.$$

To simplify this relation so we can solve for f, let us allow u to become very small so that it approaches zero (this corresponds to a glancing collision with one of the balls essentially at rest and the other moving with speed v). Then the momentum terms $f(u)mu$ are in the nonrelativistic realm and take on the classical form, simply mu, meaning that $f(u) = 1$. So the previous equation becomes

$$f(v) = \frac{1}{\sqrt{1 - v^2/c^2}}.$$

We see that $f(v)$ comes out to be the factor we used before and called γ, and here has been shown to be valid for ball A. Using Fig. 36–13b we can derive the same relation for ball B. Thus we can conclude that we need to define the relativistic momentum of a particle moving with velocity $\vec{\mathbf{v}}$ as

$$\vec{\mathbf{p}} = \frac{m\vec{\mathbf{v}}}{\sqrt{1 - v^2/c^2}} = \gamma m\vec{\mathbf{v}}.$$

With this definition the law of conservation of momentum will remain valid even in the relativistic realm. This relativistic momentum formula (Eq. 36–8) has been tested countless times on tiny elementary particles and been found valid.

*Relativistic Mass

The relativistic definition of momentum, Eq. 36–8, is sometimes interpreted as an increase in the mass of an object. In this interpretation, a particle can have a **relativistic mass**, m_{rel}, which increases with speed according to

$$m_{\text{rel}} = \frac{m}{\sqrt{1 - v^2/c^2}}.$$

In this "mass-increase" formula, m is referred to as the **rest mass** of the object. With this interpretation, *the mass of an object appears to increase as its speed increases.* But we must be careful in the use of relativistic mass. We cannot just plug it into formulas like $F = ma$ or $K = \frac{1}{2}mv^2$. For example, if we substitute it into $F = ma$, we obtain a formula that does not agree with experiment. If however, we write Newton's second law in its more general form, $\vec{\mathbf{F}} = d\vec{\mathbf{p}}/dt$, we do get a correct result (Eq. 36–9).

Also, be careful *not* to think a mass acquires more particles or more molecules as its speed becomes very large. It doesn't. In fact, many physicists believe an object has only one mass (its rest mass), and that it is only the momentum that increases with speed.

Whenever we talk about the mass of an object, we will always mean its rest mass (a fixed value).

36–10 The Ultimate Speed

A basic result of the special theory of relativity is that the speed of an object cannot equal or exceed the speed of light. That the speed of light is a natural speed limit in the universe can be seen from any of Eqs. 36–1, 36–3, 36–8, or the addition of velocities formula. It is perhaps easiest to see from Eq. 36–8. As an object is accelerated to greater and greater speeds, its momentum becomes larger and larger. Indeed, if v were to equal c, the denominator in this equation would be zero, and the momentum would be infinite. To accelerate an object up to $v = c$ would thus require infinite energy, and so is not possible.

36–11 $E = mc^2$; Mass and Energy

If momentum needs to be modified to fit with relativity as we just saw in Eq. 36–8, then we might expect energy too would need to be rethought. Indeed, Einstein not only developed a new formula for kinetic energy, but also found a new relation between mass and energy, and the startling idea that mass is a form of energy.

We start with the work-energy principle (Chapter 7), hoping it is still valid in relativity and will give verifiable results. That is, we assume the net work done on a particle is equal to its change in kinetic energy (K). Using this principle, Einstein showed that at high speeds the formula $K = \frac{1}{2}mv^2$ is not correct. Instead, as we show in the optional Subsection on page 978, the kinetic energy of a particle of mass m traveling at speed v is given by

$$K = \frac{mc^2}{\sqrt{1 - v^2/c^2}} - mc^2. \quad \textbf{(36–10a)}$$

In terms of $\gamma = 1/\sqrt{1 - v^2/c^2}$ we can rewrite Eq. 36–10a as

$$K = \gamma mc^2 - mc^2 = (\gamma - 1)mc^2. \quad \textbf{(36–10b)}$$

Equations 36–10 require some interpretation. The first term increases with the speed v of the particle. The second term, mc^2, is constant; it is called the **rest energy** of the particle, and represents a form of energy that a particle has even when at rest. Note that if a particle is at rest $(v = 0)$ the first term in Eq. 36–10a becomes mc^2, so $K = 0$ as it should.

We can rearrange Eq. 36–10b to get

$$\gamma mc^2 = mc^2 + K.$$

We call γmc^2 the *total energy* E of the particle (assuming no potential energy), because it equals the rest energy plus the kinetic energy:

$$E = K + mc^2. \tag{36–11a}$$

The total energy can also be written, using Eqs. 36–10, as

$$E = \gamma mc^2 = \frac{mc^2}{\sqrt{1 - v^2/c^2}}. \tag{36–11b}$$

For a particle at rest in a given reference frame, K is zero in Eq. 36–11a, so the total energy is its rest energy:

MASS RELATED TO ENERGY

$$E = mc^2. \tag{36–12}$$

Here we have Einstein's famous formula, $E = mc^2$. This formula mathematically relates the concepts of energy and mass. But if this idea is to have any physical meaning, then mass ought to be convertible to other forms of energy and vice versa. Einstein suggested that this might be possible, and indeed changes of mass to other forms of energy, and vice versa, have been experimentally confirmed countless times in nuclear and elementary particle physics. For example, an electron and a positron (= a positive electron, Section 37–5) have often been observed to collide and disappear, producing pure electromagnetic radiation. The amount of electromagnetic energy produced is found to be exactly equal to that predicted by Einstein's formula, $E = mc^2$. The reverse process is also commonly observed in the laboratory: electromagnetic radiation under certain conditions can be converted into material particles such as electrons (see Section 37–5 on pair production). On a larger scale, the energy produced in nuclear power plants is a result of the loss in mass of the uranium fuel as it undergoes the process called fission. Even the radiant energy we receive from the Sun is an example of $E = mc^2$; the Sun's mass is continually decreasing as it radiates electromagnetic energy outward.

The relation $E = mc^2$ is now believed to apply to all processes, although the changes are often too small to measure. That is, when the energy of a system changes by an amount ΔE, the mass of the system changes by an amount Δm given by

$$\Delta E = (\Delta m)(c^2).$$

In a nuclear reaction where an energy E is required or released, the masses of the reactants and the products will be different by $\Delta m = \Delta E/c^2$.

EXAMPLE 36–10 Pion's kinetic energy. A π^0 meson $(m = 2.4 \times 10^{-28}\,\text{kg})$ travels at a speed $v = 0.80c = 2.4 \times 10^8\,\text{m/s}$. What is its kinetic energy? Compare to a classical calculation.

APPROACH We use Eq. 36–10 and compare to $\frac{1}{2}mv^2$.

SOLUTION We substitute values into Eq. 36–10a or b

$$K = (\gamma - 1)mc^2$$

where

$$\gamma = \frac{1}{\sqrt{1 - v^2/c^2}} = \frac{1}{\sqrt{1 - (0.80)^2}} = 1.67.$$

Then

$$K = (1.67 - 1)(2.4 \times 10^{-28}\,\text{kg})(3.0 \times 10^8\,\text{m/s})^2$$

$$= 1.4 \times 10^{-11}\,\text{J}.$$

PROBLEM SOLVING *Relativistic kinetic energy*

Notice that the units of mc^2 are $\text{kg}\cdot\text{m}^2/\text{s}^2$, which is the joule.

NOTE If we were to do a classical calculation we would obtain $K = \frac{1}{2}mv^2 = \frac{1}{2}(2.4 \times 10^{-28}\,\text{kg})\,(2.4 \times 10^8\,\text{m/s})^2 = 6.9 \times 10^{-12}\,\text{J}$, about half as much, but this is not a correct result. Note that $\frac{1}{2}\gamma mv^2$ also does not work.

EXERCISE G A proton is traveling in an accelerator with a speed of $1.0 \times 10^8\,\text{m/s}$. By what factor does the proton's kinetic energy increase if its speed is doubled? (*a*) 1.3, (*b*) 2.0, (*c*) 4.0, (*d*) 5.6.

EXAMPLE 36–11 **Energy from nuclear decay.** The energy required or released in nuclear reactions and decays comes from a change in mass between the initial and final particles. In one type of radioactive decay, an atom of uranium ($m = 232.03714\,\mathrm{u}$) decays to an atom of thorium ($m = 228.02873\,\mathrm{u}$) plus an atom of helium ($m = 4.00260\,\mathrm{u}$) where the masses given are in atomic mass units ($1\,\mathrm{u} = 1.6605 \times 10^{-27}\,\mathrm{kg}$). Calculate the energy released in this decay.

APPROACH The initial mass minus the total final mass gives the mass loss in atomic mass units (u); we convert that to kg, and multiply by c^2 to find the energy released, $\Delta E = \Delta m c^2$.

SOLUTION The initial mass is 232.03714 u, and after the decay the mass is $228.02873\,\mathrm{u} + 4.00260\,\mathrm{u} = 232.03133\,\mathrm{u}$, so there is a decrease in mass of 0.00581 u. This mass, which equals $(0.00581\,\mathrm{u})(1.66 \times 10^{-27}\,\mathrm{kg}) = 9.64 \times 10^{-30}\,\mathrm{kg}$, is changed into energy. By $\Delta E = \Delta m c^2$, we have

$$\Delta E = (9.64 \times 10^{-30}\,\mathrm{kg})(3.0 \times 10^{8}\,\mathrm{m/s})^2 = 8.68 \times 10^{-13}\,\mathrm{J}.$$

Since $1\,\mathrm{MeV} = 1.60 \times 10^{-13}\,\mathrm{J}$ (Section 23–8), the energy released is 5.4 MeV.

In the tiny world of atoms and nuclei, it is common to quote energies in eV (electron volts) or multiples such as MeV (10^6 eV). Momentum (see Eq. 36–8) can be quoted in units of eV/c (or MeV/c). And mass can be quoted (from $E = mc^2$) in units of eV/c^2 (or MeV/c^2). Note the use of c to keep the units correct. The rest masses of the electron and the proton are readily shown to be 0.511 MeV/c^2 and 938 MeV/c^2, respectively. See also the Table inside the front cover.

EXAMPLE 36–12 **A 1-TeV proton.** The Tevatron accelerator at Fermilab in Illinois can accelerate protons to a kinetic energy of 1.0 TeV (10^{12} eV). What is the speed of such a proton?

APPROACH We solve the kinetic energy formula, Eq. 36–10a, for v.

SOLUTION The rest energy of a proton is $mc^2 = 938\,\mathrm{MeV}$ or $9.38 \times 10^8\,\mathrm{eV}$. Compared to the kinetic energy of 10^{12} eV, the rest energy can be neglected, so we simplify Eq. 36–10a to

$$K \approx \frac{mc^2}{\sqrt{1 - v^2/c^2}}.$$

We solve this for v in the following steps:

$$\sqrt{1 - \frac{v^2}{c^2}} = \frac{mc^2}{K};$$

$$1 - \frac{v^2}{c^2} = \left(\frac{mc^2}{K}\right)^2;$$

$$\frac{v^2}{c^2} = 1 - \left(\frac{mc^2}{K}\right)^2 = 1 - \left(\frac{9.38 \times 10^8\,\mathrm{eV}}{1.0 \times 10^{12}\,\mathrm{eV}}\right)^2;$$

$$v = \sqrt{1 - (9.38 \times 10^{-4})^2}\,c = 0.99999956\,c.$$

So the proton is traveling at a speed very nearly equal to c.

At low speeds, $v \ll c$, the relativistic formula for kinetic energy reduces to the classical one, as we now show by using the binomial expansion, $(1 \pm x)^n = 1 \pm nx + n(n-1)x^2/2! + \cdots$. With $n = -\frac{1}{2}$, we expand the square root in Eq. 36–10a

$$K = mc^2\left(\frac{1}{\sqrt{1 - v^2/c^2}} - 1\right)$$

so that

$$K \approx mc^2\left(1 + \frac{1}{2}\frac{v^2}{c^2} + \cdots - 1\right)$$

$$\approx \tfrac{1}{2}mv^2.$$

The dots in the first expression represent very small terms in the expansion which we neglect since we assumed that $v \ll c$. Thus at low speeds, the relativistic form for kinetic energy reduces to the classical form, $K = \frac{1}{2}mv^2$. This makes relativity a viable theory in that it can predict accurate results at low speed as well as at high. Indeed, the other equations of special relativity also reduce to their classical equivalents at ordinary speeds: length contraction, time dilation, and modifications to momentum as well as kinetic energy, all disappear for $v \ll c$ since $\sqrt{1 - v^2/c^2} \approx 1$.

A useful relation between the total energy E of a particle and its momentum p can also be derived. The momentum of a particle of mass m and speed v is given by Eq. 36–8

$$p = \gamma mv = \frac{mv}{\sqrt{1 - v^2/c^2}}.$$

The total energy is

$$E = K + mc^2$$

or

$$E = \gamma mc^2 = \frac{mc^2}{\sqrt{1 - v^2/c^2}}.$$

We square this equation (and we insert "$v^2 - v^2$" which is zero, but will help us):

$$E^2 = \frac{m^2c^2(v^2 - v^2 + c^2)}{1 - v^2/c^2}$$

$$= p^2c^2 + \frac{m^2c^4(1 - v^2/c^2)}{1 - v^2/c^2}$$

or

$$E^2 = p^2c^2 + m^2c^4. \qquad \textbf{(36–13)}$$

Thus, the total energy can be written in terms of the momentum p, or in terms of the kinetic energy (Eqs. 36–11), where we have assumed there is no potential energy.

We can rewrite Eq. 36–13 as $E^2 - p^2c^2 = m^2c^4$. Since the mass m of a given particle is the same in any reference frame, we see that the quantity $E^2 - p^2c^2$ must also be the same in any reference frame. Thus, at any given moment the total energy E and momentum p of a particle will be different in different reference frames, but the quantity $E^2 - p^2c^2$ will have the same value in all inertial reference frames. We say that the quantity $E^2 - p^2c^2$ is **invariant** under a Lorentz transformation.

*When Do We Use Relativistic Formulas?

From a practical point of view, we do not have much opportunity in our daily lives to use the mathematics of relativity. For example, the γ factor, $\gamma = 1/\sqrt{1 - v^2/c^2}$, which appears in many relativistic formulas, has a value of 1.005 when $v = 0.10c$. Thus, for speeds even as high as $0.10c = 3.0 \times 10^7$ m/s, the factor $\sqrt{1 - v^2/c^2}$ in relativistic formulas gives a numerical correction of less than 1%. For speeds less than $0.10c$, or unless mass and energy are interchanged, we don't usually need to use the more complicated relativistic formulas, and can use the simpler classical formulas.

If you are given a particle's mass m and its kinetic energy K, you can do a quick calculation to determine if you need to use relativistic formulas or if classical ones are good enough. You simply compute the ratio K/mc^2 because (Eq. 36–10b)

$$\frac{K}{mc^2} = \gamma - 1 = \frac{1}{\sqrt{1 - v^2/c^2}} - 1.$$

If this ratio comes out to be less than, say, 0.01, then $\gamma \leq 1.01$ and relativistic equations will correct the classical ones by about 1%. If your expected precision is no better than 1%, classical formulas are good enough. But if your precision is 1 part in 1000 (0.1%) then you would want to use relativistic formulas. If your expected precision is only 10%, you need relativity if $(K/mc^2) \gtrsim 0.1$.

EXERCISE H For 1% accuracy, does an electron with $K = 100$ eV need to be treated relativistically? [*Hint*: The mass of an electron is 0.511 MeV.]

*Deriving Relativistic Energy

To find the mathematical relationship between mass and energy, we assume that the work-energy theorem is still valid in relativity for a particle, and we take the motion of the particle to be along the x axis. The work done to increase a particle's speed from zero to v is

$$W = \int_i^f F\,dx = \int_i^f \frac{dp}{dt}\,dx = \int_i^f \frac{dp}{dt}\,v\,dt = \int_i^f v\,dp$$

where i and f refer to the initial $(v = 0)$ and final $(v = v)$ states. Since $d(pv) = p\,dv + v\,dp$ we can write

$$v\,dp = d(pv) - p\,dv$$

so

$$W = \int_i^f d(pv) - \int_i^f p\,dv.$$

The first term on the right of the equal sign is

$$\int_i^f d(pv) = pv\Big|_i^f = (\gamma mv)v = \frac{mv^2}{\sqrt{1 - v^2/c^2}}.$$

The second term in our equation for W above is easily integrated since

$$\frac{d}{dv}\left(\sqrt{1 - v^2/c^2}\right) = -(v/c^2)/\sqrt{1 - v^2/c^2},$$

and so becomes

$$-\int_i^f p\,dv = -\int_0^v \frac{mv}{\sqrt{1 - v^2/c^2}}\,dv = mc^2\sqrt{1 - v^2/c^2}\,\Big|_0^v$$
$$= mc^2\sqrt{1 - v^2/c^2} - mc^2.$$

Finally, we have for W:

$$W = \frac{mv^2}{\sqrt{1 - v^2/c^2}} + mc^2\sqrt{1 - v^2/c^2} - mc^2.$$

We multiply the second term on the right by $\sqrt{1 - v^2/c^2}/\sqrt{1 - v^2/c^2} = 1$, and obtain

$$W = \frac{mc^2}{\sqrt{1 - v^2/c^2}} - mc^2.$$

By the work-energy theorem, the work done on the particle must equal its final kinetic energy K since the particle started from rest. Therefore

$$K = \frac{mc^2}{\sqrt{1 - v^2/c^2}} - mc^2$$
$$= \gamma mc^2 - mc^2 = (\gamma - 1)mc^2,$$

which are Eqs. 36–10.

*36–12 Doppler Shift for Light

In Section 16–7 we discussed how the frequency and wavelength of sound are altered if the source of the sound and the observer are moving toward or away from each other. When a source is moving toward us, the frequency is higher than when the source is at rest. If the source moves away from us, the frequency is lower. We obtained four different equations for the Doppler shift (Eqs. 16–9a and b, Eqs. 16–10a and b), depending on the direction of the relative motion and whether the source or the observer is moving. The Doppler effect occurs also for light; but the shifted frequency or wavelength is given by slightly different equations, and there are only two of them, because for light—according to special relativity—we can make no distinction between motion of the source and motion of the observer. (Recall that sound travels in a medium such as air, whereas light does not—there is no evidence for an ether.)

To derive the Doppler shift for light, let us consider a light source and an observer that move toward each other, and let their relative velocity be v as measured in the reference frame of either the source or the observer. Figure 36–14a shows a source at rest emitting light waves of frequency f_0 and wavelength $\lambda_0 = c/f_0$. Two wavecrests are shown, a distance λ_0 apart, the second crest just having been emitted. In Fig. 36–14b, the source is shown moving at speed v toward a stationary observer who will see the wavelength λ being somewhat less than λ_0. (This is much like Fig. 16–19 for sound.) Let Δt represent the time between crests as detected by the observer, whose reference frame is shown in Fig. 36–14b. From Fig. 36–14b we see that

$$\lambda = c\,\Delta t - v\,\Delta t,$$

where $c\,\Delta t$ is the distance crest 1 has moved in the time Δt after it was emitted, and $v\,\Delta t$ is the distance the source has moved in time Δt. So far our derivation has not differed from that for sound (Section 16–7). Now we invoke the theory of relativity. The time between emission of wavecrests has undergone time dilation:

$$\Delta t = \Delta t_0/\sqrt{1 - v^2/c^2}$$

where Δt_0 is the time between emissions of wavecrests in the reference frame where the source is at rest (the "proper" time). In the source's reference frame (Fig. 36–14a), we have

$$\Delta t_0 = \frac{1}{f_0} = \frac{\lambda_0}{c}$$

(Eqs. 5–2 and 31–14). Thus

$$\lambda = (c - v)\,\Delta t = (c - v)\frac{\Delta t_0}{\sqrt{1 - v^2/c^2}} = \frac{(c - v)}{\sqrt{c^2 - v^2}}\lambda_0$$

or

$$\lambda = \lambda_0\sqrt{\frac{c - v}{c + v}}. \qquad \begin{bmatrix}\text{source and observer}\\ \text{moving toward}\\ \text{each other}\end{bmatrix} \qquad \textbf{(36–14a)}$$

The frequency f is (recall $\lambda_0 = c/f_0$)

$$f = \frac{c}{\lambda} = f_0\sqrt{\frac{c + v}{c - v}}. \qquad \begin{bmatrix}\text{source and observer}\\ \text{moving toward}\\ \text{each other}\end{bmatrix} \qquad \textbf{(36–14b)}$$

Here f_0 is the frequency of the light as seen in the source's reference frame, and f is the frequency as measured by an observer moving toward the source or toward whom the source is moving. Equations 36–14 depend only on the relative velocity v. For relative motion *away* from each other we set $v < 0$ in Eqs. 36–14, and obtain

$$\lambda = \lambda_0\sqrt{\frac{c + v}{c - v}} \qquad \textbf{(36–15a)}$$

$$f = f_0\sqrt{\frac{c - v}{c + v}}. \qquad \textbf{(36–15b)}$$

$$\begin{bmatrix}\text{source and observer}\\ \text{moving away from}\\ \text{each other}\end{bmatrix}$$

From Eqs. 36–14 and 36–15 we see that light from a source moving toward us will have a higher frequency and shorter wavelength, whereas if a light source moves away from us, we will see a lower frequency and a longer wavelength. In the latter case, visible light will have its wavelength lengthened toward the red end of the visible spectrum (Fig. 32–26), an effect called a **redshift**. As we will see in the next Chapter, all atoms have their own distinctive signature in terms of the frequencies of the light they emit. In 1929 the American astronomer Edwin Hubble (1889–1953) found that radiation from atoms in many galaxies is redshifted. That is, the frequencies of light emitted are lower than those emitted by stationary atoms on Earth, suggesting that the galaxies are receding from us. This is the origin of the idea that the universe is expanding.

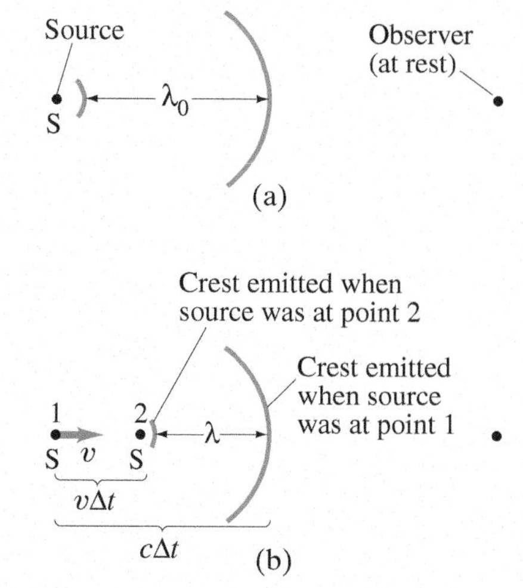

FIGURE 36–14 Doppler shift for light. (a) Source and observer at rest. (b) Source moving toward stationary observer.

EXAMPLE 36–13 **Speeding through a red light.** A driver claims that he did not go through a red light because the light was Doppler shifted and appeared green. Calculate the speed of a driver in order for a red light to appear green.

APPROACH We apply the Doppler shift equation for red light ($\lambda_0 \approx 650$ nm) and green light ($\lambda \approx 500$ nm).

SOLUTION Equation 36–14a holds for the source and the object moving toward each other:

$$\lambda = \lambda_0 \sqrt{\frac{c - v}{c + v}}.$$

We square this equation:

$$\frac{c - v}{c + v} = \left(\frac{\lambda}{\lambda_0}\right)^2$$

where $(\lambda/\lambda_0)^2 = (500\ \text{nm}/650\ \text{nm})^2 = 0.59$. We solve for v:

$$v = c\left[\frac{1 - (\lambda/\lambda_0)^2}{1 + (\lambda/\lambda_0)^2}\right] = 0.26c.$$

With this defense, the driver might not be guilty of running a red light, but he would clearly be guilty of speeding.

36–13 The Impact of Special Relativity

A great many experiments have been performed to test the predictions of the special theory of relativity. Within experimental error, no contradictions have been found. Scientists have therefore accepted relativity as an accurate description of nature.

At speeds much less than the speed of light, the relativistic formulas reduce to the old classical ones, as we have discussed. We would, of course, hope—or rather, insist—that this be true since Newtonian mechanics works so well for objects moving with speeds $v \ll c$. This insistence that a more general theory (such as relativity) give the same results as a more restricted theory (such as classical mechanics which works for $v \ll c$) is called the **correspondence principle**. The two theories must correspond where their realms of validity overlap. Relativity thus does not contradict classical mechanics. Rather, it is a more general theory, of which classical mechanics is now considered to be a limiting case.

The importance of relativity is not simply that it gives more accurate results, especially at very high speeds. Much more than that, it has changed the way we view the world. The concepts of space and time are now seen to be relative, and intertwined with one another, whereas before they were considered absolute and separate. Even our concepts of matter and energy have changed: either can be converted to the other. The impact of relativity extends far beyond physics. It has influenced the other sciences, and even the world of art and literature; it has, indeed, entered the general culture.

From a practical point of view, we do not have much opportunity in our daily lives to use the mathematics of relativity. For example, the γ factor $1/\sqrt{1 - v^2/c^2}$, which appears in relativistic formulas, has a value of only 1.005 even for a speed as high as $0.10c = 3.0 \times 10^7\ \text{m/s}$, giving a correction of less than 1%. For speeds less than $0.10c$, or unless mass and energy are interchanged, we don't usually need to use the more complicated relativistic formulas, and can use the simpler classical formulas.

Summary

An **inertial reference frame** is one in which Newton's law of inertia holds. Inertial reference frames can move at constant velocity relative to one another; accelerating reference frames are **noninertial**.

The **special theory of relativity** is based on two principles: the **relativity principle**, which states that the laws of physics are the same in all inertial reference frames, and the principle of the **constancy of the speed of light**, which states that the speed of light in empty space has the same value in all inertial reference frames.

One consequence of relativity theory is that two events that are simultaneous in one reference frame may not be simultaneous in another. Other effects are **time dilation**: moving clocks are measured to run slow; and **length contraction**: the length of a moving object is measured to be shorter (in its direction of motion) than when it is at rest. Quantitatively,

$$\Delta t = \frac{\Delta t_0}{\sqrt{1 - v^2/c^2}} = \gamma\, \Delta t_0 \qquad \textbf{(36–1)}$$

$$\ell = \ell_0\sqrt{1 - v^2/c^2} = \frac{\ell_0}{\gamma} \qquad \textbf{(36–3)}$$

where ℓ and Δt are the length and time interval of objects (or events) observed as they move by at the speed v; ℓ_0 and Δt_0 are the **proper length** and **proper time**—that is, the same quantities as measured in the rest frame of the objects or events. The quantity γ is shorthand for

$$\gamma = \frac{1}{\sqrt{1 - v^2/c^2}}. \qquad \textbf{(36–2)}$$

The theory of relativity has changed our notions of space and time, and of momentum, energy, and mass. Space and time are seen to be intimately connected, with time being the fourth dimension in addition to the three dimensions of space.

The **Lorentz transformations** relate the positions and times of events in one inertial reference frame to their positions and times in a second inertial reference frame.

$$\begin{aligned} x &= \gamma(x' + vt') \\ y &= y' \\ z &= z' \\ t &= \gamma\left(t' + \frac{vx'}{c^2}\right) \end{aligned} \qquad \textbf{(36–6)}$$

where $\gamma = 1/\sqrt{1 - v^2/c^2}$.

Velocity addition also must be done in a special way. All these relativistic effects are significant only at high speeds, close to the speed of light, which itself is the ultimate speed in the universe.

The **momentum** of an object is given by

$$p = \gamma m v = \frac{mv}{\sqrt{1 - v^2/c^2}}. \qquad \textbf{(36–8)}$$

Mass and energy are interconvertible. The equation

$$E = mc^2 \qquad \textbf{(36–12)}$$

tells how much energy E is needed to create a mass m, or vice versa. Said another way, $E = mc^2$ is the amount of energy an object has because of its mass m. The law of conservation of energy must include mass as a form of energy.

The kinetic energy K of an object moving at speed v is given by

$$K = (\gamma - 1)mc^2 = \frac{mc^2}{\sqrt{1 - v^2/c^2}} - mc^2 \qquad \textbf{(36–10)}$$

where m is the mass of the object. The total energy E, if there is no potential energy, is

$$\begin{aligned} E &= K + mc^2 \\ &= \gamma mc^2. \end{aligned} \qquad \textbf{(36–11)}$$

The momentum p of an object is related to its total energy E (assuming no potential energy) by

$$E^2 = p^2c^2 + m^2c^4. \qquad \textbf{(36–13)}$$

Questions

1. You are in a windowless car in an exceptionally smooth train moving at constant velocity. Is there any physical experiment you can do in the train car to determine whether you are moving? Explain.
2. You might have had the experience of being at a red light when, out of the corner of your eye, you see the car beside you creep forward. Instinctively you stomp on the brake pedal, thinking that you are rolling backward. What does this say about absolute and relative motion?
3. A worker stands on top of a moving railroad car, and throws a heavy ball straight up (from his point of view). Ignoring air resistance, will the ball land back in his hand or behind him?
4. Does the Earth really go around the Sun? Or is it also valid to say that the Sun goes around the Earth? Discuss in view of the relativity principle (that there is no best reference frame). Explain.
5. If you were on a spaceship traveling at $0.5c$ away from a star, at what speed would the starlight pass you?
6. The time dilation effect is sometimes expressed as "moving clocks run slowly." Actually, this effect has nothing to do with motion affecting the functioning of clocks. What then does it deal with?
7. Does time dilation mean that time actually passes more slowly in moving reference frames or that it only *seems* to pass more slowly?
8. A young-looking woman astronaut has just arrived home from a long trip. She rushes up to an old gray-haired man and in the ensuing conversation refers to him as her son. How might this be possible?
9. If you were traveling away from Earth at speed $0.5c$, would you notice a change in your heartbeat? Would your mass, height, or waistline change? What would observers on Earth using telescopes say about you?
10. Do time dilation and length contraction occur at ordinary speeds, say 90 km/h?
11. Suppose the speed of light were infinite. What would happen to the relativistic predictions of length contraction and time dilation?
12. Discuss how our everyday lives would be different if the speed of light were only 25 m/s.
13. Explain how the length contraction and time dilation formulas might be used to indicate that c is the limiting speed in the universe.

14. The drawing at the start of this Chapter shows the street as seen by Mr Tompkins, where the speed of light is $c = 20$ mi/h. What does Mr Tompkins look like to the people standing on the street (Fig. 36–15)? Explain.

FIGURE 36–15 Question 14. Mr Tompkins as seen by people on the sidewalk. See also Chapter-Opening figure on page 951.

15. An electron is limited to travel at speeds less than c. Does this put an upper limit on the momentum of an electron? If so, what is this upper limit? If not, explain.

16. Can a particle of nonzero mass attain the speed of light?

17. Does the equation $E = mc^2$ conflict with the conservation of energy principle? Explain.

18. If mass is a form of energy, does this mean that a spring has more mass when compressed than when relaxed?

19. It is not correct to say that "matter can neither be created nor destroyed." What must we say instead?

20. Is our intuitive notion that velocities simply add, as in Section 3–9, completely wrong?

Problems

36–5 and 36–6 Time Dilation, Length Contraction

1. (I) A spaceship passes you at a speed of $0.850c$. You measure its length to be 38.2 m. How long would it be when at rest?
2. (I) A certain type of elementary particle travels at a speed of 2.70×10^8 m/s. At this speed, the average lifetime is measured to be 4.76×10^{-6} s. What is the particle's lifetime at rest?
3. (II) According to the special theory of relativity, the factor γ that determines the length contraction and the time dilation is given by $\gamma = 1/\sqrt{1 - v^2/c^2}$. Determine the numerical values of γ for an object moving at speed $v = 0.01c, 0.05c, 0.10c, 0.20c, 0.30c, 0.40c, 0.50c, 0.60c, 0.70c, 0.80c, 0.90c$, and $0.99c$. Make a graph of γ versus v.
4. (II) If you were to travel to a star 135 light-years from Earth at a speed of 2.80×10^8 m/s, what would you measure this distance to be?
5. (II) What is the speed of a pion if its average lifetime is measured to be 4.40×10^{-8} s? At rest, its average lifetime is 2.60×10^{-8} s.
6. (II) In an Earth reference frame, a star is 56 light-years away. How fast would you have to travel so that to you the distance would be only 35 light-years?
7. (II) Suppose you decide to travel to a star 65 light-years away at a speed that tells you the distance is only 25 light-years. How many years would it take you to make the trip?
8. (II) At what speed v will the length of a 1.00-m stick look 10.0% shorter (90.0 cm)?
9. (II) Escape velocity from the Earth is 11.2 km/s. What would be the percent decrease in length of a 65.2-m-long spacecraft traveling at that speed as seen from Earth?
10. (II) A friend speeds by you in her spacecraft at a speed of $0.760c$. It is measured in your frame to be 4.80 m long and 1.35 m high. (*a*) What will be its length and height at rest? (*b*) How many seconds elapsed on your friend's watch when 20.0 s passed on yours? (*c*) How fast did you appear to be traveling according to your friend? (*d*) How many seconds elapsed on your watch when she saw 20.0 s pass on hers?
11. (II) At what speed do the relativistic formulas for (*a*) length and (*b*) time intervals differ from classical values by 1.00%? (This is a reasonable way to estimate when to do relativistic calculations rather than classical.)
12. (II) A certain star is 18.6 light-years away. How long would it take a spacecraft traveling $0.950c$ to reach that star from Earth, as measured by observers: (*a*) on Earth, (*b*) on the spacecraft? (*c*) What is the distance traveled according to observers on the spacecraft? (*d*) What will the spacecraft occupants compute their speed to be from the results of (*b*) and (*c*)?
13. (II) Suppose a news report stated that starship *Enterprise* had just returned from a 5-year voyage while traveling at $0.74c$. (*a*) If the report meant 5.0 years of *Earth time*, how much time elapsed on the ship? (*b*) If the report meant 5.0 years of *ship time*, how much time passed on Earth?
14. (II) An unstable particle produced in an accelerator experiment travels at constant velocity, covering 1.00 m in 3.40 ns in the lab frame before changing ("decaying") into other particles. In the rest frame of the particle, determine (*a*) how long it lived before decaying, (*b*) how far it moved before decaying.
15. (II) When it is stationary, the half-life of a certain subatomic particle is T_0. That is, if N of these particles are present at a certain time, then a time T_0 later only $N/2$ particles will be present, assuming the particles are at rest. A beam carrying N such particles per second is created at position $x = 0$ in a high-energy physics laboratory. This beam travels along the x axis at speed v in the laboratory reference frame and it is found that only $N/2$ particles per second travel in the beam at $x = 2cT_0$, where c is the speed of light. Find the speed v of the particles within the beam.
16. (II) In its own reference frame, a box has the shape of a cube 2.0 m on a side. This box is loaded onto the flat floor of a spaceship and the spaceship then flies past us with a horizontal speed of $0.80c$. What is the volume of the box as we observe it?
17. (II) When at rest, a spaceship has the form of an isosceles triangle whose two equal sides have length 2ℓ and whose base has length ℓ. If this ship flies past an observer with a relative velocity of $v = 0.95c$ directed along its base, what are the lengths of the ship's three sides according to the observer?
18. (II) How fast must a pion be moving on average to travel 25 m before it decays? The average lifetime, at rest, is 2.6×10^{-8} s.

36–8 Lorentz Transformations

19. (I) An observer on Earth sees an alien vessel approach at a speed of 0.60c. The *Enterprise* comes to the rescue (Fig. 36–16), overtaking the aliens while moving directly toward Earth at a speed of 0.90c relative to Earth. What is the relative speed of one vessel as seen by the other?

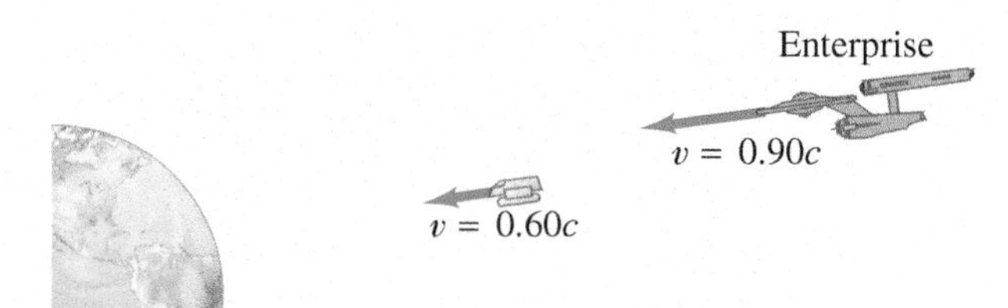

FIGURE 36–16 Problem 19.

20. (I) Suppose in Fig. 36–11 that the origins of S and S′ overlap at $t = t' = 0$ and that S′ moves at speed $v = 30\,\mathrm{m/s}$ with respect to S. In S′, a person is resting at a point whose coordinates are $x' = 25\,\mathrm{m}$, $y' = 20\,\mathrm{m}$, and $z' = 0$. Calculate this person's coordinates in S (x, y, z) at (*a*) $t = 3.5\,\mathrm{s}$, (*b*) $t = 10.0\,\mathrm{s}$. Use the Galilean transformation.
21. (I) Repeat Problem 20 using the Lorentz transformation and a relative speed $v = 1.80 \times 10^8\,\mathrm{m/s}$, but choose the time t to be (*a*) 3.5 μs and (*b*) 10.0 μs.
22. (II) In Problem 21, suppose that the person moves with a velocity whose components are $u'_x = u'_y = 1.10 \times 10^8\,\mathrm{m/s}$, What will be her velocity with respect to S? (Give magnitude and direction.)
23. (II) Two spaceships leave Earth in opposite directions, each with a speed of 0.60c with respect to Earth. (*a*) What is the velocity of spaceship 1 relative to spaceship 2? (*b*) What is the velocity of spaceship 2 relative to spaceship 1?
24. (II) Reference frame S′ moves at speed $v = 0.92c$ in the $+x$ direction with respect to reference frame S. The origins of S and S′ overlap at $t = t' = 0$. An object is stationary in S′ at position $x' = 100\,\mathrm{m}$. What is the position of the object in S when the clock in S reads 1.00 μs according to the (*a*) Galilean and (*b*) Lorentz transformation equations?
25. (II) A spaceship leaves Earth traveling at 0.61c. A second spaceship leaves the first at a speed of 0.87c with respect to the first. Calculate the speed of the second ship with respect to Earth if it is fired (*a*) in the same direction the first spaceship is already moving, (*b*) directly backward toward Earth.
26. (II) Your spaceship, traveling at 0.90c, needs to launch a probe out the forward hatch so that its speed relative to the planet that you are approaching is 0.95c. With what speed must it leave your ship?
27. (II) A spaceship traveling at 0.76c away from Earth fires a module with a speed of 0.82c at right angles to its own direction of travel (as seen by the spaceship). What is the speed of the module, and its direction of travel (relative to the spaceship's direction), as seen by an observer on Earth?
28. (II) If a particle moves in the xy plane of system S (Fig. 36–11) with speed u in a direction that makes an angle θ with the x axis, show that it makes an angle θ' in S′ given by $\tan\theta' = (\sin\theta)\sqrt{1 - v^2/c^2}/(\cos\theta - v/u)$.
29. (II) A stick of length ℓ_0, at rest in reference frame S, makes an angle θ with the x axis. In reference frame S′, which moves to the right with velocity $\vec{\mathbf{v}} = v\hat{\mathbf{i}}$ with respect to S, determine (*a*) the length ℓ of the stick, and (*b*) the angle θ' it makes with the x' axis.
30. (III) In the old West, a marshal riding on a train traveling 35.0 m/s sees a duel between two men standing on the Earth 55.0 m apart parallel to the train. The marshal's instruments indicate that in his reference frame the two men fired simultaneously. (*a*) Which of the two men, the first one the train passes (A) or the second one (B) should be arrested for firing the first shot? That is, in the gunfighter's frame of reference, who fired first? (*b*) How much earlier did he fire? (*c*) Who was struck first?
31. (III) Two lightbulbs, A and B, are placed at rest on the x axis at positions $x_A = 0$ and $x_B = +\ell$. In this reference frame, the bulbs are turned on simultaneously. Use the Lorentz transformations to find an expression for the time interval between when the bulbs are turned on as measured by an observer moving at velocity v in the $+x$ direction. According to this observer, which bulb is turned on first?
32. (III) An observer in reference frame S notes that two events are separated in space by 220 m and in time by 0.80 μs. How fast must reference frame S′ be moving relative to S in order for an observer in S′ to detect the two events as occurring at the same location in space?
33. (III) A farm boy studying physics believes that he can fit a 12.0-m long pole into a 10.0-m long barn if he runs fast enough, carrying the pole. Can he do it? Explain in detail. How does this fit with the idea that when he is running the barn looks even shorter than 10.0 m?

36–9 Relativistic Momentum

34. (I) What is the momentum of a proton traveling at $v = 0.75c$?
35. (II) (*a*) A particle travels at $v = 0.10c$. By what percentage will a calculation of its momentum be wrong if you use the classical formula? (*b*) Repeat for $v = 0.60c$.
36. (II) A particle of mass m travels at a speed $v = 0.26c$. At what speed will its momentum be doubled?
37. (II) An unstable particle is at rest and suddenly decays into two fragments. No external forces act on the particle or its fragments. One of the fragments has a speed of 0.60c and a mass of 6.68×10^{-27} kg, while the other has a mass of 1.67×10^{-27} kg. What is the speed of the less massive fragment?
38. (II) What is the percent change in momentum of a proton that accelerates (*a*) from 0.45c to 0.80c, (*b*) from 0.80c to 0.98c?

36–11 Relativistic Energy

39. (I) Calculate the rest energy of an electron in joules and in MeV $(1\,\mathrm{MeV} = 1.60 \times 10^{-13}\,\mathrm{J})$.
40. (I) When a uranium nucleus at rest breaks apart in the process known as fission in a nuclear reactor, the resulting fragments have a total kinetic energy of about 200 MeV. How much mass was lost in the process?
41. (I) The total annual energy consumption in the United States is about 8×10^{19} J. How much mass would have to be converted to energy to fuel this need?
42. (I) Calculate the mass of a proton in MeV/c^2.
43. (II) Suppose there was a process by which two photons, each with momentum 0.50 MeV/c, could collide and make a single particle. What is the maximum mass that the particle could possess?
44. (II) (*a*) How much work is required to accelerate a proton from rest up to a speed of 0.998c? (*b*) What would be the momentum of this proton?

45. (II) How much energy can be obtained from conversion of 1.0 gram of mass? How much mass could this energy raise to a height of 1.0 km above the Earth's surface?

46. (II) To accelerate a particle of mass m from rest to speed $0.90c$ requires work W_1. To accelerate the particle from speed $0.90c$ to $0.99c$, requires work W_2. Determine the ratio W_2/W_1.

47. (II) What is the speed of a particle when its kinetic energy equals its rest energy?

48. (II) What is the momentum of a 950-MeV proton (that is, its kinetic energy is 950 MeV)?

49. (II) Calculate the kinetic energy and momentum of a proton traveling 2.80×10^8 m/s.

50. (II) What is the speed of an electron whose kinetic energy is 1.25 MeV?

51. (II) What is the speed of a proton accelerated by a potential difference of 125 MV?

52. (II) Two identical particles of mass m approach each other at equal and opposite speeds, v. The collision is completely inelastic and results in a single particle at rest. What is the mass of the new particle? How much energy was lost in the collision? How much kinetic energy was lost in this collision?

53. (II) What is the speed of an electron just before it hits a television screen after being accelerated from rest by the 28,000 V of the picture tube?

54. (II) The kinetic energy of a particle is 45 MeV. If the momentum is 121 MeV/c, what is the particle's mass?

55. (II) Calculate the speed of a proton $(m = 1.67 \times 10^{-27}\,\text{kg})$ whose kinetic energy is exactly half (*a*) its total energy, (*b*) its rest energy.

56. (II) Calculate the kinetic energy and momentum of a proton $(m = 1.67 \times 10^{-27}\,\text{kg})$ traveling 8.15×10^7 m/s. By what percentages would your calculations have been in error if you had used classical formulas?

57. (II) Suppose a spacecraft of mass 17,000 kg is accelerated to $0.18c$. (*a*) How much kinetic energy would it have? (*b*) If you used the classical formula for kinetic energy, by what percentage would you be in error?

*58. (II) What magnetic field B is needed to keep 998-GeV protons revolving in a circle of radius 1.0 km (at, say, the Fermilab synchrotron)? Use the relativistic mass. The proton's rest mass is 0.938 GeV/c^2. $(1\,\text{GeV} = 10^9\,\text{eV}.)$ [*Hint*: In relativity, $m_{\text{rel}}v^2/r = qvB$ is still valid in a magnetic field, where $m_{\text{rel}} = \gamma m$.]

59. (II) The americium nucleus, $^{241}_{95}\text{Am}$, decays to a neptunium nucleus, $^{237}_{93}\text{Np}$, by emitting an alpha particle of mass 4.00260 u and kinetic energy 5.5 MeV. Estimate the mass of the neptunium nucleus, ignoring its recoil, given that the americium mass is 241.05682 u.

60. (II) Make a graph of the kinetic energy versus momentum for (*a*) a particle of nonzero mass, and (*b*) a particle with zero mass.

61. (II) A negative muon traveling at 43% the speed of light collides head on with a positive muon traveling at 55% the speed of light. The two muons (each of mass 105.7 MeV/c^2) annihilate, and produce how much electromagnetic energy?

62. (II) Show that the kinetic energy K of a particle of mass m is related to its momentum p by the equation

$$p = \sqrt{K^2 + 2Kmc^2}/c.$$

63. (III) (*a*) In reference frame S, a particle has momentum $\vec{\mathbf{p}} = p_x\hat{\mathbf{i}}$ along the positive x axis. Show that in frame S′, which moves with speed v as in Fig. 36–11, the momentum has components

$$p'_x = \frac{p_x - vE/c^2}{\sqrt{1 - v^2/c^2}}$$
$$p'_y = p_y$$
$$p'_z = p_z$$
$$E' = \frac{E - p_x v}{\sqrt{1 - v^2/c^2}}.$$

(These transformation equations hold, actually, for any direction of $\vec{\mathbf{p}}$, as long as the motion of S′ is along the x axis.) (*b*) Show that $p_x, p_y, p_z, E/c$ transform according to the Lorentz transformation in the same way as x, y, z, ct.

36–12 Doppler Shift for Light

64. (II) A certain galaxy has a Doppler shift given by $f_0 - f = 0.0987f_0$. Estimate how fast it is moving away from us.

65. (II) A spaceship moving toward Earth at $0.70c$ transmits radio signals at 95.0 MHz. At what frequency should Earth receivers be tuned?

66. (II) Starting from Eq. 36–15a, show that the Doppler shift in wavelength is

$$\frac{\Delta\lambda}{\lambda} = \frac{v}{c}$$

if $v \ll c$.

67. (III) A radar "speed gun" emits microwaves of frequency $f_0 = 36.0$ GHz. When the gun is pointed at an object moving toward it at speed v, the object senses the microwaves at the Doppler-shifted frequency f. The moving object reflects these microwaves at this same frequency f. The stationary radar apparatus detects these reflected waves at a Doppler-shifted frequency f'. The gun combines its emitted wave at f_0 and its detected wave at f'. These waves interfere, creating a beat pattern whose beat frequency is $f_{\text{beat}} = f' - f_0$. (*a*) Show that

$$v \approx \frac{cf_{\text{beat}}}{2f_0},$$

if $f_{\text{beat}} \ll f_0$. If $f_{\text{beat}} = 6670$ Hz, what is v (km/h)? (*b*) If the object's speed is different by Δv, show that the difference in beat frequency Δf_{beat} is given by

$$\Delta f_{\text{beat}} = \frac{2f_0\,\Delta v}{c}.$$

If the accuracy of the speed gun is to be 1 km/h, to what accuracy must the beat frequency be measured?

68. (III) A certain atom emits light of frequency f_0 when at rest. A monatomic gas composed of these atoms is at temperature T. Some of the gas atoms move toward and others away from an observer due to their random thermal motion. Using the rms speed of thermal motion, show that the fractional difference between the Doppler-shifted frequencies for atoms moving directly toward the observer and directly away from the observer is $\Delta f/f_0 \approx 2\sqrt{3kT/mc^2}$; assume $mc^2 \gg 3kT$. Evaluate $\Delta f/f_0$ for a gas of hydrogen atoms at 550 K. [This "Doppler-broadening" effect is commonly used to measure gas temperature, such as in astronomy.]

General Problems

69. An atomic clock is taken to the North Pole, while another stays at the Equator. How far will they be out of synchronization after 2.0 years has elapsed? [*Hint*: Use the binomial expansion, Appendix A.]

70. A spaceship in distress sends out two escape pods in opposite directions. One travels at a speed $v_1 = -0.60c$ in one direction, and the other travels at a speed $v_2 = +0.50c$ in the other direction, as observed from the spaceship. What speed does the first escape pod measure for the second escape pod?

71. An airplane travels 1300 km/h around the Earth in a circle of radius essentially equal to that of the Earth, returning to the same place. Using special relativity, estimate the difference in time to make the trip as seen by Earth and airplane observers. [*Hint*: Use the binomial expansion, Appendix A.]

72. The nearest star to Earth is Proxima Centauri, 4.3 light-years away. (*a*) At what constant velocity must a spacecraft travel from Earth if it is to reach the star in 4.6 years, as measured by travelers on the spacecraft? (*b*) How long does the trip take according to Earth observers?

73. A quasar emits familiar hydrogen lines whose wave-lengths are 7.0% longer than what we measure in the laboratory. (*a*) Using the Doppler formula for light, estimate the speed of this quasar. (*b*) What result would you obtain if you used the "classical" Doppler shift discussed in Chapter 16?

74. A healthy astronaut's heart rate is 60 beats/min. Flight doctors on Earth can monitor an astronaut's vital signs remotely while in flight. How fast would an astronaut have to be flying away from Earth in order for the doctor to measure her having a heart rate of 30 beats/min?

75. A spacecraft (reference frame S′) moves past Earth (reference frame S) at velocity $\vec{\mathbf{v}}$, which points along the x and x' axes. The spacecraft emits a light beam (speed c) along its y' axis as shown in Fig. 36–17. (*a*) What angle θ does this light beam make with the x axis in the Earth's reference frame? (*b*) Use velocity transformations to show that the light moves with speed c also in the Earth's reference frame. (*c*) Compare these relativistic results to what you would have obtained classically (Galilean transformations).

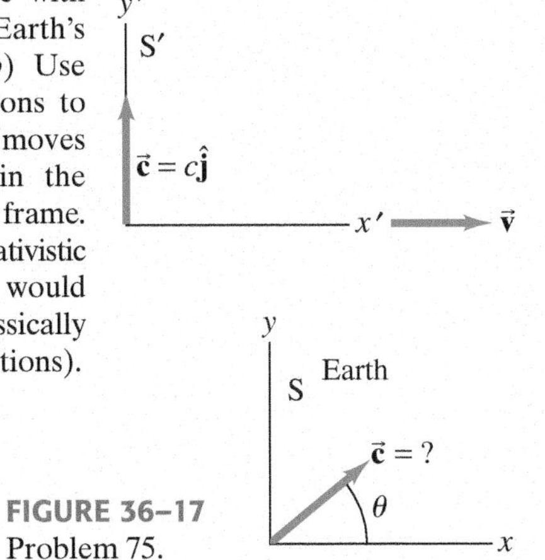

FIGURE 36–17 Problem 75.

76. Rocket A passes Earth at a speed of 0.65c. At the same time, rocket B passes Earth moving 0.85c relative to Earth in the same direction. How fast is B moving relative to A when it passes A?

77. (*a*) What is the speed v of an electron whose kinetic energy is 14,000 times its rest energy? You can state the answer as the difference $c - v$. Such speeds are reached in the Stanford Linear Accelerator, SLAC. (*b*) If the electrons travel in the lab through a tube 3.0 km long (as at SLAC), how long is this tube in the electrons' reference frame? [*Hint*: Use the binomial expansion.]

78. As a rough rule, anything traveling faster than about 0.1c is called *relativistic*—that is, special relativity is a significant effect. Determine the speed of an electron in a hydrogen atom (radius 0.53×10^{-10} m) and state whether or not it is relativistic. (Treat the electron as though it were in a circular orbit around the proton.)

79. What minimum amount of electromagnetic energy is needed to produce an electron and a positron together? A positron is a particle with the same mass as an electron, but has the opposite charge. (Note that electric charge is conserved in this process. See Section 37–5.)

80. How many grams of matter would have to be totally destroyed to run a 75-W lightbulb for 1.0 year?

81. If E is the total energy of a particle with zero potential energy, show that $dE/dp = v$, where p and v are the momentum and velocity of the particle, respectively.

82. A free neutron can decay into a proton, an electron, and a neutrino. Assume the neutrino's mass is zero; the other masses can be found in the Table inside the front cover. Determine the total kinetic energy shared among the three particles when a neutron decays at rest.

83. The Sun radiates energy at a rate of about 4×10^{26} W. (*a*) At what rate is the Sun's mass decreasing? (*b*) How long does it take for the Sun to lose a mass equal to that of Earth? (*c*) Estimate how long the Sun could last if it radiated constantly at this rate.

84. An unknown particle is measured to have a negative charge and a speed of 2.24×10^{8} m/s. Its momentum is determined to be 3.07×10^{-22} kg·m/s. Identify the particle by finding its mass.

85. How much energy would be required to break a helium nucleus into its constituents, two protons and two neutrons? The rest masses of a proton (including an electron), a neutron, and neutral helium are, respectively, 1.00783 u, 1.00867 u, and 4.00260 u. (This energy difference is called the *total binding energy* of the $^4_2$He nucleus.)

86. Show analytically that a particle with momentum p and energy E has a speed given by

$$v = \frac{pc^2}{E} = \frac{pc}{\sqrt{m^2c^2 + p^2}}.$$

87. Two protons, each having a speed of 0.985c in the laboratory, are moving toward each other. Determine (*a*) the momentum of each proton in the laboratory, (*b*) the total momentum of the two protons in the laboratory, and (*c*) the momentum of one proton as seen by the other proton.

88. When two moles of hydrogen molecules (H_2) and one mole of oxygen molecules (O_2) react to form two moles of water (H_2O), the energy released is 484 kJ. How much does the mass decrease in this reaction? What % of the total original mass of the system does this mass change represent?

89. The fictional starship *Enterprise* obtains its power by combining matter and antimatter, achieving complete conversion of mass into energy. If the mass of the *Enterprise* is approximately 6×10^{9} kg, how much mass must be converted into kinetic energy to accelerate it from rest to one-tenth the speed of light?

90. A spaceship and its occupants have a total mass of 180,000 kg. The occupants would like to travel to a star that is 35 light-years away at a speed of $0.70c$. To accelerate, the engine of the spaceship changes mass directly to energy. How much mass will be converted to energy to accelerate the spaceship to this speed? Assume the acceleration is rapid, so the speed for the entire trip can be taken to be $0.70c$, and ignore decrease in total mass for the calculation. How long will the trip take according to the astronauts on board?

91. In a nuclear reaction two identical particles are created, traveling in opposite directions. If the speed of each particle is $0.85c$, relative to the laboratory frame of reference, what is one particle's speed relative to the other particle?

92. A 32,000-kg spaceship is to travel to the vicinity of a star 6.6 light-years from Earth. Passengers on the ship want the (one-way) trip to take no more than 1.0 year. How much work must be done on the spaceship to bring it to the speed necessary for this trip?

93. Suppose a 14,500-kg spaceship left Earth at a speed of $0.98c$. What is the spaceship's kinetic energy? Compare with the total U.S. annual energy consumption (about 10^{20} J).

94. A pi meson of mass m_π decays at rest into a muon (mass m_μ) and a neutrino of negligible or zero mass. Show that the kinetic energy of the muon is $K_\mu = (m_\pi - m_\mu)^2c^2/(2m_\pi)$.

95. Astronomers measure the distance to a particular star to be 6.0 light-years (1 ly = distance light travels in 1 year). A spaceship travels from Earth to the vicinity of this star at steady speed, arriving in 2.50 years as measured by clocks on the spaceship. (*a*) How long does the trip take as measured by clocks in Earth's reference frame (assumed inertial)? (*b*) What distance does the spaceship travel as measured in its own reference frame?

96. A 1.88-kg mass oscillates on the end of a spring whose spring stiffness constant is $k = 84.2$ N/m. If this system is in a spaceship moving past Earth at $0.900c$, what is its period of oscillation according to (*a*) observers on the ship, and (*b*) observers on Earth?

97. Show that the space–time interval, $(c\,\Delta t)^2 - (\Delta x)^2$, is invariant, meaning that all observers in all inertial reference frames calculate the same number for this quantity for any pair of events.

98. A slab of glass with index of refraction n moves to the right with speed v. A flash of light is emitted at point A (Fig. 36–18) and passes through the glass arriving at point B a distance ℓ away. The glass has thickness d in the reference frame where it is at rest, and the speed of light in the glass is c/n. How long does it take the light to go from point A to point B according to an observer at rest with respect to points A and B? Check your answer for the cases $v = c$, $v = 0$, and $n = 1$.

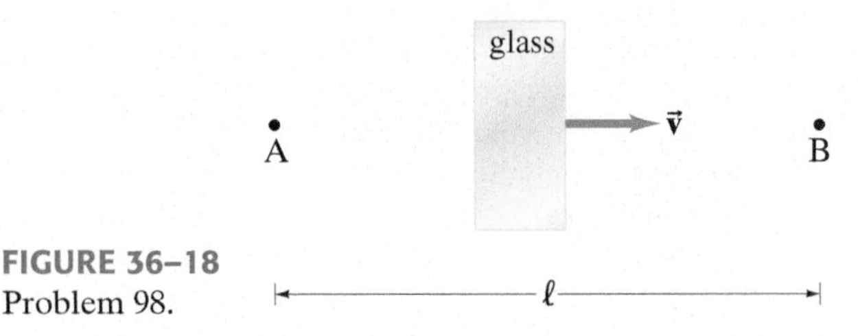

FIGURE 36–18 Problem 98.

*Numerical/Computer

* 99. (II) For a 1.0-kg mass, make a plot of the kinetic energy as a function of speed for speeds from 0 to $0.9c$, using both the classical formula $\left(K = \frac{1}{2}mv^2\right)$ and the correct relativistic formula $\left(K = (\gamma - 1)mc^2\right)$.

*100. (III) A particle of mass m is projected horizontally at a relativistic speed v_0 in the $+x$ direction. There is a constant downward force F acting on the particle. Using the definition of relativistic momentum $\vec{\mathbf{p}} = \gamma m\vec{\mathbf{v}}$ and Newton's second law $\vec{\mathbf{F}} = d\vec{\mathbf{p}}/dt$, (a) show that the x and y components of the velocity of the particle at time t are given by

$$v_x(t) = p_0c/\left(m^2c^2 + p_0^2 + F^2t^2\right)^{\frac{1}{2}}$$

$$v_y(t) = -Fct/\left(m^2c^2 + p_0^2 + F^2t^2\right)^{\frac{1}{2}}$$

where p_0 is the initial momentum of the particle. (b) Assume the particle is an electron $(m = 9.11 \times 10^{-31}\text{ kg})$, with $v_0 = 0.50c$ and $F = 1.00 \times 10^{-15}$ N. Calculate the values of v_x and v_y of the electron as a function of time t from $t = 0$ to $t = 5.00\ \mu$s in intervals of $0.05\ \mu$s. Graph the values to show how the velocity components change with time during this interval. (c) Is the path parabolic, as it would be in classical mechanics (Sections 3–7 and 3–8)? Explain.

Answers to Exercises

A: Yes.

B: (*c*).

C: (*a*) No; (*b*) yes.

D: 80 m.

E: $0.030c$, same as classical, to an accuracy of better than 0.1%.

F: (*d*).

G: (*d*).

H: No.

Electron microscopes produce images using electrons which have wave properties just as light does. Since the wavelength of electrons can be much smaller than that of visible light, much greater resolution and magnification can be obtained. A scanning electron microscope (SEM) can produce images with a three-dimensional quality, as for these *Giardia* cells inside a human small intestine. Magnification here is about 2000×. *Giardia* is on the minds of backpackers because it has become too common in untreated water, even in the high mountains, and causes an unpleasant intestinal infection not easy to get rid of.

CHAPTER 37

Early Quantum Theory and Models of the Atom

CHAPTER-OPENING QUESTION—**Guess now!**

It has been found experimentally that

(a) light behaves as a wave.
(b) light behaves as a particle.
(c) electrons behave as particles.
(d) electrons behave as waves.
(e) all of the above are true.
(f) none of the above are true.

CONTENTS

The second aspect of the revolution that shook the world of physics in the early part of the twentieth century was the quantum theory (the other was Einstein's theory of relativity). Unlike the special theory of relativity, the revolution of quantum theory required almost three decades to unfold, and many scientists contributed to its development. It began in 1900 with Planck's quantum hypothesis, and culminated in the mid-1920s with the theory of quantum mechanics of Schrödinger and Heisenberg which has been so effective in explaining the structure of matter.

37–1 Planck's Quantum Hypothesis; Blackbody Radiation

Blackbody Radiation

One of the observations that was unexplained at the end of the nineteenth century was the spectrum of light emitted by hot objects. We saw in Section 19–10 that all objects emit radiation whose total intensity is proportional to the fourth power of the Kelvin (absolute) temperature (T^4). At normal temperatures ($\approx 300\,\mathrm{K}$), we are not aware of this electromagnetic radiation because of its low intensity.

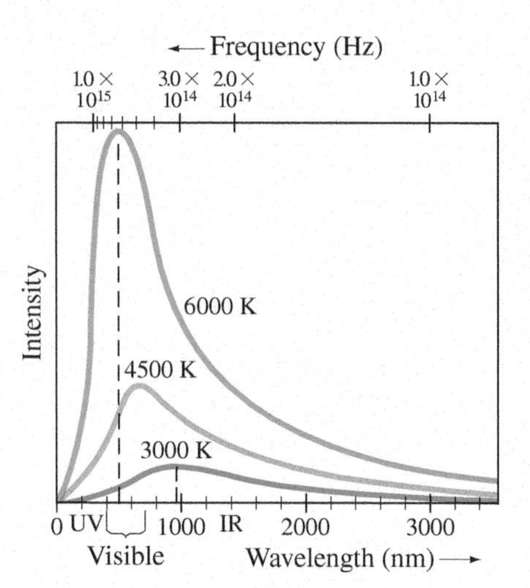

FIGURE 37–1 Measured spectra of wavelengths and frequencies emitted by a blackbody at three different temperatures.

At higher temperatures, there is sufficient infrared radiation that we can feel heat if we are close to the object. At still higher temperatures (on the order of 1000 K), objects actually glow, such as a red-hot electric stove burner or the heating element in a toaster. At temperatures above 2000 K, objects glow with a yellow or whitish color, such as white-hot iron and the filament of a lightbulb. The light emitted is of a continuous range of wavelengths or frequencies, and the *spectrum* is a plot of intensity vs. wavelength or frequency. As the temperature increases, the electromagnetic radiation emitted by objects not only increases in total intensity but reaches a peak at higher and higher frequencies.

The spectrum of light emitted by a hot dense object is shown in Fig. 37–1 for an idealized **blackbody**. A blackbody is a body that would absorb all the radiation falling on it (and so would appear black under reflection when illuminated by other sources). The radiation such an idealized blackbody would emit when hot and luminous, called **blackbody radiation** (though not necessarily black in color), approximates that from many real objects. The 6000-K curve in Fig. 37–1, corresponding to the temperature of the surface of the Sun, peaks in the visible part of the spectrum. For lower temperatures, the total radiation drops considerably and the peak occurs at longer wavelengths (or lower frequencies). (This is why objects glow with a red color at around 1000 K.) It is found experimentally that the wavelength at the peak of the spectrum, λ_{P}, is related to the Kelvin temperature T by

$$\lambda_{\mathrm{P}} T = 2.90 \times 10^{-3}\,\mathrm{m \cdot K}. \qquad \textbf{(37–1)}$$

This is known as **Wien's law**.

EXAMPLE 37–1 The Sun's surface temperature. Estimate the temperature of the surface of our Sun, given that the Sun emits light whose peak intensity occurs in the visible spectrum at around 500 nm.

APPROACH We assume the Sun acts as a blackbody, and use $\lambda_{\mathrm{P}} = 500\,\mathrm{nm}$ in Wien's law (Eq. 37–1).

SOLUTION Wien's law gives

$$T = \frac{2.90 \times 10^{-3}\,\mathrm{m \cdot K}}{\lambda_{\mathrm{P}}} = \frac{2.90 \times 10^{-3}\,\mathrm{m \cdot K}}{500 \times 10^{-9}\,\mathrm{m}} \approx 6000\,\mathrm{K}.$$

EXAMPLE 37–2 Star color. Suppose a star has a surface temperature of 32,500 K. What color would this star appear?

APPROACH We assume the star emits radiation as a blackbody, and solve for λ_{P} in Wien's law, Eq. 37–1.

SOLUTION From Wien's law we have

$$\lambda_{\mathrm{P}} = \frac{2.90 \times 10^{-3}\,\mathrm{m \cdot K}}{T} = \frac{2.90 \times 10^{-3}\,\mathrm{m \cdot K}}{3.25 \times 10^{4}\,\mathrm{K}} = 89.2\,\mathrm{nm}.$$

The peak is in the UV range of the spectrum, and will be way to the left in Fig. 37–1. In the visible region, the curve will be descending, so the shortest visible wavelengths will be strongest. Hence the star will appear bluish (or blue-white).

NOTE This example helps us to understand why stars have different colors (reddish for the coolest stars, orangish, yellow, white, bluish for "hotter" stars.)

FIGURE 37–2 Comparison of the Wien and the Rayleigh–Jeans theories to that of Planck, which closely follows experiment. The dashed lines show lack of agreement of older theories.

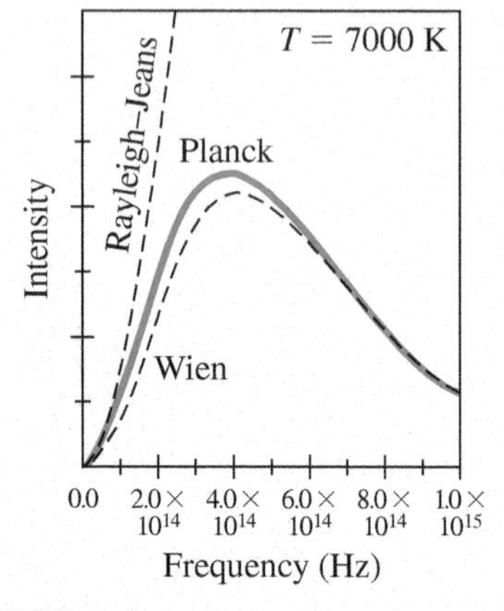

Planck's Quantum Hypothesis

A major problem facing scientists in the 1890s was to explain blackbody radiation. Maxwell's electromagnetic theory had predicted that oscillating electric charges produce electromagnetic waves, and the radiation emitted by a hot object could be due to the oscillations of electric charges in the molecules of the material. Although this would explain where the radiation came from, it did not correctly predict the observed spectrum of emitted light. Two important theoretical curves based on classical ideas were those proposed by W. Wien (in 1896) and by Lord Rayleigh (in 1900). The latter was modified later by J. Jeans and since then has been known as the Rayleigh–Jeans theory. As experimental data came in, it became clear that neither Wien's nor the Rayleigh–Jeans formulations were in accord with experiment (see Fig. 37–2).

In the year 1900 Max Planck (1858–1947) proposed an empirical formula that nicely fit the data (now often called *Planck's radiation formula*):

$$I(\lambda, T) = \frac{2\pi hc^2\lambda^{-5}}{e^{hc/\lambda kT} - 1}.$$

$I(\lambda, T)$ is the radiation intensity as a function of wavelength λ at the temperature T; k is Boltzman's constant, c is the speed of light, and h is a new constant, now called **Planck's constant**. The value of h was estimated by Planck by fitting his formula for the blackbody radiation curve to experiment. The value accepted today is

$$h = 6.626 \times 10^{-34}\,\mathrm{J\cdot s}.$$

To provide a theoretical basis for his formula, Planck made a new and radical assumption: that the energy of the oscillations of atoms within molecules cannot have just any value; instead each has energy which is a multiple of a minimum value related to the frequency of oscillation by

$$E = hf.$$

Planck's assumption suggests that the energy of any molecular vibration could be only a whole number multiple of the minimum energy hf:

$$E = nhf, \qquad n = 1, 2, 3, \cdots, \qquad \textbf{(37–2)}$$

where n is called a **quantum number** ("quantum" means "discrete amount" as opposed to "continuous"). This idea is often called **Planck's quantum hypothesis**, although little attention was brought to this point at the time. In fact, it appears that Planck considered it more as a mathematical device to get the "right answer" rather than as an important discovery in its own right. Planck himself continued to seek a classical explanation for the introduction of h. The recognition that this was an important and radical innovation did not come until later, after about 1905 when others, particularly Einstein, entered the field.

The quantum hypothesis, Eq. 37–2, states that the energy of an oscillator can be $E = hf$, or $2hf$, or $3hf$, and so on, but there cannot be vibrations with energies between these values. That is, energy would not be a continuous quantity as had been believed for centuries; rather it is **quantized**—it exists only in discrete amounts. The smallest amount of energy possible (hf) is called the **quantum of energy**. Recall from Chapter 14 that the energy of an oscillation is proportional to the amplitude squared. Another way of expressing the quantum hypothesis is that not just any amplitude of vibration is possible. The possible values for the amplitude are related to the frequency f.

A simple analogy may help. Compare a ramp, on which a box can be placed at any height, to a flight of stairs on which the box can have only certain discrete amounts of potential energy, as shown in Fig. 37–3.

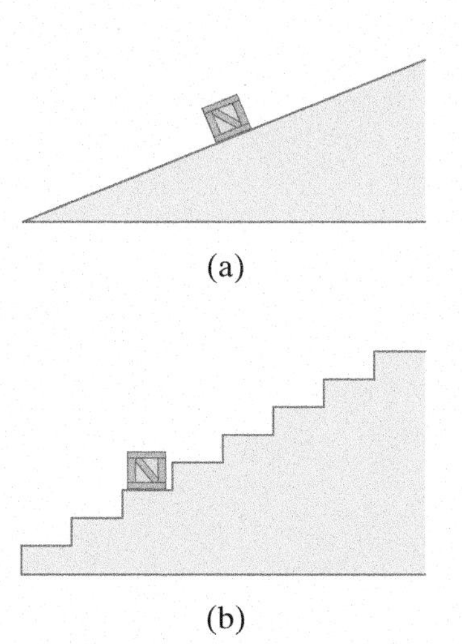

(a)

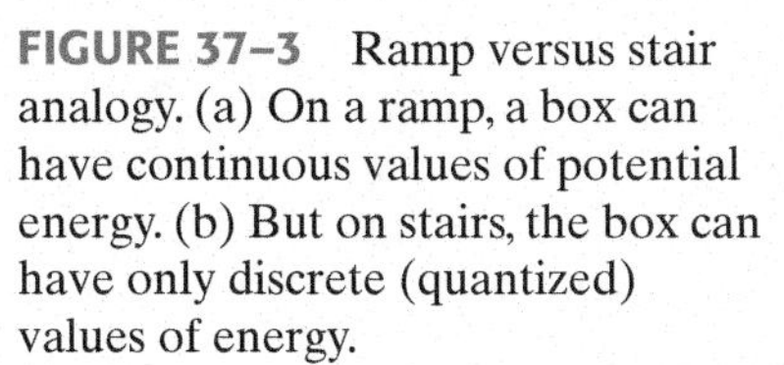

(b)

FIGURE 37–3 Ramp versus stair analogy. (a) On a ramp, a box can have continuous values of potential energy. (b) But on stairs, the box can have only discrete (quantized) values of energy.

37–2 Photon Theory of Light and the Photoelectric Effect

In 1905, the same year that he introduced the special theory of relativity, Einstein made a bold extension of the quantum idea by proposing a new theory of light. Planck's work had suggested that the vibrational energy of molecules in a radiating object is quantized with energy $E = nhf$, where n is an integer and f is the frequency of molecular vibration. Einstein argued that when light is emitted by a molecular oscillator, the molecule's vibrational energy of nhf must decrease by an amount hf (or by $2hf$, etc.) to another integer times hf, such as $(n - 1)hf$. Then to conserve energy, the light ought to be emitted in packets, or *quanta*, each with an energy

$$E = hf, \qquad \textbf{(37–3)}$$

where f is here the frequency of the emitted light.

Again h is Planck's constant. Since all light ultimately comes from a radiating source, this suggests that perhaps *light is transmitted as tiny particles*, or **photons**, as they are now called, as well as via waves predicted by Maxwell's electromagnetic theory. The photon theory of light was a radical departure from classical ideas. Einstein proposed a test of the quantum theory of light: quantitative measurements on the photoelectric effect.

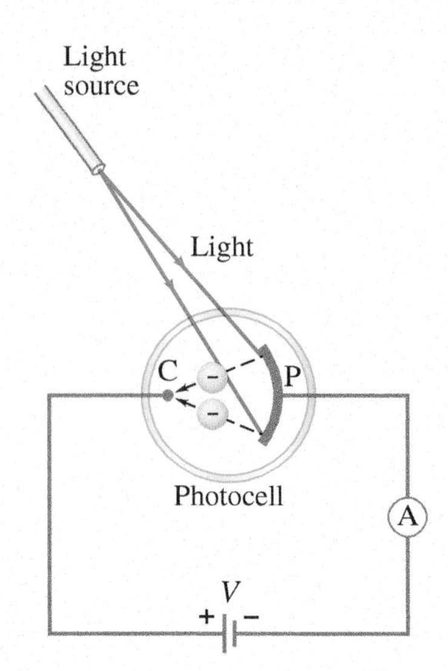

FIGURE 37–4 The photoelectric effect.

When light shines on a metal surface, electrons are found to be emitted from the surface. This effect is called the **photoelectric effect** and it occurs in many materials, but is most easily observed with metals. It can be observed using the apparatus shown in Fig. 37–4. A metal plate P and a smaller electrode C are placed inside an evacuated glass tube, called a **photocell**. The two electrodes are connected to an ammeter and a source of emf, as shown. When the photocell is in the dark, the ammeter reads zero. But when light of sufficiently high frequency illuminates the plate, the ammeter indicates a current flowing in the circuit. We explain completion of the circuit by imagining that electrons, ejected from the plate by the impinging radiation, flow across the tube from the plate to the "collector" C as indicated in Fig. 37–4.

That electrons should be emitted when light shines on a metal is consistent with the electromagnetic (EM) wave theory of light: the electric field of an EM wave could exert a force on electrons in the metal and eject some of them. Einstein pointed out, however, that the wave theory and the photon theory of light give very different predictions on the details of the photoelectric effect. For example, one thing that can be measured with the apparatus of Fig. 37–4 is the maximum kinetic energy (K_{max}) of the emitted electrons. This can be done by using a variable voltage source and reversing the terminals so that electrode C is negative and P is positive. The electrons emitted from P will be repelled by the negative electrode, but if this reverse voltage is small enough, the fastest electrons will still reach C and there will be a current in the circuit. If the reversed voltage is increased, a point is reached where the current reaches zero—no electrons have sufficient kinetic energy to reach C. This is called the *stopping potential*, or *stopping voltage*, V_0, and from its measurement, K_{max} can be determined using conservation of energy (loss of kinetic energy = gain in potential energy):

$$K_{\text{max}} = eV_0 .$$

Now let us examine the details of the photoelectric effect from the point of view of the wave theory versus Einstein's particle theory.

First the wave theory, assuming monochromatic light. The two important properties of a light wave are its intensity and its frequency (or wavelength). When these two quantities are varied, the wave theory makes the following predictions:

Wave theory predictions

1. If the light intensity is increased, the number of electrons ejected and their maximum kinetic energy should be increased because the higher intensity means a greater electric field amplitude, and the greater electric field should eject electrons with higher speed.
2. The frequency of the light should not affect the kinetic energy of the ejected electrons. Only the intensity should affect K_{max}.

The photon theory makes completely different predictions. First we note that in a monochromatic beam, all photons have the same energy $(=hf)$. Increasing the intensity of the light beam means increasing the number of photons in the beam, but does not affect the energy of each photon as long as the frequency is not changed. According to Einstein's theory, an electron is ejected from the metal by a collision with a single photon. In the process, all the photon energy is transferred to the electron and the photon ceases to exist. Since electrons are held in the metal by attractive forces, some minimum energy W_0 is required just to get an electron out through the surface. W_0 is called the **work function**, and is a few electron volts $(1\text{ eV} = 1.6 \times 10^{-19}\text{ J})$ for most metals. If the frequency f of the incoming light is so low that hf is less than W_0, then the photons will not have enough energy to eject any electrons at all. If $hf > W_0$, then electrons will be ejected and energy will be conserved in the process. That is, the input energy (of the photon), hf, will equal the outgoing kinetic energy K of the electron plus the energy required to get it out of the metal, W:

$$hf = K + W. \qquad \textbf{(37–4a)}$$

The least tightly held electrons will be emitted with the most kinetic energy (K_{max}), in which case W in this equation becomes the work function W_0, and K becomes K_{max}:

$$hf = K_{max} + W_0. \qquad \text{[least bound electrons]} \quad \textbf{(37–4b)}$$

Many electrons will require more energy than the bare minimum (W_0) to get out of the metal, and thus the kinetic energy of such electrons will be less than the maximum.

From these considerations, the photon theory makes the following predictions:

Photon theory predictions

1. An increase in intensity of the light beam means more photons are incident, so more electrons will be ejected; but since the energy of each photon is not changed, the maximum kinetic energy of electrons is not changed by an increase in intensity.
2. If the frequency of the light is increased, the maximum kinetic energy of the electrons increases linearly, according to Eq. 37–4b. That is,

$$K_{max} = hf - W_0.$$

This relationship is plotted in Fig. 37–5.

3. If the frequency f is less than the "cutoff" frequency f_0, where $hf_0 = W_0$, no electrons will be ejected, no matter how great the intensity of the light.

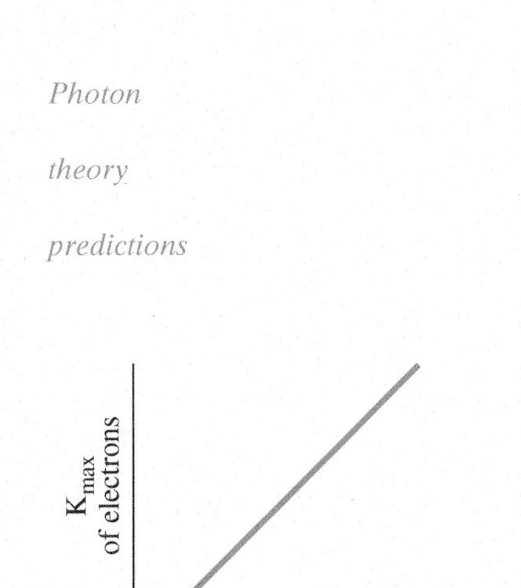

FIGURE 37–5 Photoelectric effect: the maximum kinetic energy of ejected electrons increases linearly with the frequency of incident light. No electrons are emitted if $f < f_0$.

These predictions of the photon theory are clearly very different from the predictions of the wave theory. In 1913–1914, careful experiments were carried out by R. A. Millikan. The results were fully in agreement with Einstein's photon theory.

One other aspect of the photoelectric effect also confirmed the photon theory. If extremely low light intensity is used, the wave theory predicts a time delay before electron emission so that an electron can absorb enough energy to exceed the work function. The photon theory predicts no such delay—it only takes one photon (if its frequency is high enough) to eject an electron—and experiments showed no delay. This too confirmed Einstein's photon theory.

EXAMPLE 37–3 **Photon energy.** Calculate the energy of a photon of blue light, $\lambda = 450\,\text{nm}$ in air (or vacuum).

APPROACH The photon has energy $E = hf$ (Eq. 37–3) where $f = c/\lambda$.

SOLUTION Since $f = c/\lambda$, we have

$$E = hf = \frac{hc}{\lambda} = \frac{(6.63 \times 10^{-34}\,\text{J}\cdot\text{s})(3.0 \times 10^{8}\,\text{m/s})}{(4.5 \times 10^{-7}\,\text{m})} = 4.4 \times 10^{-19}\,\text{J},$$

or $(4.4 \times 10^{-19}\,\text{J})/(1.60 \times 10^{-19}\,\text{J/eV}) = 2.8\,\text{eV}$. (See definition of eV in Section 23–8, $1\,\text{eV} = 1.60 \times 10^{-19}\,\text{J}$.)

EXAMPLE 37–4 ESTIMATE **Photons from a lightbulb.** Estimate how many visible light photons a 100-W lightbulb emits per second. Assume the bulb has a typical efficiency of about 3% (that is, 97% of the energy goes to heat).

APPROACH Let's assume an average wavelength in the middle of the visible spectrum, $\lambda \approx 500\,\text{nm}$. The energy of each photon is $E = hf = hc/\lambda$. Only 3% of the 100-W power is emitted as light, or $3\,\text{W} = 3\,\text{J/s}$. The number of photons emitted per second equals the light output of $3\,\text{J/s}$ divided by the energy of each photon.

SOLUTION The energy emitted in one second $(=3\,\text{J})$ is $E = Nhf$ where N is the number of photons emitted per second and $f = c/\lambda$. Hence

$$N = \frac{E}{hf} = \frac{E\lambda}{hc} = \frac{(3\,\text{J})(500 \times 10^{-9}\,\text{m})}{(6.63 \times 10^{-34}\,\text{J}\cdot\text{s})(3.0 \times 10^{8}\,\text{m/s})} \approx 8 \times 10^{18}$$

per second, or almost 10^{19} photons emitted per second, an enormous number.

EXERCISE A Compare a light beam that contains infrared light of a single wavelength, 1000 nm, with a beam of monochromatic UV at 100 nm, both of the same intensity. Are there more 100-nm photons or more 1000-nm photons?

EXAMPLE 37–5 **Photoelectron speed and energy.** What is the kinetic energy and the speed of an electron ejected from a sodium surface whose work function is $W_0 = 2.28\,\text{eV}$ when illuminated by light of wavelength (*a*) 410 nm, (*b*) 550 nm?

APPROACH We first find the energy of the photons $(E = hf = hc/\lambda)$. If the energy is greater than W_0, then electrons will be ejected with varying amounts of kinetic energy, with a maximum of $K_{max} = hf - W_0$.

SOLUTION (*a*) For $\lambda = 410\,\text{nm}$,

$$hf = \frac{hc}{\lambda} = 4.85 \times 10^{-19}\,\text{J} \quad \text{or} \quad 3.03\,\text{eV}.$$

The maximum kinetic energy an electron can have is given by Eq. 37–4b, $K_{max} = 3.03\,\text{eV} - 2.28\,\text{eV} = 0.75\,\text{eV}$, or $(0.75\,\text{eV})(1.60 \times 10^{-19}\,\text{J/eV}) = 1.2 \times 10^{-19}\,\text{J}$. Since $K = \frac{1}{2}mv^2$ where $m = 9.11 \times 10^{-31}\,\text{kg}$,

$$v_{max} = \sqrt{\frac{2K}{m}} = 5.1 \times 10^5\,\text{m/s}.$$

Most ejected electrons will have less kinetic energy and less speed than these maximum values.

(*b*) For $\lambda = 550\,\text{nm}$, $hf = hc/\lambda = 3.61 \times 10^{-19}\,\text{J} = 2.26\,\text{eV}$. Since this photon energy is less than the work function, no electrons are ejected.

NOTE In (*a*) we used the nonrelativistic equation for kinetic energy. If v had turned out to be more than about $0.1c$, our calculation would have been inaccurate by at least a percent or so, and we would probably prefer to redo it using the relativistic form (Eq. 36–10).

EXERCISE B Determine the lowest frequency and the longest wavelength needed to emit electrons from sodium.

It is easy to show that the energy of a photon in electron volts, when given the wavelength λ in nm, is

$$E\,(\text{eV}) = \frac{1.240 \times 10^3\,\text{eV}\cdot\text{nm}}{\lambda\,(\text{nm})}. \qquad \text{[photon energy in eV]}$$

Applications of the Photoelectric Effect

The photoelectric effect, besides playing an important historical role in confirming the photon theory of light, also has many practical applications. Burglar alarms and automatic door openers often make use of the photocell circuit of Fig. 37–4. When a person interrupts the beam of light, the sudden drop in current in the circuit activates a switch—often a solenoid—which operates a bell or opens the door. UV or IR light is sometimes used in burglar alarms because of its invisibility. Many smoke detectors use the photoelectric effect to detect tiny amounts of smoke that interrupt the flow of light and so alter the electric current. Photographic light meters use this circuit as well. Photocells are used in many other devices, such as absorption spectrophotometers, to measure light intensity. One type of film sound track is a variably shaded narrow section at the side of the film. Light passing through the film is thus "modulated," and the output electrical signal of the photocell detector follows the frequencies on the sound track. See Fig. 37–6. For many applications today, the vacuum-tube photocell of Fig. 37–4 has been replaced by a semiconductor device known as a **photodiode** (Section 40–9). In these semiconductors, the absorption of a photon liberates a bound electron, which changes the conductivity of the material, so the current through a photodiode is altered.

FIGURE 37–6 Optical sound track on movie film. In the projector, light from a small source (different from that for the picture) passes through the sound track on the moving film. The light and dark areas on the sound track vary the intensity of the transmitted light which reaches the photocell, whose output current is then a replica of the original sound. This output is amplified and sent to the loudspeakers. High-quality projectors can show movies containing several parallel sound tracks to go to different speakers around the theater.

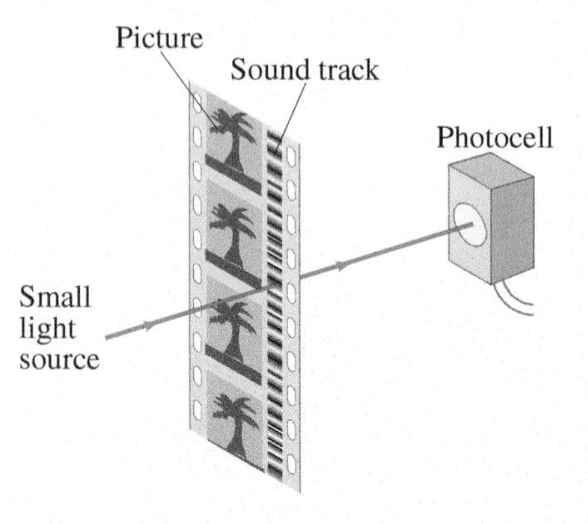

37–3 Energy, Mass, and Momentum of a Photon

We have just seen (Eq. 37–3) that the total energy of a single photon is given by $E = hf$. Because a photon always travels at the speed of light, it is truly a relativistic particle. Thus we must use relativistic formulas for dealing with its energy and momentum. The momentum of any particle of mass m is given by $p = mv/\sqrt{1 - v^2/c^2}$. Since $v = c$ for a photon, the denominator is zero. To avoid having an infinite momentum, we conclude that the photon's mass must be zero: $m = 0$. This makes sense too because a photon can never be at rest (it always moves at the speed of light). A photon's kinetic energy is its total energy:

$$K = E = hf. \qquad \text{[photon]}$$

The momentum of a photon can be obtained from the relativistic formula (Eq. 36–13) $E^2 = p^2c^2 + m^2c^4$ where we set $m = 0$, so $E^2 = p^2c^2$ or

$$p = \frac{E}{c}. \qquad \text{[photon]}$$

⚠ CAUTION
Momentum of photon is not mv

Since $E = hf$ for a photon, its momentum is related to its wavelength by

$$p = \frac{E}{c} = \frac{hf}{c} = \frac{h}{\lambda}. \qquad \textbf{(37–5)}$$

EXAMPLE 37–6 ESTIMATE **Photon momentum and force.** Suppose the 10^{19} photons emitted per second from the 100-W lightbulb in Example 37–4 were all focused onto a piece of black paper and absorbed. (*a*) Calculate the momentum of one photon and (*b*) estimate the force all these photons could exert on the paper.

APPROACH Each photon's momentum is obtained from Eq. 37–5, $p = h/\lambda$. Next, each absorbed photon's momentum changes from $p = h/\lambda$ to zero. We use Newton's second law, $F = \Delta p/\Delta t$, to get the force. Let $\lambda = 500\,\text{nm}$.

SOLUTION (*a*) Each photon has a momentum

$$p = \frac{h}{\lambda} = \frac{6.63 \times 10^{-34}\,\text{J}\cdot\text{s}}{500 \times 10^{-9}\,\text{m}} = 1.3 \times 10^{-27}\,\text{kg}\cdot\text{m/s}.$$

(*b*) Using Newton's second law for $N = 10^{19}$ photons (Example 37–4) whose momentum changes from h/λ to 0, we obtain

$$F = \frac{\Delta p}{\Delta t} = \frac{Nh/\lambda - 0}{1\,\text{s}} = N\frac{h}{\lambda} \approx (10^{19}\,\text{s}^{-1})(10^{-27}\,\text{kg}\cdot\text{m/s}) \approx 10^{-8}\,\text{N}.$$

This is a tiny force, but we can see that a very strong light source could exert a measurable force, and near the Sun or a star the force due to photons in electromagnetic radiation could be considerable. See Section 31–9.

EXAMPLE 37–7 **Photosynthesis.** In *photosynthesis*, pigments such as chlorophyll in plants capture the energy of sunlight to change CO_2 to useful carbohydrate. About nine photons are needed to transform one molecule of CO_2 to carbohydrate and O_2. Assuming light of wavelength $\lambda = 670\,\text{nm}$ (chlorophyll absorbs most strongly in the range 650 nm to 700 nm), how efficient is the photosynthetic process? The reverse chemical reaction releases an energy of 4.9 eV/molecule of CO_2.

APPROACH The efficiency is the minimum energy required (4.9 eV) divided by the actual energy absorbed, nine times the energy (hf) of one photon.

SOLUTION The energy of nine photons, each of energy $hf = hc/\lambda$ is $(9)(6.63 \times 10^{-34}\,\text{J}\cdot\text{s})(3.0 \times 10^8\,\text{m/s})/(6.7 \times 10^{-7}\,\text{m}) = 2.7 \times 10^{-18}\,\text{J}$ or 17 eV. Thus the process is $(4.9\,\text{eV}/17\,\text{eV}) = 29\%$ efficient.

37–4 Compton Effect

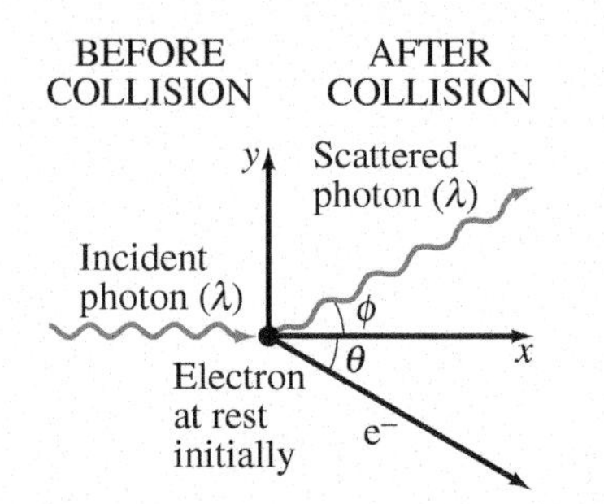

FIGURE 37–7 The Compton effect. A single photon of wavelength λ strikes an electron in some material, knocking it out of its atom. The scattered photon has less energy (some energy is given to the electron) and hence has a longer wavelength λ'.

Besides the photoelectric effect, a number of other experiments were carried out in the early twentieth century which also supported the photon theory. One of these was the **Compton effect** (1923) named after its discoverer, A. H. Compton (1892–1962). Compton scattered short-wavelength light (actually X-rays) from various materials. He found that the scattered light had a slightly longer wavelength than did the incident light, and therefore a slightly lower frequency indicating a loss of energy. He explained this result on the basis of the photon theory as incident photons colliding with electrons of the material, Fig. 37–7. Using Eq. 37–5 for momentum of a photon, Compton applied the laws of conservation of momentum and energy to the collision of Fig. 37–7 and derived the following equation for the wavelength of the scattered photons:

$$\lambda' = \lambda + \frac{h}{m_e c}(1 - \cos\phi), \qquad \textbf{(37–6a)}$$

where m_e is the mass of the electron. For $\phi = 0$, the wavelength is unchanged (there is no collision for this case of the photon passing straight through). At any other angle, λ' is longer than λ. The difference in wavelength,

$$\Delta\lambda = \lambda' - \lambda = \frac{h}{m_e c}(1 - \cos\phi), \qquad \textbf{(37–6b)}$$

is called the **Compton shift**. The quantity $h/m_e c$, which has the dimensions of length, is called the **Compton wavelength** λ_C of a free electron,

$$\lambda_C = \frac{h}{m_e c} = 2.43 \times 10^{-3}\,\text{nm} = 2.43\,\text{pm}. \qquad \text{[electron]}$$

Equations 37–6 predict that λ' depends on the angle ϕ at which the photons are detected. Compton's measurements of 1923 were consistent with this formula, confirming the value of λ_C and the dependence of λ' on ϕ. See Fig. 37–8. The wave theory of light predicts no wavelength shift: an incoming electromagnetic wave of frequency f should set electrons into oscillation at the same frequency f, and such oscillating electrons should reemit EM waves of this same frequency f (Chapter 31), and would not change with the angle ϕ. Hence the Compton effect adds to the firm experimental foundation for the photon theory of light.

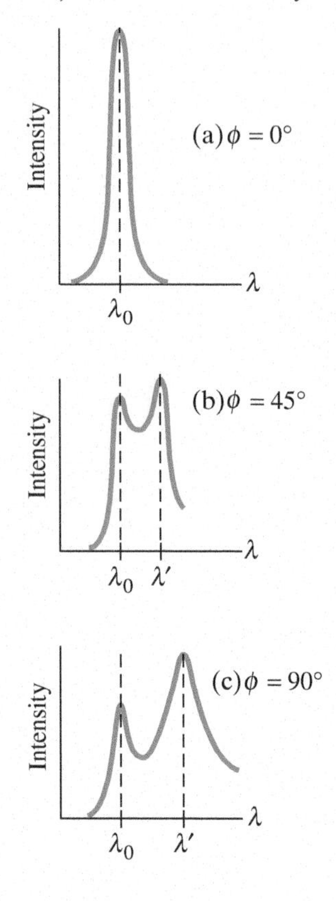

FIGURE 37–8 Plots of intensity of radiation scattered from a target such as graphite (carbon), for three different angles. The values for λ' match Eq. 37–6. For (a) $\phi = 0°$, $\lambda' = \lambda_0$. In (b) and (c) a peak is found not only at λ' due to photons scattered from free electrons (or very nearly free), but also a peak at almost precisely λ_0. The latter is due to scattering from electrons very tightly bound to their atoms so the mass in Eq. 37–6 becomes very large (mass of the atom) and $\Delta\lambda$ becomes very small.

EXERCISE C When a photon scatters off an electron by the Compton effect, which of the following increase: its energy, frequency, or wavelength?

EXAMPLE 37–8 **X-ray scattering.** X-rays of wavelength 0.140 nm are scattered from a very thin slice of carbon. What will be the wavelengths of X-rays scattered at (*a*) 0°, (*b*) 90°, (*c*) 180°?

APPROACH This is an example of the Compton effect, and we use Eq. 37–6a to find the wavelengths.

SOLUTION (*a*) For $\phi = 0°$, $\cos\phi = 1$ and $1 - \cos\phi = 0$. Then Eq. 37–6 gives $\lambda' = \lambda = 0.140\,\text{nm}$. This makes sense since for $\phi = 0°$, there really isn't any collision as the photon goes straight through without interacting.
(*b*) For $\phi = 90°$, $\cos\phi = 0$, and $1 - \cos\phi = 1$. So

$$\lambda' = \lambda + \frac{h}{m_e c} = 0.140\,\text{nm} + \frac{6.63 \times 10^{-34}\,\text{J}\cdot\text{s}}{(9.11 \times 10^{-31}\,\text{kg})(3.00 \times 10^{8}\,\text{m/s})}$$
$$= 0.140\,\text{nm} + 2.4 \times 10^{-12}\,\text{m} = 0.142\,\text{nm};$$

that is, the wavelength is longer by one Compton wavelength (= 0.0024 nm for an electron).

(c) For $\phi = 180°$, which means the photon is scattered backward, returning in the direction from which it came (a direct "head-on" collision), $\cos\phi = -1$, and $1 - \cos\phi = 2$. So

$$\lambda' = \lambda + 2\frac{h}{m_e c} = 0.140\,\text{nm} + 2(0.0024\,\text{nm}) = 0.145\,\text{nm}.$$

NOTE The maximum shift in wavelength occurs for backward scattering, and it is twice the Compton wavelength.

The Compton effect has been used to diagnose bone disease such as osteoporosis. Gamma rays, which are photons of even shorter wavelength than X-rays, coming from a radioactive source are scattered off bone material. The total intensity of the scattered radiation is proportional to the density of electrons, which is in turn proportional to the bone density. Changes in the density of bone can indicate the onset of osteoporosis.

PHYSICS APPLIED
Measuring bone density

*Derivation of Compton Shift

If the incoming photon in Fig. 37–7 has wavelength λ, then its total energy and momentum are

$$E = hf = \frac{hc}{\lambda} \quad \text{and} \quad p = \frac{h}{\lambda}.$$

After the collision of Fig. 37–7, the photon scattered at the angle ϕ has a wavelength which we call λ'. Its energy and momentum are

$$E' = \frac{hc}{\lambda'} \quad \text{and} \quad p' = \frac{h}{\lambda'}.$$

The electron, assumed at rest before the collision but free to move when struck, is scattered at an angle θ as shown in Fig. 37–8. The electron's kinetic energy is (see Eq. 36–10):

$$K_e = \left(\frac{1}{\sqrt{1 - v^2/c^2}} - 1\right)m_e c^2$$

where m_e is the mass of the electron and v is its speed. The electron's momentum is

$$p_e = \frac{1}{\sqrt{1 - v^2/c^2}}m_e v.$$

We apply conservation of energy to the collision (see Fig. 37–7):

$$\text{incoming photon} \longrightarrow \text{scattered photon} + \text{electron}$$

$$\frac{hc}{\lambda} = \frac{hc}{\lambda'} + \left(\frac{1}{\sqrt{1 - v^2/c^2}} - 1\right)m_e c^2.$$

We apply conservation of momentum to the x and y components of momentum:

$$\frac{h}{\lambda} = \frac{h}{\lambda'}\cos\phi + \frac{m_e v\cos\theta}{\sqrt{1 - v^2/c^2}}$$

$$0 = \frac{h}{\lambda'}\sin\phi - \frac{m_e v\sin\theta}{\sqrt{1 - v^2/c^2}}.$$

We can combine these three equations to eliminate v and θ, and we obtain, as Compton did, an equation for the wavelength of the scattered photon in terms of its scattering angle ϕ:

$$\lambda' = \lambda + \frac{h}{m_e c}(1 - \cos\phi),$$

which is Eq. 37–6a.

37–5 Photon Interactions; Pair Production

When a photon passes through matter, it interacts with the atoms and electrons. There are four important types of interactions that a photon can undergo:

1. The *photoelectric effect*: A photon may knock an electron out of an atom and in the process the photon disappears.
2. The photon may knock an atomic electron to a higher energy state in the atom if its energy is not sufficient to knock the electron out altogether. In this process the photon also disappears, and all its energy is given to the atom. Such an atom is then said to be in an *excited state*, and we shall discuss it more later.
3. The photon can be scattered from an electron (or a nucleus) and in the process lose some energy; this is the *Compton effect* (Fig. 37–7). But notice that the photon is not slowed down. It still travels with speed c, but its frequency will be lower because it has lost some energy.
4. *Pair production*: A photon can actually create matter, such as the production of an electron and a positron, Fig. 37–9. (A positron has the same mass as an electron, but the opposite charge, $+e$.)

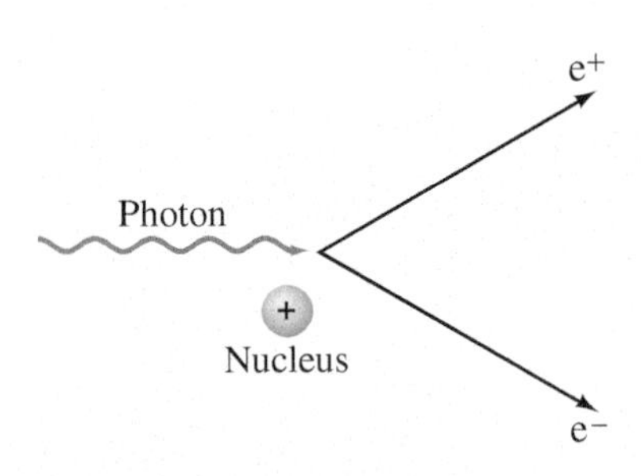

FIGURE 37–9 Pair production: a photon disappears and produces an electron and a positron.

In process 4, **pair production**, the photon disappears in the process of creating the electron–positron pair. This is an example of mass being created from pure energy, and it occurs in accord with Einstein's equation $E = mc^2$. Notice that a photon cannot create an electron alone since electric charge would not then be conserved. The inverse of pair production also occurs: if an electron collides with a positron, the two **annihilate** each other and their energy, including their mass, appears as electromagnetic energy of photons. Because of this process, positrons usually do not last long in nature.

EXAMPLE 37–9 **Pair production.** (*a*) What is the minimum energy of a photon that can produce an electron–positron pair? (*b*) What is this photon's wavelength?

APPROACH The minimum photon energy E equals the rest energy (mc^2) of the two particles created, via Einstein's famous equation $E = mc^2$ (Eq. 36–12). There is no energy left over, so the particles produced will have zero kinetic energy. The wavelength is $\lambda = c/f$ where $E = hf$ for the original photon.

SOLUTION (*a*) Because $E = mc^2$, and the mass created is equal to two electron masses, the photon must have energy

$$E = 2(9.11 \times 10^{-31}\,\mathrm{kg})(3.0 \times 10^{8}\,\mathrm{m/s})^2 = 1.64 \times 10^{-13}\,\mathrm{J} = 1.02\,\mathrm{MeV}$$

$(1\,\mathrm{MeV} = 10^{6}\,\mathrm{eV} = 1.60 \times 10^{-13}\,\mathrm{J})$. A photon with less energy cannot undergo pair production.

(*b*) Since $E = hf = hc/\lambda$, the wavelength of a 1.02-MeV photon is

$$\lambda = \frac{hc}{E} = \frac{(6.63 \times 10^{-34}\,\mathrm{J \cdot s})(3.0 \times 10^{8}\,\mathrm{m/s})}{(1.64 \times 10^{-13}\,\mathrm{J})} = 1.2 \times 10^{-12}\,\mathrm{m},$$

which is 0.0012 nm. Such photons are in the gamma-ray (or very short X-ray) region of the electromagnetic spectrum (Fig. 31–12).

NOTE Photons of higher energy (shorter wavelength) can also create an electron–positron pair, with the excess energy becoming kinetic energy of the particles.

Pair production cannot occur in empty space, for momentum could not be conserved. In Example 37–9, for instance, energy is conserved, but only enough energy was provided to create the electron–positron pair at rest, and thus with zero momentum, which could not equal the initial momentum of the photon. Indeed, it can be shown that at any energy, an additional massive object, such as an atomic nucleus, must take part in the interaction to carry off some of the momentum.

37–6 Wave–Particle Duality; the Principle of Complementarity

The photoelectric effect, the Compton effect, and other experiments have placed the particle theory of light on a firm experimental basis. But what about the classic experiments of Young and others (Chapters 34 and 35) on interference and diffraction which showed that the wave theory of light also rests on a firm experimental basis?

We seem to be in a dilemma. Some experiments indicate that light behaves like a wave; others indicate that it behaves like a stream of particles. These two theories seem to be incompatible, but both have been shown to have validity. Physicists finally came to the conclusion that this duality of light must be accepted as a fact of life. It is referred to as the **wave–particle duality**. Apparently, light is a more complex phenomenon than just a simple wave or a simple beam of particles.

To clarify the situation, the great Danish physicist Niels Bohr (1885–1962, Fig. 37–10) proposed his famous **principle of complementarity**. It states that to understand an experiment, sometimes we find an explanation using wave theory and sometimes using particle theory. Yet we must be aware of both the wave and particle aspects of light if we are to have a full understanding of light. Therefore these two aspects of light complement one another.

FIGURE 37–10 Niels Bohr (right), walking with Enrico Fermi along the Appian Way outside Rome. This photo shows one important way physics is done.

It is not easy to "visualize" this duality. We cannot readily picture a combination of wave and particle. Instead, we must recognize that the two aspects of light are different "faces" that light shows to experimenters.

Part of the difficulty stems from how we think. Visual pictures (or models) in our minds are based on what we see in the everyday world. We apply the concepts of waves and particles to light because in the macroscopic world we see that energy is transferred from place to place by these two methods. We cannot see directly whether light is a wave or particle, so we do indirect experiments. To explain the experiments, we apply the models of waves or of particles to the nature of light. But these are abstractions of the human mind. When we try to conceive of what light really "is," we insist on a visual picture. Yet there is no reason why light should conform to these models (or visual images) taken from the macroscopic world. The "true" nature of light—if that means anything—is not possible to visualize. The best we can do is recognize that our knowledge is limited to the indirect experiments, and that in terms of everyday language and images, light reveals both wave and particle properties.

CAUTION

Not correct to say light is a wave and/or a particle. Light can **act** *like a wave or like a particle.*

It is worth noting that Einstein's equation $E = hf$ itself links the particle and wave properties of a light beam. In this equation, E refers to the energy of a particle; and on the other side of the equation, we have the frequency f of the corresponding wave.

37–7 Wave Nature of Matter

In 1923, Louis de Broglie (1892–1987) extended the idea of the wave–particle duality. He appreciated the *symmetry* in nature, and argued that if light sometimes behaves like a wave and sometimes like a particle, then perhaps those things in nature thought to be particles—such as electrons and other material objects—might also have wave properties. De Broglie proposed that the wavelength of a material particle would be related to its momentum in the same way as for a photon, Eq. 37–5, $p = h/\lambda$. That is, for a particle having linear momentum $p = mv$, the wavelength λ is given by

$$\lambda = \frac{h}{p}, \tag{37–7}$$

and is valid classically $(p = mv \text{ for } v \ll c)$ and relativistically $\left(p = \gamma mv = mv/\sqrt{1 - v^2/c^2}\right)$. This is sometimes called the **de Broglie wavelength** of a particle.

EXAMPLE 37–10 **Wavelength of a ball.** Calculate the de Broglie wavelength of a 0.20-kg ball moving with a speed of 15 m/s.

APPROACH We use Eq. 37–7.

SOLUTION $\lambda = \dfrac{h}{p} = \dfrac{h}{mv} = \dfrac{(6.6 \times 10^{-34}\,\mathrm{J\cdot s})}{(0.20\,\mathrm{kg})(15\,\mathrm{m/s})} = 2.2 \times 10^{-34}\,\mathrm{m}.$

Ordinary size objects, such as the ball of Example 37–10, have unimaginably small wavelengths. Even if the speed is extremely small, say 10^{-4} m/s, the wavelength would be about 10^{-29} m. Indeed, the wavelength of any ordinary object is much too small to be measured and detected. The problem is that the properties of waves, such as interference and diffraction, are significant only when the size of objects or slits is not much larger than the wavelength. And there are no known objects or slits to diffract waves only 10^{-30} m long, so the wave properties of ordinary objects go undetected.

But tiny elementary particles, such as electrons, are another matter. Since the mass m appears in the denominator of Eq. 37–7, a very small mass should have a much larger wavelength.

EXAMPLE 37–11 **Wavelength of an electron.** Determine the wavelength of an electron that has been accelerated through a potential difference of 100 V.

APPROACH If the kinetic energy is much less than the rest energy, we can use the classical formula, $K = \frac{1}{2}mv^2$ (see Section 36–11). For an electron, $mc^2 = 0.511$ MeV. We then apply conservation of energy: the kinetic energy acquired by the electron equals its loss in potential energy. After solving for v, we use Eq. 37–7 to find the de Broglie wavelength.

SOLUTION Gain in kinetic energy equals loss in potential energy: $\Delta U = eV - 0$. Thus $K = eV$, so $K = 100$ eV. The ratio $K/mc^2 = 100\,\mathrm{eV}/(0.511 \times 10^6\,\mathrm{eV}) \approx 10^{-4}$, so relativity is not needed. Thus

$$\frac{1}{2}mv^2 = eV$$

and

$$v = \sqrt{\frac{2eV}{m}} = \sqrt{\frac{(2)(1.6 \times 10^{-19}\,\mathrm{C})(100\,\mathrm{V})}{(9.1 \times 10^{-31}\,\mathrm{kg})}} = 5.9 \times 10^6\,\mathrm{m/s}.$$

Then

$$\lambda = \frac{h}{mv} = \frac{(6.63 \times 10^{-34}\,\mathrm{J\cdot s})}{(9.1 \times 10^{-31}\,\mathrm{kg})(5.9 \times 10^6\,\mathrm{m/s})} = 1.2 \times 10^{-10}\,\mathrm{m},$$

or 0.12 nm.

EXERCISE D As a particle travels faster, does its de Broglie wavelength decrease, increase, or remain the same?

FIGURE 37–11 Diffraction pattern of electrons scattered from aluminum foil, as recorded on film.

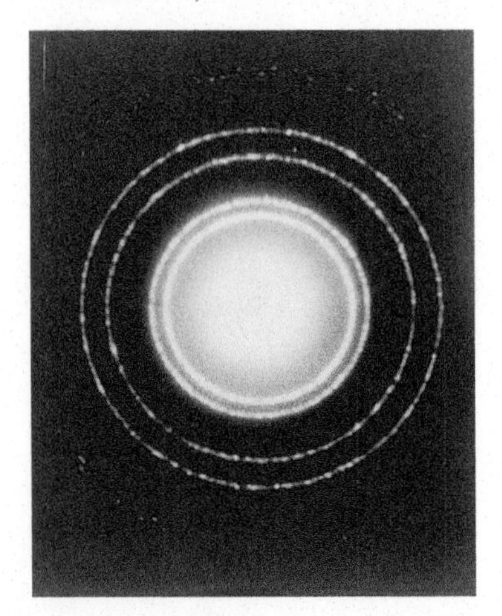

From Example 37–11, we see that electrons can have wavelengths on the order of 10^{-10} m, and even smaller. Although small, this wavelength can be detected: the spacing of atoms in a crystal is on the order of 10^{-10} m and the orderly array of atoms in a crystal could be used as a type of diffraction grating, as was done earlier for X-rays (see Section 35–10). C. J. Davisson and L. H. Germer performed the crucial experiment; they scattered electrons from the surface of a metal crystal and, in early 1927, observed that the electrons were scattered into a pattern of regular peaks. When they interpreted these peaks as a diffraction pattern, the wavelength of the diffracted electron wave was found to be just that predicted by de Broglie, Eq. 37–7. In the same year, G. P. Thomson (son of J. J. Thomson) used a different experimental arrangement and also detected diffraction of electrons. (See Fig. 37–11. Compare it to X-ray diffraction, Section 35–10.) Later experiments showed that protons, neutrons, and other particles also have wave properties.

Thus the wave–particle duality applies to material objects as well as to light. The principle of complementarity applies to matter as well. That is, we must be aware of both the particle and wave aspects in order to have an understanding of matter, including electrons. But again we must recognize that a visual picture of a "wave–particle" is not possible.

EXAMPLE 37–12 **Electron diffraction.** The wave nature of electrons is manifested in experiments where an electron beam interacts with the atoms on the surface of a solid. By studying the angular distribution of the diffracted electrons, one can indirectly measure the geometrical arrangement of atoms. Assume that the electrons strike perpendicular to the surface of a solid (see Fig. 37–12), and that their energy is low, $K = 100\,\text{eV}$, so that they interact only with the surface layer of atoms. If the smallest angle at which a diffraction maximum occurs is at 24°, what is the separation d between the atoms on the surface?

SOLUTION Treating the electrons as waves, we need to determine the condition where the difference in path traveled by the wave diffracted from adjacent atoms is an integer multiple of the de Broglie wavelength, so that constructive interference occurs. The path length difference is $d \sin\theta$; so for the smallest value of θ we must have

$$d \sin\theta = \lambda.$$

However, λ is related to the (non-relativistic) kinetic energy K by

$$K = \frac{p^2}{2m_e} = \frac{h^2}{2m_e\lambda^2}.$$

Thus

$$\lambda = \frac{h}{\sqrt{2m_e K}} = \frac{(6.63 \times 10^{-34}\,\text{J}\cdot\text{s})}{\sqrt{2(9.11 \times 10^{-31}\,\text{kg})(100\,\text{eV})(1.6 \times 10^{-19}\,\text{J/eV})}} = 0.123\,\text{nm}.$$

The surface inter-atomic spacing is

$$d = \frac{\lambda}{\sin\theta} = \frac{0.123\,\text{nm}}{\sin 24^\circ} = 0.30\,\text{nm}.$$

PHYSICS APPLIED
Electron diffraction

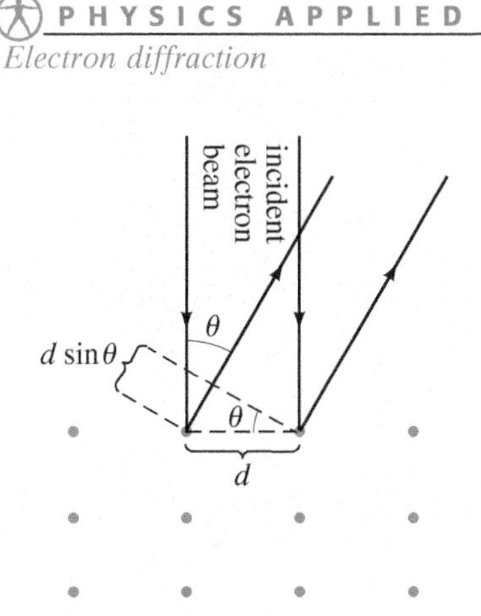

FIGURE 37–12 Example 37–12. The red dots represent atoms in an orderly array in a solid.

EXERCISE E Return to the Chapter-Opening Question, page 987, and answer it again now. Try to explain why you may have answered differently the first time.

What Is an Electron?

We might ask ourselves: "What is an electron?" The early experiments of J. J. Thomson (Section 27–7) indicated a glow in a tube, and that glow moved when a magnetic field was applied. The results of these and other experiments were best interpreted as being caused by tiny negatively charged particles which we now call electrons. No one, however, has actually seen an electron directly. The drawings we sometimes make of electrons as tiny spheres with a negative charge on them are merely convenient pictures (now recognized to be inaccurate). Again we must rely on experimental results, some of which are best interpreted using the particle model and others using the wave model. These models are mere pictures that we use to extrapolate from the macroscopic world to the tiny microscopic world of the atom. And there is no reason to expect that these models somehow reflect the reality of an electron. We thus use a wave or a particle model (whichever works best in a situation) so that we can talk about what is happening. But we should not be led to believe that an electron *is* a wave or a particle. Instead we could say that an electron is the set of its properties that we can measure. Bertrand Russell said it well when he wrote that an electron is "a logical construction."

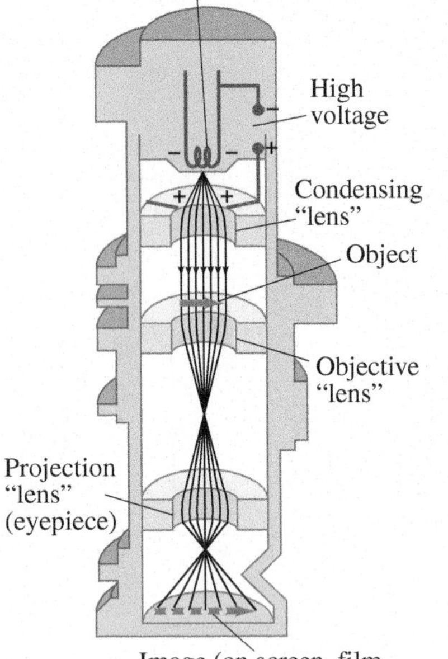

FIGURE 37–13 Transmission electron microscope. The magnetic field coils are designed to be "magnetic lenses," which bend the electron paths and bring them to a focus, as shown.

PHYSICS APPLIED
Electron microscope

*37–8 Electron Microscopes

The idea that electrons have wave properties led to the development of the **electron microscope**, which can produce images of much greater magnification than does a light microscope. Figures 37–13 and 37–14 are diagrams of two types, developed around the middle of the twentieth century: the **transmission electron microscope**, which produces a two-dimensional image, and the **scanning electron microscope** (SEM), which produces images with a three-dimensional quality. In both types, the objective and eyepiece lenses are actually magnetic fields that exert forces on the electrons to bring them to a focus. The fields are produced by carefully designed current-carrying coils of wire. Photographs using each type are shown in Fig. 37–15.

As discussed in Section 35–5, the maximum resolution of details on an object is about the size of the wavelength of the radiation used to view it. Electrons accelerated by voltages on the order of 10^5 V have wavelengths of about 0.004 nm. The maximum resolution obtainable would be on this order, but in practice, aberrations in the magnetic lenses limit the resolution in transmission electron microscopes to at best about 0.1 to 0.5 nm. This is still 10^3 times better than that attainable with a visible-light microscope, and corresponds to a useful magnification of about a million. Such magnifications are difficult to attain, and more common magnifications are 10^4 to 10^5. The maximum resolution attainable with a scanning electron microscope is somewhat less, typically 5 to 10 nm although new high-resolution SEMs approach 1 nm.

We discuss other sophisticated electron microscopes in the next Chapter, Section 38–10.

FIGURE 37–14 Scanning electron microscope. Scanning coils move an electron beam back and forth across the specimen. Secondary electrons produced when the beam strikes the specimen are collected and modulate the intensity of the beam in the CRT to produce a picture.

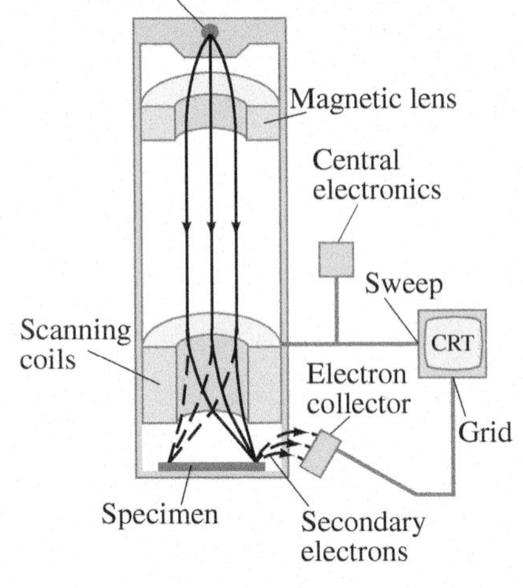

FIGURE 37–15 Electron micrographs (in false color) of viruses attacking a cell of the bacterium *Escherichia coli*: (a) transmission electron micrograph ($\approx$ 50,000×); (b) scanning electron micrograph ($\approx$ 35,000×).

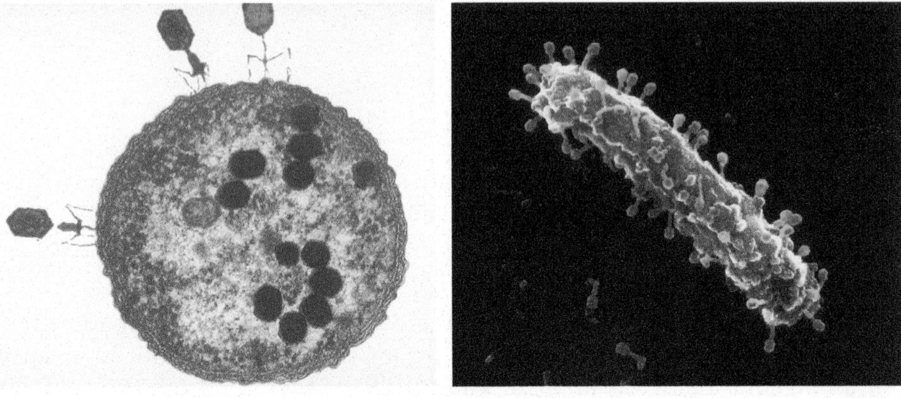

37–9 Early Models of the Atom

The idea that matter is made up of atoms was accepted by most scientists by 1900. With the discovery of the electron in the 1890s, scientists began to think of the atom itself as having a structure with electrons as part of that structure. We now introduce our modern approach to the atom and the quantum theory with which it is intertwined.†

†Some readers may say: "Tell us the facts as we know them today, and don't bother us with the historical background and its outmoded theories." Such an approach would ignore the creative aspect of science and thus give a false impression of how science develops. Moreover, it is not really possible to understand today's view of the atom without insight into the concepts that led to it.

A typical model of the atom in the 1890s visualized the atom as a homogeneous sphere of positive charge inside of which there were tiny negatively charged electrons, a little like plums in a pudding, Fig. 37–16.

Around 1911, Ernest Rutherford (1871–1937) and his colleagues performed experiments whose results contradicted the plum-pudding model of the atom. In these experiments a beam of positively charged alpha (α) particles was directed at a thin sheet of metal foil such as gold, Fig. 37–17a. (These newly discovered α particles were emitted by certain radioactive materials and were soon shown to be doubly ionized helium atoms—that is, having a charge of $+2e$.) It was expected from the plum-pudding model that the alpha particles would not be deflected significantly because electrons are so much lighter than alpha particles, and the alpha particles should not have encountered any massive concentration of positive charge to strongly repel them. The experimental results completely contradicted these predictions. It was found that most of the alpha particles passed through the foil unaffected, as if the foil were mostly empty space. And of those deflected, a few were deflected at very large angles—some even backward, nearly in the direction from which they had come. This could happen, Rutherford reasoned, only if the positively charged alpha particles were being repelled by a massive positive charge concentrated in a very small region of space (see Fig. 37–17b).

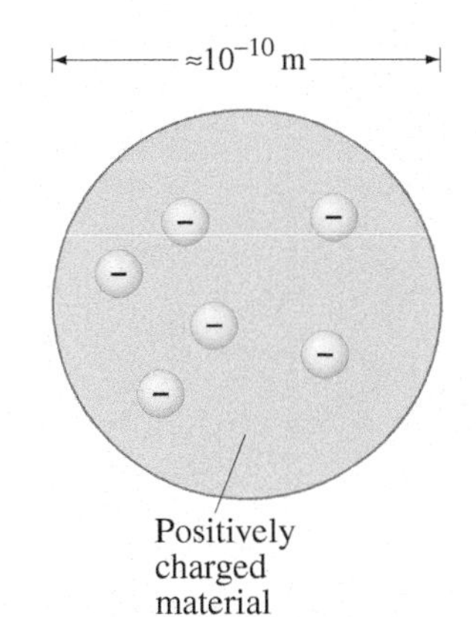

FIGURE 37–16 Plum-pudding model of the atom.

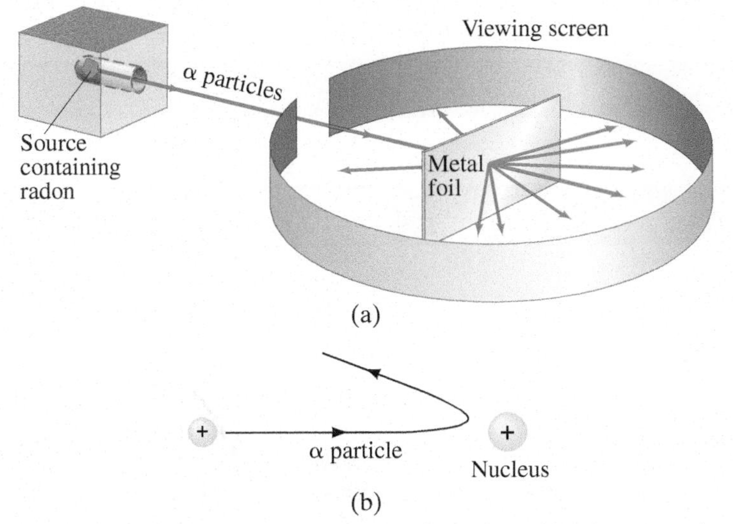

FIGURE 37–17 (a) Experimental setup for Rutherford's experiment: α particles emitted by radon are deflected by a thin metallic foil and a few rebound backward; (b) backward rebound of α particles explained as the repulsion from a heavy positively charged nucleus.

He hypothesized that the atom must consist of a tiny but massive positively charged nucleus, containing over 99.9% of the mass of the atom, surrounded by electrons some distance away. The electrons would be moving in orbits about the nucleus—much as the planets move around the Sun—because if they were at rest, they would fall into the nucleus due to electrical attraction, Fig. 37–18. Rutherford's experiments suggested that the nucleus must have a radius of about 10^{-15} to 10^{-14} m. From kinetic theory, and especially Einstein's analysis of Brownian motion (see Section 17–1), the radius of atoms was estimated to be about 10^{-10} m. Thus the electrons would seem to be at a distance from the nucleus of about 10,000 to 100,000 times the radius of the nucleus itself. (If the nucleus were the size of a baseball, the atom would have the diameter of a big city several kilometers across.) So an atom would be mostly empty space.

Rutherford's "planetary" model of the atom (also called the "nuclear model of the atom") was a major step toward how we view the atom today. It was not, however, a complete model and presented some major problems, as we shall see.

FIGURE 37–18 Rutherford's model of the atom, in which electrons orbit a tiny positive nucleus (not to scale). The atom is visualized as mostly empty space.

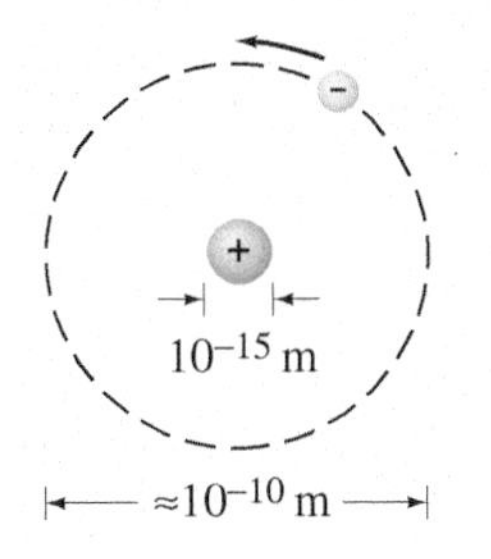

37–10 Atomic Spectra: Key to the Structure of the Atom

Earlier in this Chapter we saw that heated solids (as well as liquids and dense gases) emit light with a continuous spectrum of wavelengths. This radiation is assumed to be due to oscillations of atoms and molecules, which are largely governed by the interaction of each atom or molecule with its neighbors.

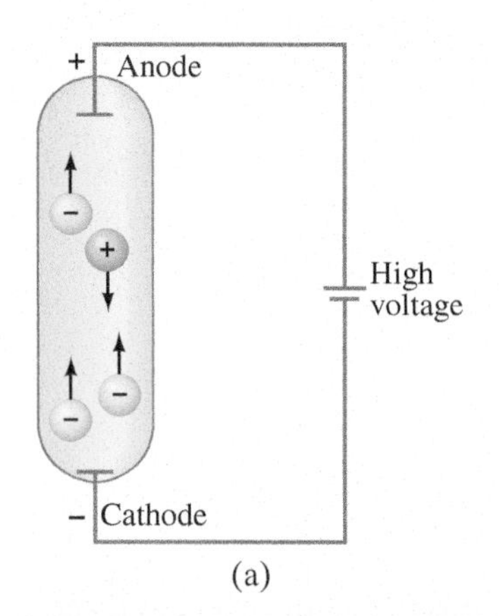

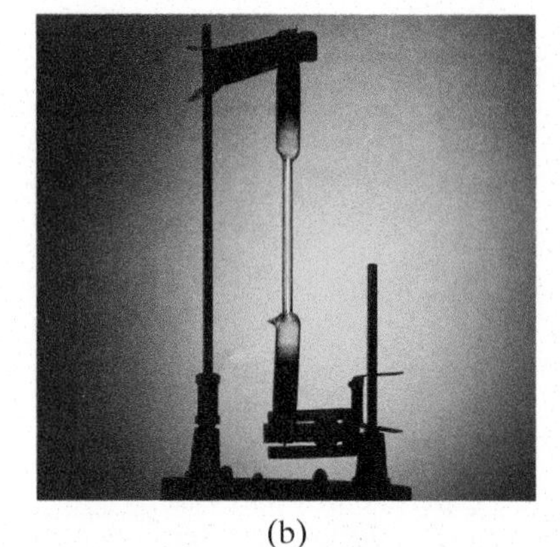
(b)

FIGURE 37–19 Gas-discharge tube: (a) diagram; (b) photo of an actual discharge tube for hydrogen.

Rarefied gases can also be excited to emit light. This is done by intense heating, or more commonly by applying a high voltage to a "discharge tube" containing the gas at low pressure, Fig. 37–19. The radiation from excited gases had been observed early in the nineteenth century, and it was found that the spectrum was not continuous, but *discrete*. Since excited gases emit light of only certain wavelengths, when this light is analyzed through the slit of a spectroscope or spectrometer, a **line spectrum** is seen rather than a continuous spectrum. The line spectra in the visible region emitted by a number of elements are shown below in Fig. 37–20, and in Chapter 35, Fig. 35–22. The **emission spectrum** is characteristic of the material and can serve as a type of "fingerprint" for identification of the gas.

We also saw (Chapter 35) that if a continuous spectrum passes through a rarefied gas, dark lines are observed in the emerging spectrum, at wavelengths corresponding to lines normally emitted by the gas. This is called an **absorption spectrum** (Fig. 37–20c), and it became clear that gases can absorb light at the same frequencies at which they emit. Using film sensitive to ultraviolet and to infrared light, it was found that gases emit and absorb discrete frequencies in these regions as well as in the visible.

FIGURE 37–20 Emission spectra of the gases (a) atomic hydrogen, (b) helium, and (c) the *solar absorption* spectrum.

FIGURE 37–21 Balmer series of lines for hydrogen.

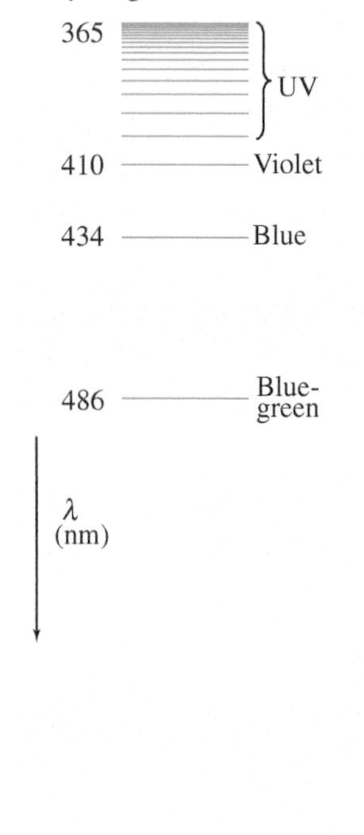

In low-density gases, the atoms are far apart on the average and hence the light emitted or absorbed is assumed to be by *individual atoms* rather than through interactions between atoms, as in a solid, liquid, or dense gas. Thus the line spectra serve as a key to the structure of the atom: any theory of atomic structure must be able to explain why atoms emit light only of discrete wavelengths, and it should be able to predict what these wavelengths are.

Hydrogen is the simplest atom—it has only one electron. It also has the simplest spectrum. The spectrum of most atoms shows little apparent regularity. But the spacing between lines in the hydrogen spectrum decreases in a regular way, Fig. 37–21. Indeed, in 1885, J. J. Balmer (1825–1898) showed that the four lines in the visible portion of the hydrogen spectrum (with measured wavelengths 656 nm, 486 nm, 434 nm, and 410 nm) have wavelengths that fit the formula

$$\frac{1}{\lambda} = R\left(\frac{1}{2^2} - \frac{1}{n^2}\right), \qquad n = 3, 4, \cdots. \tag{37–8}$$

Here n takes on the values 3, 4, 5, 6 for the four visible lines, and R, called the **Rydberg constant**, has the value $R = 1.0974 \times 10^7\,\mathrm{m}^{-1}$. Later it was found that this **Balmer series** of lines extended into the UV region, ending at $\lambda = 365$ nm, as shown in Fig. 37–21. Balmer's formula, Eq. 37–8, also worked for these lines with higher integer values of n. The lines near 365 nm became too close together to distinguish, but the limit of the series at 365 nm corresponds to $n = \infty$ (so $1/n^2 = 0$ in Eq. 37–8).

Later experiments on hydrogen showed that there were similar series of lines in the UV and IR regions, and each series had a pattern just like the Balmer series, but at different wavelengths, Fig. 37–22. Each of these series was found to fit a formula with the same form as Eq. 37–8 but with the $1/2^2$ replaced by $1/1^2, 1/3^2, 1/4^2$, and so on. For example, the so-called **Lyman series** contains lines

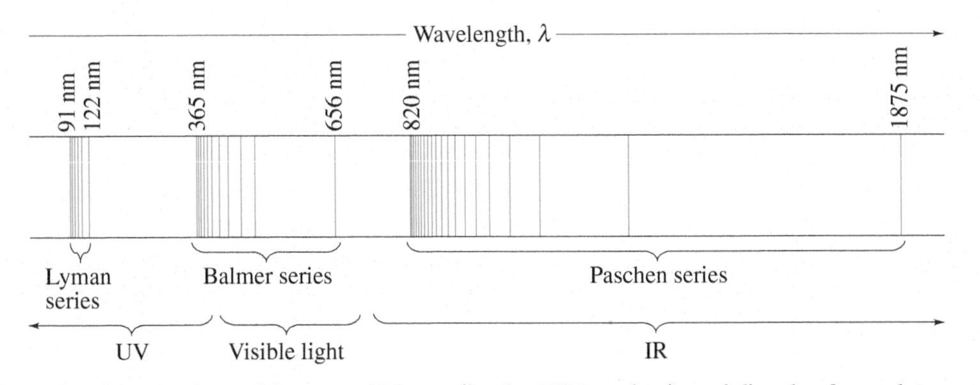

FIGURE 37–22 Line spectrum of atomic hydrogen. Each series fits the formula $\frac{1}{\lambda} = R\left(\frac{1}{n'^2} - \frac{1}{n^2}\right)$, where $n' = 1$ for the Lyman series, $n' = 2$ for the Balmer series, $n' = 3$ for the Paschen series, and so on; n can take on all integer values from $n = n' + 1$ up to infinity. The only lines in the visible region of the electromagnetic spectrum are part of the Balmer series.

with wavelengths from 91 nm to 122 nm (in the UV region) and fits the formula

$$\frac{1}{\lambda} = R\left(\frac{1}{1^2} - \frac{1}{n^2}\right), \qquad n = 2, 3, \cdots.$$

The wavelengths of the **Paschen series** (in the IR region) fit

$$\frac{1}{\lambda} = R\left(\frac{1}{3^2} - \frac{1}{n^2}\right), \qquad n = 4, 5, \cdots.$$

The Rutherford model was unable to explain why atoms emit line spectra. It had other difficulties as well. According to the Rutherford model, electrons orbit the nucleus, and since their paths are curved the electrons are accelerating. Hence they should give off light like any other accelerating electric charge (Chapter 31), with a frequency equal to its orbital frequency. Since light carries off energy and energy is conserved, the electron's own energy must decrease to compensate. Hence electrons would be expected to spiral into the nucleus. As they spiraled inward, their frequency would increase in a short time and so too would the frequency of the light emitted. Thus the two main difficulties with the Rutherford model are these: (1) it predicts that light of a continuous range of frequencies will be emitted, whereas experiment shows line spectra; (2) it predicts that atoms are unstable—electrons would quickly spiral into the nucleus—but we know that atoms in general are stable, because there is stable matter all around us.

Clearly Rutherford's model was not sufficient. Some sort of modification was needed, and Niels Bohr provided it in a model that included the quantum hypothesis. Although the Bohr model has been superceded, it did provide a crucial stepping stone to our present understanding. And some aspects of the Bohr model are still useful today, so we examine it in detail in the next Section.

37–11 The Bohr Model

Bohr had studied in Rutherford's laboratory for several months in 1912 and was convinced that Rutherford's planetary model of the atom had validity. But in order to make it work, he felt that the newly developing quantum theory would somehow have to be incorporated in it. The work of Planck and Einstein had shown that in heated solids, the energy of oscillating electric charges must change discontinuously—from one discrete energy state to another, with the emission of a quantum of light. Perhaps, Bohr argued, the electrons in an atom also cannot lose energy continuously, but must do so in quantum "jumps." In working out his model during the next year, Bohr postulated that electrons move about the nucleus in circular orbits, but that only certain orbits are allowed. He further postulated that an electron in each orbit would have a definite energy and would move in the orbit *without radiating energy* (even though this violated classical ideas since accelerating electric charges are supposed to emit EM waves; see Chapter 31). He thus called the possible orbits **stationary states**. Light is emitted, he hypothesized, only when an electron jumps from a higher (upper) stationary state to another of lower energy, Fig. 37–23. When such a transition occurs, a single photon of light is emitted whose energy, by energy conservation, is given by

$$hf = E_U - E_L, \tag{37–9}$$

where E_U refers to the energy of the upper state and E_L the energy of the lower state.

FIGURE 37–23 An atom emits a photon (energy $= hf$) when its energy changes from E_U to a lower energy E_L.

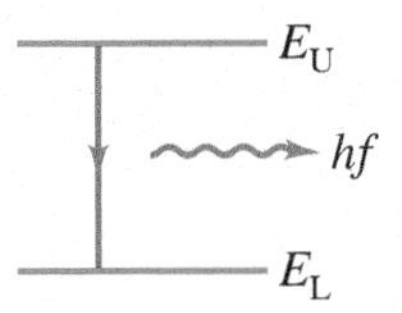

In 1912–13, Bohr set out to determine what energies these orbits would have in the simplest atom, hydrogen; the spectrum of light emitted could then be predicted from Eq. 37–9. In the Balmer formula he had the key he was looking for. Bohr quickly found that his theory would be in accord with the Balmer formula if he assumed that the electron's angular momentum L is quantized and equal to an integer n times $h/2\pi$. As we saw in Chapter 11 angular momentum is given by $L = I\omega$, where I is the moment of inertia and ω is the angular velocity. For a single particle of mass m moving in a circle of radius r with speed v, $I = mr^2$ and $\omega = v/r$; hence, $L = I\omega = (mr^2)(v/r) = mvr$. Bohr's **quantum condition** is

$$L = mvr_n = n\frac{h}{2\pi}, \qquad n = 1, 2, 3, \cdots, \tag{37–10}$$

where n is an integer and r_n is the radius of the nth possible orbit. The allowed orbits are numbered $1, 2, 3, \cdots$, according to the value of n, which is called the **principal quantum number** of the orbit.

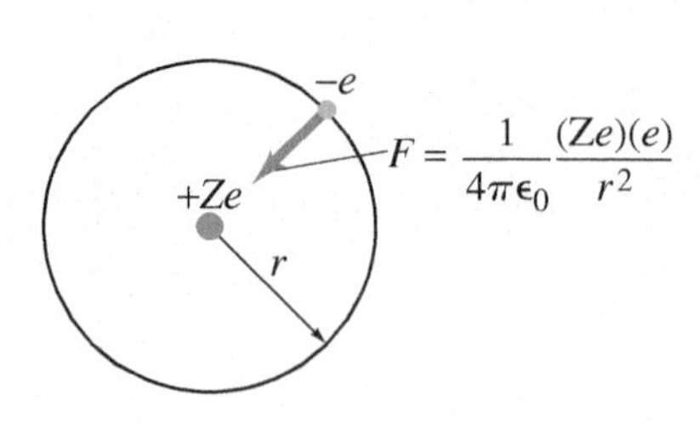

FIGURE 37–24 Electric force (Coulomb's law) keeps the negative electron in orbit around the positively charged nucleus.

Equation 37–10 did not have a firm theoretical foundation. Bohr had searched for some "quantum condition," and such tries as $E = hf$ (where E represents the energy of the electron in an orbit) did not give results in accord with experiment. Bohr's reason for using Eq. 37–10 was simply that it worked; and we now look at how. In particular, let us determine what the Bohr theory predicts for the measurable wavelengths of emitted light.

An electron in a circular orbit of radius r_n (Fig. 37–24) would have a centripetal acceleration v^2/r_n produced by the electrical force of attraction between the negative electron and the positive nucleus. This force is given by Coulomb's law,

$$F = \frac{1}{4\pi\epsilon_0}\frac{(Ze)(e)}{r_n^2}.$$

The charge on the nucleus is $+Ze$, where Z is the number of positive charges† (i.e., protons). For the hydrogen atom, $Z = +1$.

In Newton's second law, $F = ma$, we substitute Coulomb's law for F, and $a = v^2/r_n$ for a particular allowed orbit of radius r_n, and obtain

$$F = ma$$

$$\frac{1}{4\pi\epsilon_0}\frac{Ze^2}{r_n^2} = \frac{mv^2}{r_n}.$$

We solve this for r_n, and then substitute for v from Eq. 37–10 (which says $v = nh/2\pi m r_n$):

$$r_n = \frac{Ze^2}{4\pi\epsilon_0 mv^2} = \frac{Ze^2 4\pi^2 m r_n^2}{4\pi\epsilon_0 n^2 h^2}.$$

We solve for r_n (it appears on both sides, so we cancel one of them) and find

$$r_n = \frac{n^2 h^2 \epsilon_0}{\pi m Z e^2} = \frac{n^2}{Z} r_1 \tag{37–11}$$

where

$$r_1 = \frac{h^2 \epsilon_0}{\pi m e^2}.$$

Equation 37–11 gives the radii of all possible orbits. The smallest orbit is for $n = 1$, and for hydrogen $(Z = 1)$ has the value

$$r_1 = \frac{(1)^2(6.626 \times 10^{-34}\,\text{J}\cdot\text{s})^2(8.85 \times 10^{-12}\,\text{C}^2/\text{N}\cdot\text{m}^2)}{(3.14)(9.11 \times 10^{-31}\,\text{kg})(1.602 \times 10^{-19}\,\text{C})^2}$$

†We include Z in our derivation so that we can treat other single-electron ("hydrogenlike") atoms such as the ions He^+ $(Z = 2)$ and Li^{2+} $(Z = 3)$. Helium in the neutral state has two electrons: if one electron is missing, the remaining He^+ ion consists of one electron revolving around a nucleus of charge $+2e$. Similarly, doubly ionized lithium, Li^{2+}, also has a single electron, and in this case $Z = 3$.

or

$$r_1 = 0.529 \times 10^{-10}\,\text{m}. \qquad \textbf{(37–12)}$$

The radius of the smallest orbit in hydrogen, r_1, is sometimes called the **Bohr radius**. From Eq. 37–11, we see that the radii of the larger orbits† increase as n^2, so

$$\begin{aligned} r_2 &= 4r_1 = 2.12 \times 10^{-10}\,\text{m}, \\ r_3 &= 9r_1 = 4.76 \times 10^{-10}\,\text{m}, \\ &\vdots \\ r_n &= n^2 r_1. \end{aligned}$$

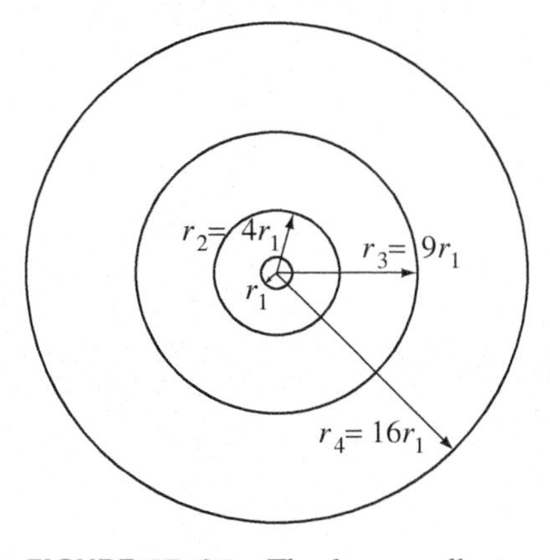

FIGURE 37–25 The four smallest orbits in the Bohr model of hydrogen; $r_1 = 0.529 \times 10^{-10}\,\text{m}$.

The first four orbits are shown in Fig. 37–25. Notice that, according to Bohr's model, an electron can exist only in the orbits given by Eq. 37–11. There are no allowable orbits in between.

For an atom with $Z \neq 1$, we can write the orbital radii, r_n, using Eq. 37–11:

$$r_n = \frac{n^2}{Z}(0.529 \times 10^{-10}\,\text{m}), \qquad n = 1, 2, 3, \cdots. \qquad \textbf{(37–13)}$$

In each of its possible orbits, the electron would have a definite energy, as the following calculation shows. The total energy equals the sum of the kinetic and potential energies. The potential energy of the electron is given by $U = qV = -eV$, where V is the potential due to a point charge $+Ze$ as given by Eq. 23–5: $V = (1/4\pi\epsilon_0)(Q/r) = (1/4\pi\epsilon_0)(Ze/r)$. So

$$U = -eV = -\frac{1}{4\pi\epsilon_0}\frac{Ze^2}{r}.$$

The total energy E_n for an electron in the nth orbit of radius r_n is the sum of the kinetic and potential energies:

$$E_n = \tfrac{1}{2}mv^2 - \frac{1}{4\pi\epsilon_0}\frac{Ze^2}{r_n}.$$

When we substitute v from Eq. 37–10 and r_n from Eq. 37–11 into this equation, we obtain

$$E_n = -\left(\frac{Z^2e^4m}{8\epsilon_0^2h^2}\right)\left(\frac{1}{n^2}\right), \qquad n = 1, 2, 3, \cdots. \qquad \textbf{(37–14a)}$$

If we evaluate the constant term in Eq. 37–14a and convert it to electron volts, as is customary in atomic physics, we obtain

$$E_n = -(13.6\,\text{eV})\frac{Z^2}{n^2}, \qquad n = 1, 2, 3, \cdots. \qquad \textbf{(37–14b)}$$

The lowest energy level $(n = 1)$ for hydrogen $(Z = 1)$ is

$$E_1 = -13.6\,\text{eV}.$$

Since n^2 appears in the denominator of Eq. 37–14b, the energies of the larger orbits in hydrogen $(Z = 1)$ are given by

$$E_n = \frac{-13.6\,\text{eV}}{n^2}.$$

For example,

$$\begin{aligned} E_2 &= \frac{-13.6\,\text{eV}}{4} = -3.40\,\text{eV}, \\ E_3 &= \frac{-13.6\,\text{eV}}{9} = -1.51\,\text{eV}. \end{aligned}$$

We see that not only are the orbit radii quantized, but from Eqs. 37–14 so is the energy. The quantum number n that labels the orbit radii also labels the energy levels. The lowest **energy level** or **energy state** has energy E_1, and is called the **ground state**. The higher states, E_2, E_3, and so on, are called **excited states**. The fixed energy levels are also called **stationary states**.

†Be careful not to believe that these well-defined orbits actually exist. Today electrons are better thought of as forming "clouds," as discussed in Chapter 39.

Notice that although the energy for the larger orbits has a smaller numerical value, all the energies are less than zero. Thus, $-3.4\,\text{eV}$ is a higher energy than $-13.6\,\text{eV}$. Hence the orbit closest to the nucleus (r_1) has the lowest energy. The reason the energies have negative values has to do with the way we defined the zero for potential energy (U). For two point charges, $U = (1/4\pi\epsilon_0)(q_1 q_2/r)$ corresponds to zero potential energy when the two charges are infinitely far apart. Thus, an electron that can just barely be free from the atom by reaching $r = \infty$ (or, at least, far from the nucleus) with zero kinetic energy will have $E = 0$, corresponding to $n = \infty$ in Eqs. 37–14. If an electron is free and has kinetic energy, then $E > 0$. To remove an electron that is part of an atom requires an energy input (otherwise atoms would not be stable). Since $E \geq 0$ for a free electron, then an electron bound to an atom needs to have $E < 0$. That is, energy must be added to bring its energy up, from a negative value, to at least zero in order to free it.

The minimum energy required to remove an electron from an atom initially in the ground state is called the **binding energy** or **ionization energy**. The ionization energy for hydrogen has been measured to be 13.6 eV, and this corresponds precisely to removing an electron from the lowest state, $E_1 = -13.6\,\text{eV}$, up to $E = 0$ where it can be free.

Spectra Lines Explained

It is useful to show the various possible energy values as horizontal lines on an energy-level diagram. This is shown for hydrogen in Fig. 37–26.[†] The electron in a hydrogen atom can be in any one of these levels according to Bohr's theory. But it could never be in between, say at $-9.0\,\text{eV}$. At room temperature, nearly all H atoms will be in the ground state $(n = 1)$. At higher temperatures, or during an electric discharge when there are many collisions between free electrons and atoms, many atoms can be in excited states $(n > 1)$. Once in an excited state, an atom's electron can jump down to a lower state, and give off a photon in the process. This is, according to the Bohr model, the origin of the emission spectra of excited gases.

[†]Note that above $E = 0$, an electron is free and can have any energy (E is not quantized). Thus there is a continuum of energy states above $E = 0$, as indicated in the energy-level diagram of Fig. 37–26.

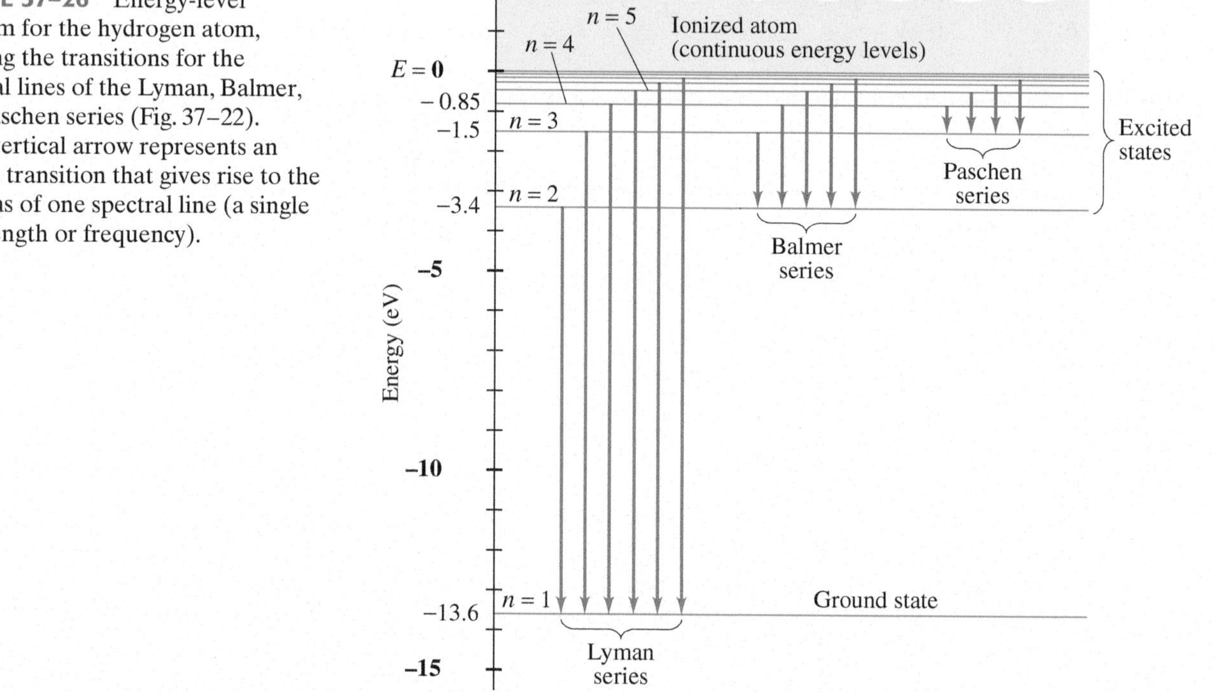

FIGURE 37–26 Energy-level diagram for the hydrogen atom, showing the transitions for the spectral lines of the Lyman, Balmer, and Paschen series (Fig. 37–22). Each vertical arrow represents an atomic transition that gives rise to the photons of one spectral line (a single wavelength or frequency).

The vertical arrows in Fig. 37–26 represent the transitions or jumps that correspond to the various observed spectral lines. For example, an electron jumping from the level $n = 3$ to $n = 2$ would give rise to the 656-nm line in the Balmer series, and the jump from $n = 4$ to $n = 2$ would give rise to the 486-nm line (see Fig. 37–21). We can predict wavelengths of the spectral lines emitted by combining Eq. 37–9 with Eq. 37–14a. Since $hf = hc/\lambda$, we have from Eq. 37–9

$$\frac{1}{\lambda} = \frac{hf}{hc} = \frac{1}{hc}(E_n - E_{n'}),$$

where n refers to the upper state and n' to the lower state. Then using Eq. 37–14a,

$$\frac{1}{\lambda} = \frac{Z^2 e^4 m}{8\epsilon_0^2 h^3 c}\left(\frac{1}{(n')^2} - \frac{1}{(n)^2}\right). \tag{37–15}$$

This theoretical formula has the same form as the experimental Balmer formula, Eq. 37–8, with $n' = 2$. Thus we see that the Balmer series of lines corresponds to transitions or "jumps" that bring the electron down to the second energy level. Similarly, $n' = 1$ corresponds to the Lyman series and $n' = 3$ to the Paschen series (see Fig. 37–26).

When the constant in Eq. 37–15 is evaluated with $Z = 1$, it is found to have the measured value of the Rydberg constant, $R = 1.0974 \times 10^7\,\mathrm{m}^{-1}$ in Eq. 37–8, in accord with experiment (see Problem 58).

The great success of Bohr's model is that it gives an explanation for why atoms emit line spectra, and accurately predicts the wavelengths of emitted light for hydrogen. The Bohr model also explains absorption spectra: photons of just the right wavelength can knock an electron from one energy level to a higher one. To conserve energy, only photons that have just the right energy will be absorbed. This explains why a continuous spectrum of light entering a gas will emerge with dark (absorption) lines at frequencies that correspond to emission lines (Fig. 37–20c).

The Bohr theory also ensures the stability of atoms. It establishes stability by decree: the ground state is the lowest state for an electron and there is no lower energy level to which it can go and emit more energy. Finally, as we saw above, the Bohr theory accurately predicts the ionization energy of 13.6 eV for hydrogen. However, the Bohr model was not so successful for other atoms, and has been superseded as we shall discuss in the next Chapter. We discuss the Bohr model because it *was* an important start, and because we still use the concept of stationary states, the ground state, and transitions between states. Also, the terminology used in the Bohr model is still used by chemists and spectroscopists.

EXAMPLE 37–13 **Wavelength of a Lyman line.** Use Fig. 37–26 to determine the wavelength of the first Lyman line, the transition from $n = 2$ to $n = 1$. In what region of the electromagnetic spectrum does this lie?

APPROACH We use Eq. 37–9, $hf = E_U - E_L$, with the energies obtained from Fig. 37–26 to find the energy and the wavelength of the transition. The region of the electromagnetic spectrum is found using the EM spectrum in Fig. 31–12.

SOLUTION In this case, $hf = E_2 - E_1 = \{-3.4\,\mathrm{eV} - (-13.6\,\mathrm{eV})\} = 10.2\,\mathrm{eV}$ $= (10.2\,\mathrm{eV})(1.60 \times 10^{-19}\,\mathrm{J/eV}) = 1.63 \times 10^{-18}\,\mathrm{J}$. Since $\lambda = c/f$, we have

$$\lambda = \frac{c}{f} = \frac{hc}{E_2 - E_1} = \frac{(6.63 \times 10^{-34}\,\mathrm{J \cdot s})(3.00 \times 10^8\,\mathrm{m/s})}{1.63 \times 10^{-18}\,\mathrm{J}} = 1.22 \times 10^{-7}\,\mathrm{m},$$

or 122 nm, which is in the UV region of the EM spectrum, Fig. 31–12. See also Fig. 37–22.

NOTE An alternate approach would be to use Eq. 37–15 to find λ, and it gives the same result.

EXAMPLE 37–14 **Wavelength of a Balmer line.** Determine the wavelength of light emitted when a hydrogen atom makes a transition from the $n = 6$ to the $n = 2$ energy level according to the Bohr model.

APPROACH We can use Eq. 37–15 or its equivalent, Eq. 37–8, with $R = 1.097 \times 10^7\,\mathrm{m}^{-1}$.

SOLUTION We find

$$\frac{1}{\lambda} = \left(1.097 \times 10^7\,\mathrm{m}^{-1}\right)\left(\frac{1}{4} - \frac{1}{36}\right) = 2.44 \times 10^6\,\mathrm{m}^{-1}.$$

So $\lambda = 1/\left(2.44 \times 10^6\,\mathrm{m}^{-1}\right) = 4.10 \times 10^{-7}\,\mathrm{m}$ or 410 nm. This is the fourth line in the Balmer series, Fig. 37–21, and is violet in color.

EXERCISE F The energy of the photon emitted when a hydrogen atom goes from the $n = 6$ state to the $n = 3$ state is (*a*) 0.378 eV; (*b*) 0.503 eV; (*c*) 1.13 eV; (*d*) 3.06 eV; (*e*) 13.6 eV.

EXAMPLE 37–15 **Absorption wavelength.** Use Fig. 37–26 to determine the maximum wavelength that hydrogen in its ground state can absorb. What would be the next smaller wavelength that would work?

APPROACH Maximum wavelength corresponds to minimum energy, and this would be the jump from the ground state up to the first excited state (Fig. 37–26). The next smaller wavelength occurs for the jump from the ground state to the second excited state. In each case, the energy difference can be used to find the wavelength.

SOLUTION The energy needed to jump from the ground state to the first excited state is $13.6\,\mathrm{eV} - 3.4\,\mathrm{eV} = 10.2\,\mathrm{eV}$; the required wavelength, as we saw in Example 37–13, is 122 nm. The energy to jump from the ground state to the second excited state is $13.6\,\mathrm{eV} - 1.5\,\mathrm{eV} = 12.1\,\mathrm{eV}$, which corresponds to a wavelength

$$\lambda = \frac{c}{f} = \frac{hc}{hf} = \frac{hc}{E_3 - E_1}$$

$$= \frac{\left(6.63 \times 10^{-34}\,\mathrm{J\cdot s}\right)\left(3.00 \times 10^8\,\mathrm{m/s}\right)}{\left(12.1\,\mathrm{eV}\right)\left(1.60 \times 10^{-19}\,\mathrm{J/eV}\right)} = 103\,\mathrm{nm}.$$

EXAMPLE 37–16 **He^+ ionization energy.** (*a*) Use the Bohr model to determine the ionization energy of the He^+ ion, which has a single electron. (*b*) Also calculate the maximum wavelength a photon can have to cause ionization.

APPROACH We want to determine the minimum energy required to lift the electron from its ground state and to barely reach the free state at $E = 0$. The ground state energy of He^+ is given by Eq. 37–14b with $n = 1$ and $Z = 2$.

SOLUTION (*a*) Since all the symbols in Eq. 37–14b are the same as for the calculation for hydrogen, except that Z is 2 instead of 1, we see that E_1 will be $Z^2 = 2^2 = 4$ times the E_1 for hydrogen:

$$E_1 = 4(-13.6\,\mathrm{eV}) = -54.4\,\mathrm{eV}.$$

Thus, to ionize the He^+ ion should require 54.4 eV, and this value agrees with experiment.

(*b*) The maximum wavelength photon that can cause ionization will have energy $hf = 54.4\,\mathrm{eV}$ and wavelength

$$\lambda = \frac{c}{f} = \frac{hc}{hf} = \frac{\left(6.63 \times 10^{-34}\,\mathrm{J\cdot s}\right)\left(3.00 \times 10^8\,\mathrm{m/s}\right)}{\left(54.4\,\mathrm{eV}\right)\left(1.60 \times 10^{-19}\,\mathrm{J/eV}\right)} = 22.8\,\mathrm{nm}.$$

NOTE If the atom absorbed a photon of greater energy (wavelength shorter than 22.8 nm), the atom could still be ionized and the freed electron would have kinetic energy of its own. If $\lambda > 22.8\,\mathrm{nm}$, the photon has too little energy to cause ionization.

In this last Example, we saw that E_1 for the He^+ ion is four times more negative than that for hydrogen. Indeed, the energy-level diagram for He^+ looks just like that for hydrogen, Fig. 37–26, except that the numerical values for each energy level are four times larger. Note, however, that we are talking here about the He^+ *ion*. Normal (neutral) helium has two electrons and its energy level diagram is entirely different.

CONCEPTUAL EXAMPLE 37–17 **Hydrogen at 20°C.** Estimate the average kinetic energy of whole hydrogen atoms (not just the electrons) at room temperature, and use the result to explain why nearly all H atoms are in the ground state at room temperature, and hence emit no light.

RESPONSE According to kinetic theory (Chapter 18), the average kinetic energy of atoms or molecules in a gas is given by Eq. 18–4:

$$\overline{K} = \tfrac{3}{2}kT,$$

where $k = 1.38 \times 10^{-23}\,\mathrm{J/K}$ is Boltzmann's constant, and T is the kelvin (absolute) temperature. Room temperature is about $T = 300\,\mathrm{K}$, so

$$\overline{K} = \tfrac{3}{2}(1.38 \times 10^{-23}\,\mathrm{J/K})(300\,\mathrm{K}) = 6.2 \times 10^{-21}\,\mathrm{J},$$

or, in electron volts:

$$\overline{K} = \frac{6.2 \times 10^{-21}\,\mathrm{J}}{1.6 \times 10^{-19}\,\mathrm{J/eV}} = 0.04\,\mathrm{eV}.$$

The average kinetic energy of an atom as a whole is thus very small compared to the energy between the ground state and the next higher energy state ($13.6\,\mathrm{eV} - 3.4\,\mathrm{eV} = 10.2\,\mathrm{eV}$). Any atoms in excited states quickly fall to the ground state and emit light. Once in the ground state, collisions with other atoms can transfer energy of only 0.04 eV on the average. A small fraction of atoms can have much more energy (see Section 18–2 on the distribution of molecular speeds), but even a kinetic energy that is 10 times the average is not nearly enough to excite atoms into states above the ground state. Thus, at room temperature, nearly all atoms are in the ground state. Atoms can be excited to upper states by very high temperatures, or by passing a current of high energy electrons through the gas, as in a discharge tube (Fig. 37–19).

Correspondence Principle

We should note that Bohr made some radical assumptions that were at variance with classical ideas. He assumed that electrons in fixed orbits do not radiate light even though they are accelerating (moving in a circle), and he assumed that angular momentum is quantized. Furthermore, he was not able to say how an electron moved when it made a transition from one energy level to another. On the other hand, there is no real reason to expect that in the tiny world of the atom electrons would behave as ordinary-sized objects do. Nonetheless, he felt that where quantum theory overlaps with the macroscopic world, it should predict classical results. This is the **correspondence principle**, already mentioned in regard to relativity (Section 36–13). This principle does work for Bohr's theory of the hydrogen atom. The orbit sizes and energies are quite different for $n = 1$ and $n = 2$, say. But orbits with $n = 100{,}000{,}000$ and 100,000,001 would be very close in radius and energy (see Fig. 37–26). Indeed, jumps between such large orbits (which would approach macroscopic sizes), would be imperceptible. Such orbits would thus appear to be continuously spaced, which is what we expect in the everyday world.

Finally, it must be emphasized that the well-defined orbits of the Bohr model do not actually exist. The Bohr model is only a model, not reality. The idea of electron orbits was rejected a few years later, and today electrons are thought of (Chapter 39) as forming "probability clouds."

37–12 de Broglie's Hypothesis Applied to Atoms

Bohr's theory was largely of an *ad hoc* nature. Assumptions were made so that theory would agree with experiment. But Bohr could give no reason why the orbits were quantized, nor why there should be a stable ground state. Finally, ten years later, a reason was proposed by Louis de Broglie. We saw in Section 37–7 that in 1923, de Broglie proposed that material particles, such as electrons, have a wave nature; and that this hypothesis was confirmed by experiment several years later.

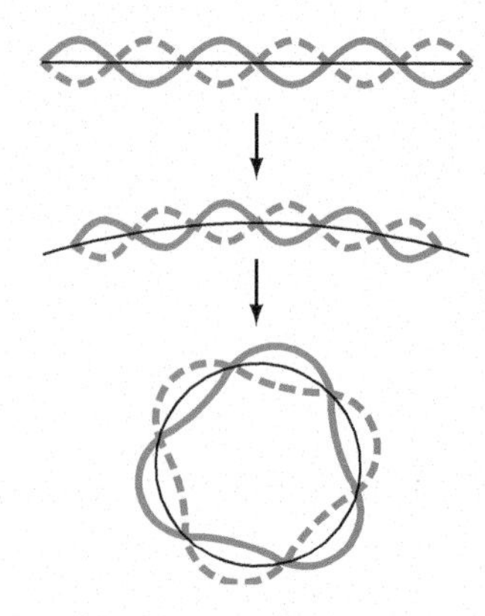

FIGURE 37–27 An ordinary standing wave compared to a circular standing wave.

FIGURE 37–28 When a wave does not close (and hence interferes destructively with itself), it rapidly dies out.

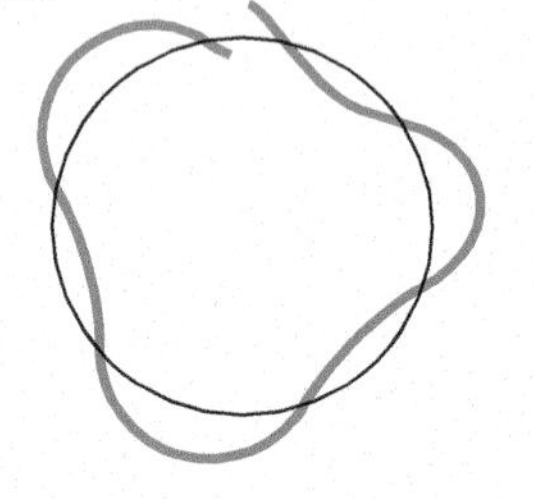

FIGURE 37–29 Standing circular waves for two, three, and five wavelengths on the circumference; n, the number of wavelengths, is also the quantum number.

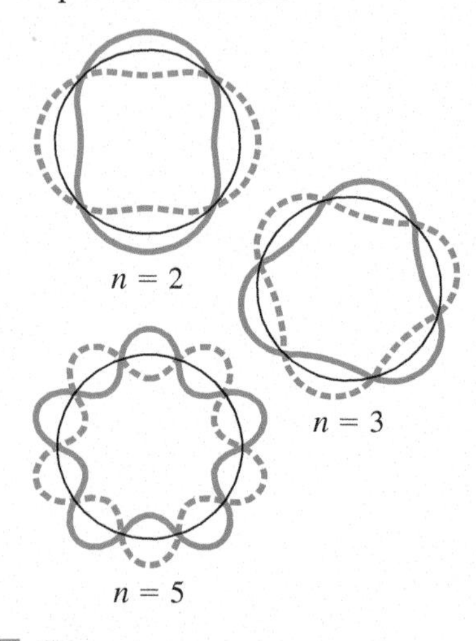

One of de Broglie's original arguments in favor of the wave nature of electrons was that it provided an explanation for Bohr's theory of the hydrogen atom. According to de Broglie, a particle of mass m moving with a nonrelativistic speed v would have a wavelength (Eq. 37–7) of

$$\lambda = \frac{h}{mv}.$$

Each electron orbit in an atom, he proposed, is actually a standing wave. As we saw in Chapter 15, when a violin or guitar string is plucked, a vast number of wavelengths are excited. But only certain ones—those that have nodes at the ends—are sustained. These are the *resonant* modes of the string. Waves with other wavelengths interfere with themselves upon reflection and their amplitudes quickly drop to zero. With electrons moving in circles, according to Bohr's theory, de Broglie argued that the electron wave was a *circular* standing wave that closes on itself, Fig. 37–27. If the wavelength of a wave does not close on itself, as in Fig. 37–28, destructive interference takes place as the wave travels around the loop, and the wave quickly dies out. Thus, the only waves that persist are those for which the circumference of the circular orbit contains a whole number of wavelengths, Fig. 37–29. The circumference of a Bohr orbit of radius r_n is $2\pi r_n$, so to have constructive interference, we need

$$2\pi r_n = n\lambda, \qquad n = 1, 2, 3, \cdots.$$

When we substitute $\lambda = h/mv$, we get $2\pi r_n = nh/mv$, or

$$mvr_n = \frac{nh}{2\pi}.$$

This is just the *quantum condition* proposed by Bohr on an *ad hoc* basis, Eq. 37–10. It is from this equation that the discrete orbits and energy levels were derived. Thus we have a first explanation for the quantized orbits and energy states in the Bohr model: they are due to the wave nature of the electron, and only resonant "standing" waves can persist.† This implies that the *wave–particle duality* is at the root of atomic structure.

In viewing the circular electron waves of Fig. 37–29, the electron is not to be thought of as following the oscillating wave pattern. In the Bohr model of hydrogen, the electron moves in a circle. The circular wave, on the other hand, represents the *amplitude* of the electron "matter wave," and in Fig. 37–29 the wave amplitude is shown superimposed on the circular path of the particle orbit for convenience.

Bohr's theory worked well for hydrogen and for one-electron ions. But it did not prove successful for multi-electron atoms. Bohr's theory could not predict line spectra even for the next simplest atom, helium. It could not explain why some emission lines are brighter than others, nor why some lines are split into two or more closely spaced lines ("fine structure"). A new theory was needed and was indeed developed in the 1920s. This new and radical theory is called *quantum mechanics*. It finally solved the problem of atomic structure, but it gives us a very different view of the atom: the idea of electrons in well-defined orbits was replaced with the idea of electron "clouds." This new theory of quantum mechanics has given us a wholly different view of the basic mechanisms underlying physical processes.

†We note, however, that Eq. 37–10 is no longer considered valid, as discussed in Chapter 39.

Summary

Quantum theory has its origins in **Planck's quantum hypothesis** that molecular oscillations are **quantized**: their energy E can only be integer (n) multiples of hf, where h is Planck's constant and f is the natural frequency of oscillation:

$$E = nhf. \qquad \textbf{(37–2)}$$

This hypothesis explained the spectrum of radiation emitted by a **blackbody** at high temperature.

Einstein proposed that for some experiments, light could be pictured as being emitted and absorbed as **quanta** (particles), which we now call **photons**, each with energy

$$E = hf \qquad \textbf{(37–3)}$$

and momentum

$$p = \frac{E}{c} = \frac{hf}{c} = \frac{h}{\lambda}. \qquad \textbf{(37–5)}$$

He proposed the photoelectric effect as a test for the photon theory of light. In the **photoelectric effect**, the photon theory says that each incident photon can strike an electron in a material and eject it if the photon has sufficient energy. The maximum energy of ejected electrons is then linearly related to the frequency of the incident light.

The photon theory is also supported by the **Compton effect** and the observation of electron–positron **pair production**.

The **wave–particle duality** refers to the idea that light and matter (such as electrons) have both wave and particle properties. The wavelength of an object is given by

$$\lambda = \frac{h}{p}, \qquad \textbf{(37–7)}$$

where p is the momentum of the object ($p = mv$ for a particle of mass m and speed v).

The **principle of complementarity** states that we must be aware of both the particle and wave properties of light and of matter for a complete understanding of them.

Early models of the atom include the plum-pudding model, and Rutherford's planetary (or nuclear) model of an atom which consists of a tiny but massive positively charged nucleus surrounded (at a relatively great distance) by electrons.

To explain the **line spectra** emitted by atoms, as well as the stability of atoms, **Bohr's theory** postulated that: (1) electrons bound in an atom can only occupy orbits for which the angular momentum is quantized, which results in discrete values for the radius and energy; (2) an electron in such a **stationary state** emits no radiation; (3) if an electron jumps to a lower state, it emits a photon whose energy equals the difference in energy between the two states; (4) the angular momentum L of atomic electrons is quantized by the rule

$$L = \frac{nh}{2\pi}, \qquad \textbf{(37–10)}$$

where n is an integer called the **quantum number**. The $n = 1$ state is the **ground state**, which in hydrogen has an energy $E_1 = -13.6\,\text{eV}$. Higher values of n correspond to **excited states**, and their energies are

$$E_n = -(13.6\,\text{eV})\frac{Z^2}{n^2}. \qquad \textbf{(37–14b)}$$

Atoms are excited to these higher states by collisions with other atoms or electrons, or by absorption of a photon of just the right frequency.

De Broglie's hypothesis that electrons (and other matter) have a wavelength $\lambda = h/mv$ gave an explanation for Bohr's quantized orbits by bringing in the wave–particle duality: the orbits correspond to circular standing waves in which the circumference of the orbit equals a whole number of wavelengths.

Questions

1. What can be said about the relative temperatures of whitish-yellow, reddish, and bluish stars? Explain.
2. If energy is radiated by all objects, why can we not see most of them in the dark?
3. Does a lightbulb at a temperature of 2500 K produce as white a light as the Sun at 6000 K? Explain.
4. Darkrooms for developing black-and-white film were sometimes lit by a red bulb. Why red? Would such a bulb work in a darkroom for developing color photographs?
5. If the threshold wavelength in the photoelectric effect increases when the emitting metal is changed to a different metal, what can you say about the work functions of the two metals?
6. Explain why the existence of a cutoff frequency in the photoelectric effect more strongly favors a particle theory rather than a wave theory of light.
7. UV light causes sunburn, whereas visible light does not. Suggest a reason.
8. The work functions for sodium and cesium are 2.28 eV and 2.14 eV, respectively. For incident photons of a given frequency, which metal will give a higher maximum kinetic energy for the electrons?
9. (*a*) Does a beam of infrared photons always have less energy than a beam of ultraviolet photons? Explain. (*b*) Does a single photon of infrared light always have less energy than a single photon of ultraviolet light?
10. Light of 450-nm wavelength strikes a metal surface, and a stream of electrons emerges from the metal. If light of the same intensity but of wavelength 400 nm strikes the surface, are more electrons emitted? Does the energy of the emitted electrons change? Explain.
11. Explain how the photoelectric circuit of Fig. 37–4 could be used in (*a*) a burglar alarm, (*b*) a smoke detector, (*c*) a photographic light meter.
12. If an X-ray photon is scattered by an electron, does the photon's wavelength change? If so, does it increase or decrease?
13. In both the photoelectric effect and in the Compton effect, a photon collides with an electron causing the electron to fly off. What then, is the difference between the two processes?
14. Consider a point source of light. How would the intensity of light vary with distance from the source according to (*a*) wave theory, (*b*) particle (photon) theory? Would this help to distinguish the two theories?
15. If an electron and a proton travel at the same speed, which has the shorter de Broglie wavelength? Explain.
16. Why do we say that light has wave properties? Why do we say that light has particle properties?
17. Why do we say that electrons have wave properties? Why do we say that electrons have particle properties?
18. What are the differences between a photon and an electron? Be specific: make a list.
19. In Rutherford's planetary model of the atom, what keeps the electrons from flying off into space?
20. How can you tell if there is oxygen near the surface of the Sun?
21. When a wide spectrum of light passes through hydrogen gas at room temperature, absorption lines are observed that correspond only to the Lyman series. Why don't we observe the other series?
22. Explain how the closely spaced energy levels for hydrogen near the top of Fig. 37–26 correspond to the closely spaced spectral lines at the top of Fig. 37–21.

23. Is it possible for the de Broglie wavelength of a "particle" to be greater than the dimensions of the particle? To be smaller? Is there any direct connection?

24. In a helium atom, which contains two electrons, do you think that on average the electrons are closer to the nucleus or farther away than in a hydrogen atom? Why?

25. How can the spectrum of hydrogen contain so many lines when hydrogen contains only one electron?

26. The Lyman series is brighter than the Balmer series because this series of transitions ends up in the most common state for hydrogen, the ground state. Why then was the Balmer series discovered first?

27. Use conservation of momentum to explain why photons emitted by hydrogen atoms have slightly less energy than that predicted by Eq. 37–9.

28. Suppose we obtain an emission spectrum for hydrogen at very high temperature (when some of the atoms are in excited states), and an absorption spectrum at room temperature, when all atoms are in the ground state. Will the two spectra contain identical lines?

Problems

37–1 Planck's Quantum Hypothesis

1. (I) Estimate the peak wavelength for radiation from (*a*) ice at 273K, (*b*) a floodlamp at 3500 K, (*c*) helium at 4.2 K, (*d*) for the universe at $T = 2.725$ K, assuming blackbody emission. In what region of the EM spectrum is each?

2. (I) How hot is metal being welded if it radiates most strongly at 460 nm?

3. (I) An HCl molecule vibrates with a natural frequency of 8.1×10^{13} Hz. What is the difference in energy (in joules and electron volts) between successive values of the oscillation energy?

4. (II) Estimate the peak wavelength of light issuing from the pupil of the human eye (which approximates a blackbody) assuming normal body temperature.

5. (III) Planck's radiation law is given by:

$$I(\lambda, T) = \frac{2\pi hc^2\lambda^{-5}}{e^{hc/\lambda kT} - 1}$$

where $I(\lambda, T)$ is the rate energy is radiated per unit surface area per unit wavelength interval at wavelength λ and Kelvin temperature T. (*a*) Show that Wien's displacement law follows from this relationship. (*b*) Determine the value of h from the experimental value of $\lambda_P T$ given in the text. [You may want to use graphing techniques.] (*c*) Derive the T^4 dependence of the rate at which energy is radiated (as in the Stefan-Boltzmann law, Eq. 19–17), by integrating Planck's formula over all wavelengths; that is, show that

$$\int I(\lambda, T)\, d\lambda \propto T^4.$$

37–2 and 37–3 Photons and the Photoelectric Effect

6. (I) What is the energy of photons (in joules) emitted by a 104.1-MHz FM radio station?

7. (I) What is the energy range (in joules and eV) of photons in the visible spectrum, of wavelength 410 nm to 750 nm?

8. (I) A typical gamma ray emitted from a nucleus during radioactive decay may have an energy of 380 keV. What is its wavelength? Would we expect significant diffraction of this type of light when it passes through an everyday opening, such as a door?

9. (I) About 0.1 eV is required to break a "hydrogen bond" in a protein molecule. Calculate the minimum frequency and maximum wavelength of a photon that can accomplish this.

10. (I) Calculate the momentum of a photon of yellow light of wavelength 6.20×10^{-7} m.

11. (I) What minimum frequency of light is needed to eject electrons from a metal whose work function is 4.8×10^{-19} J?

12. (I) What is the longest wavelength of light that will emit electrons from a metal whose work function is 3.70 eV?

13. (II) What wavelength photon would have the same energy as a 145-gram baseball moving 30.0 m/s?

14. (II) The human eye can respond to as little as 10^{-18} J of light energy. For a wavelength at the peak of visual sensitivity, 550 nm, how many photons lead to an observable flash?

15. (II) The work functions for sodium, cesium, copper, and iron are 2.3, 2.1, 4.7, and 4.5 eV, respectively. Which of these metals will not emit electrons when visible light shines on it?

16. (II) In a photoelectric-effect experiment it is observed that no current flows unless the wavelength is less than 520 nm. (*a*) What is the work function of this material? (*b*) What is the stopping voltage required if light of wavelength 470 nm is used?

17. (II) What is the maximum kinetic energy of electrons ejected from barium $(W_0 = 2.48\,\text{eV})$ when illuminated by white light, $\lambda = 410$ to 750 nm?

18. (II) Barium has a work function of 2.48 eV. What is the maximum kinetic energy of electrons if the metal is illuminated by UV light of wavelength 365 nm? What is their speed?

19. (II) When UV light of wavelength 285 nm falls on a metal surface, the maximum kinetic energy of emitted electrons is 1.70 eV. What is the work function of the metal?

20. (II) The threshold wavelength for emission of electrons from a given surface is 320 nm. What will be the maximum kinetic energy of ejected electrons when the wavelength is changed to (*a*) 280 nm, (*b*) 360 nm?

21. (II) When 230-nm light falls on a metal, the current through a photoelectric circuit (Fig. 37–4) is brought to zero at a stopping voltage of 1.84 V. What is the work function of the metal?

22. (II) A certain type of film is sensitive only to light whose wavelength is less than 630 nm. What is the energy (eV and kcal/mol) needed for the chemical reaction to occur which causes the film to change?

23. (II) The range of visible light wavelengths extends from about 410 nm to 750 nm. (*a*) Estimate the minimum energy (eV) necessary to initiate the chemical process on the retina that is responsible for vision. (*b*) Speculate as to why, at the other end of the visible range, there is a threshold photon energy beyond which the eye registers no sensation of sight. Determine this threshold photon energy (eV).

24. (II) In a photoelectric experiment using a clean sodium surface, the maximum energy of the emitted electrons was measured for a number of different incident frequencies, with the following results.

Frequency ($\times 10^{14}$ Hz)	Energy (eV)
11.8	2.60
10.6	2.11
9.9	1.81
9.1	1.47
8.2	1.10
6.9	0.57

Plot the graph of these results and find: (*a*) Planck's constant; (*b*) the cutoff frequency of sodium; (*c*) the work function.

25. (II) A **photomultiplier tube** (a very sensitive light sensor), is based on the photoelectric effect: incident photons strike a metal surface and the resulting ejected electrons are collected. By counting the number of collected electrons, the number of incident photons (i.e., the incident light intensity) can be determined. (*a*) If a photomultiplier tube is to respond properly for incident wavelengths throughout the visible range (410 nm to 750 nm), what is the maximum value for the work function W_0 (eV) of its metal surface? (*b*) If W_0 for its metal surface is above a certain threshold value, the photomultiplier will only function for incident ultraviolet wavelengths and be unresponsive to visible light. Determine this threshold value (eV).

26. (III) A group of atoms is confined to a very small (point-like) volume in a laser-based **atom trap**. The incident laser light causes each atom to emit 1.0×10^6 photons of wavelength 780 nm every second. A sensor of area $1.0\ \mathrm{cm}^2$ measures the light intensity emanating from the trap to be 1.6 nW when placed 25 cm away from the trapped atoms. Assuming each atom emits photons with equal probability in all directions, determine the number of trapped atoms.

27. (III) Assume light of wavelength λ is incident on a metal surface, whose work function is known precisely (i.e., its uncertainty is better than 0.1% and can be ignored). Show that if the stopping voltage can be determined to an accuracy of ΔV_0, the fractional uncertainty (magnitude) in wavelength is

$$\frac{\Delta\lambda}{\lambda} = \frac{\lambda e}{hc}\Delta V_0 .$$

Determine this fractional uncertainty if $\Delta V_0 = 0.01$ V and $\lambda = 550$ nm.

37–4 Compton Effect

28. (I) A high-frequency photon is scattered off of an electron and experiences a change of wavelength of 1.5×10^{-4} nm. At what angle must a detector be placed to detect the scattered photon (relative to the direction of the incoming photon)?

29. (II) Determine the Compton wavelength for (*a*) an electron, (*b*) a proton. (*c*) Show that if a photon has wavelength equal to the Compton wavelength of a particle, the photon's energy is equal to the rest energy of the particle.

30. (II) X-rays of wavelength $\lambda = 0.120$ nm are scattered from carbon. What is the expected Compton wavelength shift for photons detected at angles (relative to the incident beam) of exactly (*a*) 60°, (*b*) 90°, (*c*) 180°?

31. (II) In the Compton effect, determine the ratio $(\Delta\lambda/\lambda)$ of the maximum change $\Delta\lambda$ in a photon's wavelength to the photon's initial wavelength λ, if the photon is (*a*) a visible-light photon with $\lambda = 550$ nm, (*b*) an X-ray photon with $\lambda = 0.10$ nm.

32. (II) A 1.0-MeV gamma-ray photon undergoes a sequence of Compton-scattering events. If the photon is scattered at an angle of 0.50° in each event, estimate the number of events required to convert the photon into a visible-light photon with wavelength 555 nm. You can use an expansion for small θ; see Appendix A. [Gamma rays created near the center of the Sun are transformed to visible wavelengths as they travel to the Sun's surface through a sequence of small-angle Compton scattering events.]

33. (III) In the Compton effect, a 0.160-nm photon strikes a free electron in a head-on collision and knocks it into the forward direction. The rebounding photon recoils directly backward. Use conservation of (relativistic) energy and momentum to determine (*a*) the kinetic energy of the electron, and (*b*) the wavelength of the recoiling photon. Use Eq. 37–5, but not Eq. 37–6.

34. (III) In the Compton effect (see Fig. 37–7), use the relativistic equations for conservation of energy and of linear momentum to show that the Compton shift in wavelength is given by Eq. 37–6.

37–5 Pair Production

35. (I) How much total kinetic energy will an electron–positron pair have if produced by a 2.67-MeV photon?

36. (II) What is the longest wavelength photon that could produce a proton–antiproton pair? (Each has a mass of 1.67×10^{-27} kg.)

37. (II) What is the minimum photon energy needed to produce a $\mu^+ - \mu^-$ pair? The mass of each μ (muon) is 207 times the mass of an electron. What is the wavelength of such a photon?

38. (II) An electron and a positron, each moving at 2.0×10^5 m/s, collide head on, disappear, and produce two photons moving in opposite directions, each with the same energy and momentum. Determine the energy and momentum of each photon.

39. (II) A gamma-ray photon produces an electron and a positron, each with a kinetic energy of 375 keV. Determine the energy and wavelength of the photon.

37–7 Wave Nature of Matter

40. (I) Calculate the wavelength of a 0.23-kg ball traveling at 0.10 m/s.

41. (I) What is the wavelength of a neutron $(m = 1.67 \times 10^{-27}\ \mathrm{kg})$ traveling at 8.5×10^4 m/s?

42. (I) Through how many volts of potential difference must an electron be accelerated to achieve a wavelength of 0.21 nm?

43. (II) What is the theoretical limit of resolution for an electron microscope whose electrons are accelerated through 85 kV? (Relativistic formulas should be used.)

44. (II) The speed of an electron in a particle accelerator is 0.98*c*. Find its de Broglie wavelength. (Use relativistic momentum.)

45. (II) Calculate the ratio of the kinetic energy of an electron to that of a proton if their wavelengths are equal. Assume that the speeds are nonrelativistic.

46. (II) Neutrons can be used in diffraction experiments to probe the lattice structure of crystalline solids. Since the neutron's wavelength needs to be on the order of the spacing between atoms in the lattice, about 0.3 nm, what should the speed of the neutrons be?

47. (II) An electron has a de Broglie wavelength $\lambda = 6.0 \times 10^{-10}$ m. (*a*) What is its momentum? (*b*) What is its speed? (*c*) What voltage was needed to accelerate it to this speed?

48. (II) What is the wavelength of an electron of energy (*a*) 20 eV, (*b*) 200 eV, (*c*) 2.0 keV?

49. (II) Show that if an electron and a proton have the same nonrelativistic kinetic energy, the proton has the shorter wavelength.

50. (II) Calculate the de Broglie wavelength of an electron in a TV picture tube if it is accelerated by 33,000 V. Is it relativistic? How does its wavelength compare to the size of the "neck" of the tube, typically 5 cm? Do we have to worry about diffraction problems blurring our picture on the screen?

51. (II) After passing through two slits separated by a distance of 3.0 μm, a beam of electrons creates an interference pattern with its second-order maximum at an angle of 55°. Find the speed of the electrons in this beam.

* 37–8 Electron Microscope

*52. (II) What voltage is needed to produce electron wavelengths of 0.28 nm? (Assume that the electrons are nonrelativistic.)

*53. (II) Electrons are accelerated by 3450 V in an electron microscope. Estimate the maximum possible resolution of the microscope.

37–10 and 37–11 Bohr Model

54. (I) For the three hydrogen transitions indicated below, with n being the initial state and n' being the final state, is the transition an absorption or an emission? Which is higher, the initial state energy or the final state energy of the atom? Finally, which of these transitions involves the largest energy photon? (*a*) $n = 1$, $n' = 3$; (*b*) $n = 6$, $n' = 2$; (*c*) $n = 4$, $n' = 5$.

55. (I) How much energy is needed to ionize a hydrogen atom in the $n = 3$ state?

56. (I) (*a*) Determine the wavelength of the second Balmer line ($n = 4$ to $n = 2$ transition) using Fig. 37–26. Determine likewise (*b*) the wavelength of the third Lyman line and (*c*) the wavelength of the first Balmer line.

57. (I) Calculate the ionization energy of doubly ionized lithium, Li^{2+}, which has $Z = 3$.

58. (I) Evaluate the Rydberg constant R using the Bohr model (compare Eqs. 37–8 and 37–15) and show that its value is $R = 1.0974 \times 10^7\ m^{-1}$.

59. (II) What is the longest wavelength light capable of ionizing a hydrogen atom in the ground state?

60. (II) In the Sun, an ionized helium (He^+) atom makes a transition from the $n = 5$ state to the $n = 2$ state, emitting a photon. Can that photon be absorbed by hydrogen atoms present in the Sun? If so, between what energy states will the hydrogen atom transition occur?

61. (II) What wavelength photon would be required to ionize a hydrogen atom in the ground state and give the ejected electron a kinetic energy of 20.0 eV?

62. (II) For what maximum kinetic energy is a collision between an electron and a hydrogen atom in its ground state definitely elastic?

63. (II) Construct the energy-level diagram for the He^+ ion (like Fig. 37–26).

64. (II) Construct the energy-level diagram (like Fig. 37–26) for doubly ionized lithium, Li^{2+}.

65. (II) Determine the electrostatic potential energy and the kinetic energy of an electron in the ground state of the hydrogen atom.

66. (II) An excited hydrogen atom could, in principle, have a diameter of 0.10 mm. What would be the value of n for a Bohr orbit of this size? What would its energy be?

67. (II) Is the use of nonrelativistic formulas justified in the Bohr atom? To check, calculate the electron's velocity, v, in terms of c, for the ground state of hydrogen, and then calculate $\sqrt{1 - v^2/c^2}$.

68. (II) A hydrogen atom has an angular momentum of $5.273 \times 10^{-34}\ kg \cdot m^2/s$. According to the Bohr model, what is the energy (eV) associated with this state?

69. (II) Assume hydrogen atoms in a gas are initially in their ground state. If free electrons with kinetic energy 12.75 eV collide with these atoms, what photon wavelengths will be emitted by the gas?

70. (II) Suppose an electron was bound to a proton, as in the hydrogen atom, but by the gravitational force rather than by the electric force. What would be the radius, and energy, of the first Bohr orbit?

71. (II) *Correspondence principle*: Show that for large values of n, the difference in radius Δr between two adjacent orbits (with quantum numbers n and $n - 1$) is given by

$$\Delta r = r_n - r_{n-1} \approx \frac{2r_n}{n},$$

so $\Delta r/r_n \to 0$ as $n \to \infty$ in accordance with the correspondence principle. [Note that we can check the correspondence principle by either considering large values of n ($n \to \infty$) or by letting $h \to 0$. Are these equivalent?]

General Problems

72. If a 75-W lightbulb emits 3.0% of the input energy as visible light (average wavelength 550 nm) uniformly in all directions, estimate how many photons per second of visible light will strike the pupil (4.0 mm diameter) of the eye of an observer 250 m away.

73. At low temperatures, nearly all the atoms in hydrogen gas will be in the ground state. What minimum frequency photon is needed if the photoelectric effect is to be observed?

74. A beam of 125-eV electrons is scattered from a crystal, as in X-ray diffraction, and a first-order peak is observed at $\theta = 38°$. What is the spacing between planes in the diffracting crystal? (See Section 35–10.)

75. A microwave oven produces electromagnetic radiation at $\lambda = 12.2$ cm and produces a power of 860 W. Calculate the number of microwave photons produced by the microwave oven each second.

76. Sunlight reaching the Earth has an intensity of about $1350\ \mathrm{W/m^2}$. Estimate how many photons per square meter per second this represents. Take the average wavelength to be 550 nm.

77. A beam of red laser light $(\lambda = 633\ \mathrm{nm})$ hits a black wall and is fully absorbed. If this light exerts a total force $F = 6.5\ \mathrm{nN}$ on the wall, how many photons per second are hitting the wall?

78. The Big Bang theory states that the beginning of the universe was accompanied by a huge burst of photons. Those photons are still present today and make up the so-called cosmic microwave background radiation. The universe radiates like a blackbody with a temperature of about 2.7 K. Calculate the peak wavelength of this radiation.

79. An electron and a positron collide head on, annihilate, and create two 0.755-MeV photons traveling in opposite directions. What were the initial kinetic energies of electron and positron?

80. By what potential difference must (*a*) a proton $(m = 1.67 \times 10^{-27}\ \mathrm{kg})$, and (*b*) an electron $(m = 9.11 \times 10^{-31}\ \mathrm{kg})$, be accelerated to have a wavelength $\lambda = 6.0 \times 10^{-12}\ \mathrm{m}$?

81. In some of Rutherford's experiments (Fig. 37–17) the α particles $(\text{mass} = 6.64 \times 10^{-27}\ \mathrm{kg})$ had a kinetic energy of 4.8 MeV. How close could they get to the center of a silver nucleus (charge $= +47e$)? Ignore the recoil motion of the nucleus.

82. Show that the magnitude of the electrostatic potential energy of an electron in any Bohr orbit of a hydrogen atom is twice the magnitude of its kinetic energy in that orbit.

83. Calculate the ratio of the gravitational force to the electric force for the electron in a hydrogen atom. Can the gravitational force be safely ignored?

84. Electrons accelerated by a potential difference of 12.3 V pass through a gas of hydrogen atoms at room temperature. What wavelengths of light will be emitted?

85. In a particular photoelectric experiment, a stopping potential of 2.70 V is measured when ultraviolet light of wavelength 380 nm is incident on the metal. If blue light of wavelength 440 nm is used, what is the new stopping potential?

86. In an X-ray tube (see Fig. 35–26 and discussion in Section 35–10), the high voltage between filament and target is V. After being accelerated through this voltage, an electron strikes the target where it is decelerated (by positively charged nuclei) and in the process one or more X-ray photons are emitted. (*a*) Show that the photon of shortest wavelength will have

$$\lambda_0 = \frac{hc}{eV}.$$

(*b*) What is the shortest wavelength of X-ray emitted when accelerated electrons strike the face of a 33-kV television picture tube?

87. The intensity of the Sun's light in the vicinity of Earth is about $1350\ \mathrm{W/m^2}$. Imagine a spacecraft with a mirrored square sail of dimension 1.0 km. Estimate how much thrust (in newtons) this craft will experience due to collisions with the Sun's photons. [*Hint:* Assume the photons bounce perpendicularly off the sail with no change in the magnitude of their momentum.]

88. Photons of energy 9.0 eV are incident on a metal. It is found that current flows from the metal until a stopping potential of 4.0 V is applied. If the wavelength of the incident photons is doubled, what is the maximum kinetic energy of the ejected electrons? What would happen if the wavelength of the incident photons was tripled?

89. Light of wavelength 360 nm strikes a metal whose work function is 2.4 eV. What is the shortest de Broglie wavelength for the electrons that are produced as photoelectrons?

90. Visible light incident on a diffraction grating with slit spacing of 0.012 mm has the first maximum at an angle of 3.5° from the central peak. If electrons could be diffracted by the same grating, what electron velocity would produce the same diffraction pattern as the visible light?

91. (*a*) Suppose an unknown element has an absorption spectrum with lines corresponding to 2.5, 4.7, and 5.1 eV above its ground state, and an ionization energy of 11.5 eV. Draw an energy level diagram for this element. (*b*) If a 5.1-eV photon is absorbed by an atom of this substance, in which state was the atom before absorbing the photon? What will be the energies of the photons that can subsequently be emitted by this atom?

92. Light of wavelength 424 nm falls on a metal which has a work function of 2.28 eV. (*a*) How much voltage should be applied to bring the current to zero? (*b*) What is the maximum speed of the emitted electrons? (*c*) What is the de Broglie wavelength of these electrons?

93. (*a*) Apply Bohr's assumptions to the Earth–Moon system to calculate the allowed energies and radii of motion. (*b*) Given the known distance between Earth and the Moon, is the quantization of the energy and radius apparent?

94. Show that the wavelength of a particle of mass m with kinetic energy K is given by the relativistic formula $\lambda = hc/\sqrt{K^2 + 2mc^2K}$.

95. A small flashlight is rated at 3.0 W. As the light leaves the flashlight in one direction, a reaction force is exerted on the flashlight in the opposite direction. Estimate the size of this reaction force.

96. At the atomic-scale, the electron volt and nanometer are well-suited units for energy and distance, respectively. (*a*) Show that the energy E in eV of a photon, whose wavelength λ is in nm, is given by

$$E = \frac{1240\ \mathrm{eV \cdot nm}}{\lambda\ (\mathrm{nm})}.$$

(*b*) How much energy (eV) does a 650-nm photon have?

*97. Three fundamental constants of nature—the gravitational constant G, Planck's constant h, and the speed of light c—have the dimensions of $[L^3/MT^2]$, $[ML^2/T]$, and $[L/T]$, respectively. (*a*) Find the mathematical combination of these fundamental constants that has the dimension of time. This combination is called the "Planck time" t_P and is thought to be the earliest time, after the creation of the universe, at which the currently known laws of physics can be applied. (*b*) Determine the numerical value of t_P. (*c*) Find the mathematical combination of these fundamental constants that has the dimension of length. This combination is called the "Planck length" λ_P and is thought to be the smallest length over which the currently known laws of physics can be applied. (*d*) Determine the numerical value of λ_P.

98. Imagine a free particle of mass m bouncing back and forth between two perfectly reflecting walls, separated by distance ℓ. Imagine that the two oppositely directed matter waves associated with this particle interfere to create a standing wave with a node at each of the walls. Show that the ground state (first harmonic) and first excited state (second harmonic) have (non-relativistic) kinetic energies $h^2/8m\ell^2$ and $h^2/2m\ell^2$, respectively.

99. (*a*) A rubidium atom ($m = 85\text{ u}$) is at rest with one electron in an excited energy level. When the electron jumps to the ground state, the atom emits a photon of wavelength $\lambda = 780\text{ nm}$. Determine the resulting (nonrelativistic) recoil speed v of the atom. (*b*) The recoil speed sets the lower limit on the temperature to which an ideal gas of rubidium atoms can be cooled in a laser-based **atom trap**. Using the kinetic theory of gases (Chapter 18), estimate this "lowest achievable" temperature.

100. A rubidium atom (atomic mass 85) is initially at room temperature and has a velocity $v = 290\text{ m/s}$ due to its thermal motion. Consider the absorption of photons by this atom from a laser beam of wavelength $\lambda = 780\text{ nm}$. Assume the rubidium atom's initial velocity v is directed into the laser beam (the photons are moving right and the atom is moving left) and that the atom absorbs a new photon every 25 ns. How long will it take for this process to completely stop ("cool") the rubidium atom? [*Note*: a more detailed analysis predicts that the atom can be slowed to about 1 cm/s by this light absorption process, but it cannot be completely stopped.]

*Numerical/Computer

*101. (III) (*a*) Graph Planck's radiation formula (top of page 989) as a function of wavelength from $\lambda = 20\text{ nm}$ to 2000 nm in 20 nm steps for two lightbulb filaments, one at 2700 K and the other at 3300 K. Plot both curves on the same set of axes. (*b*) Approximately how much more intense is the visible light from the hotter bulb? Use numerical integration.

*102. (III) Estimate what % of emitted sunlight energy is in the visible range. Use Planck's radiation formula (top of page 989) and numerical integration.

*103. (III) Potassium has one of the lowest work functions of all metals and so is useful in photoelectric devices using visible light. Light from a source is incident on a potassium surface. Data for the stopping voltage V_0 as a function of wavelength λ is shown below. (*a*) Explain why a graph of V_0 vs. $1/\lambda$ is expected to yield a straight line. What are the theoretical expectations for the slope and the *y*-intercept of this line? (*b*) Using the data below, graph V_0 vs. $1/\lambda$ and show that a straight-line plot does indeed result. (*c*) Determine the slope a and *y*-intercept b of this line. Using your values for a and b, determine (*d*) potassium's work function (eV) and (*e*) Planck's constant h (J·s).

λ (μm)	0.400	0.430	0.460	0.490	0.520
V_0 (V)	0.803	0.578	0.402	0.229	0.083

Answers to Exercises

A: More 1000-nm photons (lower frequency).

B: 5.50×10^{14} Hz, 545 nm.

C: Only λ.

D: Decrease.

E: (*e*).

F: (*c*).

A. Piccard, E. Henriot, P. Ehrenfest, Ed. Herzen, Th. De Donder, E. Schrödinger, E. Verschaffelt, W. Pauli, W. Heisenberg, R. H. Fowler, L. Brillouin, P. Debye, M. Knudsen, W. L. Bragg, H. A. Kramers, P. A. M. Dirac, A. H. Compton, L. de Broglie, M. Born, N. Bohr, I. Langmuir, M. Planck, Mme Curie, H. A. Lorentz, A. Einstein, P. Langevin, Ch. E. Guye, C. T. R. Wilson, O. W. Richardson

Participants in the 1927 Solvay Conference. Albert Einstein, seated at front center, had difficulty accepting that nature could behave according to the rules of quantum mechanics. We present a brief summary of quantum mechanics in this Chapter starting with the wave function and the uncertainty principle. We examine the Schrödinger equation and its solutions for some simple cases: free particles, the square well, and tunneling through a barrier.

Quantum Mechanics

CHAPTER-OPENING QUESTION—**Guess now!**
The uncertainty principle states that

(a) no measurement can be perfect because it is technologically impossible to make perfect measuring instruments.
(b) it is impossible to measure exactly where a particle is, unless it is at rest.
(c) it is impossible to simultaneously know both the position and the momentum of a particle with complete certainty.
(d) a particle cannot actually have a completely certain value of momentum.

Bohr's model of the atom gave us a first (though rough) picture of what an atom is like. It proposed explanations for why there is emission and absorption of light by atoms at only certain wavelengths. The wavelengths of the line spectra and the ionization energy for hydrogen (and one-electron ions) are in excellent agreement with experiment. But the Bohr theory had important limitations. It was not able to predict line spectra for more complex atoms—not even for the neutral helium atom, which has only two electrons. Nor could it explain why emission lines, when viewed with great precision, consist of two or more very closely spaced lines (referred to as *fine structure*). The Bohr model also did not explain why some spectral lines were brighter than others. And it could not explain the bonding of atoms in molecules or in solids and liquids.

From a theoretical point of view, too, the Bohr theory was not satisfactory: it was a strange mixture of classical and quantum ideas. Moreover, the wave–particle duality was not really resolved.

CONTENTS

FIGURE 38–1 Erwin Schrödinger with Lise Meitner (see Chapter 42).

FIGURE 38–2 Werner Heisenberg (center) on Lake Como with Enrico Fermi (left) and Wolfgang Pauli (right).

We mention these limitations of the Bohr theory not to disparage it—for it was a landmark in the history of science. Rather, we mention them to show why, in the early 1920s, it became increasingly evident that a new, more comprehensive theory was needed. It was not long in coming. Less than two years after de Broglie gave us his matter–wave hypothesis, Erwin Schrödinger (1887–1961; Fig. 38–1) and Werner Heisenberg (1901–1976; Fig. 38–2) independently developed a new comprehensive theory.

38–1 Quantum Mechanics—A New Theory

The new theory, called **quantum mechanics**, has been extremely successful. It unifies the wave–particle duality into a single consistent theory and has successfully dealt with the spectra emitted by complex atoms, even the fine details. It explains the relative brightness of spectral lines and how atoms form molecules. It is also a much more general theory that covers all quantum phenomena from blackbody radiation to atoms and molecules. It has explained a wide range of natural phenomena and from its predictions many new practical devices have become possible. Indeed, it has been so successful that it is accepted today by nearly all physicists as the fundamental theory underlying physical processes.

Quantum mechanics deals mainly with the microscopic world of atoms and light. But this new theory, when it is applied to macroscopic phenomena, must be able to produce the old classical laws. This, the **correspondence principle** (already mentioned in Section 37–11), is satisfied fully by quantum mechanics.

This doesn't mean we should throw away classical theories such as Newton's laws. In the everyday world, the latter are far easier to apply and they give sufficiently accurate descriptions. But when we deal with high speeds, close to the speed of light, we must use the theory of relativity; and when we deal with the tiny world of the atom, we use quantum mechanics.

Although we won't go into the detailed mathematics of quantum mechanics, we will discuss the main ideas and how they involve the wave and particle properties of matter to explain atomic structure and other applications.

38–2 The Wave Function and Its Interpretation; the Double-Slit Experiment

The important properties of any wave are its wavelength, frequency, and amplitude. For an electromagnetic wave, the frequency (or wavelength) determines whether the light is in the visible spectrum or not, and if so, what color it is. We also have seen (Eq. 37–3) that the frequency is a measure of the energy of the corresponding photons ($E = hf$). The amplitude or displacement of an electromagnetic wave at any point is the strength of the electric (or magnetic) field at that point, and is related to the intensity of the wave (the brightness of the light).

For material particles such as electrons, quantum mechanics relates the wavelength to momentum according to de Broglie's formula, $\lambda = h/p$, Eq. 37–7. But what corresponds to the *amplitude* or *displacement* of a matter wave? The amplitude of an electromagnetic wave is represented by the electric and magnetic fields, E and B. In quantum mechanics, this role is played by the **wave function**, which is given the symbol Ψ (the Greek capital letter psi, pronounced "sigh"). Thus Ψ represents the wave displacement, as a function of time and position, of a new kind of field which we might call a "matter" field or a matter wave.

To understand how to interpret the wave function Ψ, we make an analogy with light using the wave–particle duality.

We saw in Chapter 15 that the intensity I of any wave is proportional to the square of the amplitude. This holds true for light waves as well, as we saw in Chapter 31. That is,

$$I \propto E^2,$$

where E is the electric field strength. From the *particle* point of view, the intensity of a light beam (of given frequency) is proportional to the number of photons, N, that pass through a given area per unit time. The more photons there are, the greater the intensity. Thus

$$I \propto E^2 \propto N.$$

This proportion can be turned around so that we have

$$N \propto E^2.$$

That is, the number of photons (striking a page of this book, say) is proportional to the square of the electric field strength.

If the light beam is very weak, only a few photons will be involved. Indeed, it is possible to "build up" a photograph in a camera using very weak light so the effect of individual photons can be seen.

If we are dealing with only one photon, the relationship above $\left(N \propto E^2\right)$ can be interpreted in a slightly different way. At any point the square of the electric field strength, E^2, is a measure of the *probability* that a photon will be at that location. At points where E^2 is large, there is a high probability the photon will be there; where E^2 is small, the probability is low.

We can interpret matter waves in the same way, as was first suggested by Max Born (1882–1970) in 1927. The wave function Ψ may vary in magnitude from point to point in space and time. If Ψ describes a collection of many electrons, then $|\Psi|^2$ at any point will be proportional to the number of electrons expected to be found at that point.† When dealing with small numbers of electrons we can't make very exact predictions, so $|\Psi|^2$ takes on the character of a probability. If Ψ, which depends on time and position, represents a single electron (say, in an atom), then $|\Psi|^2$ is interpreted as follows: *$|\Psi|^2$ at a certain point in space and time represents the probability of finding the electron at the given position and time.* Thus $|\Psi|^2$ is often referred to as the **probability density** or **probability distribution**.

Double-Slit Interference Experiment for Electrons

To understand this better, we take as a thought experiment the familiar double-slit experiment, and consider it both for light and for electrons.

Consider two slits whose size and separation are on the order of the wavelength of whatever we direct at them, either light or electrons, Fig. 38–3. We know very well what would happen in this case for light, since this is just Young's double-slit experiment (Section 34–3): an interference pattern would be seen on the screen behind. If light were replaced by electrons with wavelength comparable to the slit size, they too would produce an interference pattern (recall Fig. 37–11). In the case of light, the pattern would be visible to the eye or could be recorded on film. For electrons, a fluorescent screen could be used (it glows where an electron strikes).

FIGURE 38–3 Parallel beam, of light or electrons, falls on two slits whose sizes are comparable to the wavelength. An interference pattern is observed.

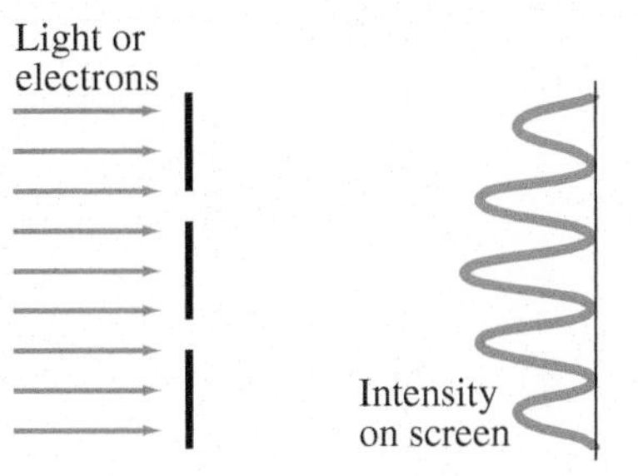

†The wave function Ψ is generally a complex quantity (that is, it involves $i = \sqrt{-1}$) and hence is not directly observable. On the other hand, $|\Psi|^2$, the absolute value of Ψ squared, is always a real quantity and it is to $|\Psi|^2$ that we can give a physical interpretation.

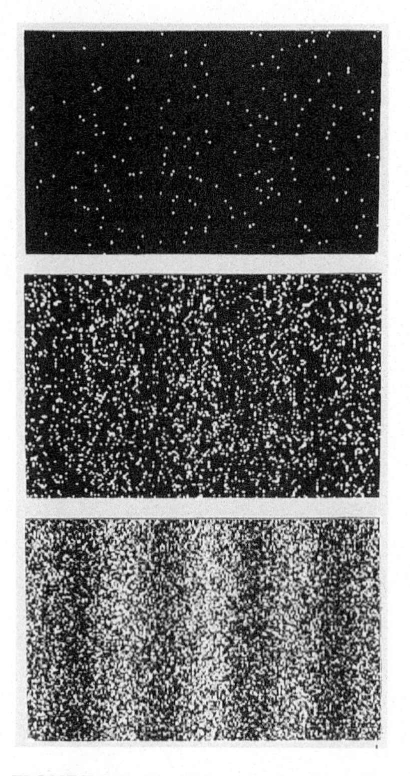

FIGURE 38–4 Young's double-slit experiment done with electrons—note that the pattern is not evident with only a few electrons (top photo), but with more and more electrons (second and third photos), the familiar double-slit interference pattern (Chapter 34) is seen.

If we reduced the flow of electrons (or photons) so they passed through the slits one at a time, we would see a flash each time one struck the screen. At first, the flashes would seem random. Indeed, there is no way to predict just where any one electron would hit the screen. If we let the experiment run for a long time, and kept track of where each electron hit the screen, we would soon see a pattern emerging—the interference pattern predicted by the wave theory; see Fig. 38–4. Thus, although we could not predict where a given electron would strike the screen, we could predict probabilities. (The same can be said for photons.) The probability, as mentioned before, is proportional to $|\Psi|^2$. Where $|\Psi|^2$ is zero, we would get a minimum in the interference pattern. And where $|\Psi|^2$ is a maximum, we would get a peak in the interference pattern.

The interference pattern would thus occur even when electrons (or photons) passed through the slits one at a time. So the interference pattern could not arise from the interaction of one electron with another. It is as if an electron passed through both slits at the same time, interfering with itself. This is possible because an electron is not precisely a particle. It is as much a wave as it is a particle, and a wave could travel through both slits at once. But what would happen if we covered one of the slits so we knew that the electron passed through the other slit, and a little later we covered the second slit so the electron had to have passed through the first slit? The result would be that no interference pattern would be seen. We would see, instead, two bright areas (or diffraction patterns) on the screen behind the slits. This confirms our idea that if both slits are open, the screen shows an interference pattern as if each electron passed through both slits, like a wave. Yet each electron would make a tiny spot on the screen as if it were a particle.

The main point of this discussion is this: if we treat electrons (and other particles) as if they were waves, then Ψ represents the wave amplitude. If we treat them as particles, then we must treat them on a *probabilistic* basis. The square of the wave function, $|\Psi|^2$, gives the probability of finding a given electron at a given point. We cannot predict—or even follow—the path of a single electron precisely through space and time.

38–3 The Heisenberg Uncertainty Principle

Whenever a measurement is made, some uncertainty is always involved. For example, you cannot make an absolutely exact measurement of the length of a table. Even with a measuring stick that has markings 1 mm apart, there will be an inaccuracy of perhaps $\frac{1}{2}$ mm or so. More precise instruments will produce more precise measurements. But there is always some uncertainty involved in a measurement, no matter how good the measuring device. We expect that by using more precise instruments, the uncertainty in a measurement can be made indefinitely small.

But according to quantum mechanics, there is actually a limit to the precision of certain measurements. This limit is not a restriction on how well instruments can be made; rather, it is inherent in nature. It is the result of two factors: the wave–particle duality, and the unavoidable interaction between the thing observed and the observing instrument. Let us look at this in more detail.

To make a measurement on an object without disturbing it, at least a little, is not possible. Consider trying to locate a Ping-Pong ball in a completely dark room. You grope about trying to find its position; and just when you touch it with your finger, you bump it and it bounces away. Whenever we measure the position of an object, whether it is a ball or an electron, we always touch it with something else that gives us the information about its position. To locate a lost Ping-Pong ball in a dark room, you could probe about with your hand or a stick; or you could shine a light and detect the light reflecting off the ball. When you search with your hand or a stick, you find the ball's position when you touch it, but at the same time you unavoidably bump it and give it some momentum.

Thus you won't know its *future* position. The same would be true if you observe the Ping-Pong ball using light. In order to "see" the ball, at least one photon must scatter from it, and the reflected photon must enter your eye or some other detector. When a photon strikes an ordinary-sized object, it does not appreciably alter the motion or position of the object. But when a photon strikes a very tiny object like an electron, it can transfer momentum to the object and thus greatly change the object's motion and position in an unpredictable way. The mere act of measuring the position of an object at one time makes our knowledge of its future position imprecise.

Now let us see where the wave–particle duality comes in. Imagine a thought experiment in which we are trying to measure the position of an object, say an electron, with photons, Fig. 38–5. (The arguments would be similar if we were using, instead, an electron microscope.) As we saw in Chapter 35, objects can be seen to a precision at best of about the wavelength of the radiation used due to diffraction. If we want a precise position measurement, we must use a short wavelength. But a short wavelength corresponds to high frequency and large momentum ($p = h/\lambda$); and the more momentum the photons have, the more momentum they can give the object when they strike it. If we use photons of longer wavelength, and correspondingly smaller momentum, the object's motion when struck by the photons will not be affected as much. But the longer wavelength means lower resolution, so the object's position will be less accurately known. Thus the act of observing produces an uncertainty in both the *position* and the *momentum* of the electron. This is the essence of the *uncertainty principle* first enunciated by Heisenberg in 1927.

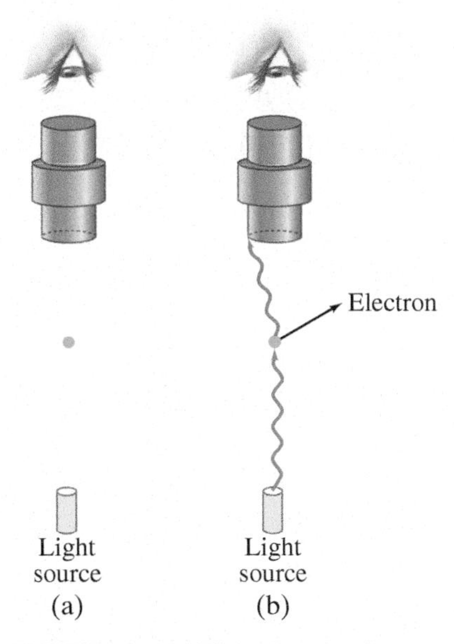

FIGURE 38–5 Thought experiment for observing an electron with a powerful light microscope. At least one photon must scatter from the electron (transferring some momentum to it) and enter the microscope.

Quantitatively, we can make an approximate calculation of the magnitude of this effect. If we use light of wavelength λ, the position can be measured at best to a precision of about λ. That is, the uncertainty in the position measurement, Δx, is approximately

$$\Delta x \approx \lambda.$$

Suppose that the object can be detected by a single photon. The photon has a momentum $p_x = h/\lambda$ (Eq. 37–5). When the photon strikes our object, it will give some or all of this momentum to the object, Fig. 38–5. Therefore, the final x momentum of our object will be uncertain in the amount

$$\Delta p_x \approx \frac{h}{\lambda}$$

since we can't tell beforehand how much momentum will be transferred. The product of these uncertainties is

$$(\Delta x)(\Delta p_x) \approx h.$$

The uncertainties could be worse than this, depending on the apparatus and the number of photons needed for detection. A more careful mathematical calculation shows the product of the uncertainties as, at best, about

$$(\Delta x)(\Delta p_x) \gtrsim \frac{h}{2\pi} \qquad \textbf{(38–1)}$$

UNCERTAINTY PRINCIPLE (Δx *and* Δp)

where Δp_x is the uncertainty of the momentum in the x direction.† This is a mathematical statement of the **Heisenberg uncertainty principle**, or, as it is sometimes called, the **indeterminancy principle**. It tells us that we cannot measure both the position *and* momentum of an object precisely at the same time. The more accurately we try to measure the position so that Δx is small, the greater will be the uncertainty in momentum, Δp_x. If we try to measure the momentum very accurately, then the uncertainty in the position becomes large.

†Note, however, that quantum mechanics does allow simultaneous precise measurements of p_x and y: that is, $(\Delta y)(\Delta p_x) \gtrsim 0$.

⚠ CAUTION

Uncertainties not due to instrument deficiency, but inherent in nature (wave–particle)

The uncertainty principle does not forbid individual precise measurements, however. For example, in principle we could measure the position of an object exactly. But then its momentum would be completely unknown. Thus, although we might know the position of the object exactly at one instant, we could have no idea at all where it would be a moment later. The uncertainties expressed here are inherent in nature, and reflect the best precision theoretically attainable even with the best instruments.

EXERCISE A Return to the Chapter-Opening Question on p. 1017, and answer it again now.

Another useful form of the uncertainty principle relates energy and time, and we examine this as follows. The object to be detected has an uncertainty in position $\Delta x \approx \lambda$. The photon that detects it travels with speed c, and it takes a time $\Delta t \approx \Delta x/c \approx \lambda/c$ to pass through the distance of uncertainty. Hence, the measured time when our object is at a given position is uncertain by about

$$\Delta t \approx \frac{\lambda}{c}.$$

Since the photon can transfer some or all of its energy $(= hf = hc/\lambda)$ to our object, the uncertainty in energy of our object as a result is

$$\Delta E \approx \frac{hc}{\lambda}.$$

The product of these two uncertainties is

$$(\Delta E)(\Delta t) \approx h.$$

A more careful calculation gives

UNCERTAINTY PRINCIPLE (ΔE and Δt)

$$(\Delta E)(\Delta t) \gtrsim \frac{h}{2\pi}. \qquad \textbf{(38–2)}$$

This form of the uncertainty principle tells us that the energy of an object can be uncertain (or can be interpreted as briefly nonconserved) by an amount ΔE for a time $\Delta t \approx h/(2\pi\,\Delta E)$.

The quantity $(h/2\pi)$ appears so often in quantum mechanics that for convenience it is given the symbol $\hbar$ ("h-bar"). That is,

$$\hbar = \frac{h}{2\pi} = \frac{6.626 \times 10^{-34}\,\mathrm{J\cdot s}}{2\pi} = 1.055 \times 10^{-34}\,\mathrm{J\cdot s}.$$

By using this notation, Eqs. 38–1 and 38–2 for the uncertainty principle can be written

$$(\Delta x)(\Delta p_x) \gtrsim \hbar \qquad \text{and} \qquad (\Delta E)(\Delta t) \gtrsim \hbar.$$

We have been discussing the position and velocity of an electron as if it were a particle. But it isn't simply a particle. Indeed, we have the uncertainty principle because an electron—and matter in general—has wave as well as particle properties. What the uncertainty principle really tells us is that if we insist on thinking of the electron as a particle, then there are certain limitations on this simplified view—namely, that the position and velocity cannot both be known precisely at the same time; and even that the electron does not *have* a precise position and momentum at the same time (because it is not simply a particle). Similarly, the energy can be uncertain in the amount ΔE for a time $\Delta t \approx \hbar/\Delta E$.

Because Planck's constant, h, is so small, the uncertainties expressed in the uncertainty principle are usually negligible on the macroscopic level. But at the level of atomic sizes, the uncertainties are significant. Because we consider ordinary objects to be made up of atoms containing nuclei and electrons, the uncertainty principle is relevant to our understanding of all of nature. The uncertainty principle expresses, perhaps most clearly, the probabilistic nature of quantum mechanics. It thus is often used as a basis for philosophic discussion.

EXAMPLE 38–1 Position uncertainty of an electron. An electron moves in a straight line with a constant speed $v = 1.10 \times 10^6\,\mathrm{m/s}$ which has been measured to a precision of 0.10%. What is the maximum precision with which its position could be simultaneously measured?

APPROACH The momentum is $p = mv$, and the uncertainty in p is $\Delta p = 0.0010p$. The uncertainty principle (Eq. 38–1) gives us the lowest Δx using the equals sign.

SOLUTION The momentum of the electron is

$$p = mv = (9.11 \times 10^{-31}\,\mathrm{kg})(1.10 \times 10^6\,\mathrm{m/s}) = 1.00 \times 10^{-24}\,\mathrm{kg \cdot m/s}.$$

The uncertainty in the momentum is 0.10% of this, or $\Delta p = 1.0 \times 10^{-27}\,\mathrm{kg \cdot m/s}$. From the uncertainty principle, the best simultaneous position measurement will have an uncertainty of

$$\Delta x \approx \frac{\hbar}{\Delta p} = \frac{1.055 \times 10^{-34}\,\mathrm{J \cdot s}}{1.0 \times 10^{-27}\,\mathrm{kg \cdot m/s}} = 1.1 \times 10^{-7}\,\mathrm{m},$$

or 110 nm.

NOTE This is about 1000 times the diameter of an atom.

EXAMPLE 38–2 Position uncertainty of a baseball. What is the uncertainty in position, imposed by the uncertainty principle, on a 150-g baseball thrown at $(93 \pm 2)\,\mathrm{mi/h} = (42 \pm 1)\,\mathrm{m/s}$?

APPROACH The uncertainty in the speed is $\Delta v = 1\,\mathrm{m/s}$. We multiply Δv by m to get Δp and then use the uncertainty principle, solving for Δx.

SOLUTION The uncertainty in the momentum is

$$\Delta p = m\,\Delta v = (0.150\,\mathrm{kg})(1\,\mathrm{m/s}) = 0.15\,\mathrm{kg \cdot m/s}.$$

Hence the uncertainty in a position measurement could be as small as

$$\Delta x = \frac{\hbar}{\Delta p} = \frac{1.055 \times 10^{-34}\,\mathrm{J \cdot s}}{0.15\,\mathrm{kg \cdot m/s}} = 7 \times 10^{-34}\,\mathrm{m}.$$

NOTE This distance is far smaller than any we could imagine observing or measuring. It is trillions of trillions of times smaller than an atom. Indeed, the uncertainty principle sets no relevant limit on measurement for macroscopic objects.

EXAMPLE 38–3 ESTIMATE **J/ψ lifetime calculated.** The J/ψ meson, discovered in 1974, was measured to have an average mass of $3100\,\mathrm{MeV}/c^2$ (note the use of energy units since $E = mc^2$) and an intrinsic width of $63\,\mathrm{keV}/c^2$. By this we mean that the masses of different J/ψ mesons were actually measured to be slightly different from one another. This mass "width" is related to the very short lifetime of the J/ψ before it decays into other particles. From the uncertainty principle, if the particle exists for only a time Δt, its mass (or rest energy) will be uncertain by $\Delta E \approx \hbar/\Delta t$. Estimate the J/ψ lifetime.

APPROACH We use the energy–time version of the uncertainty principle, Eq. 38–2.

SOLUTION The uncertainty of $63\,\mathrm{keV}/c^2$ in the J/ψ's mass is an uncertainty in its rest energy, which in joules is

$$\Delta E = (63 \times 10^3\,\mathrm{eV})(1.60 \times 10^{-19}\,\mathrm{J/eV}) = 1.01 \times 10^{-14}\,\mathrm{J}.$$

Then we expect its lifetime τ ($= \Delta t$ using Eq. 38 – 2) to be

$$\tau \approx \frac{\hbar}{\Delta E} = \frac{1.06 \times 10^{-34}\,\mathrm{J \cdot s}}{1.01 \times 10^{-14}\,\mathrm{J}} \approx 1 \times 10^{-20}\,\mathrm{s}.$$

Lifetimes this short are difficult to measure directly, and the assignment of very short lifetimes depends on this use of the uncertainty principle. (See Chapter 43.)

The uncertainty principle applies also for angular variables:

$$(\Delta L_z)(\Delta\phi) \gtrsim \hbar$$

where L is the component of angular momentum along a given axis (z) and ϕ is the angular position in a plane perpendicular to that axis.

38–4 Philosophic Implications; Probability Versus Determinism

The classical Newtonian view of the world is a deterministic one (see Section 6–5). One of its basic ideas is that once the position and velocity of an object are known at a particular time, its future position can be predicted if the forces on it are known. For example, if a stone is thrown a number of times with the same initial velocity and angle, and the forces on it remain the same, the path of the projectile will always be the same. If the forces are known (gravity and air resistance, if any), the stone's path can be precisely predicted. This mechanistic view implies that the future unfolding of the universe, assumed to be made up of particulate objects, is completely determined.

This classical deterministic view of the physical world has been radically altered by quantum mechanics. As we saw in the analysis of the double-slit experiment (Section 38–2), electrons all prepared in the same way will not all end up in the same place. According to quantum mechanics, certain probabilities exist that an electron will arrive at different points. This is very different from the classical view, in which the path of a particle is precisely predictable from the initial position and velocity and the forces exerted on it. According to quantum mechanics, the position and velocity of an object cannot even be known accurately at the same time. This is expressed in the uncertainty principle, and arises because basic entities, such as electrons, are not considered simply as particles: they have wave properties as well. Quantum mechanics allows us to calculate only the probability[†] that, say, an electron (when thought of as a particle) will be observed at various places. Quantum mechanics says there is some inherent unpredictability in nature.

Since matter is considered to be made up of atoms, even ordinary-sized objects are expected to be governed by probability, rather than by strict determinism. For example, quantum mechanics predicts a finite (but negligibly small) probability that when you throw a stone, its path will suddenly curve upward instead of following the downward-curved parabola of normal projectile motion. Quantum mechanics predicts with extremely high probability that ordinary objects will behave just as the classical laws of physics predict. But these predictions are considered probabilities, not certainties. The reason that macroscopic objects behave in accordance with classical laws with such high probability is due to the large number of molecules involved: when large numbers of objects are present in a statistical situation, deviations from the average (or most probable) approach zero. It is the average configuration of vast numbers of molecules that follows the so-called fixed laws of classical physics with such high probability, and gives rise to an apparent "determinism." Deviations from classical laws are observed when small numbers of molecules are dealt with. We can say, then, that although there are no precise deterministic laws in quantum mechanics, there are statistical laws based on probability.

It is important to note that there is a difference between the probability imposed by quantum mechanics and that used in the nineteenth century to understand thermodynamics and the behavior of gases in terms of molecules (Chapters 18 and 20). In thermodynamics, probability is used because there are far too many particles to keep track of. But the molecules are still assumed to move and interact in a deterministic way following Newton's laws. Probability in quantum mechanics is quite different; it is seen as *inherent* in nature, and not as a limitation on our abilities to calculate or to measure.

The view presented here is the generally accepted one and is called the **Copenhagen interpretation** of quantum mechanics in honor of Niels Bohr's home, since it was largely developed there through discussions between Bohr and other prominent physicists.

Because electrons are not simply particles, they cannot be thought of as following particular paths in space and time. This suggests that a description of matter in space and time may not be completely correct. This deep and far-reaching

[†]Note that these probabilities can be calculated precisely, just like exact predictions of probabilities at rolling dice or playing cards, but they are unlike predictions of probabilities at sporting events or for natural or man-made disasters, which are only estimates.

conclusion has been a lively topic of discussion among philosophers. Perhaps the most important and influential philosopher of quantum mechanics was Bohr. He argued that a space–time description of actual atoms and electrons is not possible. Yet a description of experiments on atoms or electrons must be given in terms of space and time and other concepts familiar to ordinary experience, such as waves and particles. We must not let our *descriptions* of experiments lead us into believing that atoms or electrons themselves actually move in space and time as classical particles.

38–5 The Schrödinger Equation in One Dimension—Time-Independent Form

In order to describe physical systems quantitatively using quantum mechanics, we must have a means of determining the wave function Ψ mathematically. The basic equation (in the nonrelativistic realm) for determining Ψ is the *Schrödinger equation*. We cannot, however, derive the Schrödinger equation from some higher principles, just as Newton's second law, for example, cannot be derived. The relation $\vec{\mathbf{F}} = m\vec{\mathbf{a}}$ was *invented* by Newton to describe how the motion of an object is related to the net applied force. As we saw early in this book, Newton's second law works exceptionally well. In the realm of classical physics it is the starting point for analytically solving a wide range of problems, and the solutions it yields are fully consistent with experiment. The validity of any fundamental equation resides in its agreement with experiment. The Schrödinger equation forms part of a new theory, and it too had to be *invented*—and then checked against experiment, a test that it passed splendidly.

The Schrödinger equation can be written in two forms: the time-dependent version and the time-independent version. We will mainly be interested in steady-state situations—that is, when there is no time dependence—and so we mainly deal with the time-independent version. (We briefly discuss the time-dependent version in the optional Section 38–6.) The time-independent version involves a wave function with only spatial dependence which we represent by lowercase psi, $\psi(x)$, for the simple one-dimensional case we deal with here. In three dimensions, we write $\psi(x, y, z)$ or $\psi(r, \theta, \phi)$.

In classical mechanics, we solved problems using two approaches: via Newton's laws with the concept of force, and by using the energy concept with the conservation laws. The Schrödinger equation is based on the energy approach. Even though the Schrödinger equation cannot be derived, we can suggest what form it might take by using conservation of energy and considering a very simple case: that of a free particle on which no forces act, so that its potential energy U is constant. We assume that our particle moves along the x axis, and since no force acts on it, its momentum remains constant and its wavelength $(\lambda = h/p)$ is fixed. To describe a wave for a free particle such as an electron, we expect that its wave function will satisfy a differential equation that is akin to (but not identical to) the classical wave equation. Let us see what we can infer about this equation. Consider a simple traveling wave of a single wavelength λ whose wave displacement, as we saw in Chapter 15 for mechanical waves and in Chapter 31 for electromagnetic waves, is given by $A\sin(kx - \omega t)$, or more generally as a superposition of sine and cosine: $A\sin(kx - \omega t) + B\cos(kx - \omega t)$. We are only interested in the spatial dependence, so we consider the wave at a specific moment, say $t = 0$. Thus we write as the wave function for our free particle

$$\psi(x) = A\sin kx + B\cos kx, \tag{38–3a}$$

where A and B are constants[†] and $k = 2\pi/\lambda$ (Eq. 15–11). For a particle of mass m and velocity v, the de Broglie wavelength is $\lambda = h/p$, where $p = mv$ is the particle's momentum. Hence

$$k = \frac{2\pi}{\lambda} = \frac{2\pi p}{h} = \frac{p}{\hbar}. \tag{38–3b}$$

[†]In quantum mechanics, constants can be complex (i.e., with a real and/or imaginary part).

One requirement for our wave equation, then, is that it have the wave function $\psi(x)$ as given by Eq. 38–3 as a solution for a free particle. A second requirement is that it be consistent with the conservation of energy, which we can express as

$$\frac{p^2}{2m} + U = E,$$

where E is the total energy, U is the potential energy, and (since we are considering the nonrelativistic realm) the kinetic energy K of our particle of mass m is $K = \frac{1}{2}mv^2 = p^2/2m$. Since $p = \hbar k$ (Eq. 38–3b), we can write the conservation of energy condition as

$$\frac{\hbar^2 k^2}{2m} + U = E. \qquad \textbf{(38–4)}$$

Thus we are seeking a differential equation that satisfies conservation of energy (Eq. 38–4) when $\psi(x)$ is its solution. Now, note that if we take two derivatives of our expression for $\psi(x)$, Eq. 38–3a, we get a factor $-k^2$ multiplied by $\psi(x)$:

$$\frac{d\psi(x)}{dx} = \frac{d}{dx}(A\sin kx + B\cos kx) = k(A\cos kx - B\sin kx)$$

$$\frac{d^2\psi(x)}{dx^2} = k\frac{d}{dx}(A\cos kx - B\sin kx) = -k^2(A\sin kx + B\cos kx) = -k^2\psi(x).$$

Can this last term be related to the k^2 term in Eq. 38–4? Indeed, if we multiply this last relation by $-\hbar^2/2m$, we obtain

$$-\frac{\hbar^2}{2m}\frac{d^2\psi(x)}{dx^2} = \frac{\hbar^2 k^2}{2m}\psi(x).$$

The right side is just the first term on the left in Eq. 38–4 multiplied by $\psi(x)$. If we multiply Eq. 38–4 through by $\psi(x)$, and make this substitution, we obtain

SCHRÖDINGER EQUATION (time-independent form)

$$-\frac{\hbar^2}{2m}\frac{d^2\psi(x)}{dx^2} + U(x)\psi(x) = E\psi(x). \qquad \textbf{(38–5)}$$

This is, in fact, the one-dimensional **time-independent Schrödinger equation**, where for generality we have written $U = U(x)$. It is the basis for solving problems in nonrelativistic quantum mechanics. For a particle moving in three dimensions there would be additional derivatives with respect to y and z (see Chapter 39).

Note that we have by no means derived the Schrödinger equation. Although we have made a good argument in its favor, other arguments could also be made which might or might not lead to the same equation. The Schrödinger equation as written (Eq. 38–5) is useful and valid only because it has given results in accord with experiment for a wide range of situations.

There are some requirements we impose on any wave function that is a solution of the Schrödinger equation in order that it be physically meaningful. First, we insist that it be a continuous function; after all, if $|\psi|^2$ represents the probability of finding a particle at a certain point, we expect the probability to be continuous from point to point and not to take discontinuous jumps. Second, we want the wave function to be *normalized*. By this we mean that for a single particle, the probability of finding the particle at one point or another (i.e., the probabilities summed over all space) must be exactly 1 (or 100%). For a single particle, $|\psi|^2$ represents the probability of finding the particle in unit volume. Then

$$|\psi|^2\, dV \qquad \textbf{(38–6a)}$$

is the probability of finding the particle within a volume dV, where ψ is the value of the wave function in this infinitesimal volume dV. For the one-dimensional case, $dV = dx$, so the probability of finding a particle within dx of position x is

$$|\psi(x)|^2\, dx. \qquad \textbf{(38–6b)}$$

Then the sum of the probabilities over all space—that is, the probability of finding the particle somewhere—becomes

$$\int_{\text{all space}} |\psi|^2\, dV = \int |\psi|^2\, dx = 1. \qquad \textbf{(38–6c)}$$

This is called the **normalization condition**, and the integral is taken over whatever region of space in which the particle has a chance of being found, which is often all of space, from $x = -\infty$ to $x = \infty$.

*38–6 Time-Dependent Schrödinger Equation

The more general form of the Schrödinger equation, including time dependence, for a particle of mass m moving in one dimension, is

$$-\frac{\hbar^2}{2m}\frac{\partial^2\Psi(x,t)}{\partial x^2} + U(x)\Psi(x,t) = i\hbar\frac{\partial\Psi(x,t)}{\partial t}. \qquad \textbf{(38–7)}$$

This is the **time-dependent Schrödinger equation**; here $U(x)$ is the potential energy of the particle as a function of position, and i is the imaginary number $i = \sqrt{-1}$. For a particle moving in three dimensions, there would be additional derivatives with respect to y and z, just as for the classical wave equation discussed in Section 15–5. Indeed, it is worth noting the similarity between the Schrödinger wave equation for zero potential energy $(U = 0)$ and the classical wave equation: $\partial^2 D/\partial t^2 = v^2\partial^2 D/\partial x^2$, where D is the wave displacement (equivalent of the wave function). In both equations there is the second derivative with respect to x; but in the Schrödinger equation there is only the first derivative with respect to time, whereas the classical wave equation has the second derivative for time.

As we pointed out in the preceding Section, we cannot derive the time-dependent Schrödinger equation. But we can show how the time-independent Schrödinger equation (Eq. 38–5) is obtained from it. For many problems in quantum mechanics, it is possible to write the wave function as a product of separate functions of space and time:

$$\Psi(x,t) = \psi(x)f(t).$$

Substituting this into the time-dependent Schrödinger equation (Eq. 38–7), we get:

$$-\frac{\hbar^2}{2m}f(t)\frac{d^2\psi(x)}{dx^2} + U(x)\psi(x)f(t) = i\hbar\psi(x)\frac{df(t)}{dt}.$$

We divide both sides of this equation by $\psi(x)f(t)$ and obtain an equation that involves only x on one side and only t on the other:

$$-\frac{\hbar^2}{2m}\frac{1}{\psi(x)}\frac{d^2\psi(x)}{dx^2} + U(x) = i\hbar\frac{1}{f(t)}\frac{df(t)}{dt}.$$

This **separation of variables** is very convenient. Since the left side is a function only of x, and the right side is a function only of t, the equality can be valid for all values of x and all values of t only if each side is equal to a constant (the same constant), which we call C:

$$-\frac{\hbar^2}{2m}\frac{1}{\psi(x)}\frac{d^2\psi(x)}{dx^2} + U(x) = C \qquad \textbf{(38–8a)}$$

$$i\hbar\frac{1}{f(t)}\frac{df(t)}{dt} = C. \qquad \textbf{(38–8b)}$$

We multiply the first of these (Eq. 38–8a) through by $\psi(x)$ and obtain

$$-\frac{\hbar^2}{2m}\frac{d^2\psi(x)}{dx^2} + U(x)\psi(x) = C\psi(x). \qquad \textbf{(38–8c)}$$

This we recognize immediately as the time-independent Schrödinger equation, Eq. 38–5, where the constant C equals the total energy E. Thus we have obtained the time-independent form of Schrödinger's equation from the time-dependent form.

Equation 38–8b is easy to solve. Putting $C = E,$ we rewrite Eq. 38–8b as

$$\frac{df(t)}{dt} = -i\frac{E}{\hbar}f(t)$$

(note that since $i^2 = -1,\ i = -1/i$), and then as

$$\frac{df(t)}{f(t)} = -i\frac{E}{\hbar}\,dt.$$

We integrate both sides to obtain

$$\ln f(t) = -i\frac{E}{\hbar}t$$

or

$$f(t) = e^{-i\left(\frac{E}{\hbar}\right)t}.$$

Thus the total wave function is

$$\Psi(x,t) = \psi(x)e^{-i\left(\frac{E}{\hbar}\right)t}, \tag{38–9}$$

where $\psi(x)$ satisfies Eq. 38–5. It is, in fact, the solution of the time-independent Schrödinger equation (Eq. 38–5) that is the major task of nonrelativistic quantum mechanics. Nonetheless, we should note that in general the wave function $\Psi(x,t)$ is a complex function since it involves $i = \sqrt{-1}$. It has both a real and an imaginary part.† Since $\Psi(x,t)$ is not purely real, it cannot itself be physically measurable. Rather it is only $|\Psi|^2$, which *is* real, that can be measured physically.

Note also that

$$|f(t)| = \left|e^{-i\left(\frac{E}{\hbar}\right)t}\right| = 1,$$

so $|f(t)|^2 = 1$. Hence the probability density in space does not depend on time:

$$|\Psi(x,t)|^2 = |\psi(x)|^2.$$

We thus will be interested only in the time-independent Schrödinger equation, Eq. 38–5, which we will now examine for a number of simple situations.

38–7 Free Particles; Plane Waves and Wave Packets

A **free particle** is one that is not subject to any force, and we can therefore take its potential energy to be zero. (Although we dealt with the free particle in Section 38–5 in arguing for Schrödinger's equation, here we treat it directly using Schrödinger's equation as the basis.) Schrödinger's equation (Eq. 38–5) with $U(x) = 0$ becomes

$$-\frac{\hbar^2}{2m}\frac{d^2\psi(x)}{dx^2} = E\psi(x),$$

which can be written

$$\frac{d^2\psi}{dx^2} + \frac{2mE}{\hbar^2}\psi = 0.$$

This is a familiar equation that we encountered in Chapter 14 (Eq. 14–3) in connection with the simple harmonic oscillator. The solution to this equation, but with appropriate variable changes‡ for our case here, is

$$\psi = A\sin kx + B\cos kx, \qquad \text{[free particle]} \tag{38–10}$$

where

$$k = \sqrt{\frac{2mE}{\hbar^2}}. \tag{38–11a}$$

Since $U = 0,$ the total energy E of the particle is $E = \frac{1}{2}mv^2 = p^2/2m$ (where p is the momentum); thus

$$k = \frac{p}{\hbar} = \frac{h}{\lambda\hbar} = \frac{2\pi}{\lambda}. \tag{38–11b}$$

†Recall that $e^{-i\theta} = \cos\theta - i\sin\theta.$

‡In Eq. 14–3, t becomes x and ω becomes $k = \sqrt{2mE/\hbar^2}$. (Don't confuse this k with the spring constant k of Chapter 14.) We could also write our solution (Eq. 38–10) as $\psi = A\cos(kx + \phi)$ where ϕ is a phase constant.

So a free particle of momentum p and energy E can be represented by a plane wave that varies sinusoidally. If we are not interested in the phase, we can choose $B = 0$ in Eq. 38–10, and we show this sine wave in Fig. 38–6a.

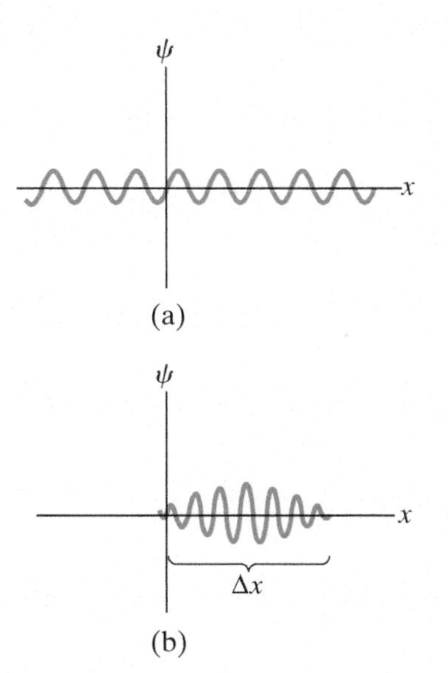

FIGURE 38–6 (a) A plane wave describing a free particle. (b) A wave packet of "width" Δx.

EXAMPLE 38–4 Free electron. An electron with energy $E = 6.3\,\text{eV}$ is in free space (where $U = 0$). Find (*a*) the wavelength λ and (*b*) the wave function ψ for the electron, assuming $B = 0$.

APPROACH The wavelength $\lambda = 2\pi/k$ (Eq. 38–11b) where the wave number k is given by Eq. 38–11a. The wave function is given by $\psi = A\sin kx$.

SOLUTION

(*a*)
$$\lambda = \frac{2\pi}{k} = \frac{2\pi\hbar}{\sqrt{2mE}}$$
$$= \frac{2\pi\left(1.055\times10^{-34}\,\text{J}\cdot\text{s}\right)}{\sqrt{2(9.11\times10^{-31}\,\text{kg})(6.3\,\text{eV})(1.60\times10^{-19}\,\text{J/eV})}}$$
$$= 4.9\times10^{-10}\,\text{m} = 0.49\,\text{nm}.$$

(*b*)
$$k = \frac{2\pi}{\lambda}$$
$$= 1.28\times10^{10}\,\text{m}^{-1},$$

so

$$\psi = A\sin kx = A\sin\left[(1.28\times10^{10}\,\text{m}^{-1})(x)\right].$$

Note in Fig. 38–6a that the sine wave will extend indefinitely† in the $+x$ and $-x$ directions. Thus, since $|\psi|^2$ represents the probability of finding the particle, the particle could be anywhere between $x = -\infty$ and $x = \infty$. This is fully consistent with the uncertainty principle (Section 38–3): the momentum of the particle was given and hence is known precisely $(p = \hbar k)$, so the particle's position must be totally unpredictable. Mathematically, if $\Delta p = 0$, $\Delta x \gtrsim \hbar/\Delta p = \infty$.

To describe a particle whose position is well localized—that is, it is known to be within a small region of space—we can use the concept of a **wave packet**. Figure 38–6b shows an example of a wave packet whose width is about Δx as shown, meaning that the particle is most likely to be found within this region of space. A well-localized particle moving through space can thus be represented by a moving wave packet.

A wave packet can be represented mathematically as the sum of many plane waves (sine waves) of slightly different wavelengths. That this will work can be seen by looking carefully at Fig. 16–17. There we combined only two nearby frequencies (to explain why there are "beats") and found that the sum of two sine waves looked like a series of wave packets. If we add additional waves with other nearby frequencies, we can eliminate all but one of the packets and arrive at Fig. 38–6b. Thus a wave packet consists of waves of a *range* of wavelengths; hence it does not have a definite momentum $p\,(= h/\lambda)$, but rather, a range of momenta. This is consistent with the uncertainty principle: we have made Δx small, so the momentum cannot be precise; that is, Δp cannot be zero. Instead, our particle can be said to have a range of momenta, Δp, or to have an uncertainty in its momentum, Δp. It is not hard to show, even for this simple situation (see Problem 20), that $\Delta p \approx h/\Delta x$, in accordance with the uncertainty principle.

†Such an infinite wave makes problems for normalization since $\int_{-\infty}^{\infty}|\psi|^2\,dx = A^2\int_{-\infty}^{\infty}\sin^2 kx\,dx$ is infinite for any nonzero value for A. For practical purposes we can usually normalize the waves $(A \neq 0)$ by assuming that the particle is in a large but finite region of space. The region can be chosen large enough so that momentum is still rather precisely fixed.

38–8 Particle in an Infinitely Deep Square Well Potential (a Rigid Box)

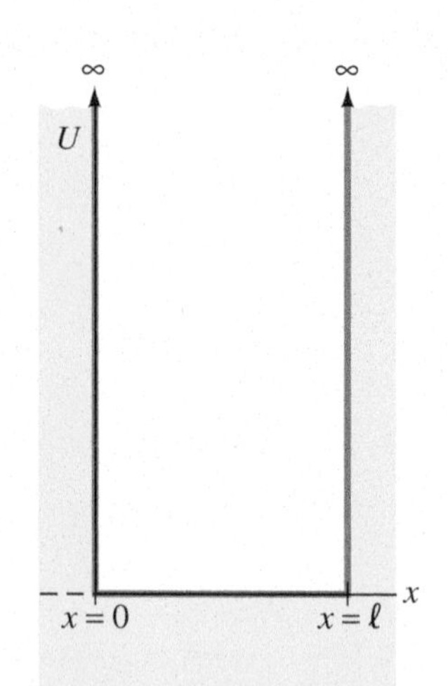

FIGURE 38–7 A plot of potential energy U vs. x for an infinitely deep square well potential.

The Schrödinger equation can be solved analytically only for a few possible forms of the potential energy U. We consider some simple cases here which at first may not seem realistic, but have simple solutions that can be used as approximations to understand a variety of phenomena.

In our first case, we assume that a particle of mass m is confined to a one-dimensional box of width ℓ whose walls are perfectly rigid. (This can serve as an approximation for an electron in a metal, for example.) The particle is trapped in this box and collisions with the walls are perfectly elastic. The potential energy for this situation, which is commonly known as an **infinitely deep square well potential** or **rigid box**, is shown in Fig. 38–7. We can write the potential energy $U(x)$ as

$$U(x) = 0 \qquad 0 < x < \ell$$
$$U(x) = \infty \qquad x \leq 0 \quad \text{and} \quad x \geq \ell.$$

For the region $0 < x < \ell$, where $U(x) = 0$, we already know the solution of the Schrödinger equation from our discussion in Section 38–7: it is just Eq. 38–10,

$$\psi(x) = A \sin kx + B \cos kx,$$

where (from Eq. 38–11a)

$$k = \sqrt{\frac{2mE}{\hbar^2}}.$$

(We could also use $\psi(x) = A \sin(kx + \phi)$ where ϕ is a phase constant.) Outside the well $U(x) = \infty$, so $\psi(x)$ must be zero. (If it weren't, the product $U\psi$ in the Schrödinger equation wouldn't be finite; besides, if $U = \infty$, we can't expect a particle of finite total energy to be in such a region.) So we are concerned only with the wave function within the well, and we must determine the constants A and B as well as any restrictions on the value of k (and hence on the energy E).

We have insisted that the wave function must be continuous. Hence, if $\psi = 0$ outside the well, it must be zero at $x = 0$ and at $x = \ell$:

$$\psi(0) = 0 \quad \text{and} \quad \psi(\ell) = 0.$$

These are the **boundary conditions** for this problem. At $x = 0$, $\sin kx = 0$ but $\cos kx = 1$, so at this point Eq. 38–10 yields

$$0 = \psi(0) = A \sin 0 + B \cos 0 = 0 + B.$$

Thus B must be zero. Our solution is reduced to

$$\psi(x) = A \sin kx.$$

Now we apply the other boundary condition, $\psi = 0$ at $x = \ell$:

$$0 = \psi(\ell) = A \sin k\ell.$$

We don't want $A = 0$ or we won't have a particle at all ($|\psi|^2 = 0$ everywhere). Therefore, we set

$$\sin k\ell = 0.$$

The sine is zero for angles of $0, \pi, 2\pi, 3\pi, \cdots$ radians, which means that $k\ell = 0, \pi, 2\pi, 3\pi, \cdots$. In other words,

$$k\ell = n\pi, \qquad n = 1, 2, 3, \cdots, \tag{38–12}$$

where n is an integer. We eliminate the case $n = 0$ since that would make $\psi = 0$ everywhere. Thus k, and hence E, cannot have just any value; rather, k is limited to values

$$k = \frac{n\pi}{\ell}.$$

Putting this expression in Eq. 38–11a (and substituting $h/2\pi$ for $\hbar$), we find that E

can have only the values

$$E = n^2 \frac{h^2}{8m\ell^2}, \qquad n = 1, 2, 3, \cdots. \tag{38–13}$$

A particle trapped in a rigid box thus can only have certain *quantized energies*. The lowest energy (the ground state) has $n = 1$ and is given by

$$E_1 = \frac{h^2}{8m\ell^2}. \qquad \text{[ground state]}$$

The next highest energy $(n = 2)$ is

$$E_2 = 4E_1,$$

and for higher energies (see Fig. 38–8),

$$E_3 = 9E_1$$
$$\vdots$$
$$E_n = n^2 E_1.$$

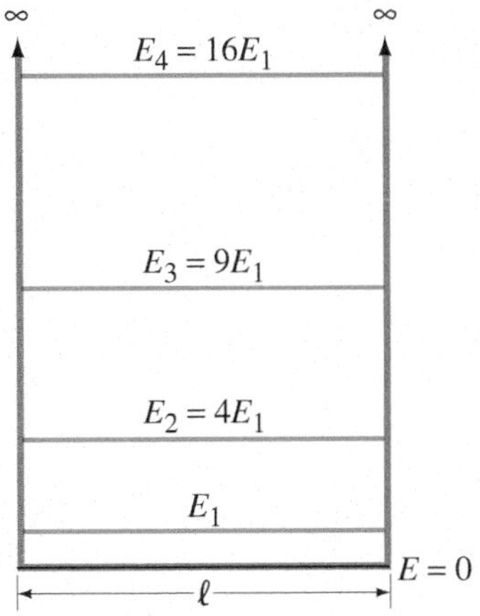

FIGURE 38–8 Possible energy levels for a particle in a box with perfectly rigid walls (infinite square well potential).

The integer n is called the **quantum number** of the state.

That the lowest energy, E_1, is not zero means that the particle in the box can never be at rest. This is contrary to classical ideas, according to which a particle can have $E = 0$. E_1 is called the **zero-point energy**. One outcome of this result is that even at a temperature of absolute zero (0 K), quantum mechanics predicts that particles in a box would not be at rest but would have a zero-point energy.

We also note that both the energy E_1 and momentum $p_1 = \hbar k = \hbar\pi/\ell$ (Eq. 38–11b) in the ground state are related inversely to the width of the box. The smaller the width ℓ, the larger the momentum (and energy). This can be considered a direct result of the uncertainty principle (see Problem 25).

The wave function $\psi = A \sin kx$ for each of the quantum states is (since $k = n\pi/\ell$)

$$\psi_n = A \sin\left(\frac{n\pi}{\ell} x\right). \tag{38–14}$$

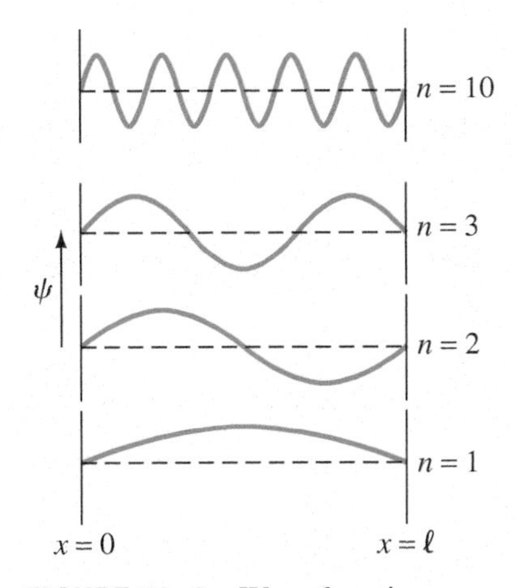

FIGURE 38–9 Wave functions corresponding to the quantum number n being 1, 2, 3, and 10 for a particle confined to a rigid box.

We can determine the constant A by imposing the normalization condition (Eq. 38–6c):

$$1 = \int_{-\infty}^{\infty} \psi^2\, dx = \int_0^{\ell} A^2 \sin^2\left(\frac{n\pi}{\ell} x\right) dx, \tag{38–15}$$

where the integral needs to be done only over the range $0 < x < \ell$ because outside these limits $\psi = 0$. The integral (see Example 38–6) is equal to $A^2\ell/2$, so we have

$$A = \sqrt{\frac{2}{\ell}} \quad \text{and} \quad \psi_n = \sqrt{\frac{2}{\ell}} \sin\left(\frac{n\pi}{\ell} x\right).$$

The amplitude A is the same for all the quantum numbers. Figure 38–9 shows the wave functions (Eq. 38–14) for $n = 1, 2, 3,$ and 10. They look just like standing waves on a string—see Fig. 15–26. This is not surprising since the wave function solutions, Eq. 38–14, are the same as for the standing waves on a string, and the condition $k\ell = n\pi$ is the same in the two cases (page 414).

Figure 38–10 shows the probability distribution, $|\psi|^2$, for the same states $(n = 1, 2, 3, 10)$ for which ψ is shown in Fig. 38–9. We see immediately that the particle is more likely to be found in some places than in others. For example, in the ground state $(n = 1)$, the electron is much more likely to be found near the center of the box than near the walls. This is clearly at variance with classical ideas, which predict a uniform probability density—the particle would be as likely to be found at one point in the box as at any other. The quantum-mechanical probability densities for higher states are even more complicated, with areas of low probability not only near the walls but also at regular intervals in between.

FIGURE 38–10 The probability distribution for a particle in a rigid box for the states with $n = 1, 2, 3,$ and 10.

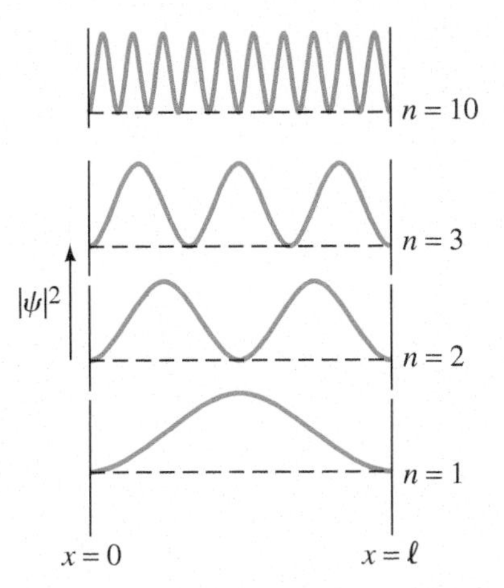

EXAMPLE 38–5 Electron in an infinite potential well. (*a*) Calculate the three lowest energy levels for an electron trapped in an infinitely deep square well potential of width $\ell = 1.00 \times 10^{-10}$ m (about the diameter of a hydrogen atom in its ground state). (*b*) If a photon were emitted when the electron jumps from the $n = 2$ state to the $n = 1$ state, what would its wavelength be?

APPROACH The energy levels are given by Eq. 38–13. In (*b*), $hf = hc/\lambda = E_2 - E_1$.

SOLUTION (*a*) The ground state $(n = 1)$ has energy

$$E_1 = \frac{h^2}{8m\ell^2} = \frac{(6.63 \times 10^{-34}\,\mathrm{J \cdot s})^2}{8(9.11 \times 10^{-31}\,\mathrm{kg})(1.00 \times 10^{-10}\,\mathrm{m})^2} = 6.03 \times 10^{-18}\,\mathrm{J}.$$

In electron volts this is

$$E_1 = \frac{6.03 \times 10^{-18}\,\mathrm{J}}{1.60 \times 10^{-19}\,\mathrm{J/eV}} = 37.7\,\mathrm{eV}.$$

Then

$$E_2 = (2)^2 E_1 = 151\,\mathrm{eV}$$
$$E_3 = (3)^2 E_1 = 339\,\mathrm{eV}.$$

(*b*) The energy difference is $E_2 - E_1 = 151\,\mathrm{eV} - 38\,\mathrm{eV} = 113\,\mathrm{eV}$ or 1.81×10^{-17} J, and this would equal the energy of the emitted photon (energy conservation). Its wavelength would be

$$\lambda = \frac{c}{f} = \frac{hc}{E} = \frac{(6.63 \times 10^{-34}\,\mathrm{J \cdot s})(3.00 \times 10^{8}\,\mathrm{m/s})}{1.81 \times 10^{-17}\,\mathrm{J}} = 1.10 \times 10^{-8}\,\mathrm{m}$$

or 11.0 nm, which is in the ultraviolet region of the spectrum.

EXERCISE B The wavelength of a photon emitted in an $n = 3$ to $n = 1$ transition is (*a*) 0.062 nm, (*b*) 620 nm, (*c*) 301 nm, (*d*) 3.2×10^{-15} m, (*e*) 4.1 nm.

EXAMPLE 38–6 Calculating a normalization constant. Show that the normalization constant A for all wave functions describing a particle in an infinite potential well of width ℓ has a value of $A = \sqrt{2/\ell}$.

APPROACH The wave functions for various n are

$$\psi = A \sin\frac{n\pi x}{\ell}.$$

To normalize ψ, we must have (Eq. 38–15)

$$1 = \int_0^{\ell} |\psi|^2\,dx = \int_0^{\ell} A^2 \sin^2\frac{n\pi x}{\ell}\,dx.$$

SOLUTION We need integrate only from 0 to ℓ since $\psi = 0$ for all other values of x. To evaluate this integral we let $\theta = n\pi x/\ell$ and use the trigonometric identity $\sin^2\theta = \frac{1}{2}(1 - \cos 2\theta)$. Then, with $dx = \ell\,d\theta/n\pi$, we have

$$1 = A^2 \int_0^{n\pi} \sin^2\theta\left(\frac{\ell}{n\pi}\right)d\theta = \frac{A^2\ell}{2n\pi}\int_0^{n\pi}(1 - \cos 2\theta)\,d\theta$$
$$= \frac{A^2\ell}{2n\pi}\left(\theta - \tfrac{1}{2}\sin 2\theta\right)\Big|_0^{n\pi}$$
$$= \frac{A^2\ell}{2}.$$

Thus $A^2 = 2/\ell$ and

$$A = \sqrt{\frac{2}{\ell}}.$$

EXAMPLE 38–7 ESTIMATE **Probability near center of rigid box.** An electron is in an infinitely deep square well potential of width $\ell = 1.00 \times 10^{-10}$ m. If the electron is in the ground state, what is the probability of finding it in a region of width $\Delta x = 1.0 \times 10^{-12}$ m at the center of the well (at $x = 0.50 \times 10^{-10}$ m)?

APPROACH The probability of finding a particle in a small region of width dx is $|\psi|^2\,dx$ (Eq. 38–6b). Using A from Example 38–6, the wave function for the ground state is

$$\psi(x) = \sqrt{\frac{2}{\ell}}\sin\frac{\pi x}{\ell}.$$

SOLUTION The $n = 1$ curve in Fig. 38–9 shows that ψ is roughly constant near the center of the well. So we can avoid doing an integral over dx and just set $dx \approx \Delta x$ and find

$$|\psi|^2\,\Delta x = \frac{2}{\ell}\sin^2\left[\frac{\pi x}{\ell}\right]\Delta x$$

$$= \frac{2}{(1.00 \times 10^{-10}\,\mathrm{m})}\sin^2\left[\frac{\pi(0.50 \times 10^{-10}\,\mathrm{m})}{(1.00 \times 10^{-10}\,\mathrm{m})}\right](1.0 \times 10^{-12}\,\mathrm{m}) = 0.02.$$

The probability of finding the electron in this region at the center of the well is thus 2%.

NOTE Since $\Delta x = 1.0 \times 10^{-12}$ m is 1% of the well width of 1.00×10^{-10} m, our result of 2% probability is not what would be expected classically. Classically, the electron would be equally likely to be anywhere in the box, and we would expect the probability to be 1% instead of 2%.

EXAMPLE 38–8 **Probability of e^- in $\frac{1}{4}$ of box.** Determine the probability of findng an electron in the left quarter of a rigid box—i.e., between one wall at $x = 0$ and position $x = \ell/4$. Assume the electron is in the ground state.

APPROACH We cannot make the assumption we did in Example 38–7 that $|\psi|^2 \approx$ constant and Δx is small. Here we need to integrate $|\psi|^2\,dx$ from $x = 0$ to $x = \ell/4$, which is equal to the area under the curve shown colored in Fig. 38–11.

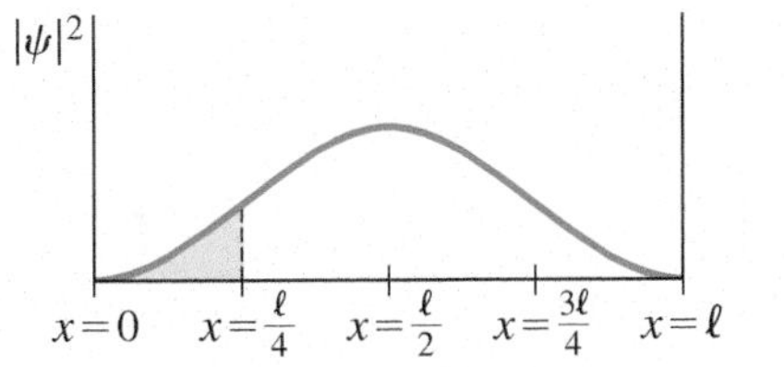

SOLUTION The wave function in the ground state is $\psi_1 = \sqrt{2/\ell}\,\sin(\pi x/\ell)$. To find the probability of the electron in the left quarter of the box, we integrate just as in Example 38–6 but with different limits on the integral (and now we know that $A = \sqrt{2/\ell}$). That is, we set $\theta = \pi x/\ell$ (then $x = \ell/4$ corresponds to $\theta = \pi/4$) and use the identity $\sin^2\theta = \frac{1}{2}(1 - \cos 2\theta)$. Thus, with $dx = (\ell/\pi)d\theta$,

$$\int_0^{\ell/4}|\psi|^2\,dx = \frac{2}{\ell}\int_0^{\ell/4}\sin^2\left(\frac{\pi}{\ell}x\right)dx$$

$$= \frac{1}{\ell}\int_0^{\pi/4}(1 - \cos 2\theta)\left(\frac{\ell}{\pi}\right)d\theta$$

$$= \frac{1}{\pi}\left(\theta - \frac{1}{2}\sin 2\theta\right)\Bigg|_0^{\pi/4}$$

$$= \frac{1}{4} - \frac{1}{2\pi} = 0.091.$$

NOTE The electron spends only 9.1% of its time in the left quarter of the box. Classically it would spend 25%.

$|\psi|^2$

$x=0$ $x=\frac{\ell}{4}$ $x=\frac{\ell}{2}$ $x=\frac{3\ell}{4}$ $x=\ell$

FIGURE 38–11 Ground-state probability distribution $|\psi|^2$ for an electron in a rigid box. Same as $n = 1$ graph of Fig. 38–10; but here we show the area under the curve from $x = 0$ to $x = \ell/4$ which represents the probability of finding the electron in that region.

EXERCISE C What is the probability of finding the electron between $x = \ell/4$ and $x = \ell/2$? (Do you need to integrate?) (*a*) 9.1%; (*b*) 18.2%; (*c*) 25%; (*d*) 33%; (*e*) 40.9%.

EXAMPLE 38–9 Most likely and average positions. Two interesting quantities are the most likely position of a particle and the average position of the particle. Consider an electron in a rigid box of width 1.00×10^{-10} m in the first excited state $n = 2$. (*a*) What is its most likely position? (*b*) What is its average position?

APPROACH To find (*a*) the most likely position (or positions), we find the maximum value(s) of the probability distribution $|\psi|^2$ by taking its derivative and setting it equal to zero. For (*b*), the average position, we integrate $\bar{x} = \int_0^\ell x\,|\psi|^2\,dx$.

SOLUTION (*a*) The wave function for $n = 2$ is $\psi(x) = \sqrt{\frac{2}{\ell}}\sin\left(\frac{2\pi}{\ell}x\right)$, so $|\psi(x)|^2 = \frac{2}{\ell}\sin^2\left(\frac{2\pi}{\ell}x\right)$. To find maxima and minima, we set $d|\psi|^2/dx = 0$:

$$\frac{d}{dx}|\psi|^2 = \frac{2}{\ell}(2)\frac{2\pi}{\ell}\sin\left(\frac{2\pi}{\ell}x\right)\cos\left(\frac{2\pi}{\ell}x\right).$$

This quantity is zero when either the sine is zero ($2\pi x/\ell = 0, \pi, 2\pi, \cdots$), or the cosine is zero ($2\pi x/\ell = \pi/2, 3\pi/2, \cdots$). The maxima and minima occur at $x = 0$, $\ell/2$, ℓ, and $x = \ell/4, 3\ell/4$. The latter ($\ell/4, 3\ell/4$) are the maxima—see the $n = 2$ curve of Fig. 38–10; the others are minima. [To confirm, you can take the second derivative, $d^2|\psi|^2/dx^2$, which is >0 for minima and <0 for maxima.]
(*b*) The average position is (again we use $\sin^2\theta = \frac{1}{2}(1 - \cos 2\theta)$):

$$\bar{x} = \int_0^\ell x|\psi|^2\,dx = \int_0^\ell \frac{2}{\ell}x\sin^2\left(\frac{2\pi}{\ell}x\right)dx = \frac{1}{\ell}\int_0^\ell x\left[1 - \cos\left(\frac{4\pi}{\ell}x\right)\right]dx,$$

which gives (integrating by parts, Appendix B: $u = x$, $dv = \cos(4\pi x/\ell)\,dx$):

$$\bar{x} = \frac{1}{\ell}\left[\frac{x^2}{2} - \frac{x\ell}{4\pi}\sin\left(\frac{4\pi}{\ell}x\right) - \frac{\ell^2}{16\pi^2}\cos\left(\frac{4\pi}{\ell}x\right)\right]_0^\ell = \frac{\ell}{2}.$$

Since the curves for $|\psi|^2$ are symmetric about the center of the box, we expect this answer. But note that for $n = 2$, the probability of finding the particle at the point $x = \ell/2$ is actually zero (Fig. 38–10).

EXAMPLE 38–10 ESTIMATE **Confined bacterium.** A tiny bacterium with a mass of about 10^{-14} kg is confined between two rigid walls 0.1 mm apart. (*a*) Estimate its minimum speed. (*b*) If, instead, its speed is about 1 mm in 100 s, estimate the quantum number of its state.

APPROACH We assume $U = 0$ inside the potential well, so $E = \frac{1}{2}mv^2$. In (*a*) the minimum speed occurs in the ground state, $n = 1$, so $v = \sqrt{2E/m}$ where E is the ground-state energy. In (*b*) we solve Eq. 38–13 for n.

SOLUTION (*a*) With $n = 1$, Eq. 38–13 gives $E = h^2/8m\ell^2$ so

$$v = \sqrt{\frac{2E}{m}} = \sqrt{\frac{h^2}{4m^2\ell^2}} = \frac{h}{2m\ell} = \frac{6.6 \times 10^{-34}\,\mathrm{J\cdot s}}{2(10^{-14}\,\mathrm{kg})(10^{-4}\,\mathrm{m})} \approx 3 \times 10^{-16}\,\mathrm{m/s}.$$

This is a speed so small that we could not measure it and the object would seem at rest, consistent with classical physics.
(*b*) Given $v = 10^{-3}\,\mathrm{m}/100\,\mathrm{s} = 10^{-5}\,\mathrm{m/s}$, the kinetic energy of the bacterium is

$$E = \tfrac{1}{2}mv^2 = \tfrac{1}{2}(10^{-14}\,\mathrm{kg})(10^{-5}\,\mathrm{m/s})^2 = 0.5 \times 10^{-24}\,\mathrm{J}.$$

From Eq. 38–13, the quantum number of this state is

$$n = \sqrt{E\left(\frac{8m\ell^2}{h^2}\right)} = \sqrt{\frac{(0.5 \times 10^{-24}\,\mathrm{J})(8)(10^{-14}\,\mathrm{kg})(10^{-4}\,\mathrm{m})^2}{(6.6 \times 10^{-34}\,\mathrm{J\cdot s})^2}} \approx 3 \times 10^{10}.$$

NOTE This number is so large that we could never distinguish between adjacent energy states (between $n = 3 \times 10^{10}$ and $3 \times 10^{10} + 1$). The energy states would appear to form a continuum. Thus, even though the energies involved here are small ($\ll 1$ eV), we are still dealing with a macroscopic object (though visible only under a microscope) and the quantum result is not distinguishable from a classical one. This is in accordance with the correspondence principle.

38–9 Finite Potential Well

Let us now look at a particle in a box whose walls are not perfectly rigid. That is, the potential energy outside the box or well is not infinite, but rises to some level U_0, as shown in Fig. 38–12. This is called a **finite potential well**. It can serve as an approximation for, say, a neutron in a nucleus. There are some significant new features that arise for the finite well as compared to the infinite well. We divide the well into three regions as shown in Fig. 38–12. In region II, inside the well, the Schrödinger equation is the same as before $(U = 0)$, although the boundary conditions will be different. So we write the solution for region II as

$$\psi_{\mathrm{II}} = A \sin kx + B \cos kx \qquad (0 < x < \ell)$$

but we don't immediately set $B = 0$ or assume that k is given by Eq. 38–12.

In regions I and III, the Schrödinger equation, now with $U(x) = U_0$, is

$$-\frac{\hbar^2}{2m}\frac{d^2\psi}{dx^2} + U_0\psi = E\psi.$$

We rewrite this as

$$\frac{d^2\psi}{dx^2} - \left[\frac{2m(U_0 - E)}{\hbar^2}\right]\psi = 0.$$

Let us assume that E is less than U_0, so the particle is "trapped" in the well (at least classically). There might be only one such **bound state**, or several, or even none, as we shall discuss later. We define the constant G by

$$G^2 = \frac{2m(U_0 - E)}{\hbar^2} \qquad \textbf{(38–16)}$$

and rewrite the Schrödinger equation as

$$\frac{d^2\psi}{dx^2} - G^2\psi = 0.$$

This equation has the general solution

$$\psi_{\mathrm{I,\,III}} = Ce^{Gx} + De^{-Gx},$$

which can be confirmed by direct substitution, since

$$\frac{d^2}{dx^2}\left(e^{\pm Gx}\right) = G^2 e^{\pm Gx}.$$

In region I, x is always negative, so D must be zero (otherwise, $\psi \to \infty$ as $x \to -\infty$, giving an unacceptable result). Similarly in region III, where x is always positive, C must be zero. Hence

$$\psi_{\mathrm{I}} = Ce^{Gx} \qquad (x < 0)$$

$$\psi_{\mathrm{III}} = De^{-Gx} \qquad (x > \ell).$$

In regions I and III, the wave function decreases exponentially with distance from the well. The mathematical forms of the wave function inside and outside the well are different, but we insist that the wave function be continuous even at the two walls. We also insist that the slope of ψ, which is its first derivative, be continuous at the walls. Hence we have the boundary conditions:

$$\psi_{\mathrm{I}} = \psi_{\mathrm{II}} \quad \text{and} \quad \frac{d\psi_{\mathrm{I}}}{dx} = \frac{d\psi_{\mathrm{II}}}{dx} \quad \text{at } x = 0$$

$$\psi_{\mathrm{II}} = \psi_{\mathrm{III}} \quad \text{and} \quad \frac{d\psi_{\mathrm{II}}}{dx} = \frac{d\psi_{\mathrm{III}}}{dx} \quad \text{at } x = \ell.$$

At the left-hand wall $(x = 0)$ these boundary conditions become

$$Ce^0 = A \sin 0 + B \cos 0 \qquad \text{or} \qquad C = B$$

and

$$GCe^0 = kA \cos 0 - kB \sin 0 \qquad \text{or} \qquad GC = kA.$$

These are two of the relations that link the constants A, B, C, D and the energy E.

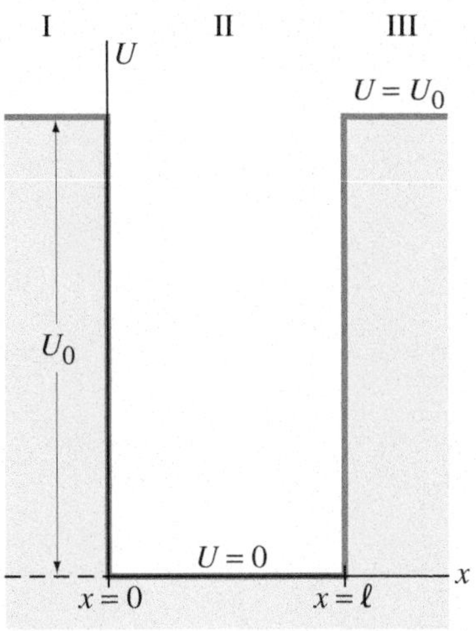

FIGURE 38–12 Potential energy U vs. x for a finite one-dimensional square well.

We get two more relations from the boundary conditions at $x = \ell$, and a fifth relation from normalizing the wave functions over all space, $\int_{-\infty}^{\infty}|\psi|^2\,dx = 1$. These five relations allow us to solve for the five unknowns, including the energy E. We will not go through the detailed mathematics, but we will discuss some of the results.

Figure 38–13a shows the wave function ψ for the three lowest possible states, and Fig. 38–13b shows the probability distributions $|\psi|^2$. We see that the wave functions are smooth at the walls of the well. Within the well ψ has the form of a sinusoidal wave; for the ground state, there is less than a half wavelength. Compare this to the infinite well (Fig. 38–9), where the ground-state wave function is exactly a half wavelength: $\lambda = 2\ell$. For our finite well, $\lambda > 2\ell$. Thus for a finite well the momentum of a particle $(p = h/\lambda)$, and hence its ground-state energy, will be less than for an infinite well of the same width ℓ.

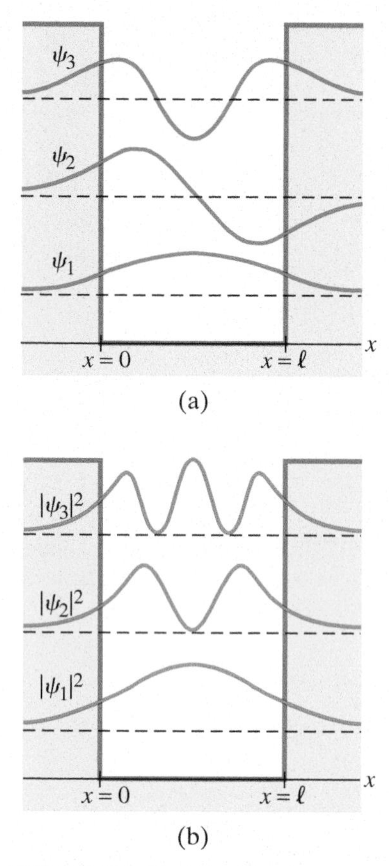

FIGURE 38–13 The wave functions (a), and probability distributions (b), for the three lowest possible states of a particle in a finite potential well. Each of the ψ and $|\psi|^2$ curves has been superposed on its energy level (dashed lines) for convenience.

Outside the finite well we see that the wave function drops off exponentially on either side of the walls. That ψ is not zero beyond the walls means that the particle can sometimes be found outside the well. This completely contradicts classical ideas. Outside the well, the potential energy of the particle is greater than its total energy: $U_0 > E$. This violates conservation of energy. But we clearly see in Fig. 38–13b that the particle can spend some time outside the well, where $U_0 > E$ (although the penetration into this classically forbidden region is generally not far since $|\psi|^2$ decreases exponentially with distance from either wall). The penetration of a particle into a classically forbidden region is a very important result of quantum mechanics. But how can it be? How can we accept this nonconservation of energy? We can look to the uncertainty principle, in the form

$$\Delta E\,\Delta t \gtrsim \hbar.$$

It tells us that the energy can be uncertain, and can even be nonconserved, by an amount ΔE for very short times $\Delta t \sim \hbar/\Delta E$.

Now let us consider the situation when the total energy E of the particle is greater than U_0. In this case the particle is a free particle and everywhere its wave function is sinusoidal, Fig. 38–14. Its wavelength is different outside the well than inside, as shown. Since $K = \frac{1}{2}mv^2 = p^2/2m$, the wavelength in region II $(U = 0)$ is

$$\lambda = \frac{h}{p} = \frac{h}{\sqrt{2mE}} \qquad 0 < x < \ell,$$

whereas in regions I and III, where $p^2/2m = K = E - U_0$, the wavelength is longer and is given by

$$\lambda = \frac{h}{p} = \frac{h}{\sqrt{2m(E - U_0)}} \qquad x < 0 \quad \text{and} \quad x > \ell.$$

For $E > U_0$, any energy E is possible. But for $E < U_0$, as we saw above, the energy is quantized and only certain states are possible.

FIGURE 38–14 The wave function of a particle of energy E traveling over a potential well whose depth U_0 is less than E (measured in the well).

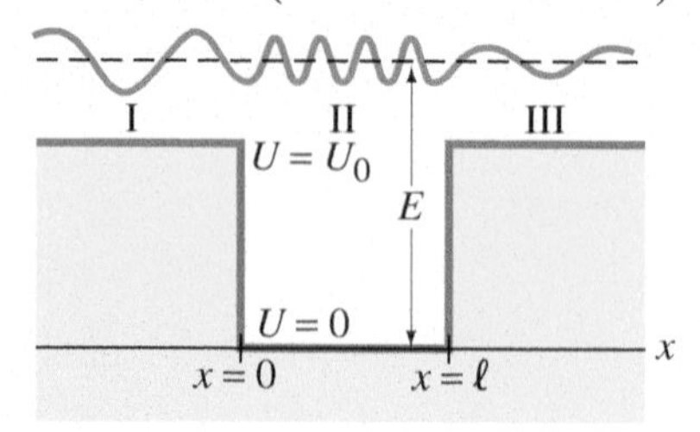

EXERCISE D An electron with energy $E = 6.0\,\text{eV}$ is near a potential well of depth $U_0 = 4.5\,\text{eV}$ and width $\ell = 10.0\,\text{nm}$. What is the wavelength of the electron when it is inside the well? (*a*) 0.50 nm; (*b*) 0.58 nm; (*c*) 1.0 nm; (*d*) 10 nm; (*e*) 20 nm.

[Another very interesting but more complicated well is the **simple harmonic oscillator** which has $U(x) = \frac{1}{2}Cx^2$ as we discussed in Chapter 14. Some quantum mechanical results, such as a zero-point energy of $\frac{1}{2}\hbar\omega$ (E cannot be zero), are treated briefly in Problem 52.]

38–10 Tunneling through a Barrier

We saw in Section 38–9 that, according to quantum mechanics, a particle such as an electron can penetrate a barrier into a region forbidden by classical mechanics. There are a number of important applications of this phenomenon, particularly as applied to penetration of a thin barrier.

We consider a particle of mass m and energy E traveling to the right along the x axis in free space where the potential energy $U = 0$ so the energy is all kinetic energy $(E = K)$. The particle encounters a narrow potential barrier whose height U_0 (in energy units) is greater than E, and whose thickness is ℓ (distance units); see Fig. 38–15a. Since $E < U_0$, we would expect from classical physics that the particle could not penetrate the barrier but would simply be reflected and would return in the opposite direction. Indeed, this is what happens for macroscopic objects. But quantum mechanics predicts a nonzero probability for finding the particle on the other side of the barrier. We can see how this can happen in part (b) of Fig. 38–15, which shows the wave function. The approaching particle has a sinusoidal wave function. Within the barrier the solution to the Schrödinger equation is a decaying exponential just as for the finite well of Section 38–9. However, before the exponential dies away to zero, the barrier ends (at $x = \ell$), and for $x > \ell$ there is again a sinusoidal wave function, since $U = 0$ and $E = K > 0$. But it is a sine wave of greatly reduced amplitude. Nonetheless, because $|\psi|^2$ is nonzero beyond the barrier, we see that there is a nonzero probability that the particle penetrates the barrier. This process is called **tunneling** through the barrier, or **barrier penetration**. Although we cannot observe the particle within the barrier (it would violate conservation of energy), we can detect it after it has penetrated the barrier.

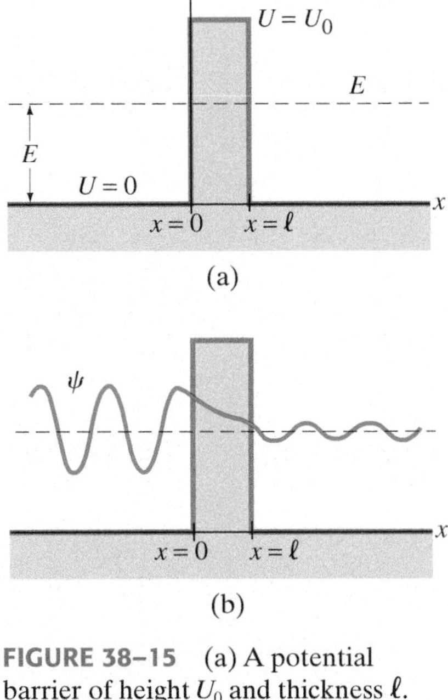

FIGURE 38–15 (a) A potential barrier of height U_0 and thickness ℓ. (b) The wave function for a particle of energy E ($< U_0$) that approaches from the left. The curve for ψ is superposed, for convenience, on the energy level line (dashed).

Quantitatively, we can describe the tunneling probability with a *transmission coefficient*, T, and a *reflection coefficient*, R. Suppose, for example, that $T = 0.03$ and $R = 0.97$; then if 100 particles struck the barrier, on the average 3 would tunnel through and 97 would be reflected. Note that $T + R = 1$, since an incident particle must either reflect or tunnel through. The transmission coefficient can be determined by writing the wave function for each of the three regions, just as we did for the finite well, and then applying the boundary conditions that ψ and $d\psi/dx$ must be continuous at the edges of the barrier ($x = 0$ and $x = \ell$). The calculation shows (see Problem 44) that if T is small ($\ll 1$), then

$$T \approx e^{-2G\ell}, \tag{38–17a}$$

where

$$G = \sqrt{\frac{2m(U_0 - E)}{\hbar^2}}. \tag{38–17b}$$

(This is the same G as in Section 38–9, Eq. 38–16.) We note that increasing the height of the barrier, U_0, or increasing its thickness, ℓ, will drastically reduce T. Indeed, for macroscopic situations, T is extremely small, in accord with classical physics, which predicts no tunneling (again the correspondence principle).

EXAMPLE 38–11 **Barrier penetration.** A 50-eV electron approaches a square barrier 70 eV high and (*a*) 1.0 nm thick, (*b*) 0.10 nm thick. What is the probability that the electron will tunnel through?

APPROACH We convert eV to joules and use Eqs. 38–17.

SOLUTION (*a*) Inside the barrier $U_0 - E = (70\,\text{eV} - 50\,\text{eV})(1.6 \times 10^{-19}\,\text{J/eV}) = 3.2 \times 10^{-18}\,\text{J}$. Then, using Eqs. 38–17, we have

$$2G\ell = 2\sqrt{\frac{2(9.11 \times 10^{-31}\,\text{kg})(3.2 \times 10^{-18}\,\text{J})}{(1.055 \times 10^{-34}\,\text{J}\cdot\text{s})^2}}\,(1.0 \times 10^{-9}\,\text{m}) = 46$$

and

$$T = e^{-2G\ell} = e^{-46} \approx 1 \times 10^{-20},$$

which is extremely small.

(*b*) For $\ell = 0.10\,\text{nm}$, $2G\ell = 4.6$ and

$$T = e^{-4.6} = 0.010.$$

Thus the electron has a 1% chance of penetrating a 0.1-nm-thick barrier, but only 1 chance in 10^{20} to penetrate a 1-nm barrier. By reducing the barrier thickness by a factor of 10, the probability of tunneling through increases 10^{18} times! Clearly the transmission coefficient is extremely sensitive to the values of ℓ, $U_0 - E$, and m.

Tunneling of Light Wave

Tunneling is a result of the wave properties of material particles, and also occurs for classical waves. For example, we saw in Section 32–7 that, when light traveling in glass strikes a glass–air boundary at an angle greater than the critical angle, the light is 100% totally reflected. We studied this phenomenon of total internal reflection from the point of view of ray optics, and we show it here in Fig. 38–16a. The wave theory, however, predicts that waves actually penetrate the air for a few wavelengths—almost as if they "needed" to pass the interface to find out there is air beyond and hence need to be totally reflected. Indeed, if a second piece of glass is brought near the first as shown in Fig. 38–16b, a transmitted wave that has tunneled through the air gap can be experimentally observed. You can actually observe this for yourself by looking down into a glass of water at an angle such that light entering your eye has been totally internally reflected from the (outer) glass surface (it will look silvery). If you press a moistened fingertip against the glass, you can see the whorls of the ridges on your fingerprints, because at the ridges you have interfered with the total internal reflection at the outer surface of the glass. So you see light that has penetrated the gap and reflected off the ridges on your finger.

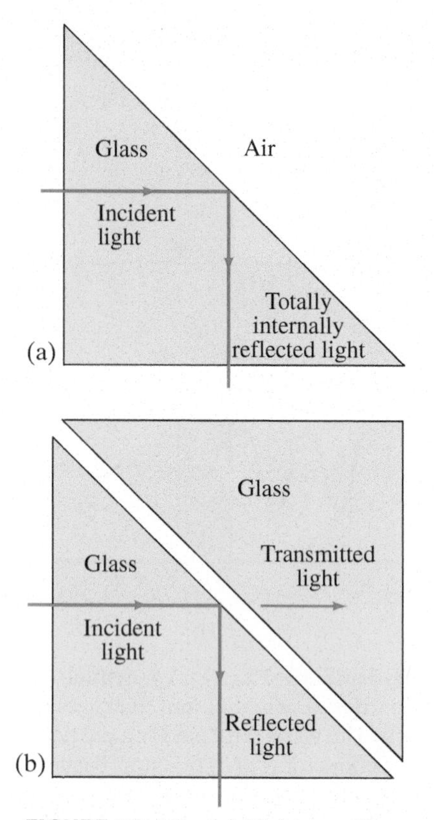

FIGURE 38–16 (a) Light traveling in glass strikes the interface with air at an angle greater than the critical angle, and is totally internally reflected. (b) A small amount of light tunnels through a narrow air gap between two pieces of glass.

Applications of Tunneling

Tunneling thus occurs even for classical waves. What is new in quantum mechanics is that material particles have wave properties and hence can tunnel. Tunneling has provided the basis for a number of useful devices, as well as helped to explain a number of important phenomena, some of which we mention briefly now.

Some atomic nuclei undergo **radioactive decay** by the emission of an alpha (α) particle, which consists of two protons and two neutrons. Inside a radioactive nucleus, we can imagine that the protons and neutrons are moving about, and sometimes two of each come together and form this stable entity, the alpha particle. We will study alpha decay in more detail in Chapter 41, but for now we note that the potential energy diagram for the alpha particle inside this type of nucleus looks something like Fig. 38–17. The square well represents the attractive nuclear force that holds the nucleus together. To this is added the $1/r$ Coulomb potential energy of repulsion between the positive alpha particle and the remaining positively charged nucleus. The barrier that results is called the **Coulomb barrier**. The wave function for the tunneling particle shown must have energy greater than zero (or the barrier would be infinitely wide and tunneling could not occur), but less than the height of the barrier. If the alpha particle had energy higher than the barrier, it would always be free and the original nucleus wouldn't exist. Thus the barrier keeps the nucleus together, but occasionally a nucleus of this type can decay by the tunneling of an alpha particle. The probability of an α particle escaping, and hence the "lifetime" of a nucleus, depends on the height and width of the barrier, and can take on a very wide range of values for only a limited change in barrier width as we saw in Example 38–11. Lifetimes of α-decaying radioactive nuclei range from less than $1\,\mu\text{s}$ to 10^{10} yr.

FIGURE 38–17 Potential energy seen by an alpha particle (charge q) in presence of nucleus (charge Q), showing the wave function for tunneling out.

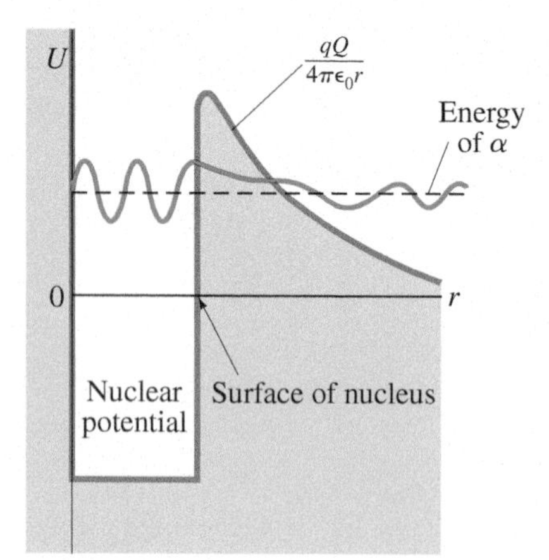

A so-called **tunnel diode** is an electronic device made of two types of semiconductor carrying opposite-sign charge carriers, separated by a very thin neutral region. Current can tunnel through this thin barrier and can be controlled by the voltage applied to it, which affects the height of the barrier.

The **scanning tunneling electron microscope** (STM), developed in the 1980s, makes use of tunneling through a vacuum. A tiny probe, whose tip may be only one (or a few) atoms wide, is moved across the specimen to be examined in a series of linear passes, like those made by the electron beam in a TV tube or CRT. The tip, as it scans, remains very close to the surface of the specimen, about 1 nm above it, Fig. 38–18. A small voltage applied between the probe and the surface causes electrons to tunnel through the vacuum between them. This tunneling current is very sensitive to the gap width (see Example 38–11), so that a feedback mechanism can be used to raise and lower the probe to maintain a constant electron tunneling current. The probe's vertical motion, following the surface of the specimen, is then plotted as a function of position, scan after scan, producing a three-dimensional image of the surface. Surface features as fine as the size of an atom can be resolved: a resolution better than 0.1 nm laterally and 10^{-2} to 10^{-3} nm vertically. This kind of resolution was not available previously and has given a great impetus to the study of the surface structure of materials. The "topographic" image of a surface actually represents the distribution of electron charge (electron wave probability distributions) on the surface.

The new **atomic force microscope** (AFM) is in many ways similar to an STM, but can be used on a wider range of sample materials. Instead of detecting an electric current, the AFM measures the force between a cantilevered tip and the sample, a force which depends strongly on the tip–sample separation at each point. The tip is moved as for the STM.

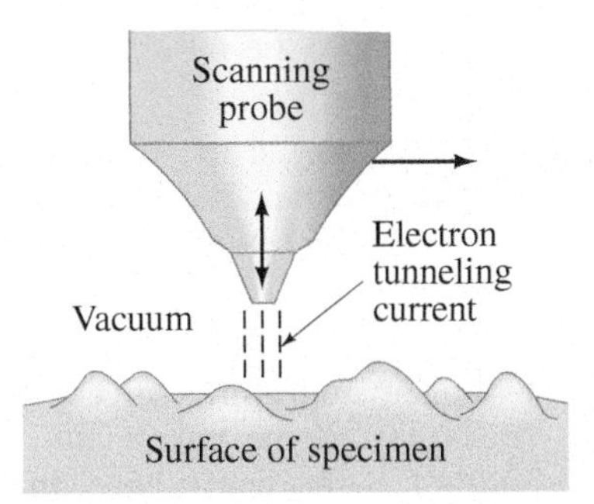

FIGURE 38–18 The probe tip of a scanning tunneling microscope, as it is moved horizontally, automatically moves up and down to maintain a constant tunneling current, thus producing an image of the surface.

Summary

In 1925, Schrödinger and Heisenberg separately worked out a new theory, **quantum mechanics**, which is now considered to be the fundamental theory at the atomic level. It is a statistical theory rather than a deterministic one.

An important aspect of quantum mechanics is the **Heisenberg uncertainty principle.** It results from the wave–particle duality and the unavoidable interaction between the observed object and the observer.

One form of the uncertainty principle states that the position x and momentum p_x of an object cannot both be measured precisely at the same time. The products of the uncertainties, $(\Delta x)(\Delta p_x)$, can be no less than $\hbar\ (= h/2\pi)$:

$$(\Delta x)(\Delta p_x) \gtrsim \hbar. \qquad \text{(38–1)}$$

Another form states that the energy can be uncertain, or nonconserved, by an amount ΔE for a time Δt where

$$(\Delta E)(\Delta t) \gtrsim \hbar. \qquad \text{(38–2)}$$

A particle such as an electron is represented by a **wave function** ψ. The square of the wave function, $|\psi|^2$, at any point in space represents the **probability** of finding the particle at that point. The wave function must be **normalized**, meaning that $\int |\psi|^2\, dV$ over all space must equal 1, since the particle must be found at one place or another.

In nonrelativistic quantum mechanics, ψ satisfies the **Schrödinger equation**:

$$-\frac{\hbar^2}{2m}\frac{d^2\psi}{dx^2} + U\psi = E\psi, \qquad \text{(38–5)}$$

here in its one-dimensional time-independent form, where U is the potential energy as a function of position and E is the total energy of the particle.

A **free particle** subject to no forces has a sinusoidal wave function $\psi = A \sin kx + B \cos kx$ with $k = p/\hbar$ and p is the particle's momentum. Such a wave of fixed momentum is spread out indefinitely in space as a plane wave.

A **wave packet**, localized in space, is a superposition of sinusoidal waves with a range of momenta.

For a particle confined to an **infinitely deep square well potential**, or **rigid box**, the Schrödinger equation gives the wave functions as

$$\psi = A \sin k\ell,$$

where ℓ is the width of the box, $A = \sqrt{2/\ell}$, and $k = n\pi/\ell$ with n an integer, as solutions inside the well. The energy is quantized,

$$E = \frac{\hbar^2 k^2}{2m} = \frac{n^2 h^2}{8m\ell^2}. \qquad \text{(38–13)}$$

In a **finite potential well**, the wave function extends into the classically forbidden region where the total energy is less than the potential energy. That this is possible is consistent with the uncertainty principle. The solutions to the Schrödinger equation in these areas are decaying exponentials.

Because quantum-mechanical particles can penetrate such classically forbidden areas, they can **tunnel** through thin barriers even though the potential energy in the barrier is greater than the total energy of the particle.

Questions

1. Compare a matter wave ψ to (*a*) a wave on a string, (*b*) an EM wave. Discuss similarities and differences.
2. Explain why Bohr's theory of the atom is not compatible with quantum mechanics, particularly the uncertainty principle.
3. Explain why it is that the more massive an object is, the easier it becomes to predict its future position.
4. In view of the uncertainty principle, why does a baseball seem to have a well-defined position and speed, whereas an electron does not?

5. Would it ever be possible to balance a very sharp needle precisely on its point? Explain.
6. When you check the pressure in a tire, doesn't some air inevitably escape? Is it possible to avoid this escape of air altogether? What is the relation to the uncertainty principle?
7. It has been said that the ground-state energy in the hydrogen atom can be precisely known but that the excited states have some uncertainty in their values (an "energy width"). Is this consistent with the uncertainty principle in its energy form? Explain.
8. If Planck's constant were much larger than it is, how would this affect our everyday life?
9. In what ways is Newtonian mechanics contradicted by quantum mechanics?
10. If you knew the position of an object precisely, with no uncertainty, how well would you know its momentum?
11. A cold thermometer is placed in a hot bowl of soup. Will the temperature reading of the thermometer be the same as the temperature of the hot soup before the measurement was made? Explain.
12. Does the uncertainty principle set a limit to how well you can make any single measurement of position?
13. Discuss the connection between the zero-point energy for a particle in a rigid box and the uncertainty principle.
14. The wave function for a particle in a rigid box is zero at points within the box (except for $n = 1$). Does this mean that the probability of finding the particle at these points is zero? Does it mean that the particle cannot pass by these points? Explain.
15. What does the probability density look like for a particle in an infinite potential well for large values of n, say $n = 100$ or $n = 1000$? As n becomes very large, do your predictions approach classical predictions in accord with the correspondence principle?
16. For a particle in an infinite potential well the separation between energy states increases as n increases (see Eq. 38–13). But doesn't the correspondence principle require closer spacing between states as n increases so as to approach a classical (nonquantized) situation? Explain.
17. A particle is trapped in an infinite potential well. Describe what happens to the particle's ground-state energy and wave function as the potential walls become finite and get lower and lower until they finally reach zero ($U = 0$ everywhere).
18. A hydrogen atom and a helium atom, each with 4 eV of kinetic energy, approach a thin barrier 6 MeV high. Which has the greater probability of tunneling through?

Problems

38–2 Wave Function, Double-Slit

1. (II) The neutrons in a parallel beam, each having kinetic energy 0.030 eV, are directed through two slits 0.60 mm apart. How far apart will the interference peaks be on a screen 1.0 m away? [*Hint*: First find the wavelength of the neutron.]
2. (II) Pellets of mass 3.0 g are fired in parallel paths with speeds of 150 m/s through a hole 3.0 mm in diameter. How far from the hole must you be to detect a 1.0-cm-diameter spread in the beam of pellets?

38–3 Uncertainty Principle

3. (I) A proton is traveling with a speed of $(7.560 \pm 0.012) \times 10^5$ m/s. With what maximum precision can its position be ascertained?
4. (I) An electron remains in an excited state of an atom for typically 10^{-8} s. What is the minimum uncertainty in the energy of the state (in eV)?
5. (I) If an electron's position can be measured to a precision of 2.6×10^{-8} m, how precisely can its speed be known?
6. (I) The lifetime of a typical excited state in an atom is about 10 ns. Suppose an atom falls from one such excited state and emits a photon of wavelength about 500 nm. Find the fractional energy uncertainty $\Delta E/E$ and wavelength uncertainty $\Delta\lambda/\lambda$ of this photon.
7. (I) A radioactive element undergoes an alpha decay with a lifetime of 12 μs. If alpha particles are emitted with 5.5-keV kinetic energy, find the uncertainty $\Delta E/E$ in the particle energy.
8. (II) A 12-g bullet leaves a rifle horizontally at a speed of 180 m/s. (*a*) What is the wavelength of this bullet? (*b*) If the position of the bullet is known to a precision of 0.65 cm (radius of the barrel), what is the minimum uncertainty in its vertical momentum?
9. (II) An electron and a 140-g baseball are each traveling 95 m/s measured to a precision of 0.085%. Calculate and compare the uncertainty in position of each.
10. (II) What is the uncertainty in the mass of a muon ($m = 105.7\ \mathrm{MeV}/c^2$), specified in eV/$c^2$, given its lifetime of 2.20 μs?
11. (II) A free neutron ($m = 1.67 \times 10^{-27}$ kg) has a mean life of 900 s. What is the uncertainty in its mass (in kg)?
12. (II) Use the uncertainty principle to show that if an electron were present in the nucleus ($r \approx 10^{-15}$ m), its kinetic energy (use relativity) would be hundreds of MeV. (Since such electron energies are not observed, we conclude that electrons are not present in the nucleus.) [*Hint*: Assume a particle can have energy as large as its uncertainty.]
13. (II) An electron in the $n = 2$ state of hydrogen remains there on average about 10^{-8} s before jumping to the $n = 1$ state. (*a*) Estimate the uncertainty in the energy of the $n = 2$ state. (*b*) What fraction of the transition energy is this? (*c*) What is the wavelength, and width (in nm), of this line in the spectrum of hydrogen?
14. (II) How accurately can the position of a 3.50-keV electron be measured assuming its energy is known to 1.00%?

15. (III) In a double-slit experiment on electrons (or photons), suppose that we use indicators to determine which slit each electron went through (Section 38–2). These indicators must tell us the y coordinate to within $d/2$, where d is the distance between slits. Use the uncertainty principle to show that the interference pattern will be destroyed. [*Note*: First show that the angle θ between maxima and minima of the interference pattern is given by $\frac{1}{2}\lambda/d$, Fig. 38–19.]

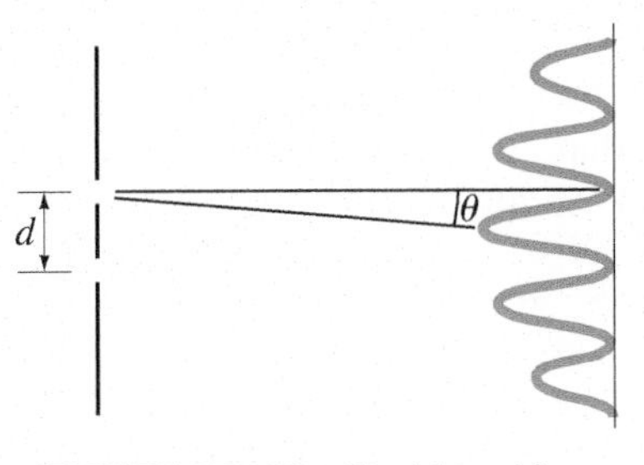

FIGURE 38–19 Problem 15.

*38–6 Time-Dependent Schrödinger Equation

*16. (II) Show that the superposition principle holds for the time-dependent Schrödinger equation. That is, show that if $\Psi_1(x,t)$ and $\Psi_2(x,t)$ are solutions, then $A\Psi_1(x,t) + B\Psi_2(x,t)$ is also a solution where A and B are arbitrary constants.

*17. (III) (*a*) Show that $\Psi(x,t) = Ae^{i(kx-\omega t)}$ is a solution to the time-dependent Schrödinger equation for a free particle $[U(x) = U_0 = \text{constant}]$ but that $\Psi(x,t) = A\cos(kx - \omega t)$ and $\Psi(x,t) = A\sin(kx - \omega t)$ are not. (*b*) Show that the valid solution of part (*a*) satisfies conservation of energy if the de Broglie relations hold; $\lambda = h/p$, $\omega = E/\hbar$. That is, show that direct substitution into Eq. 38–7 gives

$$\hbar\omega = \frac{\hbar^2 k^2}{2m} + U_0.$$

38–7 Free Particles; Plane Waves; Wave Packets

18. (I) A free electron has a wave function $\psi(x) = A\sin(2.0 \times 10^{10}x)$, where x is given in meters. Determine the electron's (*a*) wavelength, (*b*) momentum, (*c*) speed, and (*d*) kinetic energy.

19. (I) Write the wave function for (*a*) a free electron and (*b*) a free proton, each having a constant velocity $v = 3.0 \times 10^5$ m/s.

20. (III) Show that the uncertainty principle holds for a "wave packet" that is formed by two waves of similar wavelength λ_1 and λ_2. To do so, follow the argument leading up to Eq. 16–8, but use as the two waves $\psi_1 = A\sin k_1 x$ and $\psi_2 = A\sin k_2 x$. Then show that the width of each "wave packet" is $\Delta x = 2\pi/(k_1 - k_2) = 2\pi/\Delta k$ (from $t = 0.05$ s to $t = 0.15$ s in Fig. 16–17). Finally, show that $\Delta x\,\Delta p = h$ for this simple situation.

38–8 Infinite Square Well

21. (II) What is the minimum speed of an electron trapped in a 0.20-nm-wide infinitely deep square well?

22. (II) Show that for a particle in a perfectly rigid box, the wavelength of the wave function for any state is the de Broglie wavelength.

23. (II) An electron trapped in an infinitely deep square well has a ground-state energy $E = 9.0$ eV. (*a*) What is the longest wavelength photon that an excited state of this system can emit? (*b*) What is the width of the well?

24. (II) An $n = 4$ to $n = 1$ transition for an electron trapped in a rigid box produces a 340-nm photon. What is the width of the box?

25. (II) For a particle in a box with rigid walls, determine whether our results for the ground state are consistent with the uncertainty principle by calculating the product $\Delta p\,\Delta x$. Take $\Delta x \approx \ell$, since the particle is somewhere within the box. For Δp, note that although p is known $(= \hbar k)$, the direction of $\vec{\mathbf{p}}$ is not known, so the x component could vary from $-p$ to $+p$; hence take $\Delta p \approx 2p$.

26. (II) The longest-wavelength line in the spectrum emitted by an electron trapped in an infinitely deep square well is 610 nm. What is the width of the well?

27. (II) Determine the lowest four energy levels and wave functions for an electron trapped in an infinitely deep potential well of width 2.0 nm.

28. (II) Write a formula for the positions of (*a*) the maxima and (*b*) the minima in $|\psi|^2$ for a particle in the nth state in an infinite square well.

29. (II) Consider an atomic nucleus to be a rigid box of width 2.0×10^{-14} m. What would be the ground-state energy for (*a*) an electron, (*b*) a neutron, and (*c*) a proton in this nucleus?

30. (II) A proton in a nucleus can be roughly modeled as a particle in a box of nuclear dimensions. Calculate the energy released when a proton confined in a nucleus of width 1.0×10^{-14} m makes a transition from the first excited state to the ground state.

31. (II) Consider a single oxygen molecule confined in a one-dimensional rigid box of width 4.0 mm. (*a*) Treating this as a particle in a rigid box, determine the ground-state energy. (*b*) If the molecule has an energy equal to the one-dimensional average thermal energy $\frac{1}{2}kT$ at $T = 300$ K, what is the quantum number n? (*c*) What is the energy difference between the nth state and the next higher state?

32. (II) An electron is trapped in a 1.00-nm-wide rigid box. Determine the probability of finding the electron within 0.15 nm of the center of the box (on either side of center) for (*a*) $n = 1$, (*b*) $n = 5$, and (*c*) $n = 20$. (*d*) Compare to the classical prediction.

33. (III) If an infinitely deep well of width ℓ is redefined to be located from $x = -\frac{1}{2}\ell$ to $x = \frac{1}{2}\ell$ (as opposed to $x = 0$ to $x = \ell$), speculate how this will change the wave function for a particle in this well. Investigate your speculation(s) by determining the wave functions and energy levels for this newly defined well. [*Hint*: Try $\psi = A\sin(kx + \phi)$.]

38–9 Finite Potential Well

34. (II) An electron with 180 eV of kinetic energy in free space passes over a finite potential well 56 eV deep that stretches from $x = 0$ to $x = 0.50$ nm. What is the electron's wavelength (*a*) in free space, (*b*) when over the well? (*c*) Draw a diagram showing the potential energy and total energy as a function of x, and on the diagram sketch a possible wave function.

35. (II) Sketch the wave functions and the probability distributions for the $n = 4$ and $n = 5$ states for a particle trapped in a finite square well.

36. (II) Suppose that a particle of mass m is trapped in a finite potential well that has a rigid wall at $x = 0$ ($U = \infty$ for $x < 0$) and a finite wall of height $U = U_0$ at $x = \ell$, Fig. 38–20. (*a*) Sketch the wave functions for the lowest three states. (*b*) What is the form of the wave function in the ground state in the three regions $x < 0$, $0 < x < \ell$, $x > \ell$?

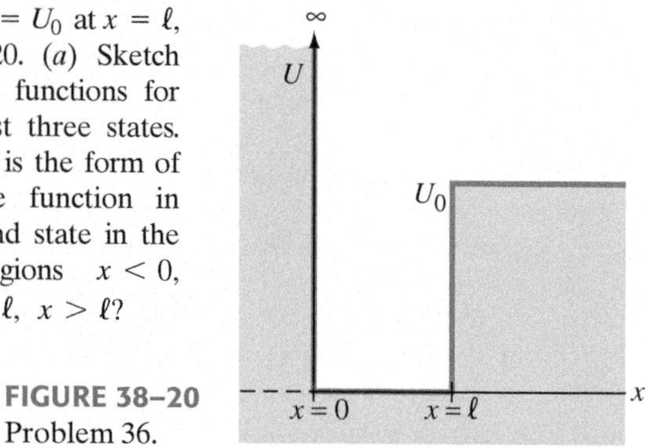

FIGURE 38–20 Problem 36.

37. (II) An electron is trapped in a 0.16-nm-wide finite square well of height $U_0 = 2.0\,\text{keV}$. Estimate at what distance outside the walls of the well the ground state wave function drops to 1.0% of its value at the walls.

38–10 Tunneling

38. (II) A potential barrier has a height $U_0 = 14\,\text{eV}$ and thickness $\ell = 0.85\,\text{nm}$. If the transmission coefficient for an incident electron is 0.00050, what is the electron's energy?

39. (II) An electron approaches a potential barrier 18 eV high and 0.55 nm wide. If the electron has a 1.0% probability of tunneling through the barrier, what is the electron's energy?

40. (II) A proton and a helium nucleus approach a 25-MeV potential energy barrier. If each has a kinetic energy of 5.0 MeV, what is the probability of each to tunnel through the barrier, assuming it is rectangular and 3.6 fm thick?

41. (II) An electron with an energy of 8.0 eV is incident on a potential barrier which is 9.2 eV high and 0.25 nm wide. (*a*) What is the probability that the electron will pass through the barrier? (*b*) What is the probability that the electron will be reflected?

42. (II) A 1.0-mA current of 1.6-MeV protons strikes a 2.6-MeV-high potential barrier 2.8×10^{-13} m thick. Estimate the transmitted current.

43. (II) For part (*b*) of Example 38–11, what effect will there be on the transmission coefficient if (*a*) the barrier height is raised 2.0%, (*b*) the barrier thickness is increased by 2.0%?

44. (II) Show that the transmission coefficient is given roughly by Eqs. 38–17 for a high or thick barrier, by calculating $|\psi(x = \ell)|^2/|\psi(0)|^2$. [*Hint*: Assume that ψ is a decaying exponential inside the barrier.]

45. (III) A uranium-238 nucleus ($Q = +92e$) lasts about 4.5×10^9 years before it decays by emission of an alpha particle ($q = +2e$, $M = 4M_{\text{proton}}$). (*a*) Assuming that the α particle is a point, and the nucleus is roughly 8 fm in radius, estimate the height of the Coulomb barrier (the peak in Fig. 38–17). (*b*) The alpha particle, when free, has kinetic energy ≈ 4 MeV. Estimate the width of the barrier. (*c*) Assuming that the square well has $U = 0$ inside (and $U = 0$ far from the nucleus), calculate the speed of the alpha particle and how often it hits the barrier inside, and from this (and Eqs. 38–17) estimate the uranium lifetime. [*Hint*: Replace the $1/r$ Coulomb barrier with an "averaged" rectangular barrier (as in Fig. 38–15) of width equal to $\frac{1}{3}$ that calculated in (*b*).]

General Problems

46. The Z^0 boson, discovered in 1985, is the mediator of the weak nuclear force, and it typically decays very quickly. Its average rest energy is 91.19 GeV, but its short lifetime shows up as an intrinsic width of 2.5 GeV. What is the lifetime of this particle?

47. Estimate the lowest possible energy of a neutron contained in a typical nucleus of radius 1.2×10^{-15} m. [*Hint*: A particle can have an energy at least as large as its uncertainty.]

48. A neutron is trapped in an infinitely deep potential well 2.5 fm in width. Determine (*a*) the four lowest possible energy states and (*b*) their wave functions. (*c*) What is the wavelength and energy of a photon emitted when the neutron makes a transition between the two lowest states? In what region of the EM spectrum does this photon lie? [*Note*: This is a rough model of an atomic nucleus.]

49. Protons are accelerated from rest across 650 V. They are then directed at two slits 0.80 mm apart. How far apart will the interference peaks be on a screen 18 m away?

50. An electron and a proton, each initially at rest, are accelerated across the same voltage. Assuming that the uncertainty in their position is given by their de Broglie wavelength, find the ratio of the uncertainty in their momentum.

51. Use the uncertainty principle to estimate the position uncertainty for the electron in the ground state of the hydrogen atom. [*Hint*: Determine the momentum using the Bohr model of Section 37–11 and assume the momentum can be anywhere between this value and zero.] How does this compare to the Bohr radius?

52. **Simple Harmonic Oscillator**. Suppose that a particle of mass m is trapped not in a square well, but in one whose potential energy is that of a simple harmonic oscillator: $U(x) = \frac{1}{2}Cx^2$. That is, if the particle is displaced from $x = 0$, a restoring force $F = -Cx$ acts on it, where C is constant. (*a*) Sketch this potential energy. (*b*) Show that $\psi = Ae^{-Bx^2}$ is a solution to the Schrödinger equation and that the energy of this state is $E = \frac{1}{2}\hbar\omega$, where $\omega = \sqrt{C/m}$ (as classically, Eq. 14–5) and $B = m\omega/2\hbar$. [*Note*: This is the ground state, and this energy $\frac{1}{2}\hbar\omega$ is the zero-point energy for a harmonic oscillator. The energies of higher states are $E_n = \left(n + \frac{1}{2}\right)\hbar\omega$, where n is an integer.]

53. Estimate the kinetic energy and speed of an alpha particle ($q = +2e$, $M = 4M_{\text{proton}}$) trapped in a nucleus 1.5×10^{-14} m wide. Assume an infinitely deep square well potential.

54. A small ball of mass 3.0×10^{-6} kg is dropped on a table from a height of 2.0 m. After each bounce the ball rises to 65% of its height before the bounce because of its inelastic collision with the table. Estimate how many bounces occur before the uncertainty principle plays a role in the problem. [*Hint*: Determine when the uncertainty in the ball's speed is comparable to its speed of impact on the table.]

55. By how much does the tunneling current through the tip of an STM change if the tip rises 0.020 nm from some initial height above a sodium surface with a work function $W_0 = 2.28$ eV? [*Hint*: Let the work function (see Section 37–2) equal the energy needed to raise the electron to the top of the barrier.]

56. Show that the function $\psi(x) = Ae^{ikx}$, where A is a constant and k is given by Eq. 38–11, is a solution of the time-independent Schrödinger equation for the case $U = 0$.

57. Show that the average value of the square of the position x of a particle in state n inside an infinite well of width ℓ is $\overline{x^2} = \int x^2|\psi_n|^2\,dx = \ell^2\left[\frac{1}{3} - \frac{1}{2}(n\pi)^{-2}\right]$. Calculate the values of $\overline{x^2}$ for $n = 1$ to 20 and make a graph of $\overline{x^2}$ versus n. [*Hint*: You may want to consult a detailed Table of integrals.]

58. Consider a particle that can exist anywhere in space with a wave function given by $\psi(x) = b^{-\frac{1}{2}}|x/b|^{\frac{1}{2}}e^{-(x/b)^2/2}$, where $b = 1.0$ nm. (*a*) Check that the wave function is normalized. (*b*) What is the most probable position for the particle in the region $x > 0$? (*c*) What is the probability of finding the particle between $x = 0$ nm and $x = 0.50$ nm?

59. A 7.0-gram pencil, 18 cm long, is balanced on its point. Classically, this is a configuration of (unstable) equilibrium, so the pencil could remain there forever if it were perfectly placed. A quantum mechanical analysis shows that the pencil must fall. (*a*) Why is this the case? (*b*) Estimate (within a factor of 2) how long it will take the pencil to hit the table if it is initially positioned as well as possible? [*Hint*: Use the uncertainty principle in its angular form to obtain an expression for the initial angle $\phi_0 \approx \Delta\phi$.]

* Numerical/Computer

*60. (III) An electron is trapped in the ground state of an infinite potential well of width $\ell = 0.10$ nm. The probability that the electron will be found in the central 1% of the well was estimated in Example 38–7 by $|\psi|^2\Delta x$. Use numerical methods to determine how large Δx could be to cause less than a 10% error in such an estimate.

*61. (III) Consider a particle of mass m and energy E traveling to the right where it encounters a narrow potential barrier of height U_0 and width ℓ as shown in Fig. 38–21. It can be shown that:

(i) for $E < U_0$, the transmission probability is

$$T = \left[1 + \frac{(e^{G\ell} - e^{-G\ell})^2}{16(E/U_0)(1 - E/U_0)}\right]^{-1}$$

where

$$G = \sqrt{\frac{2m(U_0 - E)}{\hbar^2}}$$

and the reflection probability is $R = 1 - T$;
(ii) for $E > U_0$, the transmission probability is

$$T = \left[1 + \frac{\sin^2(G'\ell)}{4(E/U_0)(E/U_0 - 1)}\right]^{-1}$$

where

$$G' = \sqrt{\frac{2m(E - U_0)}{\hbar^2}}$$

and $R = 1 - T$. Consider that the particle is an electron and it is incident on a rectangular barrier of height $U_0 = 10$ eV and width $\ell = 1.0 \times 10^{-10}$ m. (*a*) Calculate T and R for the electron from $E/U_0 = 0$ to 10, in steps of 0.1. Make a single graph showing the two curves of T and R as a function of E/U_0. (*b*) From the graph determine the energies (E/U_0) at which the electron will have transmission probabilities of 10%, 20%, 50%, and 80%.

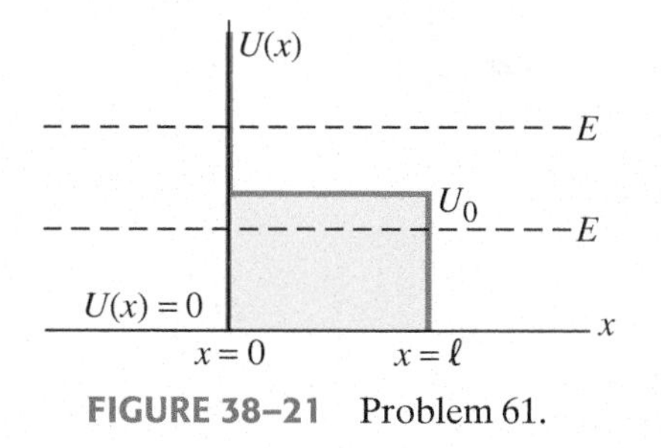

FIGURE 38–21 Problem 61.

Answers to Exercises

A: (*c*).

B: (*e*).

C: (*e*).

D: (*a*).

A neon tube is a thin glass tube, moldable into various shapes, filled with neon (or other) gas that glows with a particular color when a current at high voltage passes through it. Gas atoms, excited to upper energy levels, jump down to lower energy levels and emit light (photons) whose wavelengths (color) are characteristic of the type of gas.

In this Chapter we study what quantum mechanics tells us about atoms, their wave functions and energy levels, including the effect of the exclusion principle. We also discuss interesting applications such as lasers and holography.

Quantum Mechanics of Atoms

CONTENTS

CHAPTER-OPENING QUESTIONS—Guess now!

1. Thousands of hydrogen atoms are all in their ground state. Which statement below is true?
   (a) All of the atoms have the electron in a circular orbit at the Bohr radius.
   (b) All of the atoms have the electron in the same orbit, but it's not the same orbit for the ground state in the Bohr model.
   (c) The electron is not in an actual orbit, but the distance between the nucleus and the electron is the same in all of the atoms.
   (d) If the distance from the nucleus to the electron could be measured, it would be found at different locations in different atoms. The most probable distance would be the Bohr radius.

2. The state of an electron in an atom can be specified by a set of quantum numbers. Which of the following statements are valid?
   (a) No two electrons in the universe can be identical to each other.
   (b) All electrons in the universe are identical.
   (c) No two electrons in an atom can occupy the same quantum state.
   (d) There cannot be more than one electron in a given electron orbit.
   (e) Electrons must be excluded from the nucleus because only positive charges are allowed in the nucleus to give rise to stable electronic orbits according to Bohr's model.

At the beginning of Chapter 38 we discussed the limitations of the Bohr theory of atomic structure and why a new theory was needed. Although the Bohr theory had great success in predicting the wavelengths of light emitted and absorbed by the hydrogen atom, it could not do so for more complex atoms. Nor did it explain *fine structure*, the splitting of emission lines into two or more closely spaced lines. And, as a theory, it was an uneasy mixture of classical and quantum ideas.

Quantum mechanics came to the rescue in 1925 and 1926, and in this Chapter we examine the quantum-mechanical theory of atomic structure, which is far more complete than the old Bohr theory.

39–1 Quantum-Mechanical View of Atoms

Although the Bohr model has been discarded as an accurate description of nature, nonetheless, quantum mechanics reaffirms certain aspects of the older theory, such as that electrons in an atom exist only in discrete states of definite energy, and that a photon of light is emitted (or absorbed) when an electron makes a transition from one state to another. But quantum mechanics is a much deeper theory, and has provided us with a very different view of the atom. According to quantum mechanics, electrons do not exist in well-defined circular orbits as in the Bohr theory. Rather, the electron (because of its wave nature) can be thought of as spread out in space as if it were a "**cloud.**" The size and shape of the electron cloud can be calculated for a given state of an atom. For the ground state in the hydrogen atom, the solution of the Schrödinger equation, as we will discuss in more detail in Section 39–3, gives

$$\psi(r) = \frac{1}{\sqrt{\pi r_0^3}} e^{-\frac{r}{r_0}}.$$

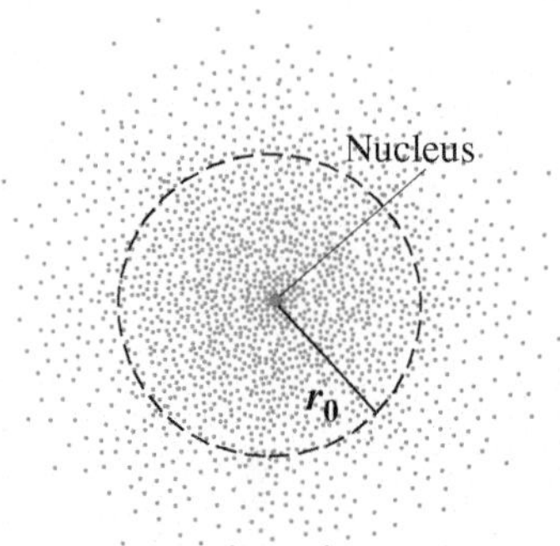

FIGURE 39–1 Electron cloud or "probability distribution" $|\psi|^2$ for the ground state of the hydrogen atom, as seen from afar. The circle represents the Bohr radius r_0. The dots represent a hypothetical detection of an electron at that point: dots closer together represent more probable presence of an electron (denser cloud).

Here $\psi(r)$ is the wave function as a function of position, and it depends only on the radial distance r from the center, and not on angular position θ or ϕ. (The constant r_0 happens to be equal to the first Bohr radius.) Thus the electron cloud, whose density is $|\psi|^2$, for the ground state of hydrogen is spherically symmetric as shown in Fig. 39–1. The extent of the electron cloud at its higher densities roughly indicates the "size" of an atom, but just as a cloud may not have a distinct border, atoms do not have a precise boundary or a well-defined size. Not all electron clouds have a spherical shape, as we shall see later in this Chapter. But note that $\psi(r)$, while becoming extremely small for large r (see the equation above), does not equal zero in any finite region. So quantum mechanics suggests that an atom is not mostly empty space. (Indeed, since $\psi \rightarrow 0$ only for $r \rightarrow \infty$, we might question the idea that there is any truly empty space in the universe.)

The electron cloud can be interpreted from either the particle or the wave viewpoint. Remember that by a particle we mean something that is localized in space—it has a definite position at any given instant. By contrast, a wave is spread out in space. The electron cloud, spread out in space as in Fig. 39–1, is a result of the wave nature of electrons. Electron clouds can also be interpreted as **probability distributions** (or **probability density**) for a particle. As we saw in Section 38–3, we cannot predict the path an electron will follow. After one measurement of its position we cannot predict exactly where it will be at a later time. We can only calculate the probability that it will be found at different points. If you were to make 500 different measurements of the position of an electron, considering it as a particle, the majority of the results would show the electron at points where the probability is high (darker area in Fig. 39–1). Only occasionally would the electron be found where the probability is low.

39–2 Hydrogen Atom: Schrödinger Equation and Quantum Numbers

The hydrogen atom is the simplest of all atoms, consisting of a single electron of charge $-e$ moving around a central nucleus (a single proton) of charge $+e$. It is with hydrogen that a study of atomic structure must begin.

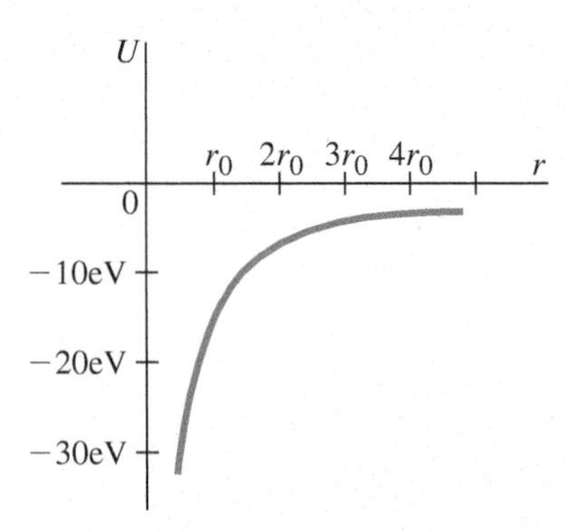

FIGURE 39–2 Potential energy $U(r)$ for the hydrogen atom. The radial distance r of the electron from the nucleus is given in terms of the Bohr radius r_0.

The Schrödinger equation (see Eq. 38–5) includes a term containing the potential energy. For the hydrogen (H) atom, the potential energy is due to the Coulomb force between electron and proton:

$$U = -\frac{1}{4\pi\epsilon_0}\frac{e^2}{r}$$

where r is the radial distance from the proton (situated at $r = 0$) to the electron. See Fig. 39–2. The (time-independent) Schrödinger equation, which must now be written in three dimensions, is then

$$-\frac{\hbar^2}{2m}\left(\frac{\partial^2\psi}{\partial x^2} + \frac{\partial^2\psi}{\partial y^2} + \frac{\partial^2\psi}{\partial z^2}\right) - \frac{1}{4\pi\epsilon_0}\frac{e^2}{r}\psi = E\psi, \qquad \textbf{(39–1)}$$

where $\partial^2\psi/\partial x^2$, $\partial^2\psi/\partial y^2$, and $\partial^2\psi/\partial z^2$ are partial derivatives with respect to x, y, and z. To solve the Schrödinger equation for the H atom, it is usual to write it in terms of spherical coordinates (r, θ, ϕ). We will not, however, actually go through the process of solving it. Instead, we look at the properties of the solutions, and (in the next Section) at the wave functions themselves.

Recall from Chapter 38 that the solutions of the Schrödinger equation in one dimension for the infinite square well were characterized by a single quantum number, which we called n, which arises from applying the boundary conditions. In the three-dimensional problem of the H atom, the solutions of the Schrödinger equation are characterized by three quantum numbers corresponding to boundary conditions applied in the three dimensions. However, four different quantum numbers are actually needed to specify each state in the H atom, the fourth coming from a relativistic treatment. We now discuss each of these quantum numbers. Much of our analysis here will also apply to more complex atoms, which we discuss starting in Section 39–4.

Quantum mechanics predicts the same energy levels (Fig. 37–26) for the H atom as does the Bohr theory. That is,

$$E_n = -\frac{13.6\,\text{eV}}{n^2} \qquad n = 1, 2, 3, \cdots, \qquad \textbf{(39–2)}$$

where n is an integer. In the simple Bohr theory, there was only one quantum number, n. In quantum mechanics, four different quantum numbers are needed to specify each state in the atom:

(1) The *quantum number*, n, from the Bohr theory is found also in quantum mechanics and is called the **principal quantum number**. It can have any integer value from 1 to ∞. The total energy of a state in the hydrogen atom depends on n, as we saw above.

(2) The **orbital quantum number**, ℓ, is related to the magnitude of the angular momentum of the electron; ℓ can take on integer values from 0 to $(n - 1)$. For the ground state $(n = 1)$, ℓ can only be zero.† For $n = 3$, ℓ can be 0, 1, or 2. The actual magnitude of the angular momentum L is related to the quantum number ℓ by

$$L = \sqrt{\ell(\ell + 1)}\,\hbar. \qquad \textbf{(39–3)}$$

The value of ℓ has almost no effect on the total energy in the hydrogen atom; only n does to any appreciable extent (but see *fine structure* below). In atoms with two or more electrons, the energy does depend on ℓ as well as n, as we shall see.

FIGURE 39–3 Quantization of angular momentum direction for $\ell = 2$. (Magnitude of $\vec{\mathbf{L}}$ is $\sqrt{6}\,\hbar$).

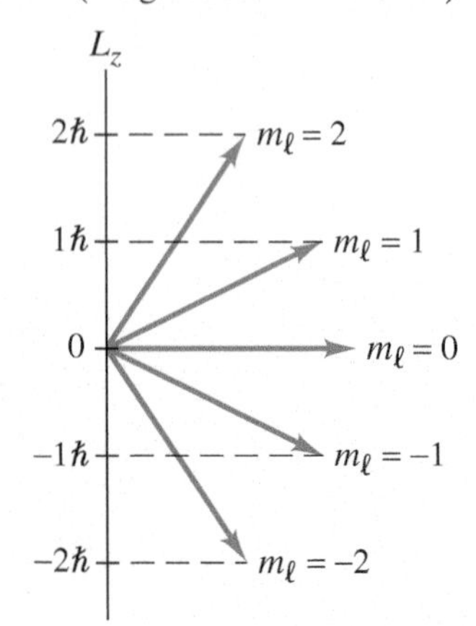

(3) The **magnetic quantum number**, m_ℓ, is related to the *direction* of the electron's angular momentum, and it can take on integer values ranging from $-\ell$ to $+\ell$. For example, if $\ell = 2$, then m_ℓ can be $-2, -1, 0, +1$, or $+2$. Since angular momentum is a vector, it is not surprising that both its magnitude and its direction would be quantized. For $\ell = 2$, the five different directions allowed can be represented by the diagram of Fig. 39–3.

†This replaces Bohr theory, which had $\ell = 1$ for the ground state (Eq. 37–10).

This limitation on the direction of $\vec{\mathbf{L}}$ is often called **space quantization**. In quantum mechanics, the direction of the angular momentum is usually specified by giving its component along the z axis (this choice is arbitrary). Then L_z is related to m_ℓ by the equation

$$L_z = m_\ell \hbar. \quad \textbf{(39–4)}$$

The values of L_x and L_y are not definite, however. The name for m_ℓ derives not from theory (which relates it to L_z), but from experiment. It was found that when a gas-discharge tube was placed in a magnetic field, the spectral lines were split into several very closely spaced lines. This splitting, known as the **Zeeman effect**, implies that the energy levels must be split (Fig. 39–4), and thus that the energy of a state depends not only on n but also on m_ℓ when a magnetic field is applied—hence the name "magnetic quantum number."

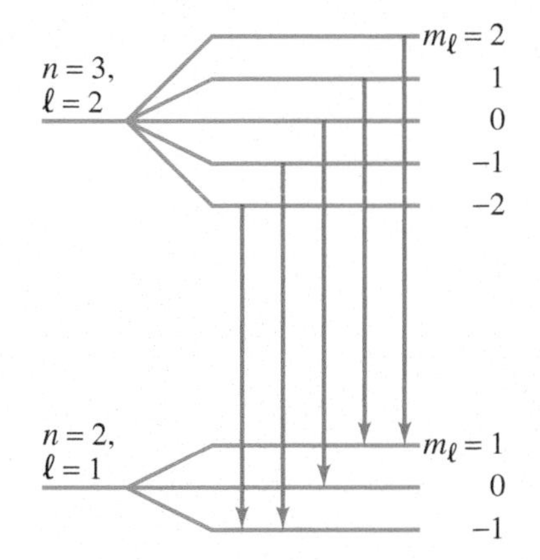

FIGURE 39–4 When a magnetic field is applied, an $n = 3, \ell = 2$ energy level is split into five separate levels (shown exaggerated—not to scale) corresponding to the five values of $m_\ell\,(2, 1, 0, -1, -2)$. An $n = 2, \ell = 1$ level is split into three levels $(m_\ell = 1, 0, -1)$. Transitions can occur between levels (not all transitions are shown), with photons of several slightly different frequencies being given off (the Zeeman effect).

(4) Finally, there is the **spin quantum number**, m_s, which for an electron can have only two values, $m_s = +\frac{1}{2}$ and $m_s = -\frac{1}{2}$. The existence of this quantum number did not come out of Schrödinger's original theory, as did n, ℓ, and m_ℓ. Instead, a subsequent modification by P. A. M. Dirac (1902–1984) explained its presence as a relativistic effect. The first hint that m_s was needed, however, came from experiment. A careful study of the spectral lines of hydrogen showed that each actually consisted of two (or more) very closely spaced lines even in the absence of an external magnetic field. It was at first hypothesized that this tiny splitting of energy levels, called **fine structure**, was due to angular momentum associated with a spinning of the electron. That is, the electron might spin on its axis as well as orbit the nucleus, just as the Earth spins on its axis as it orbits the Sun. The interaction between the tiny current of the spinning electron could then interact with the magnetic field due to the orbiting charge and cause the small observed splitting of energy levels. (The energy thus depends slightly on m_ℓ and m_s. Fine structure is said to be due to a **spin-orbit interaction**.) Today we consider the picture of a spinning electron as not legitimate. We cannot even view an electron as a localized object, much less a spinning one. What is important is that the electron can have two different states due to some intrinsic property that behaves like an angular momentum, and we still call this property "spin." The electron is said to have a spin quantum number $s = \frac{1}{2}$, which produces a spin angular momentum

$$S = \sqrt{s(s + 1)}\,\hbar = \frac{\sqrt{3}}{2}\,\hbar.$$

The z component is

$$S_z = m_s \hbar$$

where the two possible values of m_s $\left(+\frac{1}{2} \text{ and } -\frac{1}{2}\right)$ are often said to be "spin up" and "spin down," referring to the two possible directions of the spin angular momentum. See Fig. 39–5. A state with spin down $\left(m_s = -\frac{1}{2}\right)$ has slightly lower energy than a state with spin up. (Note that we include m_s, but not s, in our list of quantum numbers since s is the same for all electrons.)

FIGURE 39–5 The spin angular momentum S can take on only two directions, $m_s = +\frac{1}{2}$ or $-\frac{1}{2}$, called "spin up" and "spin down."

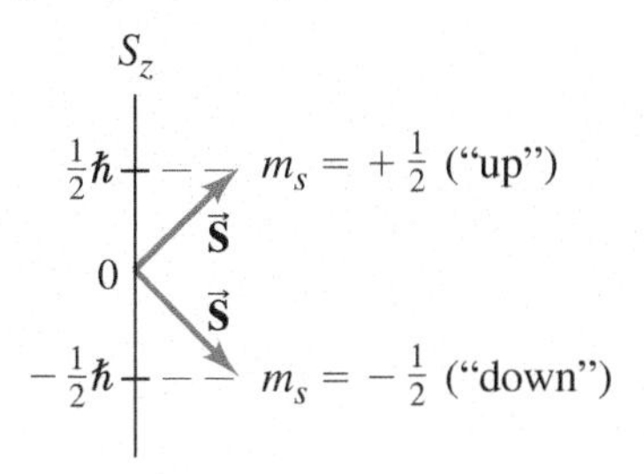

The possible values of the four quantum numbers for an electron in the hydrogen atom are summarized in Table 39–1.

TABLE 39–1 Quantum Numbers for an Electron

Name	Symbol	Possible Values
Principal	n	$1, 2, 3, \cdots, \infty$.
Orbital	ℓ	For a given n: ℓ can be $0, 1, 2, \cdots, n - 1$.
Magnetic	m_ℓ	For given n and ℓ: m_ℓ can be $\ell, \ell - 1, \cdots, 0, \cdots, -\ell$.
Spin	m_s	For each set of n, ℓ, and m_ℓ: m_s can be $+\frac{1}{2}$ or $-\frac{1}{2}$.

CONCEPTUAL EXAMPLE 39–1 **Possible states for $n = 3$.** How many different states are possible for an electron whose principal quantum number is $n = 3$?

RESPONSE For $n = 3, \ell$ can have the values $\ell = 2, 1, 0$. For $\ell = 2, m_\ell$ can be 2, 1, 0, −1, −2, which is five different possibilities. For each of these, m_s can be either up or down ($+\frac{1}{2}$ or $-\frac{1}{2}$); so for $\ell = 2$, there are $2 \times 5 = 10$ states. For $\ell = 1, m_\ell$ can be 1, 0, −1, and since m_s can be $+\frac{1}{2}$ or $-\frac{1}{2}$ for each of these, we have 6 more possible states. Finally, for $\ell = 0, m_\ell$ can only be 0, and there are only 2 states corresponding to $m_s = +\frac{1}{2}$ and $-\frac{1}{2}$. The total number of states is $10 + 6 + 2 = 18$, as detailed in the following Table:

n	ℓ	m_ℓ	m_s	n	ℓ	m_ℓ	m_s
3	2	2	$\frac{1}{2}$	3	1	1	$\frac{1}{2}$
3	2	2	$-\frac{1}{2}$	3	1	1	$-\frac{1}{2}$
3	2	1	$\frac{1}{2}$	3	1	0	$\frac{1}{2}$
3	2	1	$-\frac{1}{2}$	3	1	0	$-\frac{1}{2}$
3	2	0	$\frac{1}{2}$	3	1	−1	$\frac{1}{2}$
3	2	0	$-\frac{1}{2}$	3	1	−1	$-\frac{1}{2}$
3	2	−1	$\frac{1}{2}$	3	0	0	$\frac{1}{2}$
3	2	−1	$-\frac{1}{2}$	3	0	0	$-\frac{1}{2}$
3	2	−2	$\frac{1}{2}$				
3	2	−2	$-\frac{1}{2}$				

EXERCISE A An electron has $n = 4$, $\ell = 2$. Which of the following values of m_ℓ are possible: 4, 3, 2, 1, 0, −1, −2, −3, −4?

EXAMPLE 39–2 ***E* and *L* for $n = 3$.** Determine (*a*) the energy and (*b*) the orbital angular momentum for an electron in each of the hydrogen atom states of Example 39–1.

APPROACH The energy of a state depends only on n, except for the very small corrections mentioned above, which we will ignore. Energy is calculated as in the Bohr theory, $E_n = -13.6\,\text{eV}/n^2$. For angular momentum we use Eq. 39–3.

SOLUTION (*a*) Since $n = 3$ for all these states, they all have the same energy,

$$E_3 = -\frac{13.6\,\text{eV}}{(3)^2} = -1.51\,\text{eV}.$$

(*b*) For $\ell = 0$, Eq. 39–3 gives

$$L = \sqrt{\ell(\ell+1)}\,\hbar = 0.$$

For $\ell = 1$,

$$L = \sqrt{1(1+1)}\,\hbar = \sqrt{2}\,\hbar = 1.49 \times 10^{-34}\,\text{J}\cdot\text{s}.$$

For $\ell = 2$, $L = \sqrt{2(2+1)}\,\hbar = \sqrt{6}\,\hbar.$

NOTE Atomic angular momenta are generally given as a multiple of $\hbar$ ($\sqrt{2}\,\hbar$ or $\sqrt{6}\,\hbar$ in this case), rather than in SI units.

EXERCISE B What is the magnitude of the orbital angular momentum for orbital quantum number $\ell = 3$? (*a*) $3\hbar$; (*b*) $3.5\hbar$; (*c*) $4\hbar$; (*d*) $12\hbar$.

Selection Rules: Allowed and Forbidden Transitions

Another prediction of quantum mechanics is that when a photon is emitted or absorbed, transitions can occur only between states with values of ℓ that differ by exactly one unit:

$$\Delta\ell = \pm 1.$$

According to this **selection rule**, an electron in an $\ell = 2$ state can jump only to a state with $\ell = 1$ or $\ell = 3$. It cannot jump to a state with $\ell = 2$ or $\ell = 0$.

A transition such as $\ell = 2$ to $\ell = 0$ is called a **forbidden transition**. Actually, such a transition is not absolutely forbidden and can occur, but only with very low probability compared to **allowed transitions**—those that satisfy the selection rule $\Delta\ell = \pm 1$. Since the orbital angular momentum of an H atom must change by one unit when it emits a photon, conservation of angular momentum tells us that the photon must carry off angular momentum. Indeed, experimental evidence of many sorts shows that the photon can be assigned a spin angular momentum of $1\hbar$.

39–3 Hydrogen Atom Wave Functions

The solution of the Schrödinger equation for the ground state of hydrogen (the state with lowest energy) has an energy $E_1 = -13.6\,\text{eV}$, as we have seen. The wave function for the ground state depends only on r and so is spherically symmetric. As already mentioned in Section 39–1, its form is

$$\psi_{100} = \frac{1}{\sqrt{\pi r_0^3}} e^{-\frac{r}{r_0}} \quad \textbf{(39–5a)}$$

where $r_0 = h^2\epsilon_0/\pi m e^2 = 0.0529\,\text{nm}$ is the Bohr radius (Section 37–11). The subscript 100 on ψ represents the quantum numbers n, ℓ, m_ℓ:

$$\psi_{n\ell m_\ell}.$$

For the ground state, $n = 1, \ell = 0, m_\ell = 0$, and there is only one wave function that serves for both $m_s = +\frac{1}{2}$ and $m_s = -\frac{1}{2}$ (the value of m_s does not affect the spatial dependence of the wave function for any state, since spin is an *internal* or *intrinsic* property of the electron). The probability density for the ground state is

$$|\psi_{100}|^2 = \frac{1}{\pi r_0^3} e^{-\frac{2r}{r_0}} \quad \textbf{(39–5b)}$$

which falls off exponentially with r. Note that ψ_{100}, like all other wave functions we discuss, has been normalized:

$$\int_{\text{all space}} |\psi_{100}|^2\, dV = 1.$$

The quantity $|\psi|^2\, dV$ gives the probability of finding the electron in a volume dV about a given point. It is often more useful to specify the **radial probability distribution**, P_r, which is defined so that $P_r\, dr$ is the probability of finding the electron at a radial distance between r and $r + dr$ from the nucleus. That is, $P_r\, dr$ specifies the probability of finding the electron within a thin shell of thickness dr of inner radius r and outer radius $r + dr$, regardless of direction (see Fig. 39–6). The volume of this shell is the product of its surface area, $4\pi r^2$, and its thickness, dr:

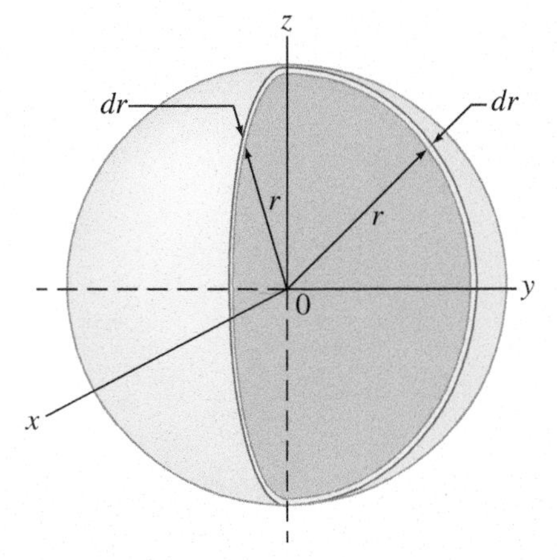

FIGURE 39–6 A spherical shell of thickness dr, inner radius r, and outer radius $r + dr$. Its volume is $dV = 4\pi r^2\, dr$.

$$dV = 4\pi r^2\, dr.$$

Hence

$$|\psi|^2\, dV = |\psi|^2 4\pi r^2\, dr$$

and the radial probability distribution is

$$P_r = 4\pi r^2 |\psi|^2. \quad \textbf{(39–6)}$$

For the ground state of hydrogen, P_r becomes

$$P_r = 4\frac{r^2}{r_0^3} e^{-\frac{2r}{r_0}} \quad \textbf{(39–7)}$$

and is plotted in Fig. 39–7. The peak of the curve is the "most probable" value of r and occurs for $r = r_0$, the Bohr radius, which we show in the following Example.

FIGURE 39–7 The radial probability distribution P_r for the ground state of hydrogen, $n = 1$, $\ell = 0$. The peak occurs at $r = r_0$, the Bohr radius.

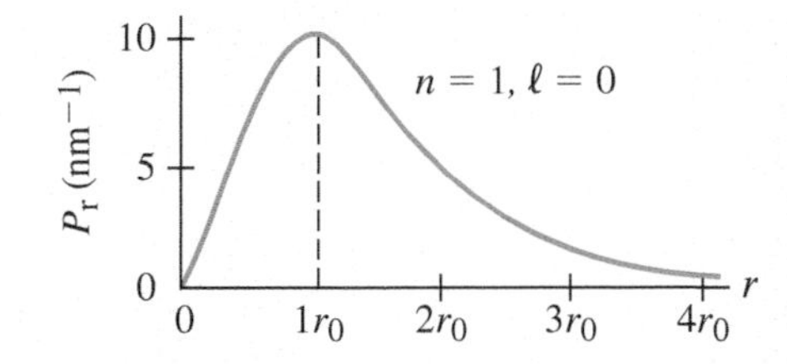

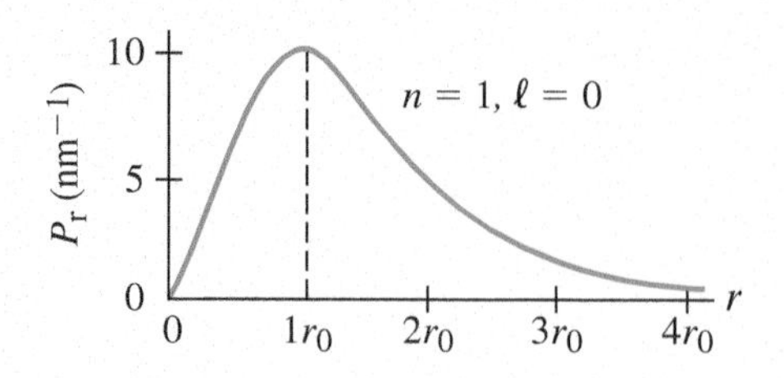

FIGURE 39–7 (Repeated.) The radial probability distribution P_r for the ground state of hydrogen, $n = 1$, $\ell = 0$. The peak occurs at $r = r_0$, the Bohr radius.

EXAMPLE 39–3 Most probable electron radius in hydrogen. Determine the most probable distance r from the nucleus at which to find the electron in the ground state of hydrogen.

APPROACH The peak of the curve in Fig. 39–7 corresponds to the most probable value of r. At this point the curve has zero slope, so we take the derivative of Eq. 39–7, set it equal to zero, and solve for r.

SOLUTION We find

$$\frac{d}{dr}\left(4\frac{r^2}{r_0^3}e^{-\frac{2r}{r_0}}\right) = 0$$

$$\left(8\frac{r}{r_0^3} - \frac{8r^2}{r_0^4}\right)e^{-\frac{2r}{r_0}} = 0.$$

Since $e^{-\frac{2r}{r_0}}$ goes to zero only at $r = \infty$, it is the term in parentheses that must be zero:

$$8\frac{r}{r_0^3} - 8\frac{r^2}{r_0^4} = 0.$$

Therefore,

$$\frac{r}{r_0^3} = \frac{r^2}{r_0^4}$$

or

$$r = r_0.$$

The most probable radial distance of the electron from the nucleus according to quantum mechanics is at the Bohr radius, an interesting coincidence.

EXERCISE C Return to the first Chapter-Opening Question, page 1044, and answer it again now. Try to explain why you may have answered differently the first time.

EXAMPLE 39–4 Calculating probability. Determine the probability of finding the electron in the ground state of hydrogen within two Bohr radii of the nucleus.

APPROACH We need to integrate P_r from $r = 0$ out to $r = 2r_0$.

SOLUTION We want to find

$$P = \int_{r=0}^{2r_0} |\psi|^2\, dV = \int_0^{2r_0} 4\frac{r^2}{r_0^3}e^{-\frac{2r}{r_0}}\, dr.$$

We first make the substitution

$$x = 2\frac{r}{r_0}$$

and then integrate by parts $\left(\int u\, dv = uv - \int v\, du\right)$ letting $u = x^2$ and $dv = e^{-x}\, dx$ (and note that $dx = 2\, dr/r_0$, and the upper limit is $x = 2(2r_0)/r_0 = 4$):

$$P = \tfrac{1}{2}\int_{x=0}^{4} x^2e^{-x}\, dx = \tfrac{1}{2}\left[-x^2e^{-x} + \int 2xe^{-x}\, dx\right]\Bigg|_0^4.$$

The second term we also integrate by parts with $u = 2x$ and $dv = e^{-x}\, dx$:

$$P = \tfrac{1}{2}\left[-x^2e^{-x} - 2xe^{-x} + 2\int e^{-x}\, dx\right]\Bigg|_0^4$$

$$= \left(-\tfrac{1}{2}x^2 - x - 1\right)e^{-x}\Big|_0^4.$$

We evaluate this at $x = 0$ and at $x = 4$:

$$P = (-8 - 4 - 1)e^{-4} + e^0 = 0.76$$

or 76%. Thus the electron would be found 76% of the time within 2 Bohr radii of the nucleus and 24% of the time farther away.

NOTE This result depends on our wave function being properly normalized, which it is, as is readily shown by letting $r \to \infty$ and integrating over all space: $\int_0^{\infty}|\psi|^2\, dV = 1$; that is, let the upper limit in the equation above be ∞.

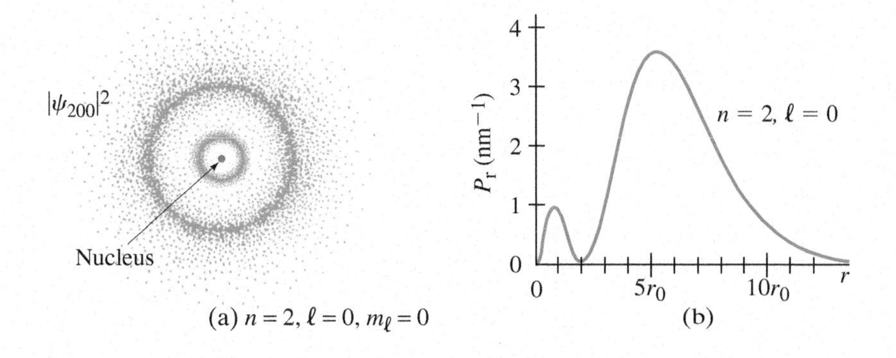

FIGURE 39–8 (a) Electron cloud, or probability distribution, for $n = 2$, $\ell = 0$ state in hydrogen. (b) The radial probability distribution P_r for the $n = 2$, $\ell = 0$ state in hydrogen.

The first excited state in hydrogen has $n = 2$. For $\ell = 0$, the solution of the Schrödinger equation (Eq. 39–1) is a wave function that is again spherically symmetric:

$$\psi_{200} = \frac{1}{\sqrt{32\pi r_0^3}}\left(2 - \frac{r}{r_0}\right)e^{-\frac{r}{2r_0}}. \qquad \textbf{(39–8)}$$

Figure 39–8a shows the probability distribution $|\psi_{200}|^2$ and Fig. 39–8b shows† a plot of the radial probability distribution

$$P_r = 4\pi r^2|\psi|^2 = \frac{1}{8}\frac{r^2}{r_0^3}\left(2 - \frac{r}{r_0}\right)^2 e^{-\frac{r}{r_0}}.$$

There are two peaks in this curve; the second, at $r \approx 5r_0$, is higher and corresponds to the most probable value for r in the $n = 2$, $\ell = 0$ state. We see that the electron tends to be somewhat farther from the nucleus in the $n = 2$, $\ell = 0$ state than in the $n = 1$, $\ell = 0$ state. (Compare to the Bohr model that gave $r_2 = 4r_0$.)

For the state with $n = 2$, $\ell = 1$, there are three possible wave functions, corresponding to $m_\ell = +1,\ 0,$ or -1:

$$\begin{aligned}\psi_{210} &= \frac{z}{\sqrt{32\pi r_0^5}}\,e^{-\frac{r}{2r_0}}\\ \psi_{211} &= \frac{x + iy}{\sqrt{64\pi r_0^5}}\,e^{-\frac{r}{2r_0}}\\ \psi_{21-1} &= \frac{x - iy}{\sqrt{64\pi r_0^5}}\,e^{-\frac{r}{2r_0}}\end{aligned} \qquad \textbf{(39–9)}$$

where i is the imaginary number $i = \sqrt{-1}$. These wave functions are *not* spherically symmetric. The probability distributions, $|\psi|^2$, are shown in Fig. 39–9a, where we can see their directional orientation.

You may wonder how such non-spherically symmetric wave functions arise when the potential energy in the Schrödinger equation has spherical symmetry. Indeed, how could an electron select one of these states? In the absence of any external influence, such as a magnetic field in a particular direction, all three of these states are equally likely, and they all have the same energy. Thus an electron can be considered to spend one-third of its time in each of these states. The net effect, then, is the sum of these three wave functions squared:

$$|\psi_{210}|^2 + |\psi_{211}|^2 + |\psi_{21-1}|^2,$$

which is spherically symmetric, since $x^2 + y^2 + z^2 = r^2$. The radial probability distribution for this sum is shown in Fig. 39–9b.

Although the spatial distributions of the electron can be calculated for the various states, it is difficult to measure them experimentally. Indeed, most of the experimental information about atoms has come from a careful examination of the emission spectra under various conditions.

FIGURE 39–9 (a) The probability distribution for the three states with $n = 2$, $\ell = 1$. (b) Radial probability distribution for the sum of the three states with $n = 2$, $\ell = 1$, and $m_\ell = +1,\ 0,$ or -1.

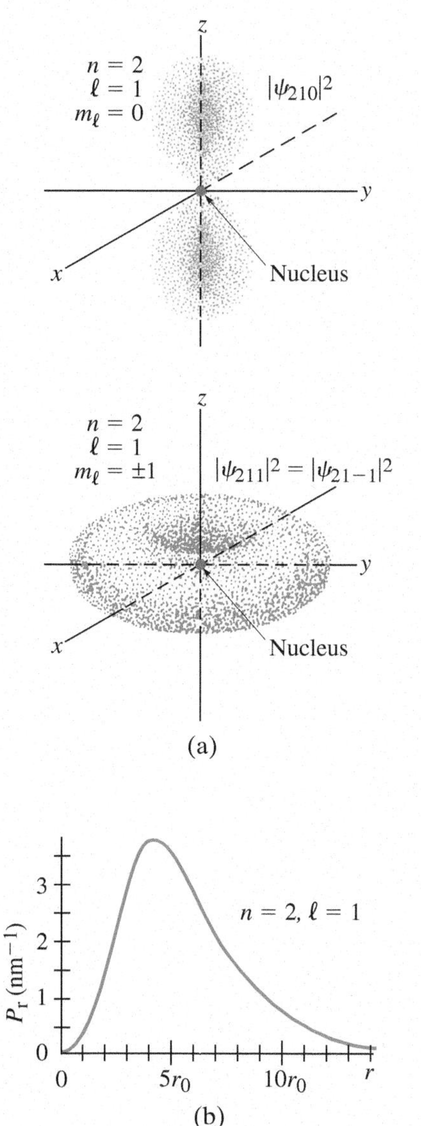

†Just as for a particle in a deep square well potential (see Figs. 38–9 and 38–10), the higher the energy, the more nodes there are in ψ and $|\psi|^2$ also for the H atom.

39–4 Complex Atoms; the Exclusion Principle

We have discussed the hydrogen atom in detail because it is the simplest to deal with. Now we briefly discuss more complex atoms, those that contain more than one electron. Their energy levels can be determined experimentally from an analysis of their emission spectra. The energy levels are *not* the same as in the H atom, since the electrons interact with each other as well as with the nucleus. Each electron in a complex atom still occupies a particular state characterized by the same quantum numbers n, ℓ, m_ℓ, and m_s. For atoms with more than one electron, the energy levels depend on both n and ℓ.

The number of electrons in a neutral atom is called its **atomic number**, Z; Z is also the number of positive charges (protons) in the nucleus, and determines what kind of atom it is. That is, Z determines the fundamental properties that distinguish one type of atom from another.

Quantum mechanics in the years after 1925 proved successful also in dealing with complex atoms. The mathematics becomes very difficult, however, since in multi-electron atoms, each electron is not only attracted to the nucleus but is also repelled by the other electrons.

To understand the possible arrangements of electrons in an atom, a new principle was needed. It was introduced by Wolfgang Pauli (1900–1958; Fig. 38–2) and is called the **Pauli exclusion principle**. It states:

No two electrons in an atom can occupy the same quantum state.

Thus, no two electrons in an atom can have exactly the same set of the quantum numbers n, ℓ, m_ℓ, and m_s. The Pauli exclusion principle forms the basis not only for understanding complex atoms, but also for understanding molecules and bonding, and other phenomena as well.

Let us now look at the structure of some of the simpler atoms when they are in the ground state. After hydrogen, the next simplest atom is *helium* with two electrons. Both electrons can have $n = 1$, since one can have spin up $(m_s = +\frac{1}{2})$ and the other spin down $(m_s = -\frac{1}{2})$, thus satisfying the exclusion principle. Since $n = 1$, then ℓ and m_ℓ must be zero (Table 39–1, p. 1047). Thus the two electrons have the quantum numbers indicated in Table 39–2.

Lithium has three electrons, two of which can have $n = 1$. But the third cannot have $n = 1$ without violating the exclusion principle. Hence the third electron must have $n = 2$. It happens that the $n = 2, \ell = 0$ level has a lower energy than $n = 2, \ell = 1$. So the electrons in the ground state have the quantum numbers indicated in Table 39–2. The quantum numbers of the third electron could also be, say, $(n, \ell, m_\ell, m_s) = (3, 1, -1, \frac{1}{2})$. But the atom in this case would be in an excited state since it would have greater energy. It would not be long before it jumped to the ground state with the emission of a photon. At room temperature, unless extra energy is supplied (as in a discharge tube), the vast majority of atoms are in the ground state.

We can continue in this way to describe the quantum numbers of each electron in the ground state of larger and larger atoms. The quantum numbers for sodium, with its eleven electrons, are shown in Table 39–2.

Figure 39–10 shows a simple energy level diagram where occupied states are shown as up or down arrows ($m_s = +\frac{1}{2}$ or $-\frac{1}{2}$), and possible empty states are shown as a small circle.

TABLE 39–2 Ground-State Quantum Numbers

Helium, $Z = 2$

n	ℓ	m_ℓ	m_s
1	0	0	$\frac{1}{2}$
1	0	0	$-\frac{1}{2}$

Lithium, $Z = 3$

n	ℓ	m_ℓ	m_s
1	0	0	$\frac{1}{2}$
1	0	0	$-\frac{1}{2}$
2	0	0	$\frac{1}{2}$

Sodium, $Z = 11$

n	ℓ	m_ℓ	m_s
1	0	0	$\frac{1}{2}$
1	0	0	$-\frac{1}{2}$
2	0	0	$\frac{1}{2}$
2	0	0	$-\frac{1}{2}$
2	1	1	$\frac{1}{2}$
2	1	1	$-\frac{1}{2}$
2	1	0	$\frac{1}{2}$
2	1	0	$-\frac{1}{2}$
2	1	−1	$\frac{1}{2}$
2	1	−1	$-\frac{1}{2}$
3	0	0	$\frac{1}{2}$

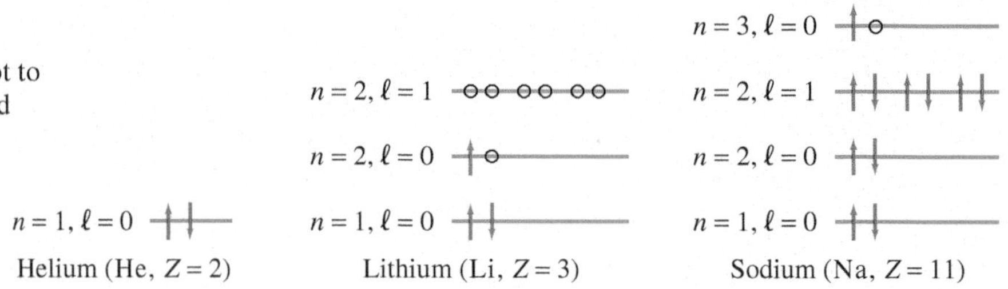

$n = 1, \ell = 0$
Helium (He, $Z = 2$)

$n = 2, \ell = 1$
$n = 2, \ell = 0$
$n = 1, \ell = 0$
Lithium (Li, $Z = 3$)

$n = 3, \ell = 0$
$n = 2, \ell = 1$
$n = 2, \ell = 0$
$n = 1, \ell = 0$
Sodium (Na, $Z = 11$)

FIGURE 39–10 Energy level diagram (not to scale) showing occupied states (arrows) and unoccupied states (o) for the ground states of He, Li, and Na. Note that we have shown the $n = 2$, $\ell = 1$ level of Li even though it is empty.

The ground-state configuration for all atoms is given in the Periodic Table, which is displayed inside the back cover of this book, and discussed in the next Section.

EXERCISE D Construct a Table of the ground-state quantum numbers for beryllium, $Z = 4$ (like those in Table 39–2).

The exclusion principle applies to identical particles whose spin quantum number is a half-integer ($\frac{1}{2}, \frac{3}{2}$, and so on), including electrons, protons, and neutrons; such particles are called **fermions**, after Enrico Fermi who derived a statistical theory describing them. A basic assumption is that all electrons are **identical**, indistinguishable one from another. Similarly, all protons are identical, all neutrons are identical, and so on. The exclusion principle does not apply to particles with integer spin (0, 1, 2, and so on), such as the photon and π meson, all of which are referred to as **bosons** (after Satyendranath Bose, who derived a statistical theory for them).

EXERCISE E Return to the second Chapter-Opening Question, page 1044, and answer it again now. Try to explain why you may have answered differently the first time.

TABLE 39–3 Value of ℓ

Value of ℓ	Letter Symbol	Maximum Number of Electrons in Subshell
0	*s*	2
1	*p*	6
2	*d*	10
3	*f*	14
4	*g*	18
5	*h*	22
⋮	⋮	⋮

39–5 Periodic Table of Elements

More than a century ago, Dmitri Mendeleev (1834–1907) arranged the (then) known elements into what we now call the **Periodic Table** of the elements. The atoms were arranged according to increasing mass, but also so that elements with similar chemical properties would fall in the same column. Today's version is shown inside the back cover of this book. Each square contains the atomic number Z, the symbol for the element, and the atomic mass (in atomic mass units). Finally, the lower left corner shows the configuration of the ground state of the atom. This requires some explanation. Electrons with the same value of n are referred to as being in the same **shell**. Electrons with $n = 1$ are in one shell (the K shell), those with $n = 2$ are in a second shell (the L shell), those with $n = 3$ are in the third (M) shell, and so on. Electrons with the same values of n and ℓ are referred to as being in the same **subshell**. Letters are often used to specify the value of ℓ as shown in Table 39–3. That is, $\ell = 0$ is the *s* subshell; $\ell = 1$ is the *p* subshell; $\ell = 2$ is the *d* subshell; beginning with $\ell = 3$, the letters follow the alphabet, *f*, *g*, *h*, *i*, and so on. (The first letters *s*, *p*, *d*, and *f* were originally abbreviations of "sharp," "principal," "diffuse," and "fundamental," experimental terms referring to the spectra.)

The Pauli exclusion principle limits the number of electrons possible in each shell and subshell. For any value of ℓ, there are $2\ell + 1$ possible m_ℓ values (m_ℓ can be any integer from 1 to ℓ, from -1 to $-\ell$, or zero), and two possible m_s values. There can be, therefore, at most $2(2\ell + 1)$ electrons in any ℓ subshell. For example, for $\ell = 2$, five m_ℓ values are possible $(2, 1, 0, -1, -2)$, and for each of these, m_s can be $+\frac{1}{2}$ or $-\frac{1}{2}$ for a total of $2(5) = 10$ states. Table 39–3 lists the maximum number of electrons that can occupy each subshell.

Since the energy levels depend almost entirely on the values of n and ℓ, it is customary to specify the electron configuration simply by giving the n value and the appropriate letter for ℓ, with the number of electrons in each subshell given as a superscript. The ground-state configuration of sodium, for example, is written as $1s^2 2s^2 2p^6 3s^1$. This is simplified in the Periodic Table by specifying the configuration only of the outermost electrons and any other nonfilled subshells (see Table 39–4 here, and the Periodic Table inside the back cover).

TABLE 39–4 Electron Configuration of Some Elements

Z (Number of Electrons)	Element†	Ground State Configuration (outer electrons)
1	H	$1s^1$
2	He	$1s^2$
3	Li	$2s^1$
4	Be	$2s^2$
5	B	$2s^2 2p^1$
6	C	$2s^2 2p^2$
7	N	$2s^2 2p^3$
8	O	$2s^2 2p^4$
9	F	$2s^2 2p^5$
10	Ne	$2s^2 2p^6$
11	Na	$3s^1$
12	Mg	$3s^2$
13	Al	$3s^2 3p^1$
14	Si	$3s^2 3p^2$
15	P	$3s^2 3p^3$
16	S	$3s^2 3p^4$
17	Cl	$3s^2 3p^5$
18	Ar	$3s^2 3p^6$
19	K	$4s^1$
20	Ca	$4s^2$
21	Sc	$3d^1 4s^2$
22	Ti	$3d^2 4s^2$
23	V	$3d^3 4s^2$
24	Cr	$3d^5 4s^1$
25	Mn	$3d^5 4s^2$
26	Fe	$3d^6 4s^2$

† Names of elements can be found in Appendix F.

CONCEPTUAL EXAMPLE 39–5 **Electron configurations.** Which of the following electron configurations are possible, and which are not: (*a*) $1s^2 2s^2 2p^6 3s^3$; (*b*) $1s^2 2s^2 2p^6 3s^2 3p^5 4s^2$; (*c*) $1s^2 2s^2 2p^6 2d^1$?

RESPONSE (*a*) This is not allowed, because too many electrons (three) are shown in the *s* subshell of the M ($n = 3$) shell. The *s* subshell has $m_\ell = 0$, with two slots only, for "spin up" and "spin down" electrons. (*b*) This is allowed, but it is an excited state. One of the electrons from the 3*p* subshell has jumped up to the 4*s* subshell. Since there are 19 electrons, the element is potassium. (*c*) This is not allowed, because there is no *d* ($\ell = 2$) subshell in the $n = 2$ shell (Table 39–1). The outermost electron will have to be (at least) in the $n = 3$ shell.

EXERCISE F Write the complete ground-state configuration for gallium, with its 31 electrons.

The grouping of atoms in the Periodic Table is according to increasing atomic number, *Z*. There is also a strong regularity according to chemical properties. Although this is treated in chemistry textbooks, we discuss it here briefly because it is a result of quantum mechanics. See the Periodic Table inside the back cover.

All the **noble gases** (in column VIII of the Periodic Table) have completely filled shells or subshells. That is, their outermost subshell is completely full, and the electron distribution is spherically symmetric. With such full spherical symmetry, other electrons are not attracted nor are electrons readily lost (ionization energy is high). This is why the noble gases are chemically inert (more on this when we discuss molecules and bonding in Chapter 40). Column VII contains the **halogens**, which lack one electron from a filled shell. Because of the shapes of the orbits (see Section 40–1), an additional electron can be accepted from another atom, and hence these elements are quite reactive. They have a valence of -1, meaning that when an extra electron is acquired, the resulting ion has a net charge of $-1e$. Column I of the Periodic Table contains the **alkali metals**, all of which have a single outer *s* electron. This electron spends most of its time outside the inner closed shells and subshells which shield it from most of the nuclear charge. Indeed, it is relatively far from the nucleus and is attracted to it by a net charge of only about $+1e$, because of the shielding effect of the other electrons. Hence this outer electron is easily removed and can spend much of its time around another atom, forming a molecule. This is why the alkali metals are chemically active and have a valence of $+1$. The other columns of the Periodic Table can be treated similarly.

The presence of the **transition elements** in the center of the Table, as well as the lanthanides (rare earths) and actinides below, is a result of incomplete inner shells. For the lowest *Z* elements, the subshells are filled in a simple order: first 1*s*, then 2*s*, followed by 2*p*, 3*s*, and 3*p*. You might expect that 3*d* ($n = 3, \ell = 2$) would be filled next, but it isn't. Instead, the 4*s* level actually has a slightly lower energy than the 3*d* (due to electrons interacting with each other), so it fills first (K and Ca). Only then does the 3*d* shell start to fill up, beginning with Sc, as can be seen in Table 39–4. (The 4*s* and 3*d* levels are close, so some elements have only one 4*s* electron, such as Cr.) Most of the chemical properties of these transition elements are governed by the relatively loosely held 4*s* electrons, and hence they usually have valences of $+1$ or $+2$. A similar effect is responsible for the *lanthanides* and *actinides*, which are shown at the bottom of the Periodic Table for convenience. All have very similar chemical properties, which are determined by their two outer 6*s* or 7*s* electrons, whereas the different numbers of electrons in the unfilled inner shells have little effect.

39–6 X-Ray Spectra and Atomic Number

The line spectra of atoms in the visible, UV, and IR regions of the EM spectrum are mainly due to transitions between states of the outer electrons. Much of the charge of the nucleus is shielded from these electrons by the negative charge on the inner electrons. But the innermost electrons in the $n = 1$ shell "see" the full charge of the nucleus. Since the energy of a level is proportional to Z^2 (see Eq. 37–14), for an atom with $Z = 50$, we would expect wavelengths about $50^2 = 2500$ times shorter than those found in the Lyman series of hydrogen (around 100 nm), or 10^{-2} to 10^{-1} nm. Such short wavelengths lie in the X-ray region of the spectrum.

X-rays are produced when electrons accelerated by a high voltage strike the metal target inside the X-ray tube (Section 35–10). If we look at the spectrum of wavelengths emitted by an X-ray tube, we see that the spectrum consists of two parts: a continuous spectrum with a cutoff at some λ_0 which depends only on the voltage across the tube, and a series of peaks superimposed. A typical example is shown in Fig. 39–11. The smooth curve and the cutoff wavelength λ_0 move to the left as the voltage across the tube increases. The sharp lines or peaks (labeled K_α and K_β in Fig. 39–11), however, remain at the same wavelength when the voltage is changed, although they are located at different wavelengths when different target materials are used. This observation suggests that the peaks are characteristic of the target material used. Indeed, we can explain the peaks by imagining that the electrons accelerated by the high voltage of the tube can reach sufficient energies that when they collide with the atoms of the target, they can knock out one of the very tightly held inner electrons. Then we explain these **characteristic X-rays** (the peaks in Fig. 39–11) as photons emitted when an electron in an upper state drops down to fill the vacated lower state. The K lines result from transitions *into* the K shell ($n = 1$). The K_α line consists of photons emitted in a transition that originates from the $n = 2$ (L) shell and drops to the $n = 1$ (K) shell, whereas the K_β line reflects a transition from the $n = 3$ (M) shell down to the K shell. An L line, on the other hand, is due to a transition into the L shell, and so on.

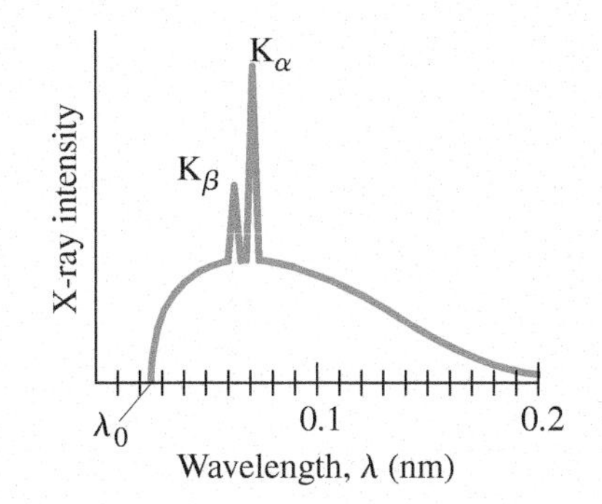

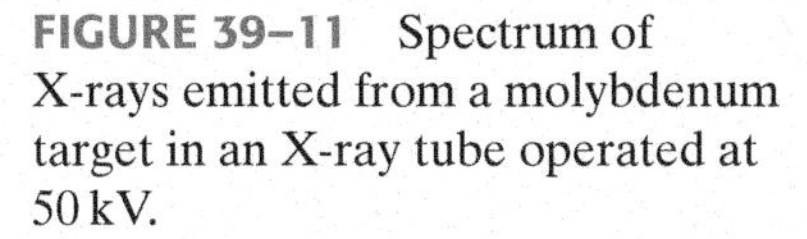
FIGURE 39–11 Spectrum of X-rays emitted from a molybdenum target in an X-ray tube operated at 50 kV.

Measurement of the characteristic X-ray spectra has allowed a determination of the inner energy levels of atoms. It has also allowed the determination of Z values for many atoms, since (as we have seen) the wavelength of the shortest characteristic X-rays emitted will be inversely proportional to Z^2. Actually, for an electron jumping from, say, the $n = 2$ to the $n = 1$ level (K_α line), the wavelength is inversely proportional to $(Z - 1)^2$ because the nucleus is shielded by the one electron that still remains in the 1s level. In 1914, H. G. J. Moseley (1887–1915) found that a plot of $\sqrt{1/\lambda}$ vs. Z produced a straight line, Fig. 39–12 where λ is the wavelength of the K_α line. The Z values of a number of elements were determined by fitting them to such a **Moseley plot**. The work of Moseley put the concept of atomic number on a firm experimental basis.

FIGURE 39–12 Plot of $\sqrt{1/\lambda}$ vs. Z for K_α X-ray lines.

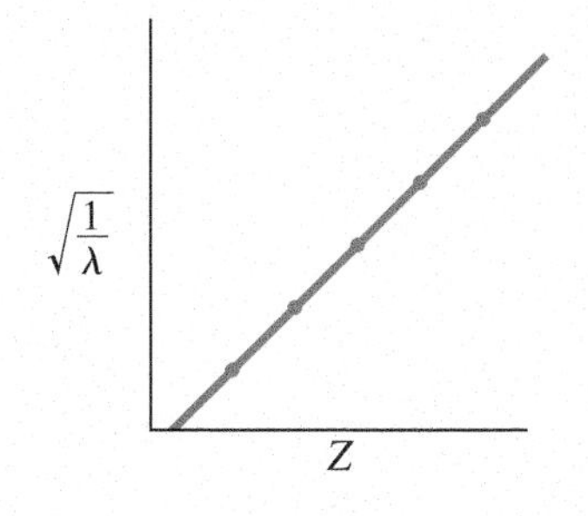

EXAMPLE 39–6 X-ray wavelength. Estimate the wavelength for an $n = 2$ to $n = 1$ transition in molybdenum ($Z = 42$). What is the energy of such a photon?

APPROACH We use the Bohr formula, Eq. 37–15 for $1/\lambda$, with Z^2 replaced by $(Z - 1)^2 = (41)^2$.

SOLUTION Equation 37–15 gives

$$\frac{1}{\lambda} = \left(\frac{e^4 m}{8\epsilon_0^2 h^3 c}\right)(Z - 1)^2\left(\frac{1}{n'^2} - \frac{1}{n^2}\right)$$

where $n = 2$ and $n' = 1$. We substitute in values:

$$\frac{1}{\lambda} = (1.097 \times 10^7\,\mathrm{m}^{-1})(41)^2\left(\frac{1}{1} - \frac{1}{4}\right) = 1.38 \times 10^{10}\,\mathrm{m}^{-1}.$$

So

$$\lambda = \frac{1}{1.38 \times 10^{10}\,\mathrm{m}^{-1}} = 0.072\,\mathrm{nm}.$$

This is close to the measured value (Fig. 39–11) of 0.071 nm. Each of these photons would have energy (in eV) of:

$$E = hf = \frac{hc}{\lambda} = \frac{(6.63 \times 10^{-34}\,\mathrm{J\cdot s})(3.00 \times 10^8\,\mathrm{m/s})}{(7.2 \times 10^{-11}\,\mathrm{m})(1.60 \times 10^{-19}\,\mathrm{J/eV})} = 17\,\mathrm{keV}.$$

The denominator includes the conversion factor from joules to eV.

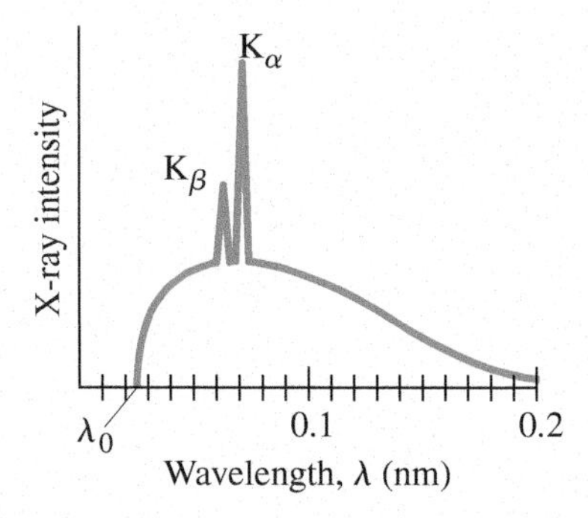

FIGURE 39–11 (Repeated.) Spectrum of X-rays emitted from a molybdenum target in an X-ray tube operated at 50 kV.

EXAMPLE 39–7 **Determining atomic number.** High-energy electrons are used to bombard an unknown material. The strongest peak is found for X-rays emitted with an energy of 7.5 keV. Guess what the material is.

APPROACH The highest intensity X-rays are generally for the K_α line (see Fig. 39–11) which occurs when high-energy external electrons knock out K shell electrons (the innermost orbit, $n = 1$) and their place is taken by electrons from the L shell $(n = 2)$. We use the Bohr model, and assume the electrons "see" a nuclear charge of $Z - 1$ (screened by one electron).

SOLUTION The hydrogen transition $n = 2$ to $n = 1$ would yield about 10.2 eV (see Fig. 37–26 or Example 37–13). Energy E is proportional to Z^2 (Eq. 37–14), or rather $(Z - 1)^2$ because the nucleus is shielded by the one electron in a 1s state (see above), so we can use ratios:

$$\frac{(Z-1)^2}{1^2} = \frac{7500\text{ eV}}{10.2\text{ eV}} = 735,$$

so $Z - 1 = \sqrt{735} = 27$, and $Z = 28$, which makes it cobalt.

Now we briefly analyze the continuous part of an X-ray spectrum (Fig. 39–11) based on the photon theory of light. When electrons strike the target, they collide with atoms of the material and give up most of their energy as heat (about 99%, so X-ray tubes must be cooled). Electrons can also give up energy by emitting a photon: an electron decelerated by interaction with atoms of the target (Fig. 39–13) emits radiation because of its deceleration (Chapter 31), and in this case it is called **bremsstrahlung** (German for "braking radiation"). Because energy is conserved, the energy of the emitted photon, hf, must equal the loss of kinetic energy of the electron, $\Delta K = K - K'$, so

$$hf = \Delta K.$$

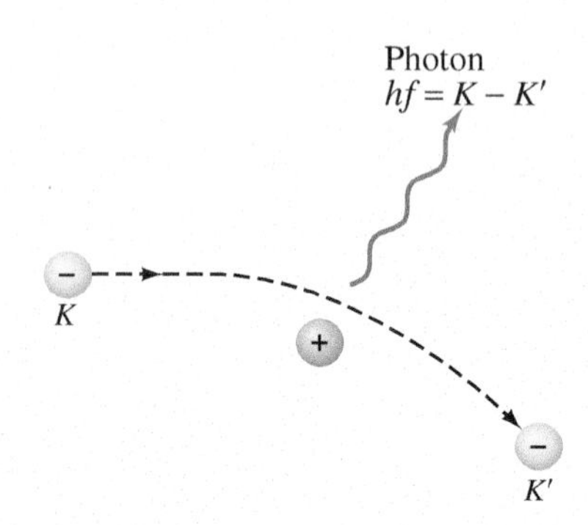

FIGURE 39–13 Bremsstrahlung photon produced by an electron decelerated by interaction with a target atom.

An electron may lose all or a part of its energy in such a collision. The continuous X-ray spectrum (Fig. 39–11) is explained as being due to such bremsstrahlung collisions in which varying amounts of energy are lost by the electrons. The shortest-wavelength X-ray (the highest frequency) must be due to an electron that gives up *all* its kinetic energy to produce one photon in a single collision. Since the initial kinetic energy of an electron is equal to the energy given it by the accelerating voltage, V, then $K = eV$. In a single collision in which the electron is brought to rest $(K' = 0)$, then $\Delta K = eV$ and

$$hf_0 = eV.$$

We set $f_0 = c/\lambda_0$ where λ_0 is the cutoff wavelength (Fig. 39–11) and find

$$\lambda_0 = \frac{hc}{eV}. \qquad \textbf{(39–10)}$$

This prediction for λ_0 corresponds precisely with that observed experimentally. This result is further evidence that X-rays are a form of electromagnetic radiation (light)† and that the photon theory of light is valid.

EXAMPLE 39–8 **Cutoff wavelength.** What is the shortest-wavelength X-ray photon emitted in an X-ray tube subjected to 50 kV?

APPROACH The electrons striking the target will have a kinetic energy of 50 keV. The shortest-wavelength photons are due to collisions in which all of the electron's kinetic energy is given to the photon so $K = eV = hf_0$.

SOLUTION From Eq. 39–10,

$$\lambda_0 = \frac{hc}{eV} = \frac{(6.63 \times 10^{-34}\,\text{J}\cdot\text{s})(3.0 \times 10^8\,\text{m/s})}{(1.6 \times 10^{-19}\,\text{C})(5.0 \times 10^4\,\text{V})} = 2.5 \times 10^{-11}\,\text{m},$$

or 0.025 nm.

NOTE This result agrees well with experiment, Fig. 39–11.

†If X-rays were not photons but rather neutral particles with mass m, Eq. 39–10 would not hold.

*39–7 Magnetic Dipole Moment; Total Angular Momentum

*Magnetic Dipole Moment and the Bohr Magneton

An electron orbiting the nucleus of an atom can be considered as a current loop, classically, and thus might be expected to have a **magnetic dipole moment** as discussed in Chapter 27. Indeed, in Example 27–12 we did a classical calculation of the magnetic dipole moment of the electron in the ground state of hydrogen based, essentially, on the Bohr model, and found it to give

$$\mu = IA = \tfrac{1}{2}evr.$$

Here v is the orbital velocity of the electron, and for a particle moving in a circle of radius r, its angular momentum is

$$L = mvr.$$

So we can write

$$\mu = \frac{1}{2}\frac{e}{m}L.$$

The direction of the angular momentum $\vec{\mathbf{L}}$ is perpendicular to the plane of the current loop. So is the direction of the magnetic dipole moment vector $\vec{\boldsymbol{\mu}}$, although in the opposite direction since the electron's charge is negative. Hence we can write the vector equation

$$\vec{\boldsymbol{\mu}} = -\frac{1}{2}\frac{e}{m}\vec{\mathbf{L}}. \qquad \textbf{(39–11)}$$

This rough semiclassical derivation was based on the Bohr theory. The same result (Eq. 39–11) is obtained using quantum mechanics. Since $\vec{\mathbf{L}}$ is quantized in quantum mechanics, the magnetic dipole moment, too, must be quantized. The magnitude of the dipole moment is given by (see Eq. 39–3)

$$\mu = \frac{e\hbar}{2m}\sqrt{\ell(\ell+1)}.$$

When a magnetic dipole moment is in a magnetic field $\vec{\mathbf{B}}$, it experiences a torque as we saw in Section 27–5, and the potential energy U of such a system depends on $\vec{\mathbf{B}}$ and the orientation of $\vec{\boldsymbol{\mu}}$ relative to $\vec{\mathbf{B}}$ (Eq. 27–12):

$$U = -\vec{\boldsymbol{\mu}} \cdot \vec{\mathbf{B}}.$$

If the magnetic field $\vec{\mathbf{B}}$ is in the z direction, then

$$U = -\mu_z B_z$$

and from Eq. 39–4 $(L_z = m_\ell \hbar)$ and Eq. 39–11, we have

$$\mu_z = -\frac{e\hbar}{2m}m_\ell.$$

(Be careful here not to confuse the electron mass m with the magnetic quantum number, m_ℓ.) It is useful to define the quantity

$$\mu_{\mathrm{B}} = \frac{e\hbar}{2m} \qquad \textbf{(39–12)}$$

which is called the **Bohr magneton** and has the value $\mu_{\mathrm{B}} = 9.27 \times 10^{-24}\,\mathrm{J/T}$ (joule/tesla). Then we can write

$$\mu_z = -\mu_{\mathrm{B}} m_\ell, \qquad \textbf{(39–13)}$$

where m_ℓ has integer values from 0 to $\pm\ell$ (see Table 39–1). An atom placed in a magnetic field would have its energy split into levels that differ by $\Delta U = \mu_{\mathrm{B}} B$; this is the *Zeeman effect*, and was shown in Fig. 39–4.

* Stern-Gerlach Experiment and the g-Factor for Electron Spin

The first evidence for the splitting of atomic energy levels, known as *space quantization* (Section 39–2), came in 1922 in a famous experiment known as the **Stern-Gerlach experiment**. Silver atoms (and later others) were heated in an oven from which they escaped as shown in Fig. 39–14. The atoms were made to pass through a

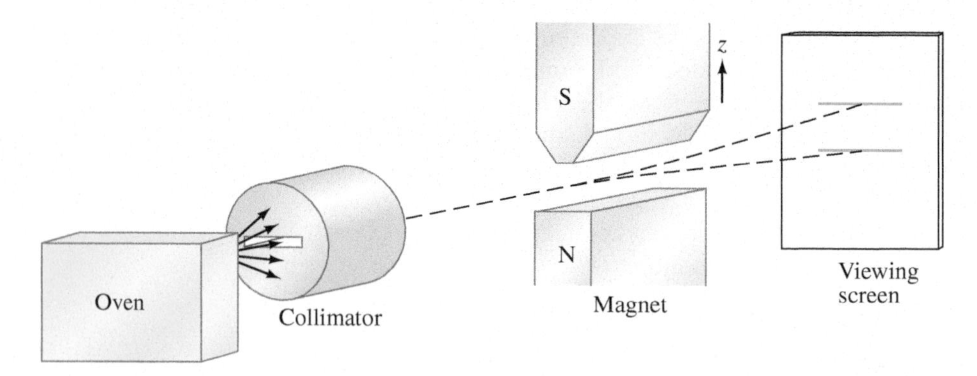

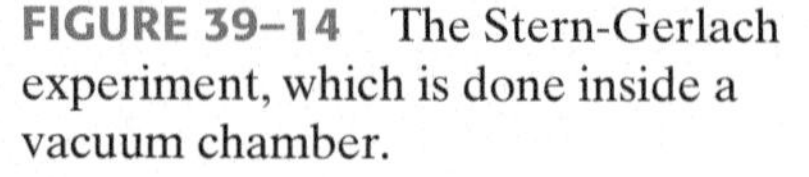

FIGURE 39–14 The Stern-Gerlach experiment, which is done inside a vacuum chamber.

collimator, which eliminated all but a narrow beam. The beam then passed into a *nonhomogeneous* magnetic field. The field was deliberately made nonhomogeneous so that it would exert a force on atomic magnetic moments: remember that the potential energy (in this case $-\vec{\boldsymbol{\mu}} \cdot \vec{\mathbf{B}}$) must change in space if there is to be a force ($F_x = -dU/dx$, etc., Section 8–2). If $\vec{\mathbf{B}}$ has a gradient along the z axis, as in Fig. 39–14, then the force is along z:

$$F_z = -\frac{dU}{dz} = \mu_z \frac{dB_z}{dz}.$$

Thus the silver atoms would be deflected up or down depending on the value of μ for each atom. Classically, we would expect to see a continuous distribution on the viewing screen, since we would expect the atoms to have randomly oriented magnetic moments. But Stern and Gerlach saw instead two distinct lines for silver (and for other atoms sometimes more than two lines). These observations were the first evidence for space quantization, though not fully explained until a few years later. If the lines were due to orbital angular momentum, there should have been an odd number of them, corresponding to the possible values of m_ℓ (since $\mu_z = -\mu_B m_\ell$). For $\ell = 0$, there is only one possibility, $m_\ell = 0$. For $\ell = 1$, m_ℓ can be $1, 0$, or -1, and we would expect three lines, and so on. Why there are only two lines was eventually explained by the concept of electron spin. With a spin of $\frac{1}{2}$, the electron spin can have only two orientations in space, as we saw in Fig. 39–5. Hence a magnetic dipole moment associated with spin would have only two positions. Thus, the two states for silver seen in the Stern-Gerlach experiment must be due to the spin of its one valence electron. Silver atoms must thus have zero orbital angular momentum but a total spin of $\frac{1}{2}$ due to this one valence electron. (Of silver's 47 electrons, the spins of the first 46 cancel.) For the H atom in its ground state, again only two lines were seen on the screen of Fig. 39–14: due to the spin $\frac{1}{2}$ of its electron since the orbital angular momentum is zero.

The Stern-Gerlach deflection is proportional to the magnetic dipole moment, μ_z, and for a spin $\frac{1}{2}$ particle we expect $\mu_z = -\mu_B m_s = -(\frac{1}{2})(e\hbar/2m)$ as for the case of orbital angular momentum, Eq. 39–13. Instead, μ_z for spin was found to be about twice as large:

$$\mu_z = -g\mu_B m_s, \qquad \text{[electron spin]} \quad \textbf{(39–14)}$$

where g, called the **g-factor** or **gyromagnetic ratio**, has been measured to be slightly larger than 2: $g = 2.0023\cdots$ for a free electron. This unexpected factor of (about) 2 clearly indicates that spin cannot be viewed as a classical angular momentum. It is a purely quantum-mechanical effect. Equation 39–14 is the same as Eq. 39–13 for orbital angular momentum with m_s replacing m_ℓ. But for the orbital case, $g = 1$.

*Total Angular Momentum $\vec{\mathbf{J}}$

An atom can have both orbital and spin angular momenta. For example, in the $2p$ state of hydrogen $\ell = 1$ and $s = \frac{1}{2}$. In the $4d$ state, $\ell = 2$ and $s = \frac{1}{2}$. The **total angular momentum** is the vector sum of the orbital angular momentum $\vec{\mathbf{L}}$ and the spin $\vec{\mathbf{S}}$:

$$\vec{\mathbf{J}} = \vec{\mathbf{L}} + \vec{\mathbf{S}}.$$

According to quantum mechanics, the magnitude of the total angular momentum $\vec{\mathbf{J}}$ is quantized:

$$J = \sqrt{j(j+1)}\,\hbar. \tag{39–15}$$

For the single electron in the H atom, quantum mechanics gives the result that j can be

$$j = \ell + s = \ell + \tfrac{1}{2}$$

or

$$j = \ell - s = \ell - \tfrac{1}{2}$$

but never less than zero, just as for ℓ and s. For the 1s state, $\ell = 0$ and $j = \frac{1}{2}$ is the only possibility. For p states, say the 2p state, $\ell = 1$ and j can be either $\frac{3}{2}$ or $\frac{1}{2}$. The z component for j is quantized in the usual way:

$$m_j = j, j-1, \cdots, -j.$$

For a 2p state with $j = \frac{1}{2}$, m_j can be $\frac{1}{2}$ or $-\frac{1}{2}$; for $j = \frac{3}{2}$, m_j can be $\frac{3}{2}, \frac{1}{2}, -\frac{1}{2}, -\frac{3}{2}$, for a total of four states. Note that the state of a single electron can be specified by giving n, ℓ, m_ℓ, m_s, or by giving n, j, ℓ, m_j (only one of these descriptions at a time).

EXERCISE G What are the possibilities for j in the 3d state of hydrogen? (*a*) $\frac{3}{2}, \frac{1}{2}$; (*b*) $\frac{5}{2}, \frac{3}{2}, \frac{1}{2}$; (*c*) $\frac{7}{2}, \frac{5}{2}, \frac{3}{2}, \frac{1}{2}$; (*d*) $\frac{5}{2}, \frac{3}{2}$; (*e*) $\frac{7}{2}, \frac{5}{2}$.

The interaction of magnetic fields with atoms, as in the Zeeman effect and the Stern-Gerlach experiment, involves the *total* angular momentum. Thus the Stern-Gerlach experiment on H atoms in the ground state shows two lines (for $m_j = +\frac{1}{2}$ and $-\frac{1}{2}$), but for the first excited state it shows four lines corresponding to the four possible m_j values ($\frac{3}{2}, \frac{1}{2}, -\frac{1}{2}, -\frac{3}{2}$).

*Spectroscopic Notation

We can specify the state of an atom, including the total angular momentum quantum number j, using the following **spectroscopic notation**. For a single electron state we can write

$$nL_j,$$

where the value of L (the orbital quantum number) is specified using the same letters as in Table 39–3, but in upper case:

L =	0	1	2	3	4	⋯
letter =	S	P	D	F	G	⋯.

So the $2P_{3/2}$ state has $n = 2$, $\ell = 1$, $j = \frac{3}{2}$, whereas $1S_{1/2}$ specifies the ground state in hydrogen.

* Fine Structure; Spin-Orbit Interaction

A magnetic effect also produces the *fine structure* splitting mentioned in Section 39–2, but it occurs in the absence of any external field. Instead, it is due to a magnetic field produced by the atom itself. We can see how it occurs by putting ourselves in the reference frame of the electron, in which case we see the nucleus revolving about us as a moving charge or electric current that produces a magnetic field, B_n. The electron has an intrinsic magnetic dipole moment μ_s (Eq. 39–14) and hence its energy will be altered by an amount (Eq. 27–12)

$$\Delta U = -\vec{\mu}_s \cdot \vec{\mathbf{B}}_n .$$

Since μ_s takes on quantized values according to the values of m_s, the energy of a single electron state will split into two closely spaced energy levels (for $m_s = +\frac{1}{2}$ and $-\frac{1}{2}$). This tiny splitting of energy levels produces a tiny splitting in spectral lines. For example, in the H atom, the $2P \rightarrow 1S$ transition is split into two lines corresponding to $2P_{1/2} \rightarrow 1S_{1/2}$ and $2P_{3/2} \rightarrow 1S_{1/2}$. The difference in energy between these two is only about $5 \times 10^{-5}\,\mathrm{eV}$, which is very small compared to the $2P \rightarrow 1S$ transition energy of $13.6\,\mathrm{eV} - 3.4\,\mathrm{eV} = 10.2\,\mathrm{eV}$.

The magnetic field $\vec{\mathbf{B}}_n$ produced by the orbital motion is proportional to the orbital angular momentum $\vec{\mathbf{L}}$, and since $\vec{\mu}_s$ is proportional to the spin $\vec{\mathbf{S}}$, then $\Delta U = -\vec{\mu}_s \cdot \vec{\mathbf{B}}_n$ can be written

$$\Delta U \propto \vec{\mathbf{L}} \cdot \vec{\mathbf{S}}.$$

This interaction, which produces the fine structure, is thus called the **spin-orbit interaction**. Its magnitude is related to a dimensionless constant known as the **fine structure constant**,

$$\alpha = \frac{e^2}{2\epsilon_0 hc} \approx \frac{1}{137},$$

which also appears elsewhere in atomic physics.

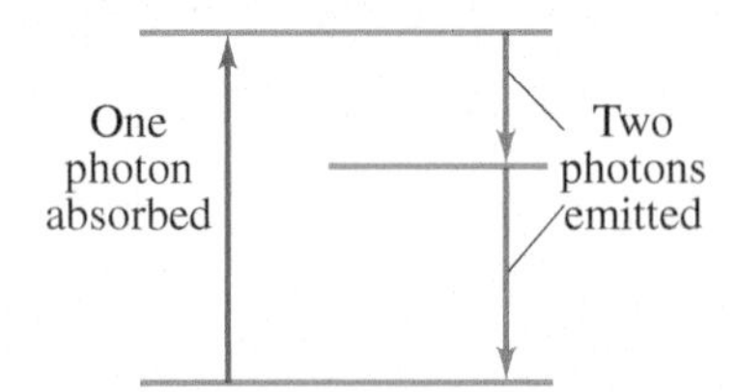

FIGURE 39–15 Fluorescence.

39–8 Fluorescence and Phosphorescence

When an atom is excited from one energy state to a higher one by the absorption of a photon, it may return to the lower level in a series of two (or more) jumps if there is an energy level in between (Fig. 39–15). The photons emitted will consequently have lower energy and frequency than the absorbed photon. When the absorbed photon is in the UV and the emitted photons are in the visible region of the spectrum, this phenomenon is called **fluorescence** (Fig. 39–16).

FIGURE 39–16 When UV light (a range of wavelengths) illuminates these various "fluorescent" rocks, they fluoresce in the visible region of the spectrum.

The wavelength for which fluorescence will occur depends on the energy levels of the particular atoms. Because the frequencies are different for different substances, and because many substances fluoresce readily, fluorescence is a powerful tool for identification of compounds. It is also used for assaying—determining how much of a substance is present—and for following substances along a natural pathway as in plants and animals. For detection of a given compound, the stimulating light must be monochromatic, and solvents or other materials present must not fluoresce in the same region of the spectrum. Sometimes the observation of fluorescent light being emitted is sufficient to detect a compound. In other cases, spectrometers are used to measure the wavelengths and intensities of the emitted light.

Fluorescent lightbulbs work in a two-step process. The applied voltage accelerates electrons that strike atoms of the gas in the tube and cause them to be excited. When the excited atoms jump down to their normal levels, they emit UV photons which strike a fluorescent coating on the inside of the tube. The light we see is a result of this material fluorescing in response to the UV light striking it.

Materials such as those used for luminous watch dials are said to be **phosphorescent**. When an atom is raised to a normal excited state, it drops back down within about 10^{-8} s. In phosphorescent substances, atoms can be excited by photon absorption to energy levels, called **metastable**, which are states that last much longer because to jump down is a "forbidden" transition as discussed in Section 39–2. Metastable states can last even a few seconds or longer. In a collection of such atoms, many of the atoms will descend to the lower state fairly soon, but many will remain in the excited state for over an hour. Hence light will be emitted even after long periods. When you put your luminous watch dial close to a bright lamp, many atoms are excited to metastable states, and you can see the glow for a long time afterward.

39–9 Lasers

A **laser** is a device that can produce a very narrow intense beam of monochromatic coherent light. (By *coherent*, we mean that across any cross section of the beam, all parts have the same phase.) The emitted beam is a nearly perfect plane wave. An ordinary light source, on the other hand, emits light in all directions (so the intensity decreases rapidly with distance), and the emitted light is incoherent (the different parts of a beam are not in phase with each other). The excited atoms that emit the light in an ordinary lightbulb act independently, so each photon emitted can be considered as a short wave train that lasts about 10^{-8} s. Different wave trains bear no phase relation to one another. Just the opposite is true of lasers.

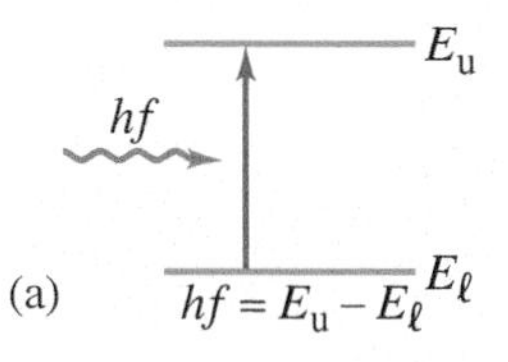

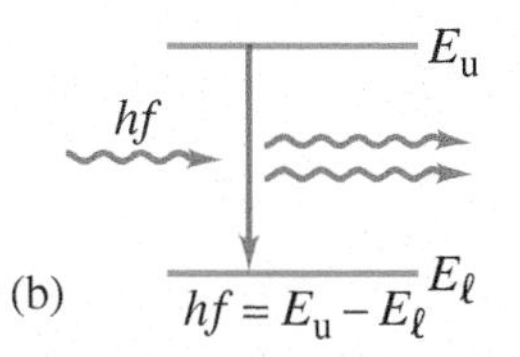

FIGURE 39–17 (a) Absorption of a photon. (b) Stimulated emission. E_u and E_ℓ refer to "upper" and "lower" energy states.

The action of a laser is based on quantum theory. We have seen that a photon can be absorbed by an atom if (and only if) the photon energy hf corresponds to the energy difference between an occupied energy level of the atom and an available excited state, Fig. 39–17a. If the atom is already in the excited state, it may of course jump down spontaneously (i.e., no apparent stimulus) to the lower state with the emission of a photon. However, if a photon with this same energy strikes the excited atom, it can stimulate the atom to make the transition sooner to the lower state, Fig. 39–17b. This phenomenon is called **stimulated emission**: not only do we still have the original photon, but also a second one of the same frequency as a result of the atom's transition. These two photons are exactly *in phase*, and they are moving in the same direction. This is how coherent light is produced in a laser. Hence the name "laser," which is an acronym for **L**ight **A**mplification by **S**timulated **E**mission of **R**adiation.

The natural population in energy states of atoms in thermal equilibrium at any temperature T (in K) is given by the **Boltzmann distribution** (or **Boltzmann factor**):

$$N_n = Ce^{-\frac{E_n}{kT}}, \tag{39–16a}$$

where N_n is the number of atoms in the state with energy E_n. For two states n and n', the ratio of the number of atoms in the two states is

$$\frac{N_n}{N_{n'}} = e^{-\left(\frac{E_n - E_{n'}}{kT}\right)}. \tag{39–16b}$$

Thus most atoms are in the ground state unless the temperature is very high. In the two-level system of Fig. 39–17, most atoms are normally in the lower state, so the majority of incident photons will be absorbed. In order to obtain the coherent light from stimulated emission, two conditions must be satisfied. First, atoms must be excited to the higher state, so that an **inverted population** is produced, one in which more atoms are in the upper state than in the lower one (Fig. 39–18). Then *emission* of photons will dominate over absorption. Hence the system will not be in thermal equilibrium. And second, the higher state must be a **metastable state**—a state in which the electrons remain longer than usual† so that the transition to the lower state occurs by stimulated emission rather than spontaneously. (How inverted populations are created will be discussed shortly.)

FIGURE 39–18 Two energy levels for a collection of atoms. Each dot represents the energy state of one atom. (a) A normal situation; (b) an inverted population.

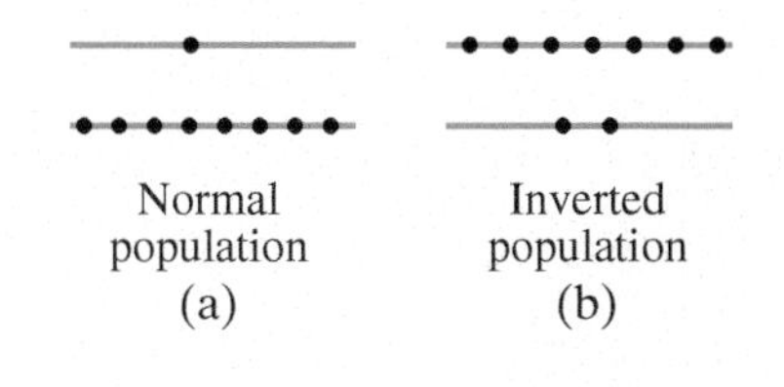

†An atom excited to such a state can jump to a lower state only by a so-called forbidden transition (discussed in Section 39–2), which is why the lifetime is longer than normal.

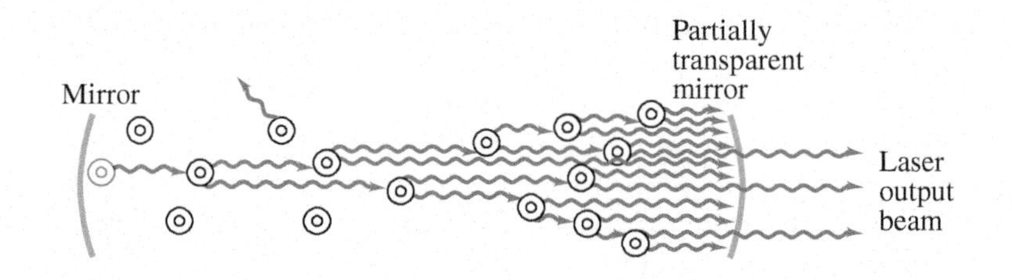

FIGURE 39–19 Laser diagram, showing excited atoms stimulated to emit light.

Figure 39–19 is a schematic diagram of a laser: the "lasing" material is placed in a long narrow tube at the ends of which are two mirrors, one of which is partially transparent (transmitting perhaps 1 or 2%). Some of the excited atoms drop down fairly soon after being excited. One of these is the blue atom shown on the far left in Fig. 39–19. If the emitted photon strikes another atom in the excited state, it stimulates this atom to emit a photon of the *same* frequency, moving in the *same* direction, and *in phase* with it. These two photons then move on to strike other atoms causing more stimulated emission. As the process continues, the number of photons multiplies. When the photons strike the end mirrors, most are reflected back, and as they move in the opposite direction, they continue to stimulate more atoms to emit photons. As the photons move back and forth between the mirrors, a small percentage passes through the partially transparent mirror at one end. These photons make up the narrow coherent external laser beam.

Inside the tube, some spontaneously emitted photons will be emitted at an angle to the axis, and these will merely go out the side of the tube and not affect the narrowness of the main beam. In a well-designed laser, the spreading of the beam is limited only by diffraction, so the angular spread is $\approx \lambda/D$ (see Eq. 35–1 or 35–10), where D is the diameter of the end mirror. The diffraction spreading can be incredibly small. The light energy, instead of spreading out in space as it does for an ordinary light source, can be a pencil-thin beam.

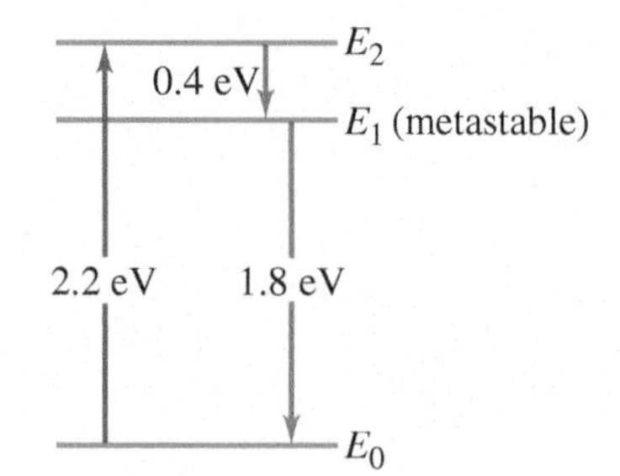

FIGURE 39–20 Energy levels of chromium in a ruby crystal. Photons of energy 2.2 eV "pump" atoms from E_0 to E_2, which then decay to metastable state E_1. Lasing action occurs by stimulated emission of photons in transition from E_1 to E_0.

Creating an Inverted Population

The excitation of the atoms in a laser can be done in several ways to produce the necessary inverted population. In a **ruby laser**, the lasing material is a ruby rod consisting of Al_2O_3 with a small percentage of aluminum (Al) atoms replaced by chromium (Cr) atoms. The Cr atoms are the ones involved in lasing. In a process called **optical pumping**, the atoms are excited by strong flashes of light of wavelength 550 nm, which corresponds to a photon energy of 2.2 eV. As shown in Fig. 39–20, the atoms are excited from state E_0 to state E_2. The atoms quickly decay either back to E_0 or to the intermediate state E_1, which is metastable with a lifetime of about 3×10^{-3} s (compared to 10^{-8} s for ordinary levels). With strong pumping action, more atoms can be found in the E_1 state than are in the E_0 state. Thus we have the inverted population needed for lasing. As soon as a few atoms in the E_1 state jump down to E_0, they emit photons that produce stimulated emission of the other atoms, and the lasing action begins. A ruby laser thus emits a beam whose photons have energy 1.8 eV and a wavelength of 694.3 nm (or "ruby-red" light).

FIGURE 39–21 Energy levels for He and Ne. He is excited in the electric discharge to the E_1 state. This energy is transferred to the E'_3 level of the Ne by collision. E'_3 is metastable and decays to E'_2 by stimulated emission.

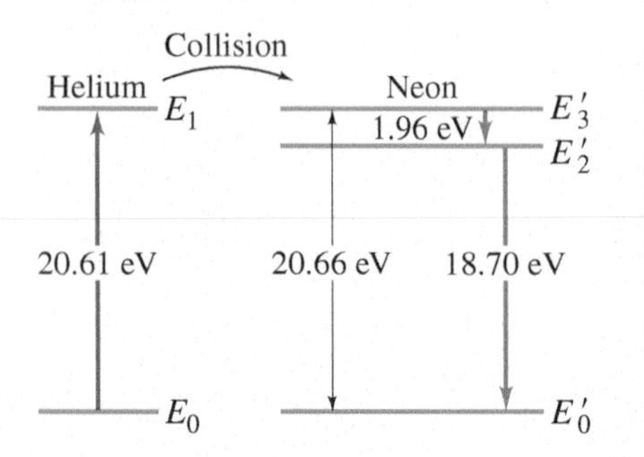

In a helium–neon (He–Ne) laser, the lasing material is a gas, a mixture of about 85% He and 15% Ne. The atoms are excited by applying a high voltage to the tube so that an electric discharge takes place within the gas. In the process, some of the He atoms are raised to the metastable state E_1 shown in Fig. 39–21, which corresponds to a jump of 20.61 eV, almost exactly equal to an excited state in neon, 20.66 eV. The He atoms do not quickly return to the ground state by spontaneous emission, but instead often give their excess energy to a Ne atom when they collide—see Fig. 39–21. In such a collision, the He drops to the ground state and the Ne atom is excited to the state E'_3 (the prime refers to neon states). The slight difference in energy (0.05 eV) is supplied by the kinetic energy of the moving atoms. In this manner, the E'_3 state in Ne—which is metastable—becomes more populated than the E'_2 level. This inverted population between E'_3 and E'_2 is what is needed for lasing.

Very common now are **semiconductor diode lasers**, also called **pn junction lasers**, which utilize an inverted population of electrons between the conduction band of the n side of the diode and the lower-energy valence band of the p side (Sections 40–7 to 40–9). When an electron jumps down, a photon can be emitted, which in turn can stimulate another electron to make the transition and emit another photon, in phase. The needed mirrors (as in Fig. 39–19) are made by the polished ends of the pn crystal. Semiconductor lasers are used in CD and DVD players (see below), and in many other applications.

Other types of laser include: *chemical lasers*, in which the energy input comes from the chemical reaction of highly reactive gases; *dye lasers*, whose frequency is tunable; CO_2 *gas lasers*, capable of high-power output in the infrared; and *rare-earth solid-state lasers* such as the high-power Nd:YAG laser.

The excitation of the atoms in a laser can be done continuously or in pulses. In a **pulsed laser**, the atoms are excited by periodic inputs of energy. In a **continuous laser**, the energy input is continuous: as atoms are stimulated to jump down to the lower level, they are soon excited back up to the upper level so the output is a continuous laser beam.

No laser is a source of energy. Energy must be put in, and the laser converts a part of it into an intense narrow beam output.

CAUTION
Laser not an energy source

*Applications

PHYSICS APPLIED
DVD and CD players, bar codes

The unique feature of light from a laser, that it is a coherent narrow beam, has found many applications. In everyday life, lasers are used as bar-code readers (at store checkout stands) and in compact disc (CD) and digital video disc (DVD) players. The laser beam reflects off the stripes and spaces of a bar code, and off the tiny pits of a CD or DVD as shown in Fig. 39–22a. The recorded information on a CD or DVD is a series of pits and spaces representing 0s and 1s (or "off" and "on") of a binary code that is decoded electronically before being sent to the audio or video system. The laser of a CD player starts reading at the inside of the disc which rotates at about 500 rpm at the start. As the disc rotates, the laser follows the spiral track (Fig. 39–22b), and as it moves outward the disc must slow down because each successive circumference $(C = 2\pi r)$ is slightly longer as r increases; at the outer edge, the disc is rotating about 200 rpm. A 1-hour CD has a track roughly 5 km long; the track width is about 1600 nm $(= 1.6\,\mu\text{m})$ and the distance between pits is about 800 nm. DVDs contain much more information. Standard DVDs use a thinner track $(0.7\,\mu\text{m})$ and shorter pit length (400 nm). New high-definition DVDs use a "blue" laser with a short wavelength (405 nm) and narrower beam, allowing a narrower track $(0.3\,\mu\text{m})$ that can store much more data for high definition. DVDs can also have two layers, one below the other. When the laser focuses on the second layer, the light passes through the semitransparent surface layer. The second layer may start reading at the outer edge instead of inside. DVDs can also have a single or double layer on *both* surfaces of the disc.

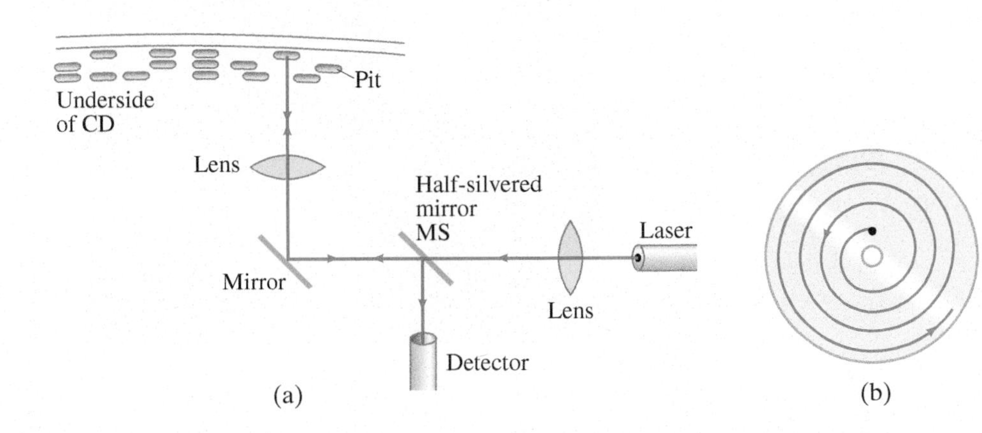

FIGURE 39–22 (a) Reading a CD (or DVD). The fine beam of a laser, focused even more finely with lenses, is directed at the undersurface of a rotating compact disc. The beam is reflected back from the areas between pits but reflects much less from pits. The reflected light is detected as shown, reflected by a half-reflecting mirror MS. The strong and weak reflections correspond to the 0s and 1s of the binary code representing the audio or video signal. (b) A laser follows the CD track which starts near the center and spirals outward.

PHYSICS APPLIED
Medical and other uses of lasers

Lasers are a useful surgical tool. The narrow intense beam can be used to destroy tissue in a localized area, or to break up gallstones and kidney stones. Because of the heat produced, a laser beam can be used to "weld" broken tissue, such as a detached retina, Fig. 39–23, or to mold the cornea of the eye (by vaporizing tiny bits of material) to correct myopia and other eye defects (LASIK surgery). The laser beam can be carried by an optical fiber (Section 32–7) to the surgical point, sometimes as an additional fiber-optic path on an endoscope (again Section 32–7). An example is the removal of plaque clogging human arteries. Tiny organelles within a living cell have been destroyed using lasers by researchers studying how the absence of that organelle affects the behavior of the cell. Laser beams are used to destroy cancerous and precancerous cells; and the heat seals off capillaries and lymph vessels, thus "cauterizing" the wound to prevent spread of the disease.

The intense heat produced in a small area by a laser beam is used for welding and machining metals and for drilling tiny holes in hard materials. Because a laser beam is coherent, monochromatic, narrow, and essentially parallel, lenses can be used to focus the light into even smaller areas. The precise straightness of a laser beam is also useful to surveyors for lining up equipment accurately, especially in inaccessible places.

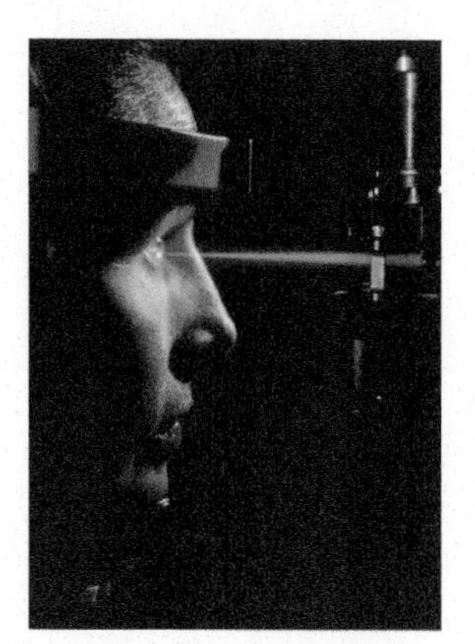

FIGURE 39–23 Laser being used in eye surgery.

*39–10 Holography

PHYSICS APPLIED
Holography

One of the most interesting applications of laser light is the production of three-dimensional images called **holograms** (see Fig. 39–24). In an ordinary photograph, the film simply records the intensity of light reaching it at each point. When the photograph or transparency is viewed, light reflecting from it or passing through it gives us a two-dimensional picture. In holography, the images are formed by interference, without lenses. When a laser hologram is made on film, a broadened laser beam is split into two parts by a half-silvered mirror, Fig. 39–24a. One part goes directly to the film; the rest passes to the object to be photographed, from which it is reflected to the film. Light from every point on the object reaches each point on the film, and the interference of the two beams allows the film to record both the intensity and relative phase of the light at each point. It is crucial that the incident light be coherent—that is, in phase at all points—which is why a laser is used. After the film is developed, it is placed again in a laser beam and a three-dimensional image of the object is created. You can walk around such an image and see it from different sides as if it were the original object (Fig. 39–24b). Yet, if you try to touch it with your hand, there will be nothing material there.

FIGURE 39–24 (a) Making a hologram. Light reflected from various points on the object interferes (at the film) with light from the direct beam. (b) A boy is looking at a hologram of a woman talking on a telephone. Holograms do not photograph well—they must be seen directly.

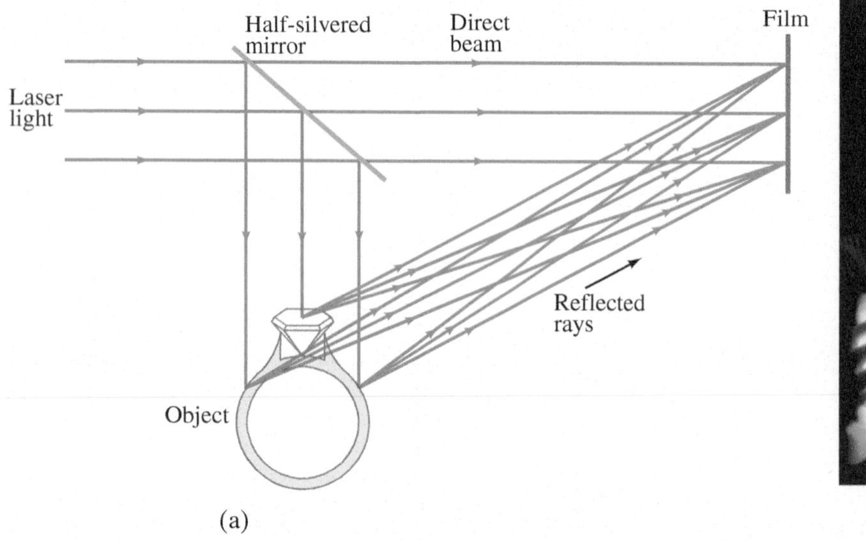

(a)

(b)

The details of how the image is formed are quite complicated. But we can get the basic idea by considering one single point on the object. In Fig. 39–25a the rays OA and OB have reflected from one point on our object. The rays CA and DB come directly from the source and interfere with OA and OB at points A and B on the film. A set of interference fringes is produced as shown in Fig. 39–25b. The spacing between the fringes changes from top to bottom as shown. Why this happens is explained in Fig. 39–26. Thus the hologram of a single point object would have the fringe pattern shown in Fig. 39–25b. The film in this case looks like a diffraction grating with variable spacing. Hence, when coherent laser light is passed back through the developed film to reconstruct the image, the diffracted rays in the first order maxima occur at slightly different angles because the spacing changes. (Remember Eq. 35–13, $\sin\theta = \lambda/d$: where the spacing d is greater, the angle θ is less.) Hence, the rays diffracted upward (in first order) seem to diverge from a single point, Fig. 39–27. This is a virtual image of the original object, which can be seen with the eye. Rays diffracted in first order *downward* converge to make a real image, which can be seen and also photographed. (Note that the straight-through undiffracted rays are of no interest.) Of course real objects consist of many points, so a hologram will be a complex interference pattern which, when laser light is incident on it, will reproduce an image of the object. Each image point will be at the correct (three-dimensional) position with respect to other points, so the image accurately represents the original object. And it can be viewed from different angles as if viewing the original object. Holograms can be made in which a viewer can walk entirely around the image (360°) and see all sides of it.

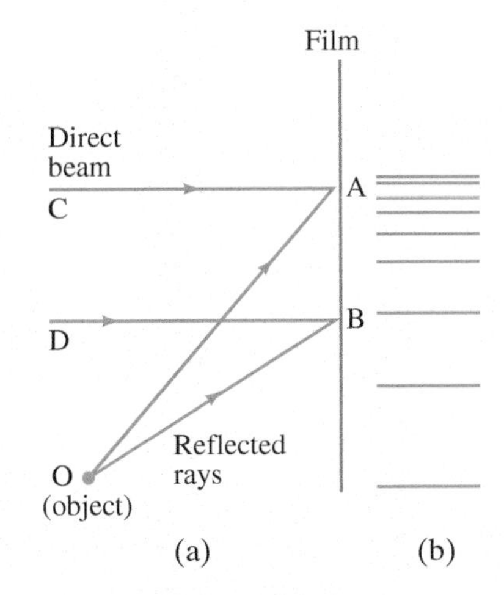

FIGURE 39–25 (a) Light from point O on the object interferes with light of the direct beam (rays CA and DB). (b) Interference fringes produced.

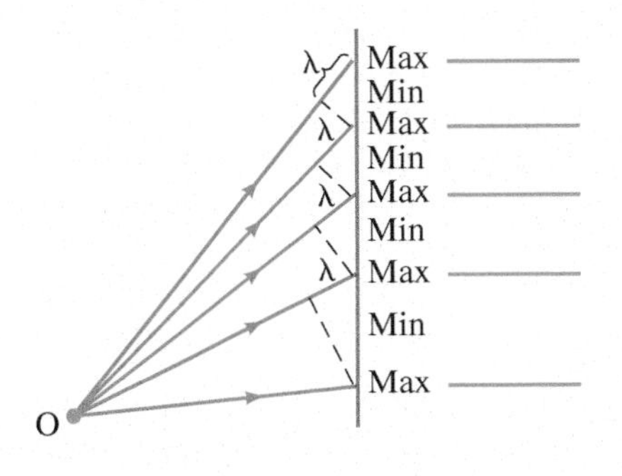

FIGURE 39–26 Each of the rays shown leaving point O is one wavelength shorter than the one above it. If the top ray is in phase with the direct beam (not shown), which has the same phase at all points on the screen, each of the rays shown produces a constructive interference fringe. From this diagram it can be seen that the fringe spacing increases toward the bottom.

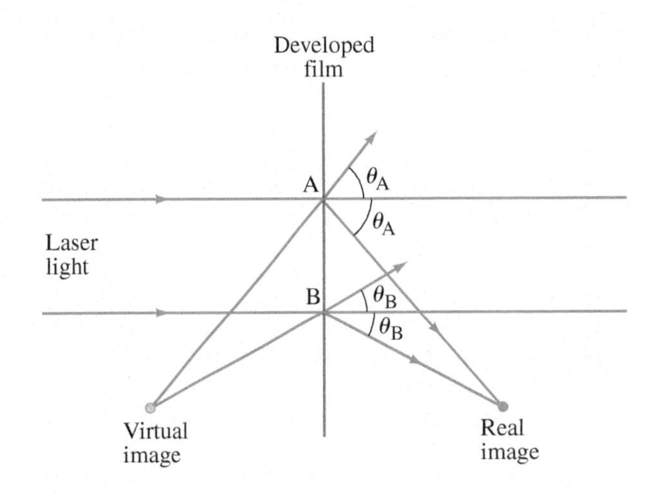

FIGURE 39–27 Reconstructing the image of one point on the object. Laser beam strikes the developed film, which is like a diffraction grating of variable spacing. Rays corresponding to the first diffraction maxima are shown emerging. The angle $\theta_A > \theta_B$ because the spacing at B is greater than at A $(\sin\theta = \lambda/d)$. Hence real and virtual images of the point are reproduced as shown.

Volume or **white-light holograms** do not require a laser to see the image, but can be viewed with ordinary white light (preferably a nearly point source, such as the Sun or a clear bulb with a small bright filament). Such holograms must be made, however, with a laser. They are made not on thin film, but on a *thick* emulsion. The interference pattern in the film emulsion can be thought of as an array of bands or ribbons where constructive interference occurred. This array, and the reconstruction of the image, can be compared to Bragg scattering of X-rays from the atoms in a crystal (see Section 35–10). White light can reconstruct the image because the Bragg condition $(m\lambda = 2d\sin\theta)$ selects out the appropriate single wavelength. If the hologram is originally produced by lasers emitting the three additive primary colors (red, green, and blue), the three-dimensional image can be seen in full color when viewed with white light.

Summary

In the quantum-mechanical view of the atom, the electrons do not have well-defined orbits, but instead exist as a "cloud." Electron clouds can be interpreted as an electron wave spread out in space, or as a **probability distribution** for electrons considered as particles.

For the simplest atom, hydrogen, the Schrödinger equation contains the potential energy

$$U = -\frac{1}{4\pi\epsilon_0}\frac{e^2}{r}.$$

The solutions give the same values of energy as the old Bohr theory.

According to quantum mechanics, the state of an electron in an atom is specified by four **quantum numbers:** n, ℓ, m_ℓ, and m_s:

1. n, the **principal quantum number**, can take on any integer value $(1, 2, 3, \cdots)$ and corresponds to the quantum number of the old Bohr theory;
2. ℓ, the **orbital quantum number**, can take on integer values from 0 up to $n - 1$;
3. m_ℓ, the **magnetic quantum number**, can take on integer values from $-\ell$ to $+\ell$;
4. m_s, the **spin quantum number**, can be $+\frac{1}{2}$ or $-\frac{1}{2}$.

The energy levels in the hydrogen atom depend on n, whereas in other atoms they depend on n and ℓ.

The orbital angular momentum of an atom has magnitude $L = \sqrt{\ell(\ell+1)}\,\hbar$ and z component $L_z = m_\ell \hbar$. Spin angular momentum has magnitude $S = \sqrt{s(s+1)}\,\hbar$ and z component $S_z = m_s \hbar$ where $s = \frac{1}{2}$ and $m_s = \pm\frac{1}{2}$.

When an external magnetic field is applied, the spectral lines are split (the **Zeeman effect**), indicating that the energy depends also on m_ℓ in this case.

Even in the absence of a magnetic field, precise measurements of spectral lines show a tiny splitting of the lines called **fine structure**, whose explanation is that the energy depends very slightly on m_ℓ and m_s.

Transitions between states that obey the **selection rule** $\Delta\ell = \pm 1$ are far more probable than other so-called **forbidden** transitions.

The ground-state wave function in hydrogen has spherical symmetry, as do other $\ell = 0$ states. States with $\ell > 0$ have some directionality in space.

The **probability density** (or probability distribution), $|\psi|^2$, and the **radial probability density**, $P_r = 4\pi r^2 |\psi|^2$, are both useful to illustrate the spatial extent of the electron cloud.

The arrangement of electrons in multi-electron atoms is governed by the **Pauli exclusion principle**, which states that no two electrons can occupy the same quantum state—that is, they cannot have the same set of quantum numbers n, ℓ, m_ℓ, and m_s.

As a result, electrons in multi-electron atoms are grouped into **shells** (according to the value of n) and **subshells** (according to ℓ).

Electron configurations are specified using the numerical values of n, and using letters for ℓ: s, p, d, f, etc., for $\ell = 0, 1, 2, 3$, and so on, plus a superscript for the number of electrons in that subshell. Thus, the ground state of hydrogen is $1s^1$, whereas that for oxygen is $1s^2 2s^2 2p^4$.

In the **Periodic Table**, the elements are arranged in horizontal rows according to increasing atomic number (number of electrons in the neutral atom). The shell structure gives rise to a periodicity in the properties of the elements, so that each vertical column can contain elements with similar chemical properties.

X-rays, which are a form of electromagnetic radiation of very short wavelength, are produced when high-speed electrons strike a target. The spectrum of X-rays so produced consists of two parts, a continuous spectrum produced when the electrons are decelerated by atoms of the target, and peaks representing photons emitted by atoms of the target after being excited by collision with the high-speed electrons. Measurement of these peaks allows determination of inner energy levels of atoms and determination of atomic number Z.

[*An atom has a **magnetic dipole moment** $\vec{\mu}$ related to its orbital angular momentum $\vec{L}$, which produces a potential energy when in a magnetic field, $U = -\vec{\mu}\cdot\vec{B}$. Electron spin also yields a magnetic moment, but the energy in a magnetic field is a factor $g = 2.0023\cdots$ times larger than expected, as determined in Stern-Gerlach experiments.]

[*Atoms have a total angular momentum $\vec{J} = \vec{L} + \vec{S}$ which is quantized as for L and S, namely $J = \sqrt{j(j+1)}\,\hbar$ where j is a half-integer equal to $\ell \pm \frac{1}{2}$ in hydrogen.]

Fluorescence occurs when absorbed UV photons are followed by emission of visible light, due to the special arrangement of energy levels of atoms of the material. **Phosphorescent** materials have **metastable** states (long-lived) that emit light seconds or minutes after absorption of light.

Lasers produce a narrow beam of monochromatic coherent light (light waves *in phase*). [***Holograms** are images with a 3-dimensional quality, formed by interference of laser light.]

Questions

1. Discuss the differences between Bohr's view of the atom and the quantum-mechanical view.
2. The probability density $|\psi|^2$ is a maximum at the center of the H atom $(r = 0)$ for the ground state, whereas the radial probability density $P_r = 4\pi r^2|\psi|^2$ is zero at this point. Explain why.
3. Which model of the hydrogen atom, the Bohr model or the quantum-mechanical model, predicts that the electron spends more time near the nucleus?
4. The size of atoms varies by only a factor of three or so, from largest to smallest, yet the number of electrons varies from one to over 100. Why?
5. Excited hydrogen and excited helium atoms both radiate light as they jump down to the $n = 1, \ell = 0, m_\ell = 0$ state. Yet the two elements have very different emission spectra. Why?
6. In Fig. 39–4, why do the upper and lower levels have different energy splittings in a magnetic field?
7. Why do three quantum numbers come out of the Schrödinger theory (rather than, say, two or four)?
8. The 589-nm yellow line in sodium is actually two very closely spaced lines. This splitting is due to an "internal" Zeeman effect. Can you explain this? [*Hint*: Put yourself in the reference frame of the electron.]

9. Which of the following electron configurations are not allowed: (*a*) $1s^2 2s^2 2p^4 3s^2 4p^2$; (*b*) $1s^2 2s^2 2p^8 3s^1$; (*c*) $1s^2 2s^2 2p^6 3s^2 3p^5 4s^2 4d^5 4f^1$? If not allowed, explain why.

10. Give the complete electron configuration for a uranium atom (careful scrutiny across the Periodic Table on the inside back cover will provide useful hints).

11. In what column of the Periodic Table would you expect to find the atom with each of the following configurations? (*a*) $1s^2 2s^2 2p^6 3s^2$; (*b*) $1s^2 2s^2 2p^6 3s^2 3p^6$; (*c*) $1s^2 2s^2 2p^6 3s^2 3p^6 4s^1$; (*d*) $1s^2 2s^2 2p^5$.

12. On what factors does the periodicity of the Periodic Table depend? Consider the exclusion principle, quantization of angular momentum, spin, and any others you can think of.

13. How would the Periodic Table look if there were no electron spin but otherwise quantum mechanics were valid? Consider the first 20 elements or so.

14. The ionization energy for neon $(Z = 10)$ is 21.6 eV and that for sodium $(Z = 11)$ is 5.1 eV. Explain the large difference.

15. Why do chlorine and iodine exhibit similar properties?

16. Explain why potassium and sodium exhibit similar properties.

17. Why are the chemical properties of the rare earths so similar?

18. Why do we not expect perfect agreement between measured values of characteristic X-ray line wavelengths and those calculated using Bohr theory, as in Example 39–6?

19. Why does the Bohr theory, which does not work at all well for normal transitions involving the outer electrons for He and more complex atoms, nevertheless predict reasonably well the atomic X-ray spectra for transitions deep inside the atom?

20. Why does the cutoff wavelength in Fig. 39–11 imply a photon nature for light?

21. How would you figure out which lines in an X-ray spectrum correspond to K_α, K_β, L, etc., transitions?

22. Why do the characteristic X-ray spectra vary in a systematic way with Z, whereas the visible spectra (Fig. 35–22) do not?

23. Why do we expect electron transitions deep within an atom to produce shorter wavelengths than transitions by outer electrons?

*24. Why is the direction of the magnetic dipole moment of an electron opposite to that of its orbital angular momentum?

*25. Why is a nonhomogeneous field used in the Stern-Gerlach experiment?

26. Compare spontaneous emission to stimulated emission.

27. Does the intensity of light from a laser fall off as the inverse square of the distance?

28. How does laser light differ from ordinary light? How is it the same?

29. Explain how a 0.0005-W laser beam, photographed at a distance, can seem much stronger than a 1000-W street lamp at the same distance.

Problems

39–2 Hydrogen Atom Quantum Numbers

1. (I) For $n = 7$, what values can ℓ have?

2. (I) For $n = 6$, $\ell = 3$, what are the possible values of m_ℓ and m_s?

3. (I) How many different states are possible for an electron whose principal quantum number is $n = 4$? Write down the quantum numbers for each state.

4. (I) If a hydrogen atom has $m_\ell = -4$, what are the possible values of n, ℓ, and m_s?

5. (I) A hydrogen atom has $\ell = 5$. What are the possible values for n, m_ℓ, and m_s?

6. (I) Calculate the magnitude of the angular momentum of an electron in the $n = 5$, $\ell = 3$ state of hydrogen.

7. (II) A hydrogen atom is in the 7*g* state. Determine (*a*) the principal quantum number, (*b*) the energy of the state, (*c*) the orbital angular momentum and its quantum number ℓ, and (*d*) the possible values for the magnetic quantum number.

8. (II) (*a*) Show that the number of different states possible for a given value of ℓ is equal to $2(2\ell + 1)$. (*b*) What is this number for $\ell = 0, 1, 2, 3, 4, 5,$ and 6?

9. (II) Show that the number of different electron states possible for a given value of n is $2n^2$. (See Problem 8.)

10. (II) An excited H atom is in a 5*d* state. (*a*) Name all the states to which the atom is "allowed" to jump with the emission of a photon. (*b*) How many different wavelengths are there (ignoring fine structure)?

11. (II) The magnitude of the orbital angular momentum in an excited state of hydrogen is $6.84 \times 10^{-34}\,\mathrm{J \cdot s}$ and the z component is $2.11 \times 10^{-34}\,\mathrm{J \cdot s}$. What are all the possible values of n, ℓ, and m_ℓ for this state?

39–3 Hydrogen Atom Wave Functions

12. (I) Show that the ground-state wave function, Eq. 39–5, is normalized. [*Hint*: See Example 39–4.]

13. (II) For the ground state of hydrogen, what is the value of (*a*) ψ, (*b*) $|\psi|^2$, and (*c*) P_r, at $r = 1.5\,r_0$?

14. (II) For the $n = 2$, $\ell = 0$ state of hydrogen, what is the value of (*a*) ψ, (*b*) $|\psi|^2$, and (*c*) P_r, at $r = 4r_0$?

15. (II) By what factor is it more likely to find the electron in the ground state of hydrogen at the Bohr radius (r_0) than at twice the Bohr radius $(2r_0)$?

16. (II) (*a*) Show that the probability of finding the electron in the ground state of hydrogen at less than one Bohr radius from the nucleus is 32%. (*b*) What is the probability of finding a 1*s* electron between $r = r_0$ and $r = 2r_0$?

17. (II) Determine the radius r of a sphere centered on the nucleus within which the probability of finding the electron for the ground state of hydrogen is (*a*) 50%, (*b*) 90%, (*c*) 99%.

18. (II) (*a*) Estimate the probability of finding an electron, in the ground state of hydrogen, within the nucleus assuming it to be a sphere of radius $r = 1.1$ fm. (*b*) Repeat the estimate assuming the electron is replaced with a muon, which is very similar to an electron (Chapter 43) except that its mass is 207 times greater.

19. (II) Show that the mean value of r for an electron in the ground state of hydrogen is $\bar{r} = \frac{3}{2}r_0$, by calculating

$$\bar{r} = \int_{\text{all space}} r|\psi_{100}|^2\,dV = \int_0^\infty r|\psi_{100}|^2 4\pi r^2\,dr.$$

20. (II) Show that ψ_{200} as given by Eq. 39–8 is normalized.

21. (II) Determine the average radial probability distribution P_r for the $n = 2, \ell = 1$ state in hydrogen by calculating

$$P_r = 4\pi r^2\left[\tfrac{1}{3}|\psi_{210}|^2 + \tfrac{1}{3}|\psi_{211}|^2 + \tfrac{1}{3}|\psi_{21-1}|^2\right].$$

22. (II) Use the result of Problem 21 to show that the most probable distance r from the nucleus for an electron in the $2p$ state of hydrogen is $r = 4r_0$, which is just the second Bohr radius (Eq. 37–11, Fig. 37–25).

23. (II) For the ground state of hydrogen, what is the probability of finding the electron within a spherical shell of inner radius $0.99r_0$ and outer radius $1.01r_0$?

24. (III) For the $n = 2, \ell = 0$ state of hydrogen, what is the probability of finding the electron within a spherical shell of inner radius $4.00\,r_0$ and outer radius $5.00\,r_0$? [*Hint*: You might integrate by parts.]

25. (III) Show that ψ_{100} (Eq. 39–5a) satisfies the Schrödinger equation (Eq. 39–1) with the Coulomb potential, for energy $E = -me^4/8\epsilon_0^2 h^2$.

26. (III) Show that the probability of finding the electron within 1 Bohr radius of the nucleus in the hydrogen atom is (*a*) 3.4% for the $n = 2, \ell = 0$ state, and (*b*) 0.37% for the $n = 2, \ell = 1$ state. (See Problem 21.)

27. (III) The wave function for the $n = 3, \ \ell = 0$ state in hydrogen is

$$\psi_{300} = \frac{1}{\sqrt{27\pi r_0^3}}\left(1 - \frac{2r}{3r_0} + \frac{2r^2}{27r_0^2}\right)e^{-\frac{r}{3r_0}}.$$

(*a*) Determine the radial probability distribution P_r for this state, and (*b*) draw the curve for it on a graph. (*c*) Determine the most probable distance from the nucleus for an electron in this state.

39–4 and 39–5 Complex Atoms

28. (I) List the quantum numbers for each electron in the ground state of oxygen $(Z = 8)$.

29. (I) List the quantum numbers for each electron in the ground state of (*a*) carbon $(Z = 6)$, (*b*) aluminum $(Z = 13)$.

30. (I) How many electrons can be in the $n = 6, \ell = 4$ subshell?

31. (II) An electron has $m_\ell = 2$ and is in its lowest possible energy state. What are the values of n and ℓ for this electron?

32. (II) If the principal quantum number n were limited to the range from 1 to 6, how many elements would we find in nature?

33. (II) What is the full electron configuration for (*a*) nickel (Ni), (*b*) silver (Ag), (*c*) uranium (U)? [*Hint*: See the Periodic Table inside the back cover.]

34. (II) Estimate the binding energy of the third electron in lithium using Bohr theory. [*Hint*: This electron has $n = 2$ and "sees" a net charge of approximately $+1e$.] The measured value is 5.36 eV.

35. (II) Using the Bohr formula for the radius of an electron orbit, estimate the average distance from the nucleus for an electron in the innermost $(n = 1)$ orbit in uranium $(Z = 92)$. Approximately how much energy would be required to remove this innermost electron?

36. (II) Let us apply the exclusion principle to an infinitely high square well (Section 38–8). Let there be five electrons confined to this rigid box whose width is ℓ. Find the lowest energy state of this system, by placing the electrons in the lowest available levels, consistent with the Pauli exclusion principle.

37. (II) Show that the total angular momentum is zero for a filled subshell.

39–6 X-rays

38. (I) If the shortest-wavelength bremsstrahlung X-rays emitted from an X-ray tube have $\lambda = 0.027$ nm, what is the voltage across the tube?

39. (I) What are the shortest-wavelength X-rays emitted by electrons striking the face of a 32.5-kV TV picture tube? What are the longest wavelengths?

40. (I) Show that the cutoff wavelength λ_0 in an X-ray spectrum is given by

$$\lambda_0 = \frac{1240}{V}\text{ nm},$$

where V is the X-ray tube voltage in volts.

41. (II) Estimate the wavelength for an $n = 2$ to $n = 1$ transition in iron $(Z = 26)$.

42. (II) Use the result of Example 39–6 to estimate the X-ray wavelength emitted when a cobalt atom $(Z = 27)$ makes a transition from $n = 2$ to $n = 1$.

43. (II) A mixture of iron and an unknown material are bombarded with electrons. The wavelength of the K_α lines are 194 pm for iron and 229 pm for the unknown. What is the unknown material?

44. (II) Use Bohr theory to estimate the wavelength for an $n = 3$ to $n = 1$ transition in molybdenum $(Z = 42)$. The measured value is 0.063 nm. Why do we not expect perfect agreement?

45. (II) Use conservation of energy and momentum to show that a moving electron cannot give off an X-ray photon unless there is a third object present, such as an atom or nucleus.

*39–7 Magnetic Dipole Moment; $\vec{J}$

*46. (I) Verify that the Bohr magneton has the value $\mu_B = 9.27 \times 10^{-24}$ J/T (see Eq. 39–12).

*47. (I) If the quantum state of an electron is specified by (n, ℓ, m_ℓ, m_s), estimate the energy difference between the states $(1, 0, 0, -\frac{1}{2})$ and $(1, 0, 0, +\frac{1}{2})$ of an electron in the 1*s* state of helium in an external magnetic field of 2.5 T.

*48. (II) Silver atoms $(\text{spin} = \frac{1}{2})$ are placed in a 1.0-T magnetic field which splits the ground state into two close levels. (*a*) What is the difference in energy between these two levels, and (*b*) what wavelength photon could cause a transition from the lower level to the upper one? (*c*) How would your answer differ if the atoms were hydrogen?

*49. (II) In a Stern-Gerlach experiment, Ag atoms exit the oven with an average speed of 780 m/s and pass through a magnetic field gradient $dB/dz = 1.8 \times 10^3$ T/m for a distance of 5.0 cm. (*a*) What is the separation of the two beams as they emerge from the magnet? (*b*) What would the separation be if the *g*-factor were 1 for electron spin?

*50. (II) For an electron in a 5*g* state, what are all the possible values of j, m_j, J, and J_z?

*51. (II) What are the possible values of j for an electron in (*a*) the 4p, (*b*) the 4f, and (*c*) the 3d state of hydrogen? (*d*) What is J in each case?

*52. (II) (*a*) Write down the quantum numbers for each electron in the gallium atom. (*b*) Which subshells are filled? (*c*) The last electron is in the 4p state; what are the possible values of the total angular momentum quantum number, j, for this electron? (*d*) Explain why the angular momentum of this last electron also represents the total angular momentum for the entire atom (ignoring any angular momentum of the nucleus). (*e*) How could you use a Stern-Gerlach experiment to determine which value of j the atom has?

*53. (III) The difference between the $2P_{3/2}$ and $2P_{1/2}$ energy levels in hydrogen is about 5×10^{-5} eV, due to the spin-orbit interaction. (*a*) Taking the electron's (orbital) magnetic moment to be 1 Bohr magneton, estimate the internal magnetic field due to the electron's orbital motion. (*b*) Estimate the internal magnetic field using a simple model of the nucleus revolving in a circle about the electron.

39–9 Lasers

54. (II) A laser used to weld detached retinas puts out 23-ms-long pulses of 640-nm light which average 0.63-W output during a pulse. How much energy can be deposited per pulse and how many photons does each pulse contain?

55. (II) Estimate the angular spread of a laser beam due to diffraction if the beam emerges through a 3.6-mm-diameter mirror. Assume that $\lambda = 694$ nm. What would be the diameter of this beam if it struck (*a*) a satellite 380 km above the Earth, (*b*) the Moon? [*Hint*: See Chapter 35.]

56. (II) A low-power laser used in a physics lab might have a power of 0.50 mW and a beam diameter of 3.0 mm. Calculate (*a*) the average light intensity of the laser beam, and (*b*) compare it to the intensity of a very powerful lightbulb emitting 100 W of light as viewed from a distance of 2.0 m.

57. (II) Calculate the wavelength of a He–Ne laser.

58. (II) Suppose that the energy level system in Fig. 39–20 is not being pumped and is in thermal equilibrium. Determine the fraction of atoms in levels E_2 and E_1 relative to E_0 at $T = 300$ K.

59. (II) To what temperature would the system in Fig. 39–20 have to be raised (see Problem 58) so that in thermal equilibrium the level E_2 would have half as many atoms as E_0? (Note that pumping mechanisms do not maintain thermal equilibrium.)

60. (II) Show that a population inversion for two levels (as in a pumped laser) corresponds to a negative Kelvin temperature in the Boltzmann distribution. Explain why such a situation does not contradict the idea that negative Kelvin temperatures cannot be reached in the normal sense of temperature.

General Problems

61. The ionization (binding) energy of the outermost electron in boron is 8.26 eV. (*a*) Use the Bohr model to estimate the "effective charge," Z_{eff}, seen by this electron. (*b*) Estimate the average orbital radius.

62. How many electrons can there be in an "h" subshell?

63. What is the full electron configuration in the ground state for elements with Z equal to (*a*) 25, (*b*) 34, (*c*) 39? [*Hint*: See the Periodic Table inside the back cover.]

64. What are the largest and smallest possible values for the angular momentum L of an electron in the $n = 6$ shell?

65. Estimate (*a*) the quantum number ℓ for the orbital angular momentum of the Earth about the Sun, and (*b*) the number of possible orientations for the plane of Earth's orbit.

66. Use the Bohr theory (especially Eq. 37–15) to show that the Moseley plot (Fig. 39–12) can be written

$$\sqrt{\frac{1}{\lambda}} = a(Z - b),$$

where $b \approx 1$, and evaluate a.

67. Determine the most probable distance from the nucleus of an electron in the $n = 2, \ell = 0$ state of hydrogen.

68. Show that the diffractive spread of a laser beam, $\approx \lambda/D$ as described in Section 39–9, is precisely what you might expect from the uncertainty principle. [*Hint*: Since the beam's width is constrained by the dimension of the aperture D, the component of the light's momentum perpendicular to the laser axis is uncertain.]

69. In the so-called **vector model** of the atom, space quantization of angular momentum (Fig. 39–3) is illustrated as shown in Fig. 39–28. The angular momentum vector of magnitude $L = \sqrt{\ell(\ell+1)}\,\hbar$ is thought of as precessing around the z axis (like a spinning top or gyroscope) in such a way that the z component of angular momentum, $L_z = m_\ell \hbar$, also stays constant. Calculate the possible values for the angle θ between $\vec{\mathbf{L}}$ and the z axis (*a*) for $\ell = 1$, (*b*) $\ell = 2$, and (*c*) $\ell = 3$. (*d*) Determine the minimum value of θ for $\ell = 100$ and $\ell = 10^6$. Is this consistent with the correspondence principle?

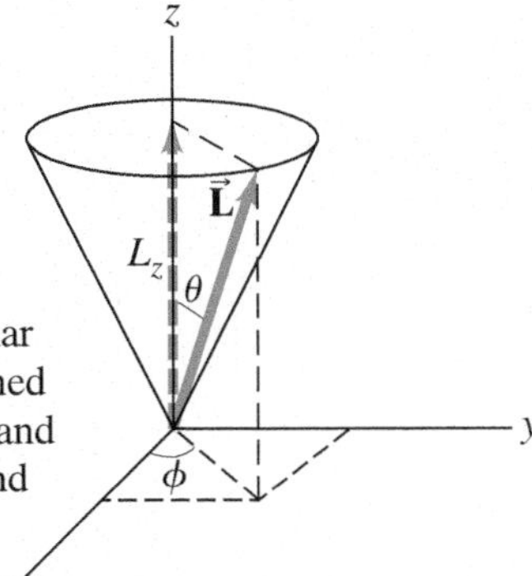

FIGURE 39–28 The vector model for orbital angular momentum. The orbital angular momentum vector $\vec{\mathbf{L}}$ is imagined to precess about the z axis; L and L_z remain constant, but L_x and L_y continually change. Problems 69 and 70.

70. The vector model (Problem 69) gives some insight into the uncertainty principle for angular momentum, which is

$$\Delta L_z \, \Delta\phi \geq \hbar$$

for the z component. Here ϕ is the angular position measured in the plane perpendicular to the z axis. Once m_ℓ for an atom is known, L_z is known precisely, so $\Delta L_z = 0$. (*a*) What does this tell us about ϕ? (*b*) What can you say about L_x and L_y, which are *not* quantized (only L and L_z are)? (*c*) Show that although L_x and L_y are not quantized, nonetheless $\left(L_x^2 + L_y^2\right)^{1/2} = \left[\ell(\ell+1) - m_\ell^2\right]^{1/2}\hbar$ is.

71. (*a*) Show that the mean value for $1/r$ of an electron in the ground state of hydrogen equals $1/r_0$, and from this conclude that the mean value of the potential energy is

$$\overline{U} = -\frac{1}{4\pi\epsilon_0}\frac{e^2}{r_0}.$$

(*b*) Using $E = \overline{U} + \overline{K}$, find a relationship between the average kinetic energy and the average potential energy in the ground state. [*Hint*: For (*a*), see Problem 19 or Example 38–9.]

72. The angular momentum in the hydrogen atom is given both by the Bohr model and by quantum mechanics. Compare the results for $n = 2$.

73. For each of the following atomic transitions, state whether the transition is *allowed* or *forbidden*, and why: (*a*) $4p \rightarrow 3p$; (*b*) $3p \rightarrow 1s$; (*c*) $4d \rightarrow 3d$; (*d*) $4d \rightarrow 3s$; (*e*) $4s \rightarrow 2p$.

74. It is possible for atoms to be excited into states with very high values of the principal quantum number. Electrons in these so-called *Rydberg states* have very small ionization energies and huge orbital radii. This makes them particularly sensitive to external perturbation, as would be the case if the atom were in an electric field. Consider the $n = 45$ state of the hydrogen atom. Determine the binding energy, the radius of the orbit, and the effective cross-sectional area of this Rydberg state.

75. Suppose that the spectrum of an unknown element shows a series of lines with one out of every four matching a line from the Lyman series of hydrogen. Assuming that the unknown element is an ion with Z protons and one electron, determine Z and the element in question.

*76. Suppose that the splitting of energy levels shown in Fig. 39–4 was produced by a 1.6-T magnetic field. (*a*) What is the separation in energy between adjacent m_ℓ levels for the same ℓ? (*b*) How many different wavelengths will there be for $3d$ to $2p$ transitions, if m_ℓ can change only by ± 1 or 0? (*c*) What is the wavelength for each of these transitions?

77. *Populations in the H atom.* Use the Boltzmann factor (Eq. 39–16) to estimate the fraction of H atoms in the $n = 2$ and $n = 3$ levels (relative to the ground state) for thermal equilibrium at (*a*) $T = 300$ K and (*b*) $T = 6000$ K. [Note: Since there are eight states with $n = 2$ and only two with $n = 1$, multiply your result for $n = 2$ by $\frac{8}{2} = 4$; do similarly for $n = 3$.] (*c*) Given 1.0 g of hydrogen, estimate the number of atoms in each state at $T = 6000$ K. (*d*) Estimate the number of $n = 3$ to $n = 1$ and $n = 2$ to $n = 1$ photons that will be emitted per second at $T = 6000$ K. Assume that the lifetime of each excited state is 10^{-8} s. [*Hint*: To evaluate a large exponent, you can use base-10 logarithms, Appendix A.]

Answers to Exercises

A: 2, 1, 0, −1, −2.

B: (*b*).

C: (*d*).

D: Add one line to Li in Table 39–2: 2, 0, 0, $-\frac{1}{2}$.

E: (*b*), (*c*).

F: $1s^2 2s^2 2p^6 3s^2 3p^6 3d^{10} 4s^2 4p^1$.

G: (*d*).

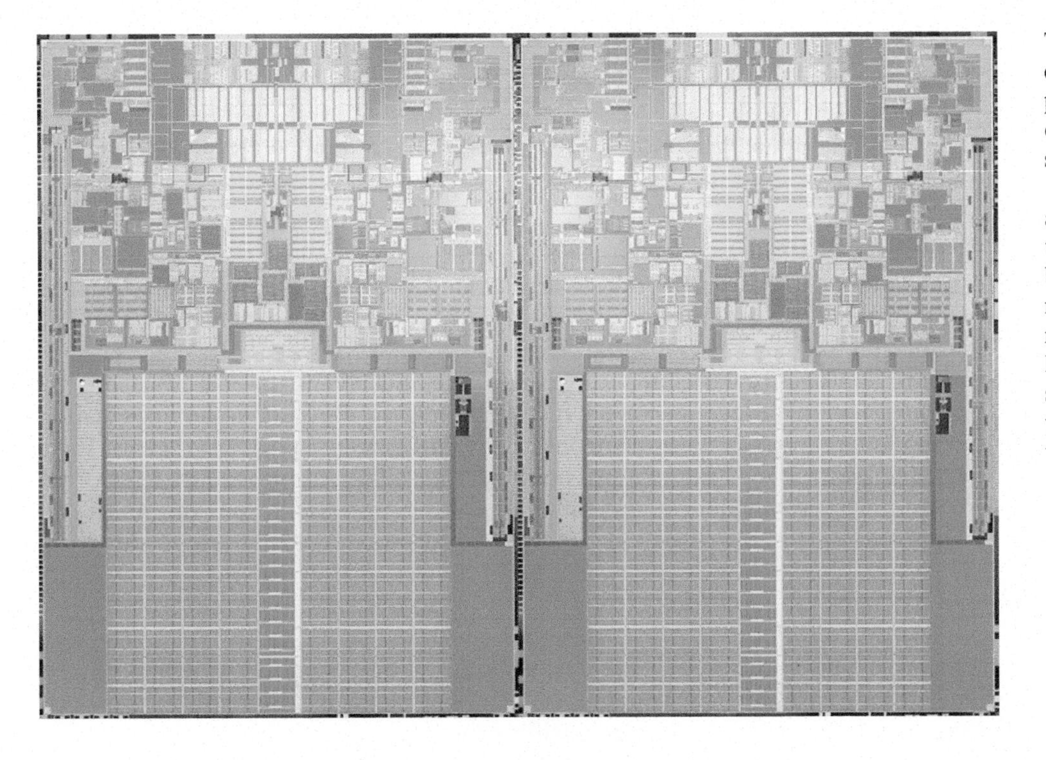

This computer processor chip contains over 800 million transistors, plus diodes and other semiconductor electronic elements, all in a space smaller than a penny.

Before discussing semiconductors and their applications, we study the quantum theory description of bonding between atoms to form molecules, and how it explains molecular behavior. We then examine how atoms and molecules form solids, with emphasis on metals as well as on semiconductors and their use in electronics.

Molecules and Solids

CHAPTER-OPENING QUESTION—**Guess now!**

As a metal is heated, how does the rms speed (v_{rms}) of the electrons change? Assume a temperature change of about 30C° near room temperature.

(a) v_{rms} increases linearly with temperature.
(b) v_{rms} decreases linearly with temperature.
(c) v_{rms} increases exponentially with temperature.
(d) v_{rms} decreases exponentially with temperature.
(e) v_{rms} changes very little as the temperature is increased.

CONTENTS

Since its development in the 1920s, quantum mechanics has had a profound influence on our lives, both intellectually and technologically. Even the way we view the world has changed, as we have seen in the last few Chapters. Now we discuss how quantum mechanics has given us an understanding of the structure of molecules and matter in bulk, as well as a number of important applications including semiconductor devices.

40–1 Bonding in Molecules

One of the great successes of quantum mechanics was to give scientists, at last, an understanding of the nature of chemical bonds. Since it is based in physics, and because this understanding is so important in many fields, we discuss it here.

By a molecule, we mean a group of two or more atoms that are strongly held together so as to function as a single unit. When atoms make such an attachment, we say that a chemical **bond** has been formed. There are two main types of strong chemical bond: covalent and ionic. Many bonds are actually intermediate between these two types.

Covalent Bonds

To understand how *covalent bonds* are formed, we take the simplest case, the bond that holds two hydrogen atoms together to form the hydrogen molecule, H_2. The mechanism is basically the same for other covalent bonds. As two H atoms approach each other, the electron clouds begin to overlap, and the electrons from each atom can "orbit" both nuclei. (This is sometimes called "sharing" electrons.) If both electrons are in the ground state $(n = 1)$ of their respective atoms, there are two possibilities: their spins can be parallel (both up or both down), in which case the total spin is $S = \frac{1}{2} + \frac{1}{2} = 1$; or their spins can be opposite ($m_s = +\frac{1}{2}$ for one, $m_s = -\frac{1}{2}$ for the other), so that the total spin $S = 0$. We shall now see that a bond is formed only for the $S = 0$ state, when the spins are opposite.

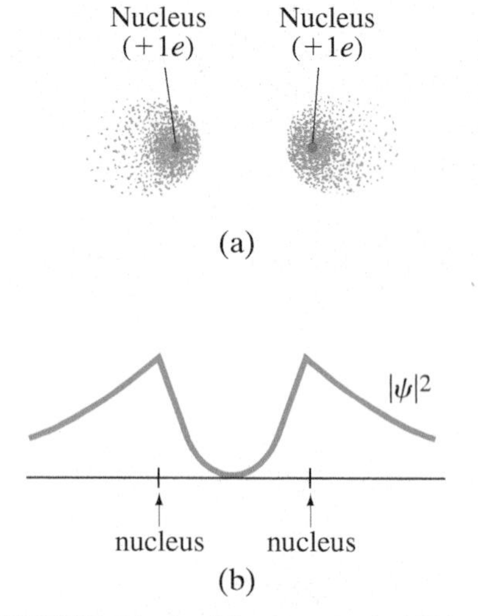

FIGURE 40–1 Electron probability distribution (electron cloud) for two H atoms when the spins are the same $(S = 1)$: (a) electron cloud; (b) projection of $|\psi|^2$ along the line through the centers of the two atoms.

First we consider the $S = 1$ state, for which the spins are the same. The two electrons cannot both be in the lowest energy state and be attached to the same atom, for then they would have identical quantum numbers in violation of the exclusion principle. The exclusion principle tells us that, because no two electrons can occupy the same quantum state, if two electrons have the same quantum numbers, they must be different in some other way—namely, by being in different places in space (for example, attached to different atoms). Thus, for $S = 1$, when the two atoms approach each other, the electrons will stay away from each other as shown by the probability distribution of Fig. 40–1. The electrons spend very little time between the two nuclei, so the positively charged nuclei repel each other and no bond is formed.

For the $S = 0$ state, on the other hand, the spins are opposite and the two electrons are consequently in different quantum states (m_s is different, $+\frac{1}{2}$ for one, $-\frac{1}{2}$ for the other). Hence the two electrons can come close together, and the probability distribution looks like Fig. 40–2: the electrons can spend much of their time between the two nuclei. The two positively charged nuclei are attracted to the negatively charged electron cloud between them, and this is the attraction that holds the two hydrogen atoms together to form a hydrogen molecule. This is a **covalent bond**.

FIGURE 40–2 Electron probability distribution for two H atoms when the spins are opposite $(S = 0)$: (a) electron cloud; (b) projection of $|\psi|^2$ along the line through the centers of the atoms. In this case a bond is formed because the positive nuclei are attracted to the concentration of negative charge between them. This is a hydrogen molecule, H_2.

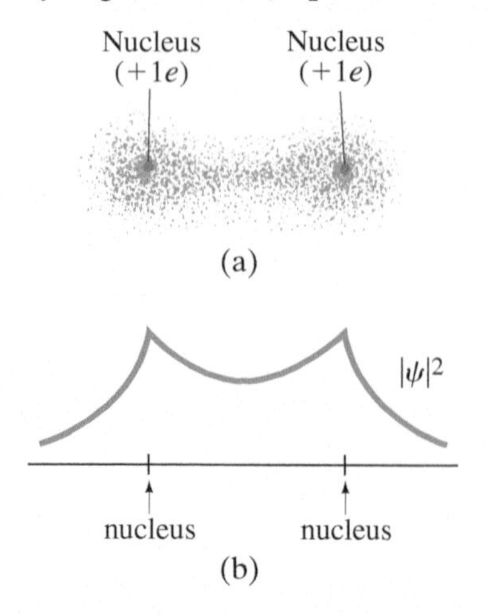

The probability distributions of Figs. 40–1 and 40–2 can perhaps be better understood on the basis of waves. What the exclusion principle requires is that when the spins are the same, there is destructive interference of the electron wave functions in the region between the two atoms. But when the spins are opposite, constructive interference occurs in the region between the two atoms, resulting in a large amount of negative charge there. Thus a covalent bond can be said to be the result of constructive interference of the electron wave functions in the space between the two atoms, and of the electrostatic attraction of the two positive nuclei for the negative charge concentration between them.

Why a bond is formed can also be understood from the energy point of view. When the two H atoms approach close to one another, if the spins of their electrons are opposite, the electrons can occupy the same space, as discussed above. This means that each electron can now move about in the space of two atoms instead of in the volume of only one. Because each electron now occupies more space, it is less well localized. From the uncertainty principle with Δx larger, we see that Δp and the minimum momentum can be less. With less momentum, each electron has less energy when the two atoms combine than when they are separate. That is, the molecule has less energy than the two separate atoms, and so is more stable. An energy input is required to break the H_2 molecule into two separate H atoms, so the H_2 molecule is a stable entity.

This is what we mean by a *bond*. The energy required to break a bond is called the **bond energy**, the **binding energy**, or the **dissociation energy**. For the hydrogen molecule, H_2, the bond energy is 4.5 eV.

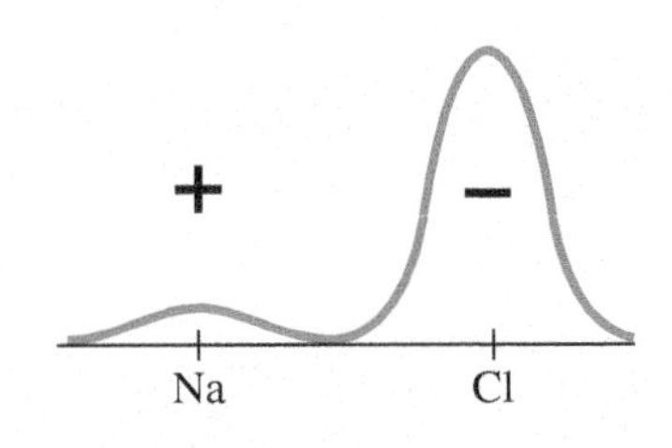

FIGURE 40–3 Probability distribution $|\psi|^2$ for the outermost electron of Na in NaCl.

Ionic Bonds

An *ionic bond* is, in a sense, a special case of the covalent bond. Instead of the electrons being shared equally, they are shared unequally. For example, in sodium chloride (NaCl), the outer electron of the sodium spends nearly all its time around the chlorine (Fig. 40–3). The chlorine atom acquires a net negative charge as a result of the extra electron, whereas the sodium atom is left with a net positive charge. The electrostatic attraction between these two charged atoms holds them together. The resulting bond is called an **ionic bond** because it is created by the attraction between the two ions (Na^+ and Cl^-). But to understand the ionic bond, we must understand why the extra electron from the sodium spends so much of its time around the chlorine. After all, the chlorine atom is neutral; why should it attract another electron?

The answer lies in the probability distributions of the two neutral atoms. Sodium contains 11 electrons, 10 of which are in spherically symmetric closed shells (Fig. 40–4). The last electron spends most of its time beyond these closed shells. Because the closed shells have a total charge of $-10e$ and the nucleus has charge $+11e$, the outermost electron in sodium "feels" a net attraction due to $+1e$. It is not held very strongly. On the other hand, 12 of chlorine's 17 electrons form closed shells, or subshells (corresponding to $1s^2 2s^2 2p^6 3s^2$). These 12 electrons form a spherically symmetric shield around the nucleus. The other five electrons are in $3p$ states whose probability distributions are not spherically symmetric and have a form similar to those for the $2p$ states in hydrogen shown in Fig. 39–9. Four of these $3p$ electrons can have "doughnut-shaped" distributions symmetric about the z axis, as shown in Fig. 40–5. The fifth can have a "barbell-shaped" distribution (as for $m_\ell = 0$ in Fig. 39–9), which in Fig. 40–5 is shown only in dashed outline because it is half empty. That is, the exclusion principle allows one more electron to be in this state (it will have spin opposite to that of the electron already there). If an extra electron—say from a Na atom—happens to be in the vicinity, it can be in this state, perhaps at point x in Fig. 40–5. It could experience an attraction due to as much as $+5e$ because the $+17e$ of the nucleus is partly shielded at this point by the 12 inner electrons. Thus, the outer electron of a sodium atom will be more strongly attracted by the $+5e$ of the chlorine atom than by the $+1e$ of its own atom. This, combined with the strong attraction between the two ions when the extra electron stays with the Cl^-, produces the charge distribution of Fig. 40–3, and hence the ionic bond.

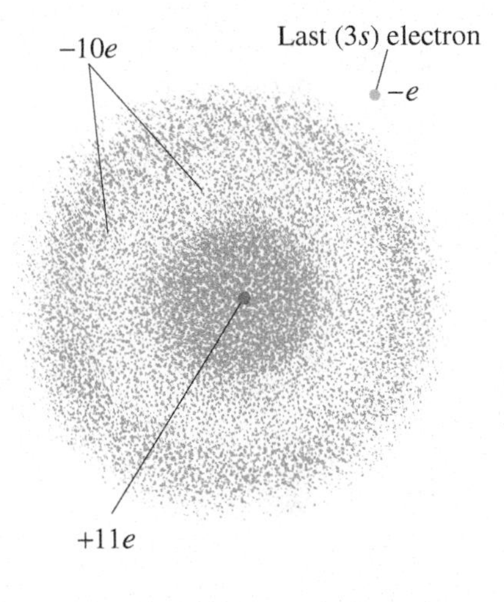

FIGURE 40–4 In a neutral sodium atom, the 10 inner electrons shield the nucleus, so the single outer electron is attracted by a net charge of $+1e$.

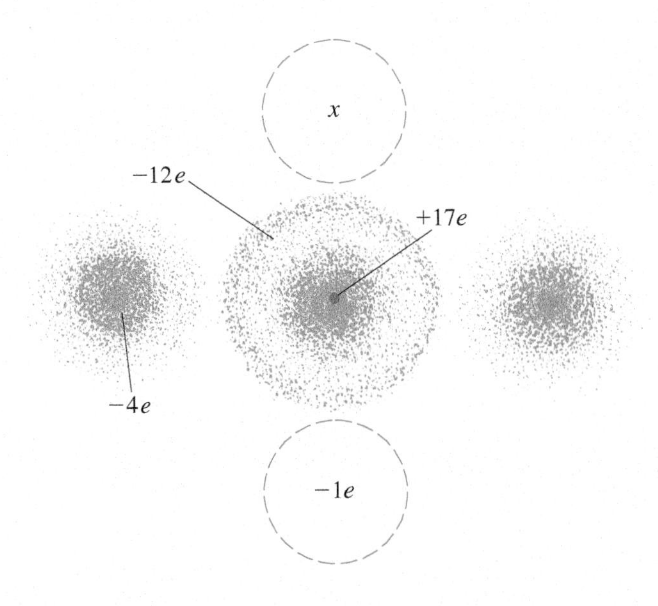

FIGURE 40–5 Neutral chlorine atom. The $+17e$ of the nucleus is shielded by the 12 electrons in the inner shells and subshells. Four of the five $3p$ electrons are shown in doughnut-shaped clouds (seen in cross section at left and right), and the fifth is in the dashed-line cloud concentrated about the z axis (vertical). An extra electron at x will be attracted by a net charge that can be as much as $+5e$.

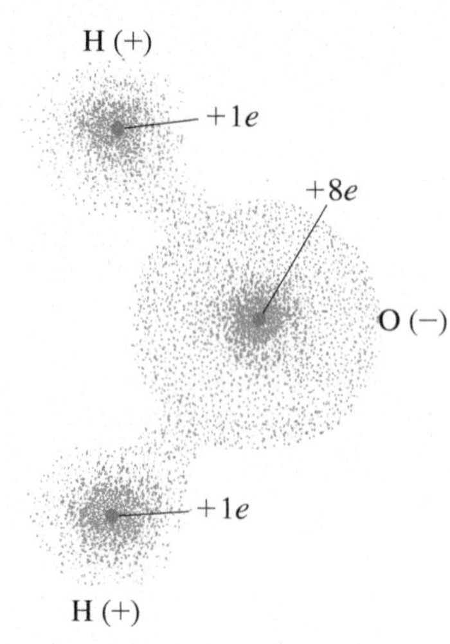

FIGURE 40–6 The water molecule H_2O is polar.

Partial Ionic Character of Covalent Bonds

A pure covalent bond in which the electrons are shared equally occurs mainly in symmetrical molecules such as H_2, O_2, and Cl_2. When the atoms involved are different from each other, it is usual to find that the shared electrons are more likely to be in the vicinity of one atom than the other. The extreme case is an ionic bond; in intermediate cases the covalent bond is said to have a *partial ionic character*. The molecules themselves are **polar**—that is, one part (or parts) of the molecule has a net positive charge and other parts a net negative charge. An example is the water molecule, H_2O (Fig. 40–6). The shared electrons are more likely to be found around the oxygen atom than around the two hydrogens. The reason is similar to that discussed above in connection with ionic bonds. Oxygen has eight electrons ($1s^2 2s^2 2p^4$), of which four form a spherically symmetric core and the other four could have, for example, a doughnut-shaped distribution. The barbell-shaped distribution on the z axis (like that shown dashed in Fig. 40–5) could be empty, so electrons from hydrogen atoms can be attracted by a net charge of $+4e$. They are also attracted by the H nuclei, so they partly orbit the H atoms as well as the O atom. The net effect is that there is a net positive charge on each H atom (less than $+1e$), because the electrons spend only part of their time there. And, there is a net negative charge on the O atom.

40–2 Potential-Energy Diagrams for Molecules

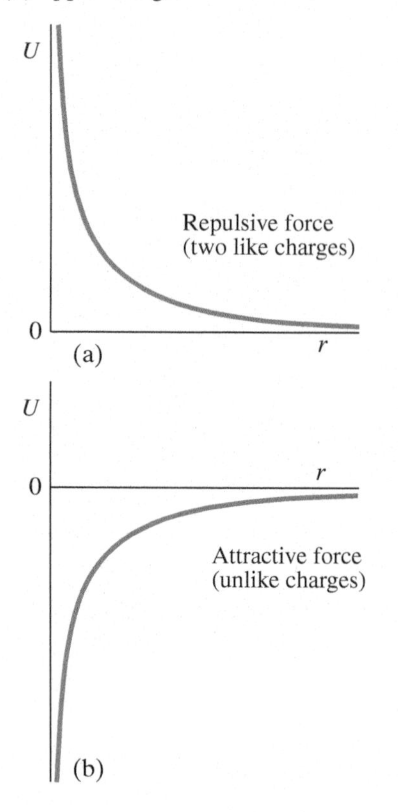

FIGURE 40–7 Potential energy U as a function of separation r for two point charges of (a) like sign and (b) opposite sign.

It is useful to analyze the interaction between two objects—say, between two atoms or molecules—with the use of a potential-energy diagram, which is a plot of the potential energy versus the separation distance.

For the simple case of two point charges, q_1 and q_2, the potential energy U is given by (see Chapter 23)

$$U = \frac{1}{4\pi\epsilon_0}\frac{q_1 q_2}{r},$$

where r is the distance between the charges, and the constant $(1/4\pi\epsilon_0)$ is equal to $9.0 \times 10^9\,\mathrm{N\cdot m^2/C^2}$. If the two charges have the same sign, the potential energy U is positive for all values of r, and a graph of U versus r in this case is shown in Fig. 40–7a. The force is repulsive (the charges have the *same* sign) and the curve rises as r decreases; this makes sense since work is done to bring the charges together, thereby increasing their potential energy. If, on the other hand, the two charges are of the *opposite* sign, the potential energy is negative because the product $q_1 q_2$ is negative. The force is attractive in this case, and the graph of $U\,(\propto -1/r)$ versus r looks like Fig. 40–7b. The potential energy becomes more *negative* as r decreases.

Now let us look at the potential-energy diagram for the formation of a covalent bond, such as for the hydrogen molecule, H_2. The potential energy U of one H atom in the presence of the other is plotted in Fig. 40–8. Starting at large r, U decreases as the atoms approach, because the electrons concentrate between the two nuclei (Fig. 40–2), so attraction occurs. However, at very short distances, the electrons would be "squeezed out"—there is no room for them between the two nuclei.

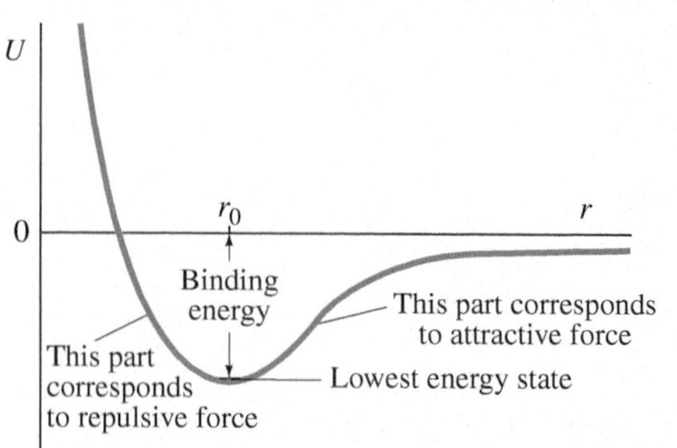

FIGURE 40–8 (right) Potential-energy diagram for H_2 molecule; r is the separation of the two H atoms. The binding energy (the energy difference between $U = 0$ and the lowest energy state near the bottom of the well) is 4.5 eV, and $r_0 = 0.074$ nm.

Without the electrons between them, each nucleus would feel a repulsive force due to the other, so the curve rises as r decreases further. There is an optimum separation of the atoms, r_0 in Fig. 40–8, at which the energy is lowest. This is the point of greatest stability for the hydrogen molecule, and r_0 is the average separation of atoms in the H_2 molecule. The depth of this "well" is the *binding energy*,† as shown. This is how much energy must be put into the system to separate the two atoms to infinity, where $U = 0$. For the H_2 molecule, the binding energy is about 4.5 eV and $r_0 = 0.074\,\text{nm}$.

In molecules made of larger atoms, say, oxygen or nitrogen, repulsion also occurs at short distances, because the closed inner electron shells begin to overlap and the exclusion principle forbids their coming too close. The repulsive part of the curve rises even more steeply than $1/r$. A reasonable approximation to the potential energy, at least in the vicinity of r_0, is

$$U = -\frac{A}{r^m} + \frac{B}{r^n}, \tag{40–1}$$

where A and B are constants associated with the attractive and repulsive parts of the potential energy and the exponents m and n are small integers. For ionic and some covalent bonds, the attractive term can often be written with $m = 1$ (Coulomb potential).

For many bonds, the potential-energy curve has the shape shown in Fig. 40–9. There is still an optimum distance r_0 at which the molecule is stable. But when the atoms approach from a large distance, the force is initially repulsive rather than attractive. The atoms thus do not interact spontaneously. Instead, some additional energy must be injected into the system to get it over the "hump" (or barrier) in the potential-energy diagram. This required energy is called the **activation energy**.

The curve of Fig. 40–9 is much more common than that of Fig. 40–8. The activation energy often reflects a need to break other bonds, before the one under discussion can be made. For example, to make water from O_2 and H_2, the H_2 and O_2 molecules must first be broken into H and O atoms by an input of energy; this is what the activation energy represents. Then the H and O atoms can combine to form H_2O with the release of a great deal more energy than was put in initially. The initial activation energy can be provided by applying an electric spark to a mixture of H_2 and O_2, breaking a few of these molecules into H and O atoms. The resulting explosive release of energy when these atoms combine to form H_2O quickly provides the activation energy needed for further reactions, so additional H_2 and O_2 molecules are broken up and recombined to form H_2O.

The potential-energy diagrams for ionic bonds can have similar shapes. In NaCl, for example, the Na^+ and Cl^- ions attract each other at distances a bit larger than some r_0, but at shorter distances the overlapping of inner electron shells gives rise to repulsion. The two atoms thus are most stable at some intermediate separation, r_0, and for many bonds there is an activation energy.

†The binding energy corresponds not quite to the bottom of the potential energy curve, but to the lowest quantum energy state, slightly above the bottom, as shown in Fig. 40–8.

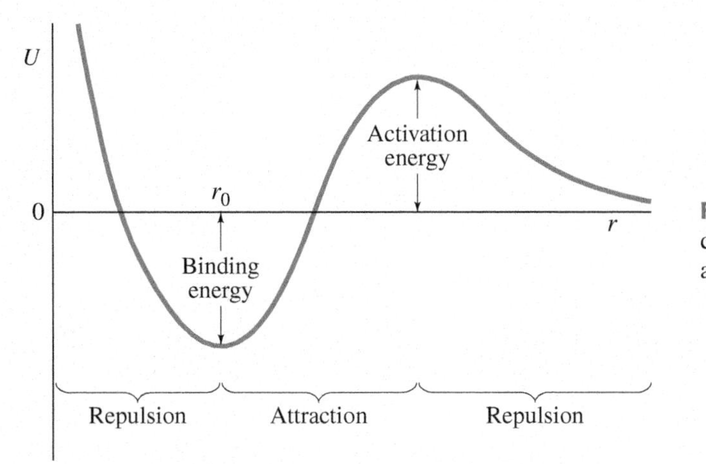

FIGURE 40–9 Potential-energy diagram for a bond requiring an activation energy.

EXAMPLE 40–1 ESTIMATE **Sodium chloride bond.** A potential-energy diagram for the NaCl ionic bond is shown in Fig. 40–10, where we have set $U = 0$ for free Na and Cl neutral atoms (which are represented on the right in Fig. 40–10). Measurements show that 5.14 eV are required to remove an electron from a neutral Na atom to produce the Na^+ ion; and 3.61 eV of energy is released when an electron is "grabbed" by a Cl atom to form the Cl^- ion. Thus, forming Na^+ and Cl^- ions from neutral Na and Cl atoms requires $5.14\,eV - 3.61\,eV = 1.53\,eV$ of energy, a form of activation energy. This is shown as the "bump" in Fig. 40–10. But note that the potential-energy diagram from here out to the right is not really a function of distance—it is drawn dashed to remind us that it only represents the energy difference between the ions and the neutral atoms (for which we have chosen $U = 0$). (*a*) Calculate the separation distance, r_1, at which the potential of the Na^+ and Cl^- ions drops to zero (measured value is $r_1 = 0.94$ nm). (*b*) Estimate the binding energy of the NaCl bond, which occurs at a separation $r_0 = 0.24$ nm. Ignore the repulsion of the overlapping electron shells that occurs at this distance (and causes the rise of the potential-energy curve for $r < r_0$, Fig. 40–10).

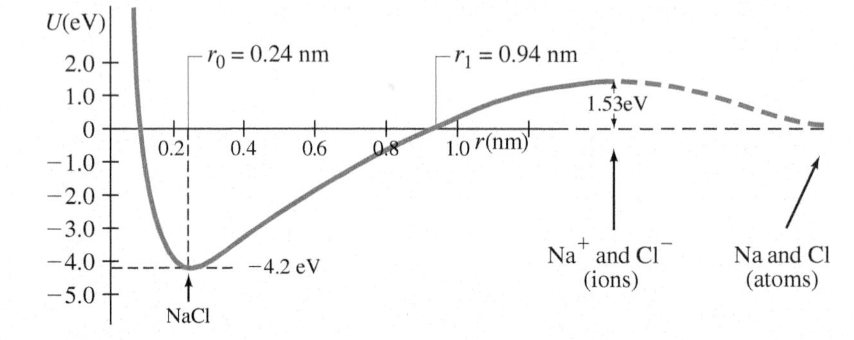

FIGURE 40–10 Example 40–1. Potential-energy diagram for the NaCl bond. Beyond about $r = 1.2$ nm, the diagram is schematic only, and represents the energy difference between ions and neutral atoms. $U = 0$ is chosen for the two separated atoms Na and Cl (not for the ions). [For the two ions, Na^+ and Cl^-, the zero of potential energy at $r = \infty$ corresponds to $U \approx 1.53$ eV on this diagram.]

SOLUTION (*a*) The potential energy of two point charges is given by Coulomb's law:

$$U' = \frac{1}{4\pi\epsilon_0}\frac{q_1 q_2}{r},$$

where we distinguish U' from the U in Fig. 40–10. This formula works for our two ions if we set $U' = 0$ at $r = \infty$, which in the plot of Fig. 40–10 corresponds to $U = +1.53$ eV (Fig. 40–10 is drawn for $U = 0$ for the free *atoms*). The point r_1 in Fig. 40–10 corresponds to $U' = -1.53$ eV relative to the two free ions. We solve the U' equation above for r, setting $q_1 = +e$ and $q_2 = -e$:

$$r_1 = \frac{1}{4\pi\epsilon_0}\frac{q_1 q_2}{U'} = \frac{(9.0\times10^9\,\mathrm{N\cdot m^2/C^2})(-1.60\times10^{-19}\,\mathrm{C})(+1.60\times10^{-19}\,\mathrm{C})}{(-1.53\,\mathrm{eV})(1.60\times10^{-19}\,\mathrm{J/eV})}$$
$$= 0.94\,\mathrm{nm},$$

which is just the measured value.

(*b*) At $r_0 = 0.24$ nm, the potential energy of the two ions (relative to $r = \infty$ for the two ions) is

$$U' = \frac{1}{4\pi\epsilon_0}\frac{q_1 q_2}{r}$$
$$= \frac{(9.0\times10^9\,\mathrm{N\cdot m^2/C^2})(-1.60\times10^{-19}\,\mathrm{C})(+1.60\times10^{-19}\,\mathrm{C})}{(0.24\times10^{-9}\,\mathrm{m})(1.60\times10^{-19}\,\mathrm{J/eV})} = -6.0\,\mathrm{eV}.$$

Thus, we estimate that 6.0 eV of energy is given up when Na^+ and Cl^- ions form a NaCl bond. Put another way, it takes 6.0 eV to break the NaCl bond and form the Na^+ and Cl^- ions. To get the binding energy—the energy to separate the NaCl into Na and Cl *atoms*—we need to subtract out the 1.53 eV (the "bump" in Fig. 40–10) needed to ionize them:

$$\text{binding energy} = 6.0\,\mathrm{eV} - 1.53\,\mathrm{eV} = 4.5\,\mathrm{eV}.$$

The measured value (shown on Fig. 40–10) is 4.2 eV. The difference can be attributed to the energy associated with the repulsion of the electron shells at this distance.

PHYSICS APPLIED
ATP and energy in the cell

Sometimes the potential energy of a bond looks like that of Fig. 40–11. In this case, the energy of the bonded molecule, at a separation r_0, is greater than when there is no bond ($r = \infty$). That is, an energy *input* is required to make the bond (hence the binding energy is negative), and there is energy release when the bond is broken.

Such a bond is stable only because there is the barrier of the activation energy. This type of bond is important in living cells, for it is in such bonds that energy can be stored efficiently in certain molecules, particularly ATP (adenosine triphosphate). The bond that connects the last phosphate group (designated Ⓟ in Fig. 40–11) to the rest of the molecule (ADP, meaning adenosine diphosphate, since it contains only two phosphates) has potential energy of the shape shown in Fig. 40–11. Energy is stored in this bond. When the bond is broken (ATP $\rightarrow$ ADP + Ⓟ), energy is released and this energy can be used to make other chemical reactions "go."

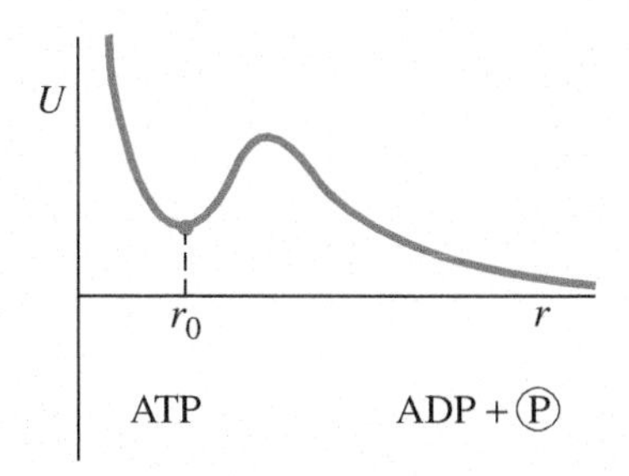

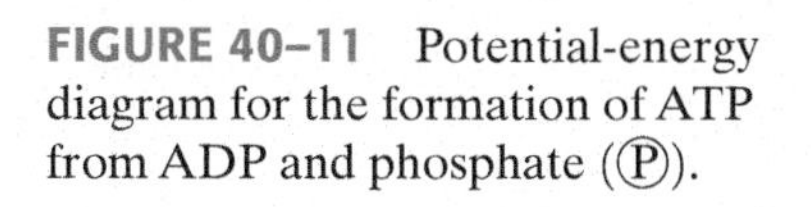

FIGURE 40–11 Potential-energy diagram for the formation of ATP from ADP and phosphate (Ⓟ).

In living cells, many chemical reactions have activation energies that are often on the order of several eV. Such energy barriers are not easy to overcome in the cell. This is where enzymes come in. They act as *catalysts*, which means they act to lower the activation energy so that reactions can occur that otherwise would not. Enzymes act via the electrostatic force to distort the bonding electron clouds, so that the initial bonds are easily broken.

EXAMPLE 40–2 **Bond length.** Suppose a diatomic molecule has a potential energy given by $U = -(1/4\pi\epsilon_0)(e^2/r) + B/r^6$ where $B = 1.0 \times 10^{-78}\,\mathrm{J\cdot m^6}$ and r is the distance between the centers of the two atoms. Determine the expected equilibrium separation of the two atoms (bond length of the molecule).

APPROACH The force between the two atoms is given by $F = -dU/dr$ (analogous to Eq. 8–7). The classical equilibrium separation is found by setting $F = 0$.

SOLUTION

$$F = -dU/dr = (1/4\pi\epsilon_0)(-e^2/r^2) - B(-6/r^7) = -(1/4\pi\epsilon_0)(e^2/r^2) + 6B/r^7 = 0.$$

Then

$$r^5 = 6B(4\pi\epsilon_0)/e^2 = 2.6 \times 10^{-50}\,\mathrm{m}^5.$$

So $r = 1.2 \times 10^{-10}\,\mathrm{m} = 0.12\,\mathrm{nm}$.

NOTE This can be shown to be a stable equilibrium point by checking the sign of d^2U/dr^2, or by evaluating F at positions slightly larger and smaller than the equilibrium position.

40–3 Weak (van der Waals) Bonds

Once a bond between two atoms or ions is made, energy must normally be supplied to break the bond and separate the atoms. As mentioned in Section 40–1, this energy is called the *bond energy* or *binding energy*. The binding energy for covalent and ionic bonds is typically 2 to 5 eV. These bonds, which hold atoms together to *form* molecules, are often called **strong bonds** to distinguish them from so-called "weak bonds." The term **weak bond**, as we use it here, refers to attachments *between* molecules due to simple electrostatic attraction—such as *between* polar molecules (and not *within* a polar molecule, which is a strong bond). The strength of the attachment is much less than for the strong bonds. Binding energies are typically in the range 0.04 to 0.3 eV—hence their name "weak bonds."

Weak bonds are generally the result of attraction between dipoles (Sections 21–11, 23–6). For example, Fig. 40–12 shows two molecules, which have permanent dipole moments, attracting one another. Besides such **dipole–dipole bonds**, there can also be **dipole–induced dipole bonds**, in which a polar molecule with a permanent dipole moment can induce a dipole moment in an otherwise electrically balanced (nonpolar) molecule, just as a single charge can induce a separation of charge in a nearby object (see Fig. 21–7). There can even be an attraction between two nonpolar molecules, because their electrons are moving about: at any instant there may be a transient separation of charge, creating a brief dipole moment and weak attraction. All these weak bonds are referred to as **van der Waals bonds**, and the forces involved **van der Waals forces**. The potential energy has the general shape shown in Fig. 40–8, with the attractive van der Waals potential energy varying as $1/r^6$.

FIGURE 40–12 The $C^+ — O^-$ and $H^+ — N^-$ dipoles attract each other. (These dipoles may be part of, for example, the nucleotide bases cytosine and guanine in DNA molecules. See Fig. 40–13.) The + and − charges typically have magnitudes of a fraction of e.

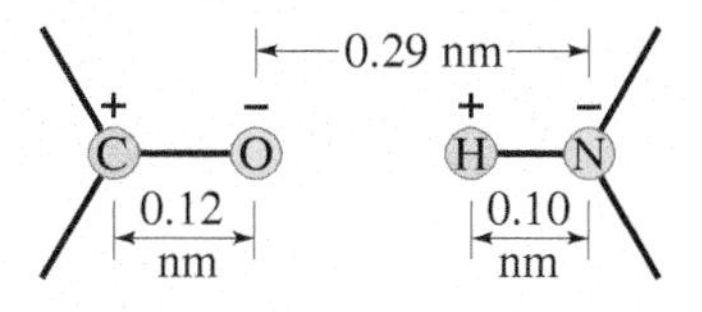

When one of the atoms in a dipole–dipole bond is hydrogen, as in Fig. 40–12, it is called a **hydrogen bond**. A hydrogen bond is generally the strongest of the weak bonds, because the hydrogen atom is the smallest atom and can be approached more closely. Hydrogen bonds also have a partial "covalent" character: that is, electrons between the two dipoles may be shared to a small extent, making a stronger, more lasting bond.

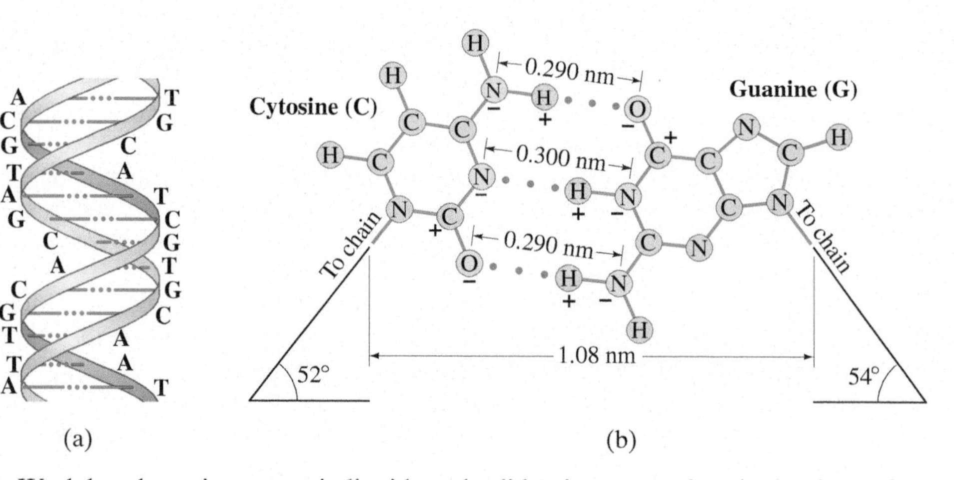

FIGURE 40–13 (a) Section of a DNA double helix. The red dots represent hydrogen bonds between the two strands. (b) "Close-up" view: cytosine (C) and guanine (G) molecules on separate strands of a DNA double helix are held together by the hydrogen bonds (red dots) involving an H^+ on one molecule attracted to an N^- or $C^+—O^-$ of a molecule on the adjacent chain. See also Section 21–12 and Figs. 21–47 and 21–48.

PHYSICS APPLIED
DNA

Weak bonds are important in liquids and solids when strong bonds are absent (see Section 40–5). They are also very important for understanding the activities of cells, such as the double helix shape of DNA (Fig. 40–13), and DNA replication. The average kinetic energy of molecules in a living cell at normal temperatures ($T \approx 300$ K) is around $\frac{3}{2}kT \approx 0.04$ eV, about the magnitude of weak bonds. This means that a weak bond can readily be broken just by a molecular collision. Hence weak bonds are not very permanent—they are, instead, brief attachments. This helps them play particular roles in the cell. On the other hand, strong bonds—those that hold molecules together—are almost never broken simply by molecular collision. Thus they are relatively permanent. They can be broken by chemical action (the making of even stronger bonds), and this usually happens in the cell with the aid of an enzyme, which is a protein molecule.

EXAMPLE 40–3 **Nucleotide energy.** Calculate the potential energy between a C=O dipole of the nucleotide base cytosine and the nearby H—N dipole of guanine, assuming that the two dipoles are lined up as shown in Fig. 40–12. Dipole moment ($= q\ell$) measurements (see Table 23–2 and Fig. 40–12) give

$$q_{\mathrm{H}} = -q_{\mathrm{N}} = \frac{3.0 \times 10^{-30}\,\mathrm{C\cdot m}}{0.10 \times 10^{-9}\,\mathrm{m}} = 3.0 \times 10^{-20}\,\mathrm{C} = 0.19e,$$

and

$$q_{\mathrm{C}} = -q_{\mathrm{O}} = \frac{8.0 \times 10^{-30}\,\mathrm{C\cdot m}}{0.12 \times 10^{-9}\,\mathrm{m}} = 6.7 \times 10^{-20}\,\mathrm{C} = 0.42e.$$

APPROACH We want to find the potential energy of the two charges in one dipole due to the two charges in the other, since this will be equal to the work needed to pull the two dipoles infinitely far apart. The potential energy U of a charge q_1 in the presence of a charge q_2 is $U = (1/4\pi\epsilon_0)(q_1 q_2/r_{12})$ where $1/4\pi\epsilon_0 = 9.0 \times 10^9\,\mathrm{N\cdot m^2/C^2}$ and r_{12} is the distance between the two charges.

SOLUTION The potential energy U consists of four terms:

$$U = U_{\mathrm{CH}} + U_{\mathrm{CN}} + U_{\mathrm{OH}} + U_{\mathrm{ON}},$$

where U_{CH} means the potential energy of C in the presence of H, and similarly for the other terms. We do not have terms corresponding to C and O, or N and H, because the two dipoles are assumed to be stable entities. Then, using the distances shown in Fig. 40–12, we get:

$$\begin{aligned} U &= \frac{1}{4\pi\epsilon_0}\left[\frac{q_{\mathrm{C}}q_{\mathrm{H}}}{r_{\mathrm{CH}}} + \frac{q_{\mathrm{C}}q_{\mathrm{N}}}{r_{\mathrm{CN}}} + \frac{q_{\mathrm{O}}q_{\mathrm{H}}}{r_{\mathrm{OH}}} + \frac{q_{\mathrm{O}}q_{\mathrm{N}}}{r_{\mathrm{ON}}}\right] \\ &= (9.0 \times 10^9\,\mathrm{N\cdot m^2/C^2})(6.7)(3.0)\left(\frac{1}{0.31} - \frac{1}{0.41} - \frac{1}{0.19} + \frac{1}{0.29}\right)\frac{(10^{-20}\,\mathrm{C})^2}{(10^{-9}\,\mathrm{m})} \\ &= -1.86 \times 10^{-20}\,\mathrm{J} = -0.12\,\mathrm{eV}. \end{aligned}$$

The potential energy is negative, meaning 0.12 eV of work (or energy input) is required to separate the dipoles. That is, the binding energy of this "weak" or hydrogen bond is 0.12 eV. This is only an estimate, of course, since other charges in the vicinity would have an influence too.

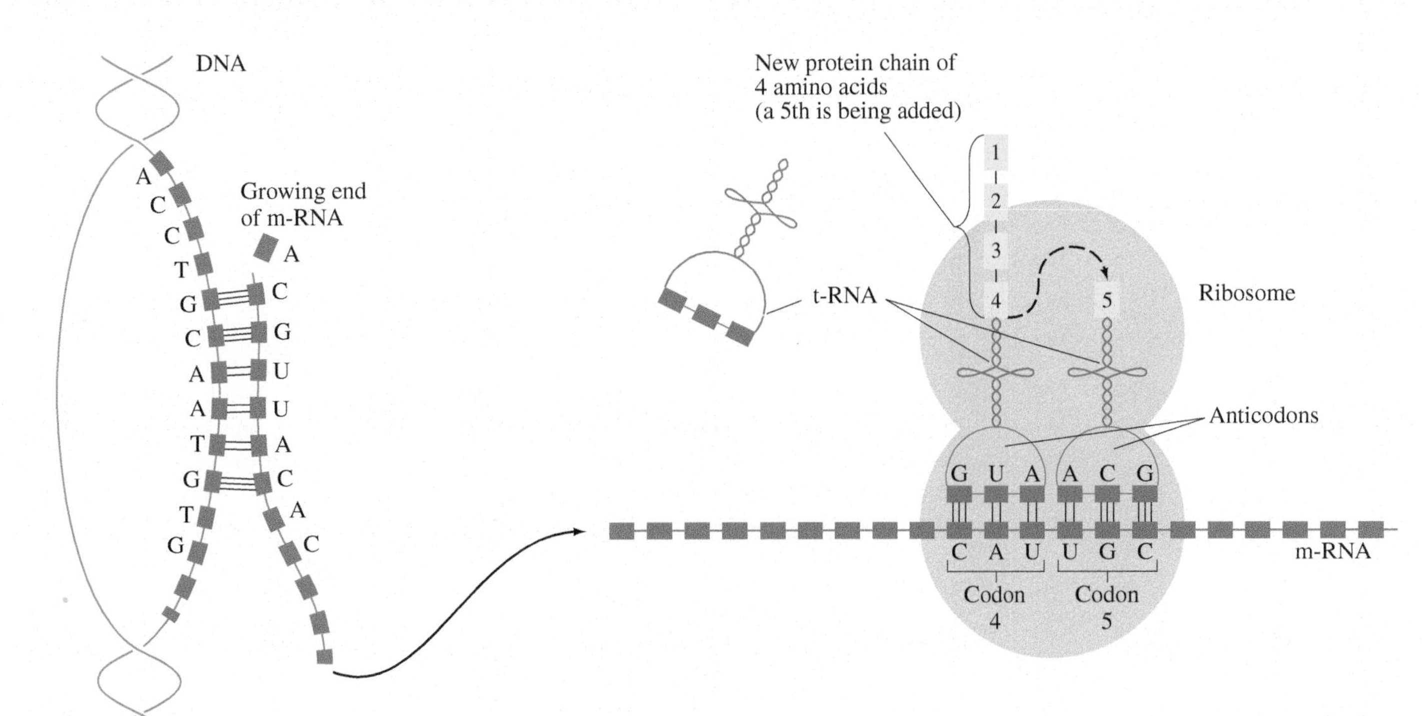

FIGURE 40–14 Protein synthesis. See text for details.

*Protein Synthesis

PHYSICS APPLIED
Protein synthesis

Weak bonds, especially hydrogen bonds, are crucial to the process of protein synthesis. Proteins serve as structural parts of the cell and as enzymes to catalyze chemical reactions needed for the growth and survival of the organism. A protein molecule consists of one or more chains of small molecules known as *amino acids*. There are 20 different amino acids, and a single protein chain may contain hundreds of them in a specific order. The standard model for how amino acids are connected together in the correct order to form a protein molecule is shown schematically in Fig. 40–14.

It begins at the DNA double helix: each gene on a chromosome contains the information for producing one protein. The ordering of the four bases, A, C, G, and T, provides the "code," the **genetic code**, for the order of amino acids in the protein. First, the DNA double helix unwinds and a new molecule called *messenger*-RNA (m-RNA) is synthesized using one strand of the DNA as a "template." m-RNA is a chain molecule containing four different bases, like those of DNA (Section 21–12) except that thymine (T) is replaced by the similar uracil molecule (U). Near the top left in Fig. 40–14, a C has just been added to the growing m-RNA chain in much the same way that DNA replicates; and an A, attracted and held close to the T on the DNA chain by the electrostatic force, will soon be attached to the C by an enzyme. The order of the bases, and thus the genetic information, is preserved in the m-RNA because the shapes of the molecules only allow the "proper" one to get close enough so the electrostatic force can act to form weak bonds.

Next, the m-RNA is buffeted about in the cell (kinetic theory) until it gets close to a tiny organelle known as a *ribosome*, to which it can attach by electrostatic attraction (on the right in Fig. 40–14), because their shapes allow the charged parts to get close enough to form weak bonds. Also held by the electrostatic force to the ribosome are one or two *transfer*-RNA (t-RNA) molecules. These t-RNA molecules "translate" the genetic code of nucleotide bases into amino acids in the following way. There is a different t-RNA molecule for each amino acid and each combination of three bases. On one end of a t-RNA molecule is an amino acid. On the other end of the t-RNA molecule is the appropriate "anticodon," a set of three nucleotide bases that "code" for that amino acid. If all three bases of an anticodon match the three bases of the "codon" on the m-RNA (in the sense of G to C and A to U), the anticodon is attracted electrostatically to the m-RNA codon and that t-RNA molecule is held there briefly. The ribosome has two particular attachment sites which hold two t-RNA molecules while enzymes link their two amino acids together to lengthen the amino acid chain (yellow in Fig. 40–14). As each amino acid is connected by an enzyme (four are already connected in Fig. 40–14, top right, and a fifth is about to be connected), the old t-RNA molecule is removed—perhaps by a random collision with some molecule in the cellular fluid. A new one soon becomes attracted as the ribosome moves along the m-RNA.

This process of protein synthesis is often presented as if it occurred in clockwork fashion—as if each molecule knew its role and went to its assigned place. But this is not the case. The forces of attraction between the electric charges of the molecules are rather weak and become significant only when the molecules can come close together and several weak bonds can be made. Indeed, if the shapes are not just right, there is almost no electrostatic attraction, which is why there are few mistakes. The fact that weak bonds are weak is very important. If they were strong, collisions with other molecules would not allow a t-RNA molecule to be released from the ribosome, or the m-RNA to be released from the DNA. If they were not temporary encounters, metabolism would grind to a halt.

As each amino acid is added to the next, the protein molecule grows in length until it is complete. Even as it is being made, this chain is being buffeted about in the cellular sea—we might think of a wiggling worm. But a protein molecule has electrically charged polar groups along its length. And as it takes on various shapes, the electric forces of attraction between different parts of the molecule will eventually lead to a particular configuration that is quite stable. Each type of protein has its own special shape, depending on the location of charged atoms. In the last analysis, the final shape depends on the order of the amino acids.

40–4 Molecular Spectra

When atoms combine to form molecules, the probability distributions of the outer electrons overlap and this interaction alters the energy levels. Nonetheless, molecules can undergo transitions between electron energy levels just as atoms do. For example, the H_2 molecule can absorb a photon of just the right frequency to excite one of its ground-state electrons to an excited state. The excited electron can then return to the ground state, emitting a photon. The energy of photons emitted by molecules can be of the same order of magnitude as for atoms, typically 1 to 10 eV, or less.

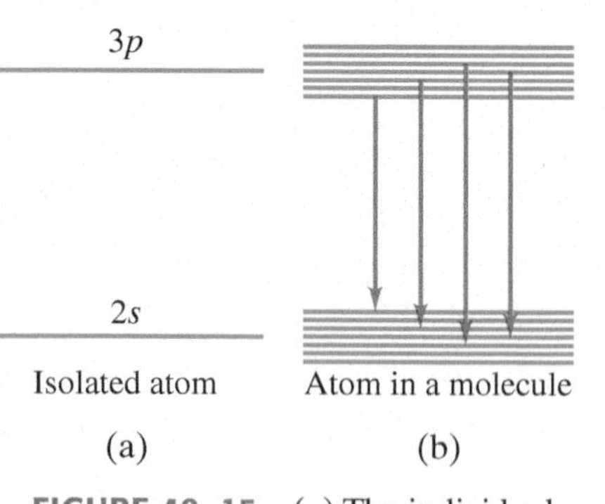

FIGURE 40–15 (a) The individual energy levels of an isolated atom become (b) bands of closely spaced levels in molecules, as well as in solids and liquids.

Additional energy levels become possible for molecules (but not for atoms) because the molecule as a whole can rotate, and the atoms of the molecule can vibrate relative to each other. The energy levels for both rotational and vibrational levels are quantized, and are generally spaced much more closely (10^{-3} to 10^{-1} eV) than the electronic levels. Each atomic energy level thus becomes a set of closely spaced levels corresponding to the vibrational and rotational motions, Fig. 40–15. Transitions from one level to another appear as many very closely spaced lines. In fact, the lines are not always distinguishable, and these spectra are called **band spectra**. Each type of molecule has its own characteristic spectrum, which can be used for identification and for determination of structure. We now look in more detail at rotational and vibrational states in molecules.

Rotational Energy Levels in Molecules

FIGURE 40–16 Diatomic molecule rotating about a vertical axis.

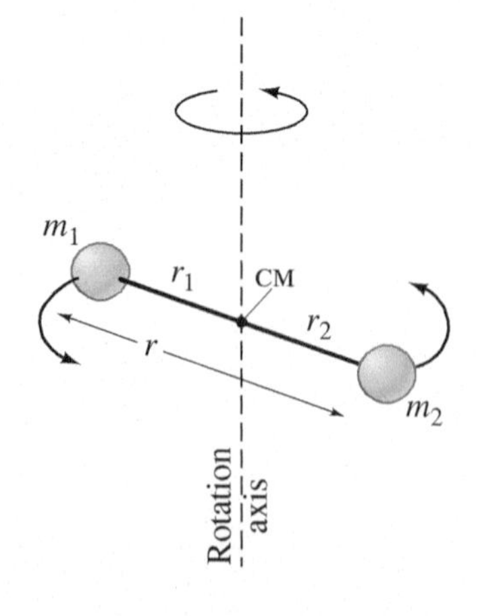

We consider only diatomic molecules, although the analysis can be extended to polyatomic molecules. When a diatomic molecule rotates about its center of mass as shown in Fig. 40–16, its kinetic energy of rotation (see Section 10–8) is

$$E_{\text{rot}} = \frac{1}{2} I\omega^2 = \frac{(I\omega)^2}{2I},$$

where $I\omega$ is the angular momentum (Section 11–1). Quantum mechanics predicts quantization of angular momentum just as in atoms (see Eq. 39–3):

$$I\omega = \sqrt{\ell(\ell+1)}\,\hbar, \qquad \ell = 0, 1, 2, \cdots,$$

where ℓ is an integer called the **rotational angular momentum quantum number**. Thus the rotational energy is quantized:

$$E_{\text{rot}} = \frac{(I\omega)^2}{2I} = \ell(\ell+1)\frac{\hbar^2}{2I}. \qquad \ell = 0, 1, 2, \cdots. \tag{40–2}$$

Transitions between rotational energy levels are subject to the **selection rule** (as in Section 39–2):

$$\Delta\ell = \pm 1.$$

The energy of a photon emitted or absorbed for a transition between rotational states with angular momentum quantum number ℓ and $\ell - 1$ will be

$$\Delta E_{\text{rot}} = E_\ell - E_{\ell-1} = \frac{\hbar^2}{2I}\ell(\ell+1) - \frac{\hbar^2}{2I}(\ell-1)(\ell)$$

$$= \frac{\hbar^2}{I}\ell. \qquad \left[\begin{array}{c}\ell \text{ is for upper} \\ \text{energy state}\end{array}\right] \qquad \textbf{(40–3)}$$

We see that the transition energy increases directly with ℓ. Figure 40–17 shows some of the allowed rotational energy levels and transitions. Measured absorption lines fall in the microwave or far-infrared regions of the spectrum, and their frequencies are generally 2, 3, 4, ⋯ times higher than the lowest one, as predicted by Eq. 40–3.

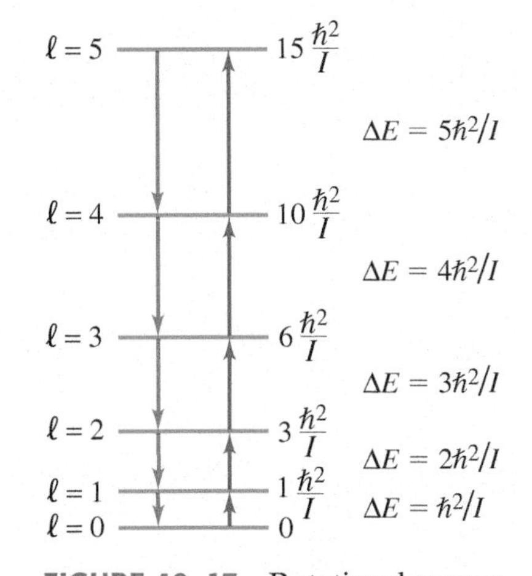

FIGURE 40–17 Rotational energy levels and allowed transitions (emission and absorption) for a diatomic molecule. Upward-pointing arrows represent absorption of a photon, and downward arrows represent emission of a photon.

EXERCISE A Determine the three lowest rotational energy states (in eV) for a nitrogen molecule which has a moment of inertia $I = 1.39 \times 10^{-46}\,\text{kg}\cdot\text{m}^2$.

The moment of inertia of the molecule in Fig. 40–16 rotating about its center of mass (Section 10–5) is

$$I = m_1 r_1^2 + m_2 r_2^2,$$

where r_1 and r_2 are the distances of each atom from their common center of mass. We can show (in Example 40–4 below) that I can be written

$$I = \frac{m_1 m_2}{m_1 + m_2} r^2 = \mu r^2, \qquad \textbf{(40–4)}$$

where $r = r_1 + r_2$ is the distance between the two atoms of the molecule and $\mu = m_1 m_2/(m_1 + m_2)$ is called the **reduced mass.** If $m_1 = m_2$, then $\mu = \frac{1}{2}m_1 = \frac{1}{2}m_2$.

EXAMPLE 40–4 Reduced mass. Show that the moment of inertia of a diatomic molecule rotating about its center of mass can be written

$$I = \mu r^2,$$

where

$$\mu = \frac{m_1 m_2}{m_1 + m_2}$$

is the reduced mass, Eq. 40–4, and r is the distance between the atoms.

SOLUTION The moment of inertia of a single particle of mass m a distance r from the rotation axis is $I = mr^2$ (Eq. 10–11 or 10–13). For our diatomic molecule (Fig. 40–16)

$$I = m_1 r_1^2 + m_2 r_2^2.$$

Now $r = r_1 + r_2$ and $m_1 r_1 = m_2 r_2$ because the axis of rotation passes through the center of mass. Hence

$$r_1 = r - r_2 = r - \frac{m_1}{m_2} r_1.$$

Solving for r_1 gives

$$r_1 = \frac{r}{1 + \frac{m_1}{m_2}} = \frac{m_2 r}{m_1 + m_2}.$$

Similarly,

$$r_2 = \frac{m_1 r}{m_1 + m_2}.$$

Then (see first equation of this Solution)

$$I = m_1\left(\frac{m_2 r}{m_1 + m_2}\right)^2 + m_2\left(\frac{m_1 r}{m_1 + m_2}\right)^2 = \frac{m_1 m_2 (m_1 + m_2) r^2}{(m_1 + m_2)^2}$$

$$= \frac{m_1 m_2}{m_1 + m_2} r^2 = \mu r^2,$$

where

$$\mu = \frac{m_1 m_2}{m_1 + m_2},$$

which is what we wished to show.

EXAMPLE 40–5 **Rotational transition.** A rotational transition $\ell = 1$ to $\ell = 0$ for the molecule CO has a measured absorption wavelength $\lambda_1 = 2.60\,\text{mm}$ (microwave region). Use this to calculate (*a*) the moment of inertia of the CO molecule, and (*b*) the CO bond length, *r*. (*c*) Calculate the wavelengths of the next three rotational transitions, and the energies of the photon emitted for each of these four transitions.

APPROACH The absorption wavelength is used to find the energy of the absorbed photon. The moment of inertia *I* is found from Eq. 40–3, and the bond length *r* from Eq. 40–4.

SOLUTION (*a*) The photon energy, $E = hf = hc/\lambda$, equals the rotational energy level difference, ΔE_{rot}. From Eq. 40–3, we can write

$$\frac{\hbar^2}{I}\ell = \Delta E = hf = \frac{hc}{\lambda_1}.$$

With $\ell = 1$ (the upper state) in this case, we solve for *I*:

$$I = \frac{\hbar^2 \ell}{hc}\lambda_1 = \frac{h\lambda_1}{4\pi^2 c} = \frac{(6.63 \times 10^{-34}\,\text{J}\cdot\text{s})(2.60 \times 10^{-3}\,\text{m})}{4\pi^2(3.00 \times 10^8\,\text{m/s})}$$
$$= 1.46 \times 10^{-46}\,\text{kg}\cdot\text{m}^2.$$

(*b*) The masses of C and O are 12.0 u and 16.0 u, respectively, where $1\,\text{u} = 1.66 \times 10^{-27}\,\text{kg}$. Thus the reduced mass is

$$\mu = \frac{m_1 m_2}{m_1 + m_2} = \frac{(12.0)(16.0)}{28.0}(1.66 \times 10^{-27}\,\text{kg}) = 1.14 \times 10^{-26}\,\text{kg}$$

or 6.86 u. Then, from Eq. 40–4, the bond length is

$$r = \sqrt{\frac{I}{\mu}} = \sqrt{\frac{1.46 \times 10^{-46}\,\text{kg}\cdot\text{m}^2}{1.14 \times 10^{-26}\,\text{kg}}} = 1.13 \times 10^{-10}\,\text{m} = 0.113\,\text{nm}.$$

(*c*) From Eq. 40–3, $\Delta E \propto \ell$. Hence $\lambda = c/f = hc/\Delta E$ is proportional to $1/\ell$. Thus, for $\ell = 2$ to $\ell = 1$ transitions, $\lambda_2 = \frac{1}{2}\lambda_1 = 1.30\,\text{mm}$. For $\ell = 3$ to $\ell = 2$, $\lambda_3 = \frac{1}{3}\lambda_1 = 0.87\,\text{mm}$. And for $\ell = 4$ to $\ell = 3$, $\lambda_4 = 0.65\,\text{mm}$. All are close to measured values. The energies of the photons, $hf = hc/\lambda$, are respectively $4.8 \times 10^{-4}\,\text{eV}$, $9.5 \times 10^{-4}\,\text{eV}$, $1.4 \times 10^{-3}\,\text{eV}$, and $1.9 \times 10^{-3}\,\text{eV}$.

Vibrational Energy Levels in Molecules

The potential energy of the two atoms in a typical diatomic molecule has the shape shown in Fig. 40–8 or 40–9, and Fig. 40–18 again shows the potential energy for the H_2 molecule. We note that the potential energy, at least in the vicinity of the equilibrium separation r_0, closely resembles the potential energy of a simple harmonic oscillator, $U = \frac{1}{2}kx^2$, which is shown superposed in dashed lines. Thus, for small displacements from r_0, each atom experiences a restoring force approximately proportional to the displacement, and the molecule vibrates as a simple harmonic oscillator (SHO). The classical frequency of vibration is

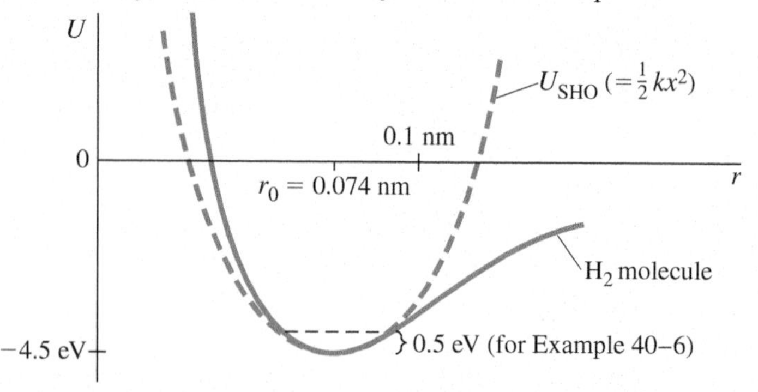

FIGURE 40–18 Potential energy for the H_2 molecule and for a simple harmonic oscillator ($U_{SHO} = \frac{1}{2}kx^2$, with $|x| = |r - r_0|$). The 0.50-eV energy height marked is for use in Example 40–6 to estimate *k*. [Note that $U_{SHO} = 0$ is not the same as $U = 0$ for the molecule.]

$$f = \frac{1}{2\pi}\sqrt{\frac{k}{\mu}}, \tag{40–5}$$

where *k* is the "stiffness constant" (as for a spring, Chapter 14) and instead of the mass *m* we must again use the reduced mass $\mu = m_1 m_2/(m_1 + m_2)$. (This is shown in Problem 19.) The Schrödinger equation for the SHO potential energy yields solutions for energy that are quantized according to

$$E_{\text{vib}} = \left(\nu + \tfrac{1}{2}\right)hf \qquad \nu = 0, 1, 2, \cdots, \tag{40–6}$$

where f is given by Eq. 40–5 and ν is an integer called the **vibrational quantum number**. The lowest energy state $(\nu = 0)$ is not zero (as for rotation) but has $E = \frac{1}{2}hf$. This is called the **zero-point energy**.† Higher states have energy $\frac{3}{2}hf$, $\frac{5}{2}hf$, and so on, as shown in Fig. 40–19. Transitions are subject to the **selection rule**

$$\Delta\nu = \pm 1,$$

so allowed transitions occur only between adjacent states and all give off photons of energy

$$\Delta E_{\text{vib}} = hf. \tag{40–7}$$

This is very close to experimental values for small ν; but for higher energies, the potential-energy curve (Fig. 40–18) begins to deviate from a perfect SHO curve, which affects the wavelengths and frequencies of the transitions. Typical transition energies are on the order of 10^{-1} eV, about 10 to 100 times larger than for rotational transitions, with wavelengths in the infrared region of the spectrum ($\approx 10^{-5}$ m).‡

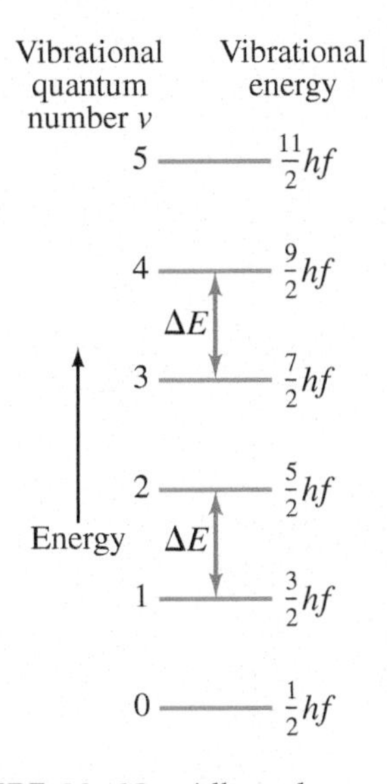

FIGURE 40–19 Allowed vibrational energies of a diatomic molecule, where f is the fundamental frequency of vibration, given by Eq. 40–5. The energy levels are equally spaced. Transitions are allowed only between adjacent levels $(\Delta\nu = \pm 1)$.

EXAMPLE 40–6 ESTIMATE **Wavelength for H_2.** (*a*) Use the curve of Fig. 40–18 to estimate the value of the stiffness constant k for the H_2 molecule, and then (*b*) estimate the fundamental wavelength for vibrational transitions.

APPROACH To find k, we arbitrarily choose an energy height of 0.50 eV which is indicated in Fig. 40–18. By measuring directly on the graph, we find that this energy corresponds to a vibration on either side of $r_0 = 0.074$ nm of about $x = \pm 0.017$ nm.

SOLUTION (*a*) For SHO, $U_{\text{SHO}} = \frac{1}{2}kx^2$ and $U_{\text{SHO}} = 0$ at $x = 0$ $(r = r_0)$; then

$$k = \frac{2U_{\text{SHO}}}{x^2} \approx \frac{2(0.50\,\text{eV})(1.6 \times 10^{-19}\,\text{J/eV})}{(1.7 \times 10^{-11}\,\text{m})^2} \approx 550\,\text{N/m}.$$

NOTE This value of k would also be reasonable for a macroscopic spring.

(*b*) The reduced mass is $\mu = m_1 m_2/(m_1 + m_2) = m_1/2 = \frac{1}{2}(1.0\,\text{u})(1.66 \times 10^{-27}\,\text{kg}) = 0.83 \times 10^{-27}$ kg. Hence, using Eq. 40–5,

$$\lambda = \frac{c}{f} = 2\pi c\sqrt{\frac{\mu}{k}} = 2\pi(3.0 \times 10^8\,\text{m/s})\sqrt{\frac{0.83 \times 10^{-27}\,\text{kg}}{550\,\text{N/m}}} = 2300\,\text{nm},$$

which is in the infrared region of the spectrum.

Experimentally, we do the inverse process: The wavelengths of vibrational transitions for a given molecule are measured, and from this the stiffness constant k can be calculated. The values of k calculated in this way are a measure of the strength of the molecular bond.

EXAMPLE 40–7 **Vibrational energy levels in hydrogen.** Hydrogen molecule vibrations emit infrared radiation of wavelength around 2300 nm. (*a*) What is the separation in energy between adjacent vibrational levels? (*b*) What is the lowest vibrational energy state?

APPROACH The energy separation between adjacent vibrational levels is $\Delta E_{\text{vib}} = hf = hc/\lambda$. The lowest energy (Eq. 40–6) has $\nu = 0$.

SOLUTION (*a*)

$$\Delta E_{\text{vib}} = hf = \frac{hc}{\lambda} = \frac{(6.63 \times 10^{-34}\,\text{J}\cdot\text{s})(3.00 \times 10^8\,\text{m/s})}{(2300 \times 10^{-9}\,\text{m})(1.60 \times 10^{-19}\,\text{J/eV})} = 0.54\,\text{eV},$$

where the denominator includes the conversion factor from joules to eV.

(*b*) The lowest vibrational energy has $\nu = 0$ in Eq. 40–6:

$$E_{\text{vib}} = \left(\nu + \tfrac{1}{2}\right)hf = \tfrac{1}{2}hf = 0.27\,\text{eV}.$$

†Recall this phenomenon for a square well, Fig. 38–8.

‡Forbidden transitions with $\Delta\nu = 2$ are emitted somewhat more weakly, but their observation can be important in some cases, such as in astronomy.

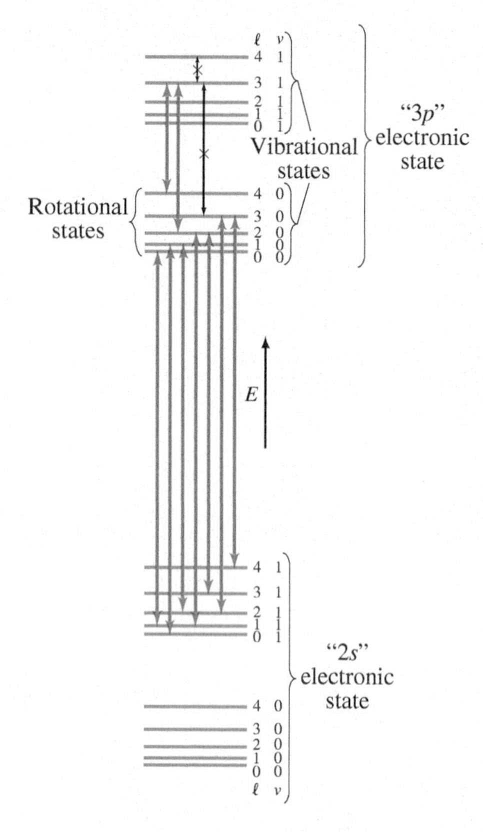

FIGURE 40–20 Combined electronic, vibrational, and rotational energy levels. Transitions marked with an × are not allowed by selection rules.

Rotational plus Vibrational Levels

When energy is imparted to a molecule, both the rotational and vibrational modes can be excited. Because rotational energies are an order of magnitude or so smaller than vibrational energies, which in turn are smaller than the electronic energy levels, we can represent the grouping of levels as shown in Fig. 40–20. Transitions between energy levels, with emission of a photon, are subject to the **selection rules**:

$$\Delta\nu = \pm 1 \qquad \text{and} \qquad \Delta\ell = \pm 1.$$

Some allowed and forbidden (marked ×) transitions are indicated in Fig. 40–20. Not all transitions and levels are shown, and the separation between vibrational levels, and (even more) between rotational levels, has been exaggerated. But we can clearly see the origin of the very closely spaced lines that give rise to the band spectra, as mentioned with reference to Fig. 40–15 earlier in this Section.

The spectra are quite complicated, so we consider briefly only transitions within the same electronic level, such as those at the top of Fig. 40–20. A transition from a state with quantum numbers ν and ℓ, to one with quantum numbers $\nu + 1$ and $\ell \pm 1$ (see the selection rules above), will absorb[†] a photon of energy:

$$\Delta E = \Delta E_{\text{vib}} + \Delta E_{\text{rot}}$$

$$= hf + (\ell + 1)\frac{\hbar^2}{I} \qquad \begin{bmatrix} \ell \rightarrow \ell + 1 \\ (\Delta\ell = +1) \end{bmatrix}, \quad \ell = 0, 1, 2, \cdots \qquad \textbf{(40–8a)}$$

$$= hf - \ell\frac{\hbar^2}{I} \qquad \begin{bmatrix} \ell \rightarrow \ell - 1 \\ (\Delta\ell = -1) \end{bmatrix}, \quad \ell = 1, 2, 3, \cdots, \qquad \textbf{(40–8b)}$$

where we have used Eqs. 40–3 and 40–7. Note that for $\ell \rightarrow \ell - 1$ transitions, ℓ cannot be zero because there is then no state with $\ell = -1$. Equations 40–8 predict an absorption spectrum like that shown schematically in Fig. 40–21, with transitions $\ell \rightarrow \ell - 1$ on the left and $\ell \rightarrow \ell + 1$ on the right. Figure 40–22 shows the molecular absorption spectrum of HCl, which follows this pattern very well.

[†]Eqs. 40–8 are for absorption; for emission of a photon, the transition would be $\nu \rightarrow \nu - 1$, $\ell \rightarrow \ell \pm 1$.

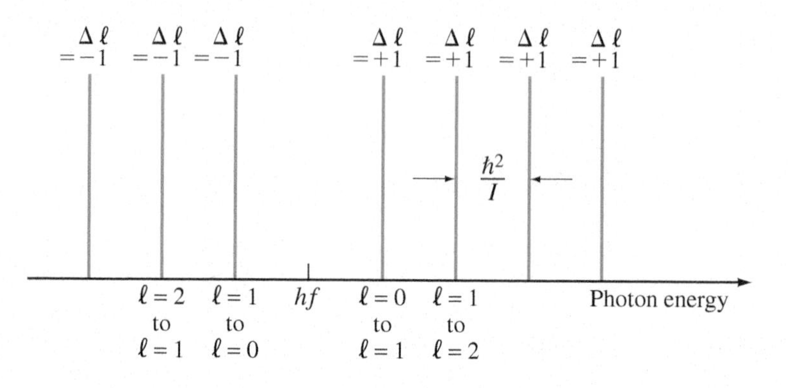

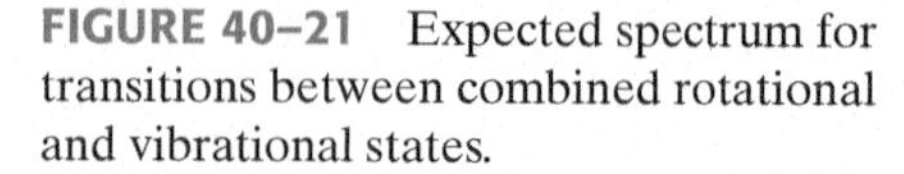
FIGURE 40–21 Expected spectrum for transitions between combined rotational and vibrational states.

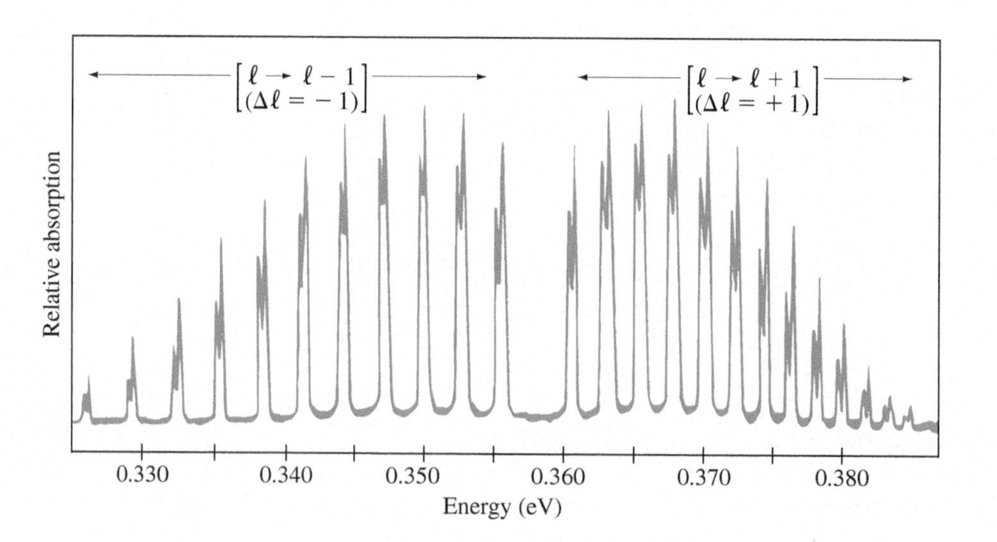

FIGURE 40–22 Absorption spectrum for HCl molecules. Lines on the left correspond to transitions where $\ell \rightarrow \ell - 1$; those on the right are for $\ell \rightarrow \ell + 1$. Each line has a double peak because chlorine has two isotopes of different mass and different moment of inertia.

(Each line in the spectrum of Fig. 40–22 is split into two because Cl consists of two isotopes of different mass; hence there are two kinds of HCl molecule with slightly different moments of inertia I.)

EXAMPLE 40–8 ESTIMATE **The HCl molecule.** Estimate the moment of inertia of the HCl molecule using the absorption spectrum shown in Fig. 40–22. For the purposes of a rough estimate you can ignore the difference between the two isotopes.

APPROACH The locations of the peaks in Fig. 40–22 should correspond to Eqs. 40–8. We don't know what value of ℓ each peak corresponds to in Fig. 40–22, but we can estimate the energy difference between peaks to be about $\Delta E' = 0.0025\,\mathrm{eV}$.

SOLUTION From Eqs. 40–8, the energy difference between two peaks is given by

$$\Delta E' = \Delta E_{\ell+1} - \Delta E_\ell = \frac{\hbar^2}{I}.$$

Then

$$I = \frac{\hbar^2}{\Delta E'} = \frac{(6.626 \times 10^{-34}\,\mathrm{J\cdot s}/2\pi)^2}{(0.0025\,\mathrm{eV})(1.6 \times 10^{-19}\,\mathrm{J/eV})} = 2.8 \times 10^{-47}\,\mathrm{kg\cdot m^2}.$$

NOTE To get an idea of what this number means, we write $I = \mu r^2$ (Eq. 40–4), where μ is the reduced mass (Example 40–4); then we calculate μ:

$$\mu = \frac{m_1 m_2}{m_1 + m_2} = \frac{(1.0\,\mathrm{u})(35\,\mathrm{u})}{36\,\mathrm{u}}(1.66 \times 10^{-27}\,\mathrm{kg/u}) = 1.6 \times 10^{-27}\,\mathrm{kg};$$

the bond length is given by (Eq. 40–4)

$$r = \left(\frac{I}{\mu}\right)^{\frac{1}{2}} = \left(\frac{2.8 \times 10^{-47}\,\mathrm{kg\cdot m^2}}{1.6 \times 10^{-27}\,\mathrm{kg}}\right)^{\frac{1}{2}} = 1.3 \times 10^{-10}\,\mathrm{m},$$

which is the expected order of magnitude for a bond length.

40–5 Bonding in Solids

Quantum mechanics has been a great tool for understanding the structure of solids. This active field of research today is called **solid-state physics**, or **condensed-matter physics** so as to include liquids as well. The rest of this Chapter is devoted to this subject, and we begin with a brief look at the structure of solids and the bonds that hold them together.

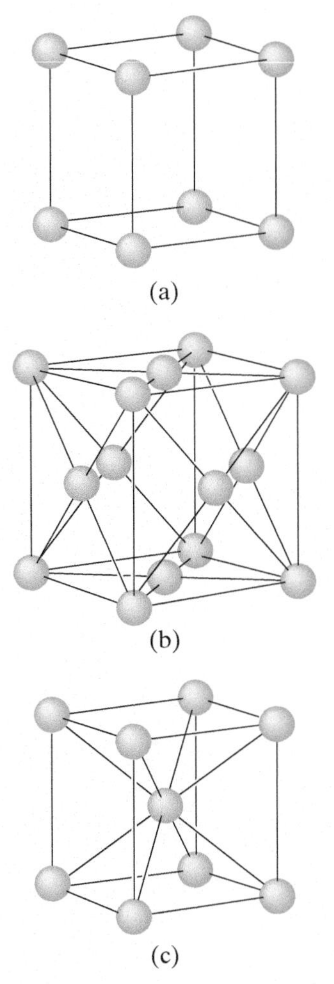

FIGURE 40–23 Arrangement of atoms in (a) a simple cubic crystal, (b) face-centered cubic crystal (note the atom at the center of each face), and (c) body-centered cubic crystal. Each shows the relationship of the bonds. Each of these "cells" is repeated in three dimensions to the edges of the macroscopic crystal.

Although some solid materials are *amorphous* in structure, in that the atoms and molecules show no long-range order, we will be interested here in the large class of *crystalline* substances whose atoms, ions, or molecules are generally believed to form an orderly array known as a **lattice**. Figure 40–23 shows three of the possible arrangements of atoms in a crystal: simple cubic, face-centered cubic, and body-centered cubic. The NaCl crystal is face-centered cubic (see Fig. 40–24), with one Na^+ ion or one Cl^- ion at each lattice point (i.e., considering Na and Cl separately).

The molecules of a solid are held together in a number of ways. The most common are by *covalent* bonding (such as between the carbon atoms of the diamond crystal) or *ionic* bonding (as in a NaCl crystal). Often the bonds are partially covalent and partially ionic. Our discussion of these bonds earlier in this Chapter for molecules applies equally well here to solids.

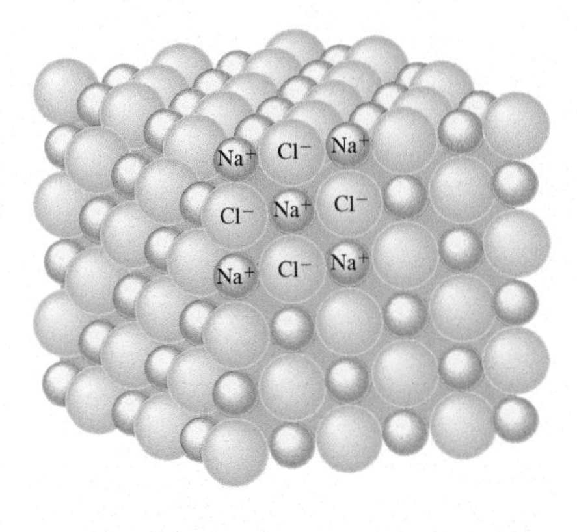

FIGURE 40–24 Diagram of an NaCl crystal, showing the "packing" of atoms.

Let us look for a moment at the NaCl crystal of Fig. 40–24. Each Na^+ ion feels an attractive Coulomb potential due to each of the six "nearest neighbor" Cl^- ions surrounding it. Note that one Na^+ does not "belong" exclusively to one Cl^-, so we must not think of ionic solids as consisting of individual molecules. Each Na^+ also feels a repulsive Coulomb potential due to other Na^+ ions, although this is weaker since the other Na^+ ions are farther away. Thus we expect a net attractive potential

$$U = -\alpha \frac{1}{4\pi\epsilon_0}\frac{e^2}{r}.$$

The factor α is called the *Madelung constant*. If each Na^+ were surrounded by only the six Cl^- ions, α would be 6, but the influence of all the other ions reduces it to a value $\alpha = 1.75$ for the NaCl crystal. The potential must also include a term representing the repulsive force when the wave functions of the inner shells and subshells overlap, and this has the form $U = B/r^m$, where m is a small integer.

The sum of these two terms suggests a potential energy

$$U = -\frac{\alpha}{4\pi\epsilon_0}\frac{e^2}{r} + \frac{B}{r^m}, \tag{40–9}$$

which has the same form as Eq. 40–1 for molecules (Section 40–2). It can be shown (Problem 25) that, at the equilibrium distance r_0,

$$U = U_0 = -\frac{\alpha}{4\pi\epsilon_0}\frac{e^2}{r_0}\left(1 - \frac{1}{m}\right).$$

This U_0 is known as the *ionic cohesive energy*; it is a sort of "binding energy"—the energy (per ion) needed to take the solid apart into separated ions, one by one.

A different type of bond occurs in metals. Metal atoms have relatively loosely held outer electrons. Present-day **metallic bond** theories propose that in a metallic solid, these outer electrons roam rather freely among all the metal atoms which, without their outer electrons, act like positive ions. The electrostatic attraction between the metal ions and this negative electron "gas" is what is believed, at least in part, to hold the solid together. The binding energy of metal bonds are typically 1 to 3 eV, somewhat weaker than ionic or covalent bonds (5 to 10 eV in solids). The "free electrons," according to this theory, are responsible for the high electrical and thermal conductivity of metals (see Sections 40–6 and 40–7). This theory also nicely accounts for the shininess of smooth metal surfaces: the electrons are free and can vibrate at any frequency, so when light of a range of frequencies falls on a metal, the electrons can vibrate in response and re-emit light of those same frequencies. Hence the reflected light will consist largely of the same frequencies as the incident light. Compare this to nonmetallic materials that have a distinct color—the atomic electrons exist only in certain energy states, and when white light falls on them, the atoms absorb at certain frequencies, and reflect other frequencies which make up the color we see.

Here is a brief summary of important strong bonds:

- ionic: an electron is stolen from one atom by another;
- covalent: electrons are shared by atoms within a single molecule;
- metallic: electrons are shared by all atoms in the metal.

The atoms or molecules of some materials, such as the noble gases, can form only *weak bonds* with each other. As we saw in Section 40–3, weak bonds have very low binding energies and would not be expected to hold atoms together as a liquid or solid at room temperature. The noble gases condense only at very low temperatures, where the atomic kinetic energy is small and the weak attraction can then hold the atoms together.

40–6 Free-Electron Theory of Metals; Fermi Energy

Let us look more closely at the free-electron theory of metals mentioned in the preceding Section. Let us imagine the electrons trapped within the metal as being in a potential well: inside the metal, the potential energy is zero, but at the edges of the metal there are high potential walls. Since very few electrons leave the metal at room temperature, we can imagine the walls as being infinitely high (as in Section 38–8). At higher temperatures, electrons do leave the metal (we know that thermionic emission occurs, Section 23–9), so we must recognize that the well is of finite depth. In this simple model, the electrons are trapped within the metal, but are free to move about inside the well whose size is macroscopic—the size of the piece of metal. The energy will be quantized, but the spacing between energy levels will be very tiny (see Eq. 38–13) because the width of the potential well ℓ is very large. Indeed, for a cube 1 cm on a side, the number of states with energy between, say, 5.0 and 5.5 eV, is on the order of 10^{22} (see Example 40–9).

To deal with such vast numbers of states, which are so closely spaced as to seem continuous, we need to use statistical methods. We define a quantity known as the **density of states**, $g(E)$, whose meaning is similar to the Maxwell distribution, Eq. 18–6 (see Section 18–2). That is, the quantity $g(E)\,dE$ represents the number of states per unit volume that have energy between E and $E + dE$. A careful calculation (see Problem 41), which must treat the potential well as three dimensional, shows that

$$g(E) = \frac{8\sqrt{2}\,\pi m^{3/2}}{h^3}\,E^{1/2} \qquad \textbf{(40–10)}$$

where m is the mass of the electron. This function is plotted in Fig. 40–25.

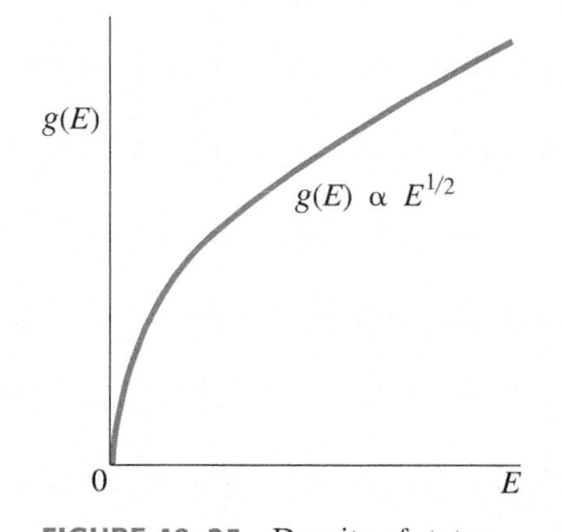

FIGURE 40–25 Density of states $g(E)$ as a function of energy E (Eq. 40–10).

EXAMPLE 40–9 ESTIMATE **Electron states in copper.** Estimate the number of states in the range 5.0 to 5.5 eV available to electrons in a 1.0-cm cube of copper metal.

APPROACH Since $g(E)$ is the number of states per unit volume per unit energy interval, the number N of states is approximately (it is approximate because ΔE is not small)

$$N \approx g(E)V\,\Delta E,$$

where the volume $V = 1.0\,\text{cm}^3 = 1.0 \times 10^{-6}\,\text{m}^3$ and $\Delta E = 0.50\,\text{eV}$.

SOLUTION We evaluate $g(E)$ at 5.25 eV, and find (Eq. 40–10):

$$\begin{aligned} N \approx g(E)V\,\Delta E &= \frac{8\sqrt{2}\,\pi(9.1 \times 10^{-31}\,\text{kg})^{\frac{3}{2}}}{(6.63 \times 10^{-34}\,\text{J}\cdot\text{s})^3}\sqrt{(5.25\,\text{eV})(1.6 \times 10^{-19}\,\text{J/eV})} \\ &\quad \times (1.0 \times 10^{-6}\,\text{m}^3)(0.50\,\text{eV})(1.6 \times 10^{-19}\,\text{J/eV}) \\ &\approx 8 \times 10^{21} \end{aligned}$$

states in $1.0\,\text{cm}^3$. Note that the type of metal did not enter the calculation.

Equation 40–10 gives us the density of states. Now we must ask: How are the states available to an electron gas actually populated? Let us first consider the situation at absolute zero, $T = 0\,\text{K}$. For a classical ideal gas, all the particles would be in the lowest state, with zero kinetic energy $\left(= \frac{3}{2}kT = 0\right)$—recall Eq. 18–4. But the situation is vastly different for an electron gas because electrons obey the exclusion principle. Electrons do not obey classical statistics but rather a quantum statistics called **Fermi–Dirac statistics**† that takes into account the exclusion principle. All particles that have spin $\frac{1}{2}$ (or other half-integral spin: $\frac{3}{2}, \frac{5}{2}$, etc.), such as electrons, protons, and neutrons, obey Fermi–Dirac statistics and are referred to as **fermions**.‡ The electron gas in a metal is often called a **Fermi gas**. According to the exclusion principle, no two electrons in the metal can have the same set of quantum numbers. Therefore, in each of the states of our potential well, there can be at most two electrons: one with spin up $\left(m_s = +\frac{1}{2}\right)$ and one with spin down $\left(m_s = -\frac{1}{2}\right)$. (This factor of 2 has already been included in Eq. 40–10.) Thus, at $T = 0\,\text{K}$, the possible energy levels will be filled, two electrons each, up to a maximum level called the **Fermi level**. This is shown in Fig. 40–26, where the vertical axis is labeled $n_o(E)$ for "density of occupied states." The energy of the state at the Fermi level is called the **Fermi energy**, E_F. To determine E_F, we integrate Eq. 40–10 from $E = 0$ to $E = E_F$ (all states up to E_F are filled at $T = 0\,\text{K}$):

$$\frac{N}{V} = \int_0^{E_F} g(E)\,dE, \qquad \textbf{(40–11)}$$

where N/V is the number of conduction electrons per unit volume in the metal.

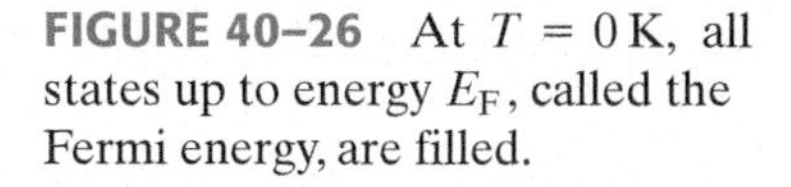
FIGURE 40–26 At $T = 0\,\text{K}$, all states up to energy E_F, called the Fermi energy, are filled.

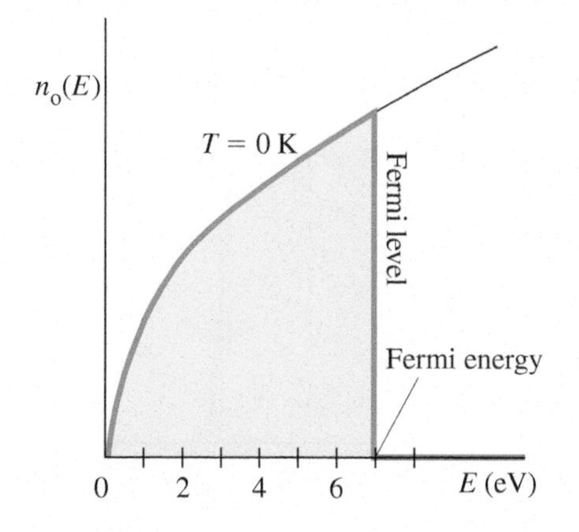

†Developed independently by Enrico Fermi (Figs. 41–8, 38–2, 37–10) in early 1926 and by P. A. M. Dirac a few months later.

‡Particles with integer spin (0, 1, 2, etc.), such as the photon, obey *Bose–Einstein* statistics and are called *bosons*, as mentioned in Section 39–4.

Then, solving for E_F, the result (see Example 40–10 below) is

$$E_F = \frac{h^2}{8m}\left(\frac{3}{\pi}\frac{N}{V}\right)^{\frac{2}{3}}. \tag{40–12}$$

The average energy in this distribution (see Problem 35) is

$$\overline{E} = \tfrac{3}{5}E_F. \tag{40–13}$$

For copper, $E_F = 7.0\,\text{eV}$ (see Example 40–10) and $\overline{E} = 4.2\,\text{eV}$. This is very much greater than the energy of thermal motion at room temperature ($\frac{3}{2}kT \approx 0.04\,\text{eV}$). Clearly, all motion does not stop at absolute zero.

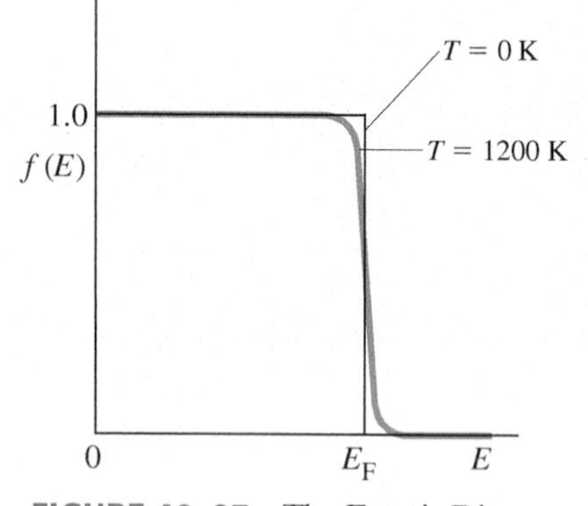

FIGURE 40–27 The Fermi–Dirac probability function for two temperatures, $T = 0\,\text{K}$ (black line) and $T = 1200\,\text{K}$ (blue curve). For $f(E) = 1$, a state with energy E is certainly occupied. For $f(E) = 0.5$, which occurs at $E = E_F$, the state with E_F has a 50% chance of being occupied.

Thus, at $T = 0$, all states with energy below E_F are occupied, and all states above E_F are empty. What happens for $T > 0$? We expect that some (at least) of the electrons will increase in energy due to thermal motion. Classically, the distribution of occupied states would be given by the Boltzmann factor, $e^{-E/kT}$ (see Eqs. 39–16). But for our electron gas, a quantum-mechanical system obeying the exclusion principle, the probability of a given state of energy E being occupied is given by the **Fermi–Dirac probability function** (or **Fermi factor**):

$$f(E) = \frac{1}{e^{(E-E_F)/kT} + 1}, \tag{40–14}$$

where E_F is the Fermi energy. This function is plotted in Fig. 40–27 for two temperatures, $T = 0\,\text{K}$ and $T = 1200\,\text{K}$ (just below the melting point of copper). At $T = 0$ (or as T approaches zero) the factor $e^{(E-E_F)/kT}$ in Eq. 40–14 is zero if $E < E_F$ and is ∞ if $E > E_F$. Thus

$$f(E) = \begin{cases} 1 & E < E_F \\ 0 & E > E_F \end{cases} \quad \text{at} \quad T = 0.$$

This is what is plotted in black in Fig. 40–27 and is consistent with Fig. 40–26: all states up to the Fermi level are occupied [probability $f(E) = 1$] and all states above are unoccupied. For $T = 1200\,\text{K}$, the Fermi factor changes only a little, as shown in Fig. 40–27 as the blue curve. Note that at any temperature T, when $E = E_F$, then Eq. 40–14 gives $f(E) = 0.50$, meaning the state at $E = E_F$ has a 50% chance of being occupied by an electron. Alternatively, at $E = E_F$, half the available states are filled and half are empty. To see how $f(E)$ affects the actual distribution of electrons in energy states, we must weight the density of possible states, $g(E)$, by the probability that those states will be occupied, $f(E)$. The product of these two functions gives the **density of occupied states**,

$$n_o(E) = g(E)f(E) = \frac{8\sqrt{2}\,\pi m^{3/2}}{h^3}\frac{E^{1/2}}{e^{(E-E_F)/kT} + 1}. \tag{40–15}$$

Then $n_o(E)\,dE$ represents the number of electrons per unit volume with energy between E and $E + dE$ in thermal equilibrium at temperature T. This is plotted in Fig. 40–28 for $T = 1200\,\text{K}$, a temperature at which a metal is so hot it would glow. We see immediately that the distribution differs very little from that at $T = 0$. We see also that the changes that do occur are concentrated about the Fermi level. A few electrons from slightly below the Fermi level move to energy states slightly above it. The average energy of the electrons increases only very slightly when the temperature is increased from $T = 0\,\text{K}$ to $T = 1200\,\text{K}$. This is very different from the behavior of an ideal gas, for which kinetic energy increases directly with T. Nonetheless, this behavior is readily understood as follows. Energy of thermal motion at $T = 1200\,\text{K}$ is about $\frac{3}{2}kT \approx 0.1\,\text{eV}$. The Fermi level, on the other hand, is on the order of several eV: for copper it is $E_F \approx 7.0\,\text{eV}$. An electron at $T = 1200\,\text{K}$ may have 7 eV of energy, but it can acquire at most only a few times 0.1 eV of energy by a (thermal) collision with the lattice. Only electrons very near the Fermi level would find vacant states close enough to make such a transition. Essentially none of the electrons could increase in energy by, say, 3 eV, so electrons farther down in the electron gas are unaffected. Only electrons near the top of the energy distribution can be thermally excited to higher states. And their new energy is on the average only slightly higher than their old energy.

FIGURE 40–28 The density of occupied states for the electron gas in copper. The width kT shown above the graph represents thermal energy at $T = 1200\,\text{K}$.

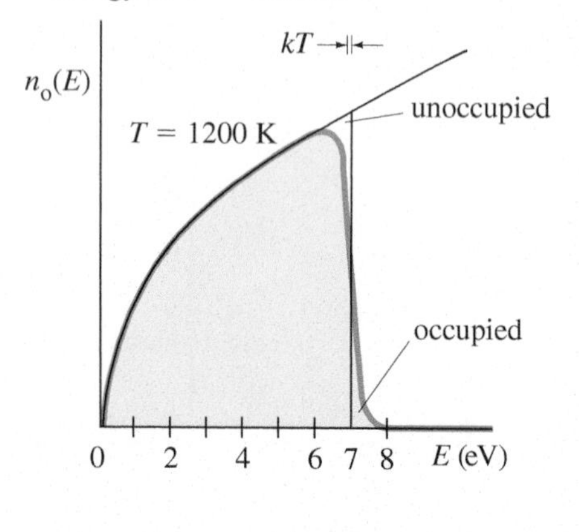

EXERCISE B Return to the Chapter-Opening Question, page 1071, and answer it again now. Try to explain why you may have answered differently the first time.

EXAMPLE 40–10 **The Fermi level.** For the metal copper, determine (*a*) the Fermi energy, (*b*) the average energy of electrons, and (*c*) the speed of electrons at the Fermi level (this is called the *Fermi speed*).

APPROACH We first derive Eq. 40–12 by combining Eqs. 40–10 and 40–11:

$$\frac{N}{V} = \frac{8\sqrt{2}\,\pi m^{3/2}}{h^3}\int_0^{E_F} E^{1/2}\,dE = \frac{8\sqrt{2}\,\pi m^{3/2}}{h^3}\,\frac{2}{3}\,E_F^{3/2}.$$

Solving for E_F, we obtain

$$E_F = \frac{h^2}{8m}\left(\frac{3}{\pi}\frac{N}{V}\right)^{\frac{2}{3}}$$

[note: $(2\sqrt{2})^{\frac{2}{3}} = (2^{\frac{3}{2}})^{\frac{2}{3}} = 2$], and this is Eq. 40–12. We calculated N/V, the number of conduction electrons per unit volume in copper, in Example 25–14 to be $N/V = 8.4 \times 10^{28}\,\mathrm{m}^{-3}$.

SOLUTION (*a*) The Fermi energy for copper is thus

$$E_F = \frac{(6.63 \times 10^{-34}\,\mathrm{J\cdot s})^2}{8(9.1 \times 10^{-31}\,\mathrm{kg})}\left[\frac{3(8.4 \times 10^{28}\,\mathrm{m}^{-3})}{\pi}\right]^{\frac{2}{3}}\frac{1}{1.6 \times 10^{-19}\,\mathrm{J/eV}} = 7.0\,\mathrm{eV}.$$

(*b*) From Eq. 40–13,

$$\overline{E} = \tfrac{3}{5}E_F = 4.2\,\mathrm{eV}.$$

(*c*) In our model, we have taken $U = 0$ inside the metal (assuming a 3-D infinite potential well, Section 38–8, Eq. 40–10, and Problem 41). Then E is only kinetic energy $= \frac{1}{2}mv^2$. Therefore, at the Fermi level, the Fermi speed is

$$v_F = \sqrt{\frac{2E_F}{m}} = \sqrt{\frac{2(7.0\,\mathrm{eV})(1.6 \times 10^{-19}\,\mathrm{J/eV})}{9.1 \times 10^{-31}\,\mathrm{kg}}} = 1.6 \times 10^6\,\mathrm{m/s},$$

a very high speed. The temperature of a classical gas would have to be extremely high to produce an average particle speed this large.

EXAMPLE 40–11 **Incorrect classical calculation.** Let us see what result we get if the electrons are treated as a classical ideal gas. That is, estimate the average kinetic energy of electrons at room temperature using the kinetic theory of gases, Chapter 18.

APPROACH The average kinetic energy of particles in an ideal gas was given in Chapter 18, Eq. 18–4, as

$$\overline{K} = \tfrac{3}{2}kT,$$

where k is Boltzmann's constant and $T \approx 300\,\mathrm{K}$.

SOLUTION The ideal gas model gives

$$\overline{K} = \frac{3}{2}(1.38 \times 10^{-23}\,\mathrm{J/K})(300\,\mathrm{K})\left(\frac{1}{1.6 \times 10^{-19}\,\mathrm{J/eV}}\right) = 0.039\,\mathrm{eV}.$$

This result is far from correct. It is off by a factor of 100: Example 40–10 gave 4.2 eV. The ideal gas model does not work for electrons which obey the exclusion principle. Indeed, we see here how important and powerful the exclusion principle is.

EXERCISE C Determine the Fermi energy for gold $(\text{density} = 19{,}300\,\mathrm{kg/m^3})$. (*a*) 5.5 eV, (*b*) 6.2 eV, (*c*) 7.2 eV, (*d*) 8.1 eV, (*e*) 8.4 eV.

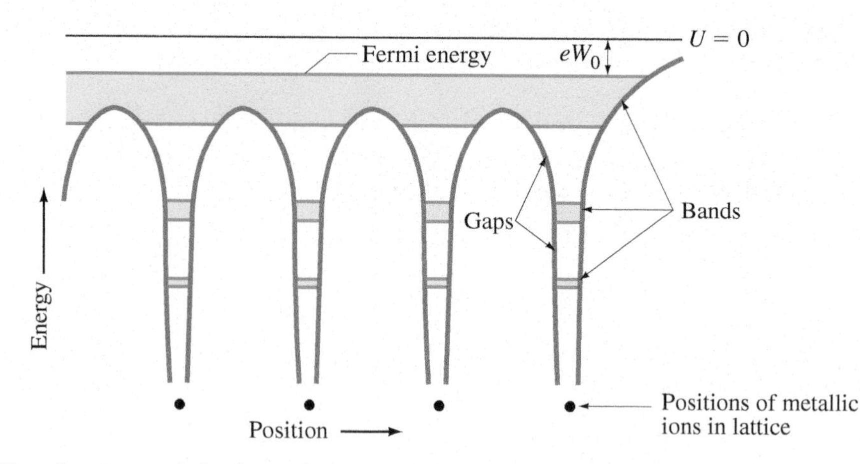

FIGURE 40–29 Potential energy for an electron in a metal crystal, with deep potential wells in the vicinity of each ion in the crystal lattice.

The simple model of an electron gas presented here provides good explanations for the electrical and thermal properties of conductors. But it does not explain why some materials are good conductors and others are good insulators. To provide an explanation, our model of electrons inside a metal moving in a uniform potential well needs to be refined to include the effect of the lattice. Figure 40–29 shows a "periodic" potential that takes into account the attraction of electrons for each atomic ion in the lattice. Here we have taken $U = 0$ for an electron free of the metal; so within the metal, electron energies are less than zero (just as for molecules, or for the H atom in which the ground state has $E = -13.6\,\text{eV}$). The quantity eW_0 represents the minimum energy to remove an electron from the metal, where W_0 is the *work function* (see Section 37–2). The crucial outcome of putting a periodic potential (more easily approximated with narrow square wells) into the Schrödinger equation is that the allowed energy states are divided into *bands*, with energy gaps in between. Only electrons in the highest band, close to the Fermi level, are able to move about freely within the metal crystal. In the next Section we will see physically why there are bands and how they explain the properties of conductors, insulators, and semiconductors.

40–7 Band Theory of Solids

We saw in Section 40–1 that when two hydrogen atoms approach each other, the wave functions overlap, and the two 1*s* states (one for each atom) divide into two states of different energy. (As we saw, only one of these states, $S = 0$, has low enough energy to give a bound H_2 molecule.) Figure 40–30a shows this situation for 1*s* and 2*s* states for two atoms: as the two atoms get closer (toward the left in Fig. 40–30a), the 1*s* and 2*s* states split into two levels. If six atoms come together, as in Fig. 40–30b, each of the states splits into six levels. If a large number of atoms come together to form a solid, then each of the original atomic levels becomes a **band** as shown in Fig. 40–30c. The energy levels are so close together in each band that they seem essentially continuous. This is why the spectrum of heated solids (Section 37–1) appears continuous. (See also Fig. 40–15 and its discussion at the start of Section 40–4.)

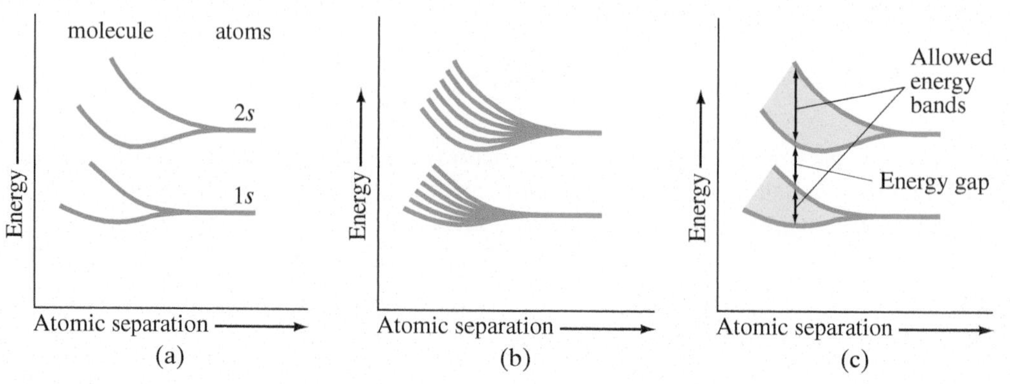

FIGURE 40–30 The splitting of 1*s* and 2*s* atomic energy levels as (a) two atoms approach each other (the atomic separation decreases toward the left on the graph), (b) the same for six atoms, and (c) for many atoms when they come together to form a solid.

The crucial aspect of a good **conductor** is that the highest energy band containing electrons is only partially filled. Consider sodium metal, for example, whose energy bands are shown in Fig. 40–31. The $1s$, $2s$, and $2p$ bands are full (just as in a sodium atom) and don't concern us. The $3s$ band, however, is only half full. To see why, recall that the exclusion principle stipulates that in an atom, only two electrons can be in the $3s$ state, one with spin up and one with spin down. These two states have slightly different energy. For a solid consisting of N atoms, the $3s$ band will contain $2N$ possible energy states. A sodium atom has a single $3s$ electron, so in a sample of sodium metal containing N atoms, there are N electrons in the $3s$ band, and N unoccupied states. When a potential difference is applied across the metal, electrons can respond by accelerating and increasing their energy, since there are plenty of unoccupied states of slightly higher energy available. Hence, a current flows readily and sodium is a good conductor. The characteristic of all good conductors is that the highest energy band is only partially filled, or two bands overlap so that unoccupied states are available. An example of the latter is magnesium, which has two $3s$ electrons, so its $3s$ band is filled. But the unfilled $3p$ band overlaps the $3s$ band in energy, so there are lots of available states for the electrons to move into. Thus magnesium, too, is a good conductor.

FIGURE 40–31 Energy bands for sodium (Na).

In a material that is a good **insulator**, on the other hand, the highest band containing electrons, called the **valence band**, is completely filled. The next highest energy band, called the **conduction band**, is separated from the valence band by a "forbidden" **energy gap** (or **band gap**), E_g, of typically 5 to 10 eV. So at room temperature (300 K), where thermal energies (that is, average kinetic energy—see Chapter 18) are on the order of $\frac{3}{2}kT \approx 0.04$ eV, almost no electrons can acquire the 5 eV needed to reach the conduction band. When a potential difference is applied across the material, no available states are accessible to the electrons, and no current flows. Hence, the material is a good insulator.

Figure 40–32 compares the relevant energy bands (a) for conductors, (b) for insulators, and also (c) for the important class of materials known as **semiconductors**. The bands for a pure (or **intrinsic**) semiconductor, such as silicon or germanium, are like those for an insulator, except that the unfilled conduction band is separated from the filled valence band by a much smaller energy gap, E_g, typically on the order of 1 eV. At room temperature, a few electrons can acquire enough thermal energy to reach the conduction band, and so a very small current may flow when a voltage is applied. At higher temperatures, more electrons have enough energy to jump the gap. Often this effect can more than offset the effects of more frequent collisions due to increased disorder at higher temperature, so the resistivity of semiconductors can *decrease* with increasing temperature (see Table 25–1). But this is not the whole story of semiconductor conduction. When a potential difference is applied to a semiconductor, the few electrons in the conduction band move toward the positive electrode. Electrons in the valence band try to do the same thing, and a few can because there are a small number of unoccupied states which were left empty by the electrons reaching the conduction band. Such unfilled electron states are called **holes**. Each electron in the valence band that fills a hole in this way as it moves toward the positive electrode leaves behind its own hole, so the holes migrate toward the negative electrode. As the electrons tend to accumulate at one side of the material, the holes tend to accumulate on the opposite side. We will look at this phenomenon in more detail in the next Section.

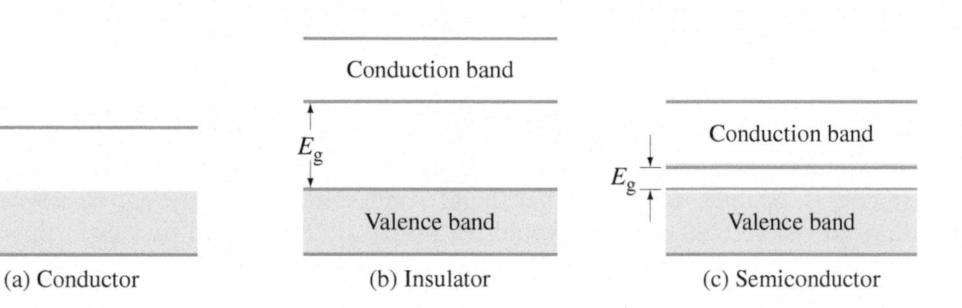

FIGURE 40–32 Energy bands for (a) a conductor, (b) an insulator, which has a large energy gap E_g, and (c) a semiconductor, which has a small energy gap E_g. Shading represents occupied states. Pale shading in (c) represents electrons that can pass from the top of the valence band to the bottom of the conduction band due to thermal agitation at room temperature (exaggerated).

EXAMPLE 40–12 **Calculating the energy gap.** It is found that the conductivity of a certain semiconductor increases when light of wavelength 345 nm or shorter strikes it, suggesting that electrons are being promoted from the valence band to the conduction band. What is the energy gap, E_g, for this semiconductor?

APPROACH The longest wavelength (lowest energy) photon to cause an increase in conductivity has $\lambda = 345$ nm, and its energy $(= hf)$ equals the energy gap.

SOLUTION The gap energy equals the energy of a $\lambda = 345$-nm photon:

$$E_g = hf = \frac{hc}{\lambda} = \frac{(6.63 \times 10^{-34}\,\mathrm{J\cdot s})(3.00 \times 10^{8}\,\mathrm{m/s})}{(345 \times 10^{-9}\,\mathrm{m})(1.60 \times 10^{-19}\,\mathrm{J/eV})} = 3.6\,\mathrm{eV}.$$

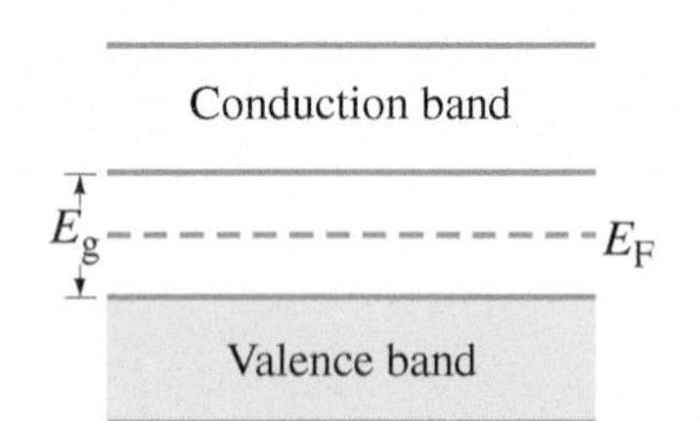

FIGURE 40–33 Example 40–13: The Fermi energy is midway between the valence band and the conduction band.

EXAMPLE 40–13 ESTIMATE **Free electrons in semiconductors and insulators.** Use the Fermi–Dirac probability function, Eq. 40–14, to estimate the order of magnitude of the numbers of free electrons in the conduction band of a solid containing 10^{21} atoms, assuming the solid is at room temperature $(T = 300\,\mathrm{K})$ and is (*a*) a semiconductor with $E_g \approx 1.1$ eV, (*b*) an insulator with $E_g \approx 5$ eV. Compare to a conductor.

APPROACH At $T = 0$, all states above the Fermi energy E_F are empty, and all those below are full. So for semiconductors and insulators we can take E_F to be about midway between the valence and conduction bands, Fig. 40–33, and it does not change significantly as we go to room temperature. We can thus use Eq. 40–14 to find the fraction of electrons in the conduction band at room temperature for the two cases.

SOLUTION (*a*) For the semiconductor, the gap $E_g \approx 1.1$ eV, so $E - E_F \approx 0.55$ eV for the lowest states in the conduction band. Since at room temperature we have $kT \approx 0.026$ eV, then $(E - E_F)/kT \approx 0.55\,\mathrm{eV}/0.026\,\mathrm{eV} \approx 21$ and

$$f(E) = \frac{1}{e^{(E-E_F)/kT} + 1} \approx \frac{1}{e^{21}} \approx 10^{-9}.$$

Thus about 1 atom in 10^9 can contribute an electron to the conductivity.

(*b*) For the insulator with $E - E_F \approx 5.0\,\mathrm{eV} - \frac{1}{2}(5.0\,\mathrm{eV}) = 2.5\,\mathrm{eV}$, we get

$$f(E) \approx \frac{1}{e^{2.5/0.026} + 1} \approx \frac{1}{e^{96}} \approx 10^{-42}.$$

Thus in an ordinary sample containing 10^{21} atoms, there would be no free electrons in an insulator $(10^{21} \times 10^{-42} = 10^{-21})$, about 10^{12} $(10^{21} \times 10^{-9})$ free electrons in a semiconductor, and about 10^{21} free electrons in a good conductor.

PHYSICS APPLIED
Transparency

CONCEPTUAL EXAMPLE 40–14 **Which is transparent?** The energy gap for silicon is 1.14 eV at room temperature, whereas that of zinc sulfide (ZnS) is 3.6 eV. Which one of these is opaque to visible light, and which is transparent?

RESPONSE Visible light photons span energies from roughly 1.8 eV to 3.1 eV $(E = hf = hc/\lambda$ where $\lambda = 400$ nm to 700 nm and $1\,\mathrm{eV} = 1.6 \times 10^{-19}\,\mathrm{J})$. Light is absorbed by the electrons in a material. Silicon's energy gap is small enough to absorb these photons, thus bumping electrons well up into the conduction band, so silicon is opaque. On the other hand, zinc sulfide's energy gap is so large that no visible light photons would be absorbed; they would pass right through the material which would thus be transparent.

40–8 Semiconductors and Doping

Nearly all electronic devices today use semiconductors. The most common are silicon (Si) and germanium (Ge). An atom of silicon or germanium has four outer electrons that act to hold the atoms in the regular lattice structure of the crystal, shown schematically in Fig. 40–34a. Germanium and silicon acquire properties useful for electronics when a tiny amount of impurity is introduced into the crystal structure (perhaps 1 part in 10^6 or 10^7). This is called **doping** the semiconductor. Two kinds of doped semiconductor can be made, depending on the type of impurity used. If the impurity is an element whose atoms have five outer electrons, such as arsenic, we have the situation shown in Fig. 40–34b, with the arsenic atoms holding positions in the crystal lattice where normally silicon atoms would be. Only four of arsenic's electrons fit into the bonding structure. The fifth does not fit in and can move relatively freely, somewhat like the electrons in a conductor. Because of this small number of extra electrons, a doped semiconductor becomes slightly conducting.

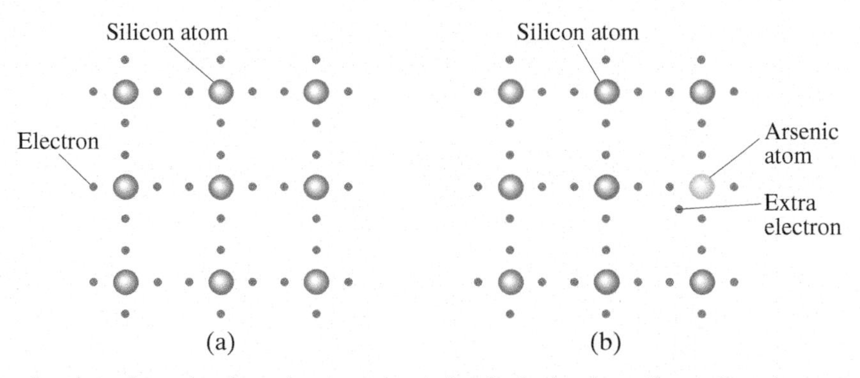

FIGURE 40–34 Two-dimensional representation of a silicon crystal. (a) Four (outer) electrons surround each silicon atom. (b) Silicon crystal doped with a small percentage of arsenic atoms: the extra electron doesn't fit into the crystal lattice and so is free to move about. This is an n-type semiconductor.

The density of conduction electrons in an intrinsic (undoped) semiconductor at room temperature is very low, usually less than 1 per 10^9 atoms. With an impurity concentration of 1 in 10^6 or 10^7 when doped, the conductivity will be much higher and it can be controlled with great precision. An arsenic-doped silicon crystal is called an ***n*-type semiconductor** because *negative* charges (electrons) carry the electric current.

In a ***p*-type semiconductor**, a small percentage of semiconductor atoms are replaced by atoms with three outer electrons—such as gallium. As shown in Fig. 40–35a, there is a "hole" in the lattice structure near a gallium atom since it has only three outer electrons. Electrons from nearby silicon atoms can jump into this hole and fill it. But this leaves a hole where that electron had previously been, Fig. 40–35b. The vast majority of atoms are silicon, so holes are almost always next to a silicon atom. Since silicon atoms require four outer electrons to be neutral, this means that there is a net positive charge at the hole. Whenever an electron moves to fill a hole, the positive hole is then at the previous position of that electron. Another electron can then fill this hole, and the hole thus moves to a new location; and so on. This type of semiconductor is called *p-type* because it is the positive holes that seem to carry the electric current. Note, however, that both *p-type* and *n-type* semiconductors have *no net charge* on them.

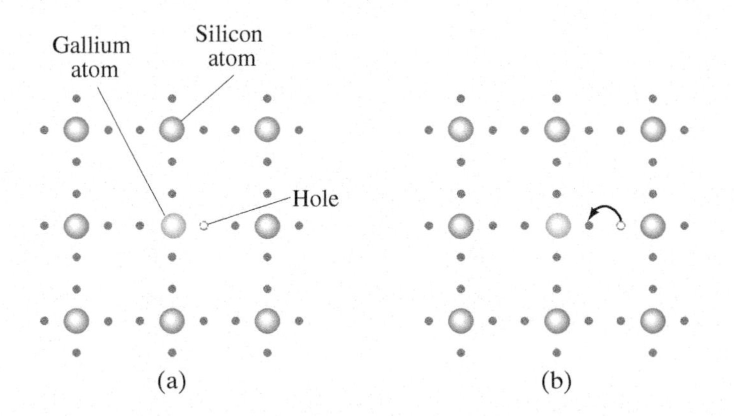

FIGURE 40–35 A p-type semiconductor, gallium-doped silicon. (a) Gallium has only three outer electrons, so there is an empty spot, or *hole* in the structure. (b) Electrons from silicon atoms can jump into the hole and fill it. As a result, the hole moves to a new location (to the right in this Figure), to where the electron used to be.

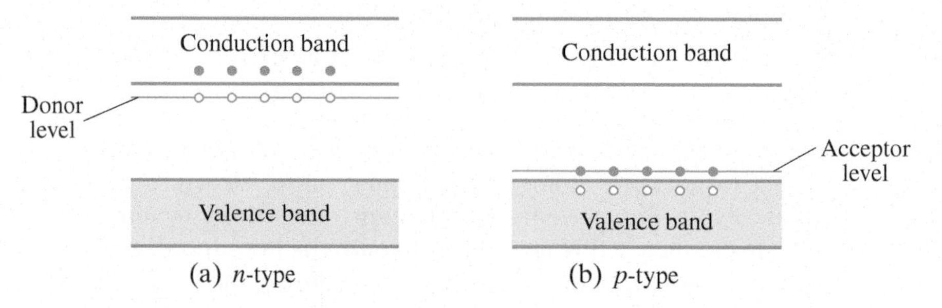

FIGURE 40–36 Impurity energy levels in doped semiconductors.

According to the band theory (Section 40–7), in a doped semiconductor the impurity provides additional energy states between the bands as shown in Fig. 40–36. In an n-type semiconductor, the impurity energy level lies just below the conduction band, Fig. 40–36a. Electrons in this energy level need only about 0.05 eV in Si (even less in Ge) to reach the conduction band; this is on the order of the thermal energy, $\frac{3}{2}kT$ (= 0.04 eV at 300 K), so transitions occur readily at room temperature. This energy level can thus supply electrons to the conduction band, so it is called a **donor** level. In p-type semiconductors, the impurity energy level is just above the valence band (Fig. 40–36b). It is called an **acceptor** level because electrons from the valence band can easily jump into it. Positive holes are left behind in the valence band, and as other electrons move into these holes, the holes move as discussed earlier.

EXERCISE D Which of the following impurity atoms would produce a p-type semiconductor? (*a*) Ge; (*b*) Ne; (*c*) Al; (*d*) As; (*e*) none of the above.

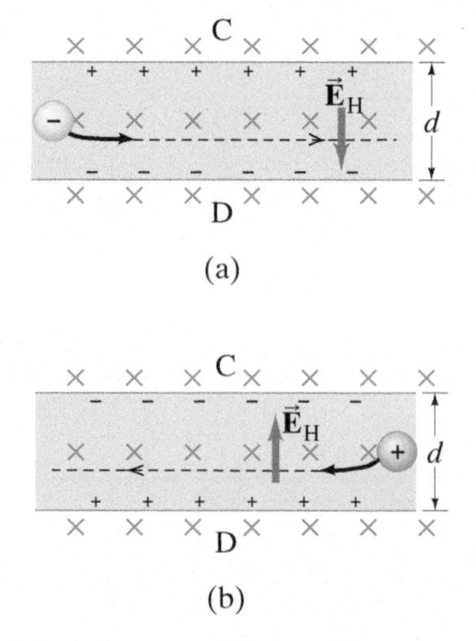

FIGURE 27–32 (Repeated.) The Hall effect. (a) Negative charges moving to the right as the current. (b) The same current, but as positive charges moving to the left.

CONCEPTUAL EXAMPLE 40–15 **Determining charge of conductors.** How can we determine if a p-type semiconductor has a current that is really due to the motion of holes? Or, is this just a convenient model?

RESPONSE Recall from Section 27–8 that the Hall effect can be used to distinguish the sign of the charges involved in a current. When placed in a magnetic field, the current in a particular direction can result in a voltage perpendicular to that current due to the magnetic force on the moving charges (Fig. 27–32, repeated here). The direction of this *Hall voltage* depends on the sign of the charges carrying the current. In this way, it has been demonstrated that it really is moving holes that are responsible for the current in a p-type semiconductor.

40–9 Semiconductor Diodes

Semiconductor diodes and transistors are essential components of modern electronic devices. The miniaturization achieved today allows many thousands of diodes, transistors, resistors, and so on, to be placed on a single *chip* less than a millimeter on a side. We now discuss, briefly and qualitatively, the operation of diodes and transistors.

When an n-type semiconductor is joined to a p-type, a ***pn* junction diode** is formed. Separately, the two semiconductors are electrically neutral. When joined, a few electrons near the junction diffuse from the n-type into the p-type semiconductor, where they fill a few of the holes. The n-type is left with a positive charge, and the p-type acquires a net negative charge. Thus a potential difference is established, with the n side positive relative to the p side, and this prevents further diffusion of electrons.

If a battery is connected to a diode with the positive terminal to the p side and the negative terminal to the n side as in Fig. 40–37a, the externally applied voltage opposes the internal potential difference and the diode is said to be **forward biased**. If the voltage is great enough (about 0.3 V for Ge, 0.6 V for Si at room temperature), a current will flow. The positive holes in the p-type semiconductor are repelled by the positive terminal of the battery, and the electrons in the n-type are repelled by the negative terminal of the battery. The holes and electrons meet at the junction, and the electrons cross over and fill the holes. A current is flowing. Meanwhile, the positive terminal of the battery is continually pulling electrons off the p end, forming new holes, and electrons are being supplied by the negative terminal at the n end. Consequently, a large current flows through the diode.

When the diode is **reverse biased**, as in Fig. 40–37b, the holes in the p end are attracted to the battery's negative terminal and the electrons in the n end are attracted to the positive terminal. The current carriers do not meet near the junction and, ideally, no current flows.

A graph of current versus voltage for a typical diode is shown in Fig. 40–38. As can be seen, a real diode does allow a small amount of reverse current to flow. For most practical purposes, it is negligible. (At room temperature, the reverse current is a few μA in Ge and a few pA in Si; but it increases rapidly with temperature, and may render a diode ineffective above 200°C.)

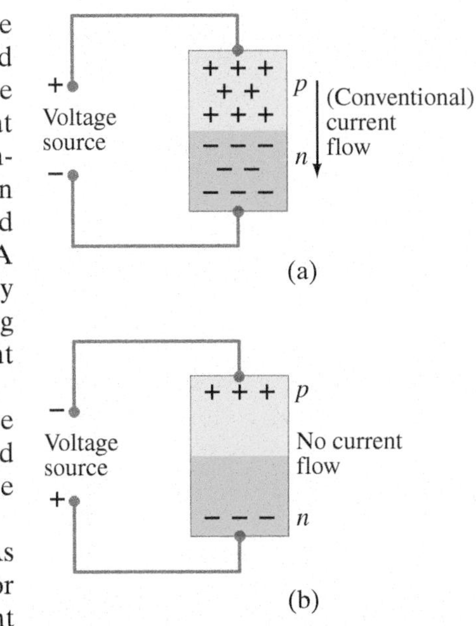

FIGURE 40–37 Schematic diagram showing how a semiconductor diode operates. Current flows when the voltage is connected in forward bias, as in (a), but not when connected in reverse bias, as in (b).

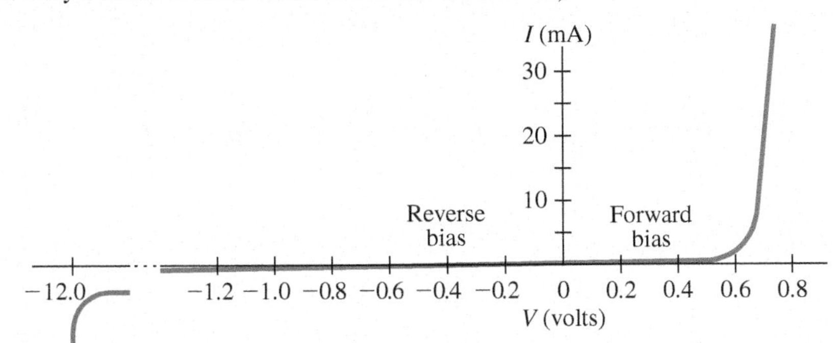

FIGURE 40–38 Current through a silicon pn diode as a function of applied voltage.

EXAMPLE 40–16 **A diode.** The diode whose current–voltage characteristics are shown in Fig. 40–38 is connected in series with a 4.0-V battery in forward bias and a resistor. If a current of 15 mA is to pass through the diode, what resistance must the resistor have?

APPROACH We use Fig. 40–38, where we see that the voltage drop across the diode is about 0.7 V when the current is 15 mA. Then we use simple circuit analysis and Ohm's law (Chapters 25 and 26).

SOLUTION The voltage drop across the resistor is $4.0\,\text{V} - 0.7\,\text{V} = 3.3\,\text{V}$, so $R = V/I = (3.3\,\text{V})/(1.5 \times 10^{-2}\,\text{A}) = 220\,\Omega$.

The symbol for a diode is

—▶|— [diode]

where the arrow represents the direction conventional (+) current flows readily.

If the voltage across a diode connected in reverse bias is increased greatly, a point is reached where breakdown occurs. The electric field across the junction becomes so large that ionization of atoms results. The electrons thus pulled off their atoms contribute to a larger and larger current as breakdown continues. The voltage remains constant over a wide range of currents. This is shown on the far left in Fig. 40–38. This property of diodes can be used to accurately regulate a voltage supply. A diode designed for this purpose is called a **zener diode**. When placed across the output of an unregulated power supply, a zener diode can maintain the voltage at its own breakdown voltage as long as the supply voltage is always above this point. Zener diodes can be obtained corresponding to voltages of a few volts to hundreds of volts.

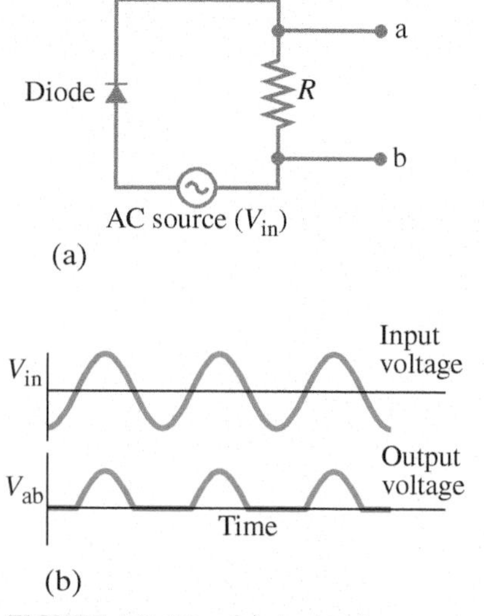

FIGURE 40–39 (a) A simple (half-wave) rectifier circuit using a semiconductor diode. (b) AC source input voltage, and output voltage across R, as functions of time.

Since a *pn* junction diode allows current to flow only in one direction (as long as the voltage is not too high), it can serve as a **rectifier**—to change ac into dc. A simple rectifier circuit is shown in Fig. 40–39a. The ac source applies a voltage across the diode alternately positive and negative. Only during half of each cycle will a current pass through the diode; only then is there a current through the resistor R. Hence, a graph of the voltage V_{ab} across R as a function of time looks like the output voltage shown in Fig. 40–39b. This **half-wave rectification** is not exactly dc, but it is unidirectional. More useful is a **full-wave rectifier** circuit, which uses two diodes (or sometimes four) as shown in Fig. 40–40a. At any given instant, either one diode or the other will conduct current to the right. Therefore, the output across the load resistor R will be as shown in Fig. 40–40b. Actually this is the voltage if the capacitor C were not in the circuit. The capacitor tends to store charge and, if the time constant RC is sufficiently long, helps to smooth out the current as shown in Fig. 40–40c. (The variation in output shown in Fig. 40–40c is called **ripple voltage**.)

Rectifier circuits are important because most line voltage in buildings is ac, and most electronic devices require a dc voltage for their operation. Hence, diodes are found in nearly all electronic devices including radio and TV sets, calculators, and computers.

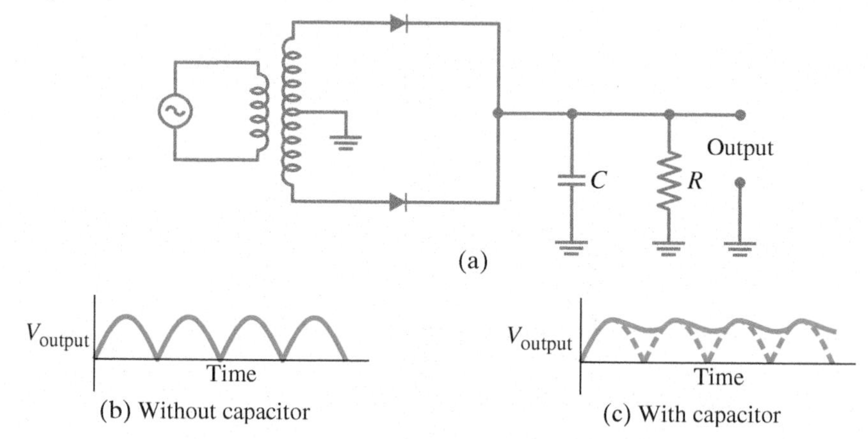

FIGURE 40–40 (a) Full-wave rectifier circuit (including a transformer so the magnitude of the voltage can be changed). (b) Output voltage in the absence of capacitor C. (c) Output voltage with the capacitor in the circuit.

PHYSICS APPLIED
LEDs and applications
Car safety (brakes)

Another useful device is a **light-emitting diode** (LED), invented in the 1960s. When a *pn* junction is forward biased, a current begins to flow. Electrons cross from the *n* region into the *p* region and combine with holes, and a photon can be emitted with an energy approximately equal to the band gap, E_g (see Figs. 40–32c and 40–36). Often the energy, and hence the wavelength, is in the red region of the visible spectrum, producing the familiar LED displays on electronic devices, car instrument panels, digital clocks, and so on. Infrared (i.e., nonvisible) LEDs are used in remote controls for TVs, DVDs, and stereos. New types of LEDs emit other colors, and LED “bulbs” are beginning to replace other types of lighting in applications such as flashlights, traffic signals, car brake lights, and outdoor signs, billboards, and theater displays. LED bulbs, sometimes called **solid-state** lighting, are costly, but they offer advantages: they are long-lived, efficient, and rugged. LED traffic lights, for example (Fig. 40–41), last 5 to 10 times longer than traditional incandescent bulbs, and use only 20% of the energy for the same light output. As car brake lights, they light up a fraction of a second sooner, allowing a driver an extra 5 or 6 meters (15–20 ft) more stopping distance at highway speeds.

FIGURE 40–41 LED traffic light.

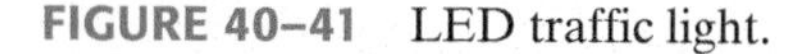

Solar cells and **photodiodes** (Section 37–2) are *pn* junctions used in the reverse way. Photons are absorbed, creating electron–hole pairs if the photon energy is greater than the band gap energy, E_g. The created electrons and holes produce a current that, when connected to an external circuit, becomes a source of emf and power. *Particle detectors* (Section 41–11) operate similarly.

A diode is called a **nonlinear device** because the current is not proportional to the voltage. That is, a graph of current versus voltage (Fig. 40–38) is not a straight line, as it is for a resistor (which ideally *is* linear). Transistors are also *nonlinear* devices.

40–10 Transistors and Integrated Circuits (Chips)

A simple **junction transistor** consists of a crystal of one type of doped semiconductor sandwiched between two crystals of the opposite type. Both *npn* and *pnp* transistors are made, and they are shown schematically in Fig. 40–42a. The three semiconductors are given the names *collector*, *base*, and *emitter*. The symbols for *npn* and *pnp* transistors are shown in Fig. 40–42b. The arrow is always placed on the emitter and indicates the direction of (conventional) current flow in normal operation.

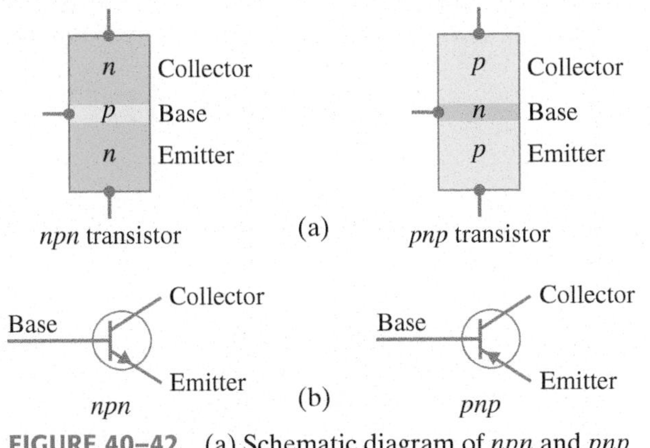

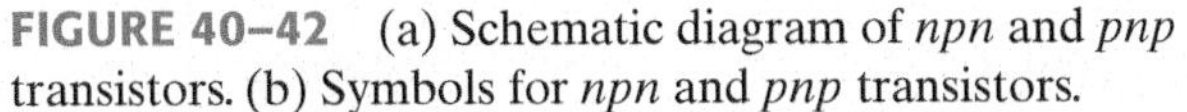

FIGURE 40–42 (a) Schematic diagram of *npn* and *pnp* transistors. (b) Symbols for *npn* and *pnp* transistors.

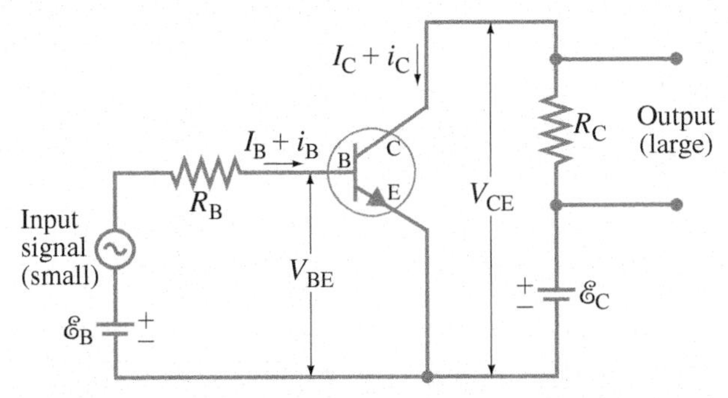

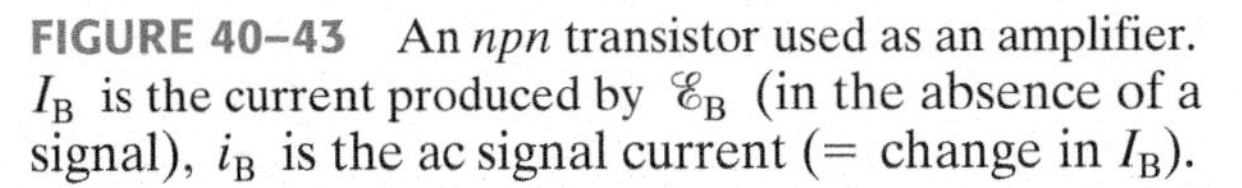

FIGURE 40–43 An *npn* transistor used as an amplifier. I_B is the current produced by $\mathscr{E}_B$ (in the absence of a signal), i_B is the ac signal current (= change in I_B).

The operation of a transistor can be analyzed qualitatively—very briefly—as follows. Consider an *npn* transistor connected as shown in Fig. 40–43. A voltage V_{CE} is maintained between the collector and emitter by the battery $\mathscr{E}_C$. The voltage applied to the base is called the *base bias voltage*, V_{BE}. If V_{BE} is positive, conduction electrons in the emitter are attracted into the base. Since the base region is very thin (less than 1 μm—much less if on a chip), most of these electrons flow right across into the collector, which is maintained at a positive voltage. A large current, I_C, flows between collector and emitter and a much smaller current, I_B, through the base. In the steady state, I_B and I_C can be considered dc. A small variation in the base voltage due to an input signal causes a large change in the collector current and therefore a large change in the voltage drop across the output resistor R_C. Hence a transistor can *amplify* a small signal into a larger one.

Typically a small ac signal (call it i_B) is to be amplified, and when added to the base bias voltage and current causes the voltage and current at the collector to vary at the same rate but magnified. Thus, what is important for amplification is the *change* in collector current for a given input *change* in base current. We label these ac signal currents (= changes in I_C and I_B) as i_C and i_B. The **current gain** is defined as the ratio

$$\beta_I = \frac{\text{output (collector) ac current}}{\text{input (base) ac current}} = \frac{i_C}{i_B}.$$

β_I is typically on the order of 10 to 100. Similarly, the **voltage gain** is

$$\beta_V = \frac{\text{output (collector) ac voltage}}{\text{input (base) ac voltage}}.$$

Transistors are the basic elements in modern electronic **amplifiers** of all sorts.

A *pnp* transistor operates like an *npn*, except that holes move instead of electrons. The collector voltage is negative, and so is the base voltage in normal operation.

In **digital circuits**, including computers, where "off" and "on" (or zero and one) make up the binary code, transistors act like a **gate** or switch. That is, they let current pass ("on") or they block it ("off").

Transistors were a great advance in miniaturization of electronic circuits. Although individual transistors are very small compared to the once-used vacuum tubes, they are huge compared to **integrated circuits** or **chips** (see photo at start of this Chapter). Tiny amounts of impurities can be placed at particular locations within a single silicon crystal. These can be arranged to form diodes, transistors, and resistors (undoped semiconductors). Capacitors and inductors can also be formed, although they are often connected separately. A tiny chip, a few millimeters on a side, may contain millions of transistors and other circuit elements. Integrated circuits are the heart of computers, televisions, calculators, cameras, and the electronic instruments that control aircraft, space vehicles, and automobiles. The "miniaturization" produced by integrated circuits not only allows extremely complicated circuits to be placed in a small space, but also has allowed a great increase in the speed of operation of, say, computers, because the distances the electronic signals travel are so tiny.

Summary

Quantum mechanics explains the bonding together of atoms to form **molecules**. In a **covalent bond**, the electron clouds of two or more atoms overlap because of constructive interference between the electron waves. The positive nuclei are attracted to this concentration of negative charge between them, forming the bond.

An **ionic bond** is an extreme case of a covalent bond in which one or more electrons from one atom spend much more time around the other atom than around their own. The atoms then act as oppositely charged ions that attract each other, forming the bond.

These **strong bonds** hold molecules together, and also hold atoms and molecules together in solids. Also important are **weak bonds** (or **van der Waals bonds**), which are generally dipole attractions between molecules.

When atoms combine to form molecules, the energy levels of the outer electrons are altered because they now interact with each other. Additional energy levels also become possible because the atoms can vibrate with respect to each other, and the molecule as a whole can rotate. The energy levels for both vibrational and rotational motion are quantized, and are very close together (typically, 10^{-1} eV to 10^{-3} eV apart). Each atomic energy level thus becomes a set of closely spaced levels corresponding to the vibrational and rotational motions. Transitions from one level to another appear as many very closely spaced lines. The resulting spectra are called **band spectra**.

The quantized rotational energy levels are given by

$$E_{\text{rot}} = \ell(\ell+1)\frac{\hbar^2}{2I}, \qquad \ell = 0, 1, 2, \cdots, \qquad \textbf{(40–2)}$$

where I is the moment of inertia of the molecule.

The energy levels for vibrational motion are given by

$$E_{\text{vib}} = \left(\nu + \tfrac{1}{2}\right)hf, \qquad \nu = 0, 1, 2, \cdots, \qquad \textbf{(40–6)}$$

where f is the classical natural frequency of vibration for the molecule. Transitions between energy levels are subject to the selection rules $\Delta\ell = \pm 1$ and $\Delta\nu = \pm 1$.

Some **solids** are bound together by covalent and ionic bonds, just as molecules are. In metals, the electrostatic force between free electrons and positive ions helps form the **metallic bond**.

In the free-electron theory of metals, electrons occupy the possible energy states according to the exclusion principle. At $T = 0\,\text{K}$, all possible states are filled up to a maximum energy level called the **Fermi energy**, E_{F}, the magnitude of which is typically a few eV. All states above E_{F} are vacant at $T = 0\,\text{K}$.

At normal temperatures (300 K) the distribution of occupied states is only slightly altered and is given by the **Fermi–Dirac probability function**

$$f(E) = \frac{1}{e^{(E-E_{\text{F}})/kT} + 1}. \qquad \textbf{(40–14)}$$

In a crystalline solid, the possible energy states for electrons are arranged in **bands**. Within each band the levels are very close together, but between the bands there may be forbidden **energy gaps**. Good conductors are characterized by the highest occupied band (the **conduction band**) being only partially full, so there are many accessible states available to electrons to move about and accelerate when a voltage is applied. In a good insulator, the highest occupied energy band (the **valence band**) is completely full, and there is a large energy gap (5 to 10 eV) to the next highest band, the *conduction band*. At room temperature, molecular kinetic energy (thermal energy) available due to collisions is only about 0.04 eV, so almost no electrons can jump from the valence to the conduction band. In a **semiconductor**, the gap between valence and conduction bands is much smaller, on the order of 1 eV, so a few electrons can make the transition from the essentially full valence band to the nearly empty conduction band.

In a **doped** semiconductor, a small percentage of impurity atoms with five or three valence electrons replace a few of the normal silicon atoms with their four valence electrons. A five-electron impurity produces an ***n*-type** semiconductor with negative electrons as carriers of current. A three-electron impurity produces a ***p*-type** semiconductor in which positive **holes** carry the current. The energy level of impurity atoms lies slightly below the conduction band in an *n*-type semiconductor, and acts as a **donor** from which electrons readily pass into the conduction band. The energy level of impurity atoms in a *p*-type semiconductor lies slightly above the valence band and acts as an **acceptor** level, since electrons from the valence band easily reach it, leaving holes behind to act as charge carriers.

A semiconductor **diode** consists of a ***pn* junction** and allows current to flow in one direction only; it can be used as a **rectifier** to change ac to dc. Common **transistors** consist of three semiconductor sections, either as ***pnp*** or ***npn***. Transistors can amplify electrical signals and in computers serve as switches or **gates** for the 1s and 0s. An integrated circuit consists of a tiny semiconductor crystal or **chip** on which many transistors, diodes, resistors, and other circuit elements have been constructed using careful placement of impurities.

Questions

1. What type of bond would you expect for (*a*) the N_2 molecule, (*b*) the HCl molecule, (*c*) Fe atoms in a solid?
2. Describe how the molecule $CaCl_2$ could be formed.
3. Does the H_2 molecule have a permanent dipole moment? Does O_2? Does H_2O? Explain.
4. Although the molecule H_3 is not stable, the ion H_3^+ is. Explain, using the Pauli exclusion principle.
5. The energy of a molecule can be divided into four categories. What are they?
6. Would you expect the molecule H_2^+ to be stable? If so, where would the single electron spend most of its time?
7. Explain why the carbon atom $(Z = 6)$ usually forms four bonds with hydrogen-like atoms.
8. Explain on the basis of energy bands why the sodium chloride crystal is a good insulator. [*Hint*: Consider the shells of Na^+ and Cl^- ions.]
9. If conduction electrons are free to roam about in a metal, why don't they leave the metal entirely?
10. Explain why the resistivity of metals increases with increasing temperature whereas the resistivity of semiconductors may decrease with increasing temperature.
11. Figure 40–44 shows a "bridge-type" full-wave rectifier. Explain how the current is rectified and how current flows during each half cycle.

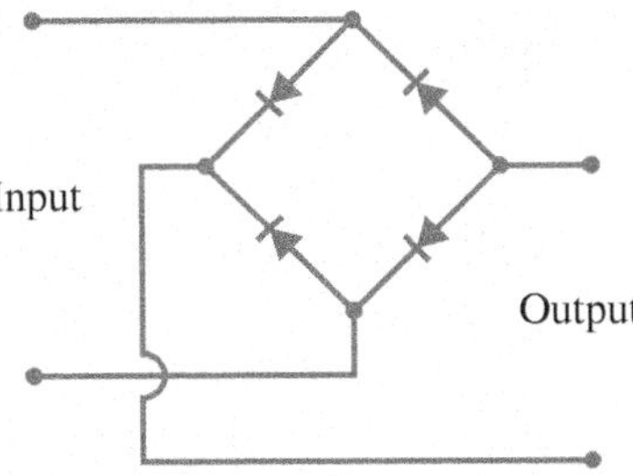

Input

Output

FIGURE 40–44 Question 11.

12. Discuss the differences between an ideal gas and a Fermi electron gas.
13. Compare the resistance of a *pn* junction diode connected in forward bias to its resistance when connected in reverse bias.
14. Which aspects of Fig. 40–28 are peculiar to copper, and which are valid in general for other metals?
15. Explain how a transistor can be used as a switch.
16. What is the main difference between *n*-type and *p*-type semiconductors?
17. Draw a circuit diagram showing how a *pnp* transistor can operate as an amplifier.
18. In a transistor, the base–emitter junction and the base–collector junction are essentially diodes. Are these junctions reverse-biased or forward-biased in the application shown in Fig. 40–43?
19. A transistor can amplify an electronic signal, meaning it can increase the power of an input signal. Where does it get the energy to increase the power?
20. A silicon semiconductor is doped with phosphorus. Will these atoms be donors or acceptors? What type of semiconductor will this be?
21. Do diodes and transistors obey Ohm's law? Explain.
22. Can a diode be used to amplify a signal? Explain.
23. If $\mathscr{E}_C$ were reversed in Fig. 40–43, how would the amplification be altered?

Problems

40–1 to 40–3 Molecular Bonds

1. (I) Estimate the binding energy of a KCl molecule by calculating the electrostatic potential energy when the K^+ and Cl^- ions are at their stable separation of 0.28 nm. Assume each has a charge of magnitude 1.0*e*.
2. (II) The measured binding energy of KCl is 4.43 eV. From the result of Problem 1, estimate the contribution to the binding energy of the repelling electron clouds at the equilibrium distance $r_0 = 0.28$ nm.
3. (II) Estimate the binding energy of the H_2 molecule, assuming the two H nuclei are 0.074 nm apart and the two electrons spend 33% of their time midway between them.
4. (II) The equilibrium distance r_0 between two atoms in a molecule is called the **bond length**. Using the bond lengths of homogeneous molecules (like H_2, O_2, and N_2), one can estimate the bond length of heterogeneous molecules (like CO, CN, and NO). This is done by summing half of each bond length of the homogenous molecules to estimate that of the heterogeneous molecule. Given the following bond lengths: H_2 (= 74 pm), N_2 (= 145 pm), O_2 (= 121 pm), C_2 (= 154 pm), estimate the bond lengths for: HN, CN, and NO.
5. (II) Estimate the energy associated with the repulsion of the electron shells of a lithium fluoride (LiF) molecule. The ionization energy of lithium is 5.39 eV, and it takes 3.41 eV to remove the extra electron from an F^- ion. The bond length is 0.156 nm, and the binding energy of LiF is 5.95 eV.
6. (II) Binding energies are often measured experimentally in kcal per mole, and then the binding energy in eV per molecule is calculated from that result. What is the conversion factor in going from kcal per mole to eV per molecule? What is the binding energy of KCl (= 4.43 eV) in kcal per mole?
7. (III) (*a*) Apply reasoning similar to that in the text for the $S = 0$ and $S = 1$ states in the formation of the H_2 molecule to show why the molecule He_2 is *not* formed. (*b*) Explain why the He_2^+ molecular ion *could* form. (Experiment shows it has a binding energy of 3.1 eV at $r_0 = 0.11$ nm.)

40–4 Molecular Spectra

8. (I) Show that the quantity $\hbar^2/I$ has units of energy.
9. (I) What is the reduced mass of the molecules (*a*) KCl; (*b*) O_2; (*c*) HCl?
10. (II) (*a*) Calculate the "characteristic rotational energy," $\hbar^2/2I$, for the O_2 molecule whose bond length is 0.121 nm. (*b*) What are the energy and wavelength of photons emitted in an $\ell = 2$ to $\ell = 1$ transition?

11. (II) The "characteristic rotational energy," $\hbar^2/2I$, for N_2 is 2.48×10^{-4} eV. Calculate the N_2 bond length.

12. (II) Estimate the longest wavelength emitted by a lithium hydride (LiH) molecule for a change in its rotational state if its equilibrium separation is 0.16 nm.

13. (II) The equilibrium separation of H atoms in the H_2 molecule is 0.074 nm (Fig. 40–8). Calculate the energies and wavelengths of photons for the rotational transitions (*a*) $\ell = 1$ to $\ell = 0$, (*b*) $\ell = 2$ to $\ell = 1$, and (*c*) $\ell = 3$ to $\ell = 2$.

14. (II) Explain why there is no transition for $\Delta E = hf$ in Fig. 40–21 (and Fig. 40–22). See Eqs. 40–8.

15. (II) The fundamental vibration frequency for the CO molecule is 6.42×10^{13} Hz. Determine (*a*) the reduced mass, and (*b*) the effective value of the "stiffness" constant k. Compare to k for the H_2 molecule.

16. (II) Li and Br form a molecule for which the lowest vibrational frequency is 1.7×10^{13} Hz. What is the effective stiffness constant k?

17. (II) Calculate the bond length for the NaCl molecule given that three successive wavelengths for rotational transitions are 23.1 mm, 11.6 mm, and 7.71 mm.

18. (II) (*a*) Use the curve of Fig. 40–18 to estimate the stiffness constant k for the H_2 molecule. (Recall that $U = \frac{1}{2}kx^2$.) (*b*) Then estimate the fundamental wavelength for vibrational transitions using the classical formula (Chapter 14), but use only $\frac{1}{2}$ the mass of an H atom (because both H atoms move).

19. (III) Imagine the two atoms of a diatomic molecule as if they were connected by a spring, Fig. 40–45. Show that the classical frequency of vibration is given by Eq. 40–5. [*Hint*: Let x_1 and x_2 be the displacements of each mass from initial equilibrium positions; then $m_1\, d^2x_1/dt^2 = -kx$, and $m_2\, d^2x_2/dt^2 = -kx$, where $x = x_1 + x_2$. Find another relationship between x_1 and x_2, assuming that the center of mass of the system stays at rest, and then show that $\mu\, d^2x/dt^2 = -kx$.]

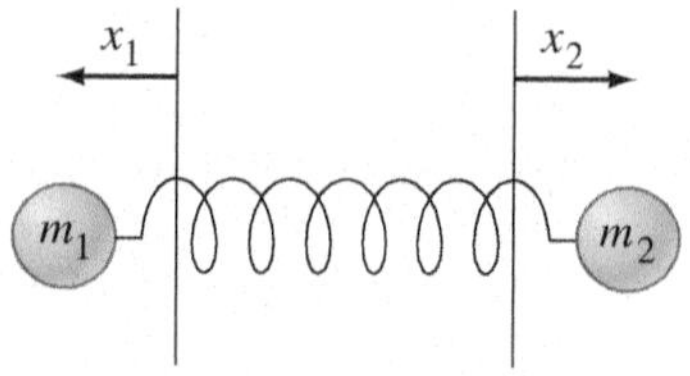

FIGURE 40–45 Problem 19.

40–5 Bonding in Solids

20. (I) Estimate the ionic cohesive energy for NaCl taking $\alpha = 1.75$, $m = 8$, and $r_0 = 0.28$ nm.

21. (II) Common salt, NaCl, has a density of $2.165\ \text{g/cm}^3$. The molecular weight of NaCl is 58.44. Estimate the distance between nearest neighbor Na and Cl ions. [*Hint*: *Each* ion can be considered to have one "cube" or "cell" of side s (our unknown) extending out from it.]

22. (II) Repeat the previous Problem for KCl whose density is $1.99\ \text{g/cm}^3$.

23. (II) The spacing between "nearest neighbor" Na and Cl ions in a NaCl crystal is 0.24 nm. What is the spacing between two nearest neighbor Na ions?

24. (III) For a long one-dimensional chain of alternating positive and negative ions, show that the Madelung constant would be $\alpha = 2 \ln 2$. [*Hint*: Use a series expansion for $\ln(1 + x)$.]

25. (III) (*a*) Starting from Eq. 40–9, show that the ionic cohesive energy is given by $U_0 = -(\alpha e^2/4\pi\epsilon_0 r_0)(1 - 1/m)$. Determine U_0 for (*b*) NaI $(r_0 = 0.33\ \text{nm})$ and (*c*) MgO $(r_0 = 0.21\ \text{nm})$. Assume $m = 10$. (*d*) If you used $m = 8$ instead, how far off would your answers be? Assume $\alpha = 1.75$.

40–6 Free-Electron Theory of Metals

26. (II) Estimate the number of possible electron states in a 1.00-cm^3 cube of silver between $0.985E_F$ and E_F $(= 5.48\ \text{eV})$.

27. (II) Estimate the number of states between 7.00 eV and 7.05 eV that are available to electrons in a 1.0-cm^3 cube of copper.

28. (II) What, roughly, is the ratio of the density of molecules in an ideal gas at 285 K and 1 atm (say O_2) to the density of free electrons (assume one per atom) in a metal (copper) also at 285 K?

29. (II) Calculate the energy which has 85.0% occupancy probability for copper at (*a*) $T = 295$ K; (*b*) $T = 750$ K.

30. (II) Calculate the energy which has 15.0% occupancy probability for copper at (*a*) $T = 295$ K; (*b*) $T = 950$ K.

31. (II) What is the occupancy probability for a conduction electron in copper at $T = 295$ K for an energy $E = 1.015E_F$?

32. (II) The atoms in zinc metal $(\rho = 7.1 \times 10^3\ \text{kg/m}^3)$ each have two free electrons. Calculate (*a*) the density of conduction electrons, (*b*) their Fermi energy, and (*c*) their Fermi speed.

33. (II) Calculate the Fermi energy and Fermi speed for sodium, which has a density of $0.97 \times 10^3\ \text{kg/m}^3$ and has one conduction electron per atom.

34. (II) Given that the Fermi energy of aluminum is 11.63 eV, (*a*) calculate the density of free electrons using Eq. 40–12, and (*b*) estimate the valence of aluminum using this model and the known density $(2.70 \times 10^3\ \text{kg/m}^3)$ and atomic mass (27.0) of aluminum.

35. (II) Show that the average energy of conduction electrons in a metal at $T = 0$ K is $\overline{E} = \frac{3}{5}E_F$ (Eq. 40–13) by calculating

$$\overline{E} = \frac{\int E\, n_o(E)\, dE}{\int n_o(E)\, dE}.$$

36. (II) The neutrons in a neutron star (Chapter 44) can be treated as a Fermi gas with neutrons in place of the electrons in our model of an electron gas. Determine the Fermi energy for a neutron star of radius 12 km and mass 2.5 times that of our Sun. Assume that the star is made entirely of neutrons and is of uniform density.

37. (II) For a one-dimensional potential well of width ℓ, start with Eq. 38–13 and show that the number of states per unit energy interval for an electron gas is given by

$$g_\ell(E) = \sqrt{\frac{8m\ell^2}{h^2E}}.$$

Remember that there can be two electrons (spin up and spin down) for each value of n. [*Hint*: Write the quantum number n in terms of E. Then $g_\ell(E) = 2\, dn/dE$ where dn is the number of energy levels between E and $E + dE$.]

38. (II) Show that the probability for the state at the Fermi energy being occupied is exactly $\frac{1}{2}$, independent of temperature.

39. (II) A very simple model of a "one-dimensional" metal consists of N electrons confined to a rigid box of width ℓ. We neglect the Coulomb interaction between the electrons. (*a*) Calculate the Fermi energy for this one-dimensional metal (E_F = the energy of the most energetic electron at $T = 0\,\text{K}$), taking into account the Pauli exclusion principle. You can assume for simplicity that N is even. (*b*) What is the smallest amount of energy, ΔE, that this 1-D metal can absorb? (*c*) Find the limit of $\Delta E/E_F$ for large N. What does this result say about how well metals can conduct?

40. (II) (*a*) For copper at room temperature ($T = 293\,\text{K}$), calculate the Fermi factor, Eq. 40–14, for an electron with energy 0.12 eV above the Fermi energy. This represents the probability that this state is occupied. Is this reasonable? (*b*) What is the probability that a state 0.12 eV below the Fermi energy is occupied? (*c*) What is the probability that the state in part (*b*) is unoccupied?

41. (III) Proceed as follows to derive the density of states, $g(E)$, the number of states per unit volume per unit energy interval, Eq. 40–10. Let the metal be a cube of side ℓ. Extend the discussion of Section 38–8 for an infinite well to three dimensions, giving energy levels

$$E = \frac{h^2}{8m\ell^2}\left(n_1^2 + n_2^2 + n_3^2\right).$$

(Explain the meaning of n_1, n_2, n_3.) Each set of values for the quantum numbers n_1, n_2, n_3 corresponds to one state. Imagine a space where n_1, n_2, n_3 are the axes, and each state is represented by a point on a cubic lattice in this space, each separated by 1 unit along an axis. Consider the octant $n_1 > 0$, $n_2 > 0$, $n_3 > 0$. Show that the number of states N within a radius $R = \left(n_1^2 + n_2^2 + n_3^2\right)^{\frac{1}{2}}$ is $2\left(\frac{1}{8}\right)\left(\frac{4}{3}\pi R^3\right)$. Then, to get Eq. 40–10, set $g(E) = (1/V)(dN/dE)$, where $V = \ell^3$ is the volume of the metal.

40–7 Band Theory of Solids

42. (I) A semiconductor is struck by light of slowly increasing frequency and begins to conduct when the wavelength of the light is 580 nm; estimate the size of the energy gap E_g.

43. (I) Calculate the longest-wavelength photon that can cause an electron in silicon ($E_g = 1.14\,\text{eV}$) to jump from the valence band to the conduction band.

44. (II) The energy gap between valence and conduction bands in germanium is 0.72 eV. What range of wavelengths can a photon have to excite an electron from the top of the valence band into the conduction band?

45. (II) We saw that there are $2N$ possible electron states in the 3*s* band of Na, where N is the total number of atoms. How many possible electron states are there in the (*a*) 2*s* band, (*b*) 2*p* band, and (*c*) 3*p* band? (*d*) State a general formula for the total number of possible states in any given electron band.

46. (II) The energy gap E_g in germanium is 0.72 eV. When used as a photon detector, roughly how many electrons can be made to jump from the valence to the conduction band by the passage of a 730-keV photon that loses all its energy in this fashion?

40–8 Semiconductors and Doping

47. (II) Suppose that a silicon semiconductor is doped with phosphorus so that one silicon atom in 1.2×10^6 is replaced by a phosphorus atom. Assuming that the "extra" electron in every phosphorus atom is donated to the conduction band, by what factor is the density of conduction electrons increased? The density of silicon is $2330\,\text{kg/m}^3$, and the density of conduction electrons in pure silicon is about $10^{16}\,\text{m}^{-3}$ at room temperature.

40–9 Diodes

48. (I) At what wavelength will an LED radiate if made from a material with an energy gap $E_g = 1.6\,\text{eV}$?

49. (I) If an LED emits light of wavelength $\lambda = 680\,\text{nm}$, what is the energy gap (in eV) between valence and conduction bands?

50. (II) A silicon diode, whose current–voltage characteristics are given in Fig. 40–38, is connected in series with a battery and an 860-Ω resistor. What battery voltage is needed to produce a 12-mA current?

51. (II) Suppose that the diode of Fig. 40–38 is connected in series to a 150-Ω resistor and a 2.0-V battery. What current flows in the circuit? [*Hint*: Draw a line on Fig. 40–38 representing the current in the resistor as a function of the voltage across the diode; the intersection of this line with the characteristic curve will give the answer.]

52. (II) Sketch the resistance as a function of current, for $V > 0$, for the diode shown in Fig. 40–38.

53. (II) An ac voltage of 120 V rms is to be rectified. Estimate very roughly the average current in the output resistor R (35 kΩ) for (*a*) a half-wave rectifier (Fig. 40–39), and (*b*) a full-wave rectifier (Fig. 40–40) without capacitor.

54. (II) A semiconductor diode laser emits 1.3-μm light. Assuming that the light comes from electrons and holes recombining, what is the band gap in this laser material?

55. (II) A silicon diode passes significant current only if the forward-bias voltage exceeds about 0.6 V. Make a rough estimate of the average current in the output resistor R of (*a*) a half-wave rectifier (Fig. 40–39), and (*b*) a full-wave rectifier (Fig. 40–40) without a capacitor. Assume that $R = 120\,\Omega$ in each case and that the ac voltage is 9.0 V rms in each case.

56. (III) A 120-V rms 60-Hz voltage is to be rectified with a full-wave rectifier as in Fig. 40–40, where $R = 28\,\text{k}\Omega$, and $C = 35\,\mu\text{F}$. (*a*) Make a rough estimate of the average current. (*b*) What happens if $C = 0.10\,\mu\text{F}$? [*Hint*: See Section 26–5.]

40–10 Transistors

57. (II) If the current gain of the transistor amplifier in Fig. 40–43 is $\beta = i_C/i_B = 95$, what value must R_C have if a 1.0-μA ac base current is to produce an ac output voltage of 0.35 V?

58. (II) Suppose that the current gain of the transistor in Fig. 40–43 is $\beta = i_C/i_B = 85$. If $R_C = 4.3\,\text{k}\Omega$, calculate the ac output voltage for an ac input current of 2.0 μA.

59. (II) An amplifier has a voltage gain of 65 and a 25-kΩ load (output) resistance. What is the peak output current through the load resistor if the input voltage is an ac signal with a peak of 0.080 V?

60. (II) A transistor, whose current gain $\beta = i_C/i_B = 75$, is connected as in Fig. 40–43 with $R_B = 3.8\,\text{k}\Omega$ and $R_C = 7.8\,\text{k}\Omega$. Calculate (*a*) the voltage gain, and (*b*) the power amplification.

61. (II) From Fig. 40–43, write an equation for the relationship between the base current (I_B), the collector current (I_C), and the emitter current (I_E, not labeled in Fig. 40–43). Assume $i_B = i_C = 0$.

General Problems

62. Use the uncertainty principle to estimate the binding energy of the H_2 molecule by calculating the difference in kinetic energy of the electrons between when they are in separate atoms and when they are in the molecule. Take Δx for the electrons in the separated atoms to be the radius of the first Bohr orbit, 0.053 nm, and for the molecule take Δx to be the separation of the nuclei, 0.074 nm. [*Hint*: Let $\Delta p \approx \Delta p_x$.]

63. The average translational kinetic energy of an atom or molecule is about $\overline{K} = \frac{3}{2}kT$ (see Chapter 18), where $k = 1.38 \times 10^{-23}\,\text{J/K}$ is Boltzmann's constant. At what temperature T will $\overline{K}$ be on the order of the bond energy (and hence the bond easily broken by thermal motion) for (*a*) a covalent bond (say H_2) of binding energy 4.0 eV, and (*b*) a "weak" hydrogen bond of binding energy 0.12 eV?

64. In the ionic salt KF, the separation distance between ions is about 0.27 nm. (*a*) Estimate the electrostatic potential energy between the ions assuming them to be point charges (magnitude 1*e*). (*b*) When F "grabs" an electron, it releases 3.41 eV of energy, whereas 4.34 eV is required to ionize K. Find the binding energy of KF relative to free K and F atoms, neglecting the energy of repulsion.

65. A diatomic molecule is found to have an activation energy of 1.4 eV. When the molecule is disassociated, 1.6 eV of energy is released. Draw a potential energy curve for this molecule.

66. One possible form for the potential energy (U) of a diatomic molecule (Fig. 40–8) is called the *Morse Potential*:

$$U = U_0\left[1 - e^{-a(r-r_0)}\right]^2.$$

(*a*) Show that r_0 represents the equilibrium distance and U_0 the dissociation energy. (*b*) Graph U from $r = 0$ to $r = 4r_0$, assuming $a = 18\,\text{nm}^{-1}$, $U_0 = 4.6\,\text{eV}$, and $r_0 = 0.13\,\text{nm}$.

67. The fundamental vibration frequency for the HCl molecule is 8.66×10^{13} Hz. Determine (*a*) the reduced mass, and (*b*) the effective value of the stiffness constant k. Compare to k for the H_2 molecule.

68. For H_2, estimate how many rotational states there are between vibrational states.

69. Explain, using the Boltzmann factor (Eq. 39–16), why the heights of the peaks in Fig. 40–22 are different from one another. Explain also why the lines are not equally spaced. [*Hint*: Does the moment of inertia necessarily remain constant?]

70. The rotational absorption spectrum of a molecule displays peaks about 8.4×10^{11} Hz apart. Determine the moment of inertia of this molecule.

71. A TV remote control emits IR light. If the detector on the TV set is *not* to react to visible light, could it make use of silicon as a "window" with its energy gap $E_g = 1.14\,\text{eV}$? What is the shortest-wavelength light that can strike silicon without causing electrons to jump from the valence band to the conduction band?

72. Do we need to consider quantum effects for everyday rotating objects? Estimate the differences between rotational energy levels for a spinning baton compared to the energy of the baton. Assume the baton consists of a uniform 32-cm-long bar with a mass of 260 g and two small end masses, each of mass 380 g, and that it rotates at 1.6 rev/s about the bar's center.

73. Consider a monatomic solid with a weakly bound cubic lattice, with each atom connected to six neighbors, each bond having a binding energy of 3.9×10^{-3} eV. When this solid melts, its latent heat of fusion goes directly into breaking the bonds between the atoms. Estimate the latent heat of fusion for this solid, in J/mol. [*Hint*: Show that in a simple cubic lattice (Fig. 40–46), there are *three* times as many bonds as there are atoms, when the number of atoms is large.]

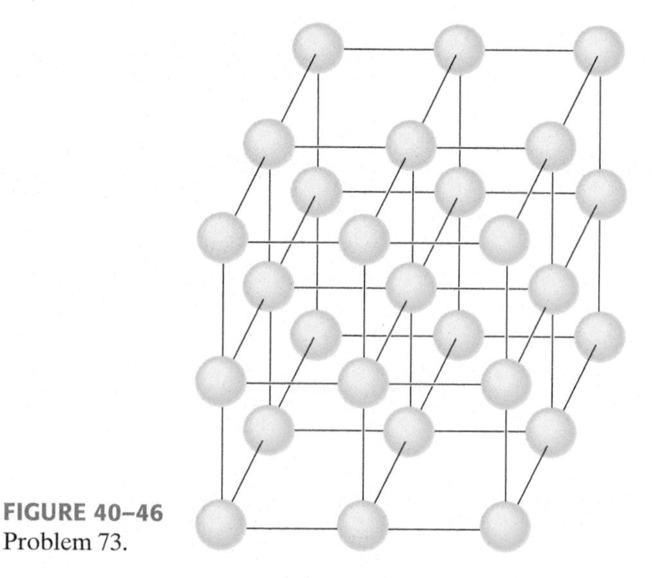

FIGURE 40–46 Problem 73.

74. The energy gap between valence and conduction bands in zinc sulfide is 3.6 eV. What range of wavelengths can a photon have to excite an electron from the top of the valence band into the conduction band?

75. When EM radiation is incident on diamond, it is found that light with wavelengths shorter than 226 nm will cause the diamond to conduct. What is the energy gap between the valence band and the conduction band for diamond?

76. The **Fermi temperature** T_F is defined as that temperature at which the thermal energy kT (without the $\frac{3}{2}$) is equal to the Fermi energy: $kT_F = E_F$. (*a*) Determine the Fermi temperature for copper. (*b*) Show that for $T \gg T_F$, the Fermi factor (Eq. 40–14) approaches the Boltzmann factor. (Note: This last result is not very useful for understanding conductors. Why?)

77. Estimate the number of states from 4.0 eV to 6.2 eV available to electrons in a 10-cm cube of iron.

78. The band gap of silicon is 1.14 eV. (*a*) For what range of wavelengths will silicon be transparent? (See Example 40–14.) In what region of the electromagnetic spectrum does this transparent range begin? (*b*) If window glass is transparent for all visible wavelengths, what is the minimum possible band gap value for glass (assume $\lambda = 450\,\mathrm{nm}$ to $750\,\mathrm{nm}$)? [*Hint*: If the photon has less energy than the band gap, the photon will pass through the solid without being absorbed.]

79. For a certain semiconductor, the longest wavelength radiation that can be absorbed is 1.92 mm. What is the energy gap in this semiconductor?

80. Assume conduction electrons in a semiconductor behave as an ideal gas. (This is not true for conduction electrons in a metal.) (*a*) Taking mass $m = 9 \times 10^{-31}$ kg and temperature $T = 300\,\mathrm{K}$, determine the de Broglie wavelength of a semiconductor's conduction electrons. (*b*) Given that the spacing between atoms in a semiconductor's atomic lattice is on the order of 0.3 nm, would you expect room-temperature conduction electrons to travel in straight lines or diffract when traveling through this lattice? Explain.

81. Most of the Sun's radiation has wavelengths shorter than 1100 nm. For a solar cell to absorb all this, what energy gap ought the material have?

82. Green and blue LEDs became available many years after red LEDs were first developed. Approximately what energy gaps would you expect to find in green (525 nm) and in blue (465 nm) LEDs?

83. For an arsenic donor atom in a doped silicon semiconductor, assume that the "extra" electron moves in a Bohr orbit about the arsenic ion. For this electron in the ground state, take into account the dielectric constant $K = 12$ of the Si lattice (which represents the weakening of the Coulomb force due to all the other atoms or ions in the lattice), and estimate (*a*) the binding energy, and (*b*) the orbit radius for this extra electron. [*Hint*: Substitute $\epsilon = K\epsilon_0$ in Coulomb's law; see Section 24–5.]

84. A strip of silicon 1.8 cm wide and 1.0 mm thick is immersed in a magnetic field of strength 1.3 T perpendicular to the strip (Fig. 40–47). When a current of 0.28 mA is run through the strip, there is a resulting Hall effect voltage of 18 mV across the strip (Section 27–8). How many electrons per silicon atom are in the conduction band? The density of silicon is $2330\,\mathrm{kg/m^3}$.

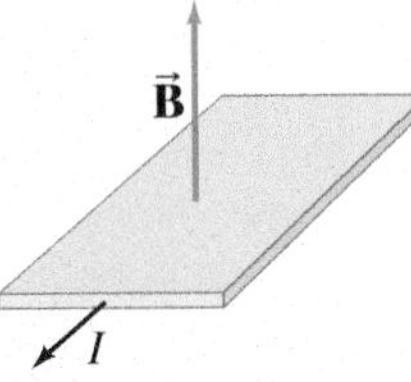

FIGURE 40–47 Problem 84.

85. A zener diode voltage regulator is shown in Fig. 40–48. Suppose that $R = 2.80\,\mathrm{k\Omega}$ and that the diode breaks down at a reverse voltage of 130 V. (The current increases rapidly at this point, as shown on the far left of Fig. 40–38 at a voltage of -12 V on that diagram.) The diode is rated at a maximum current of 120 mA. (*a*) If $R_{\mathrm{load}} = 18.0\,\mathrm{k\Omega}$, over what range of supply voltages will the circuit maintain the output voltage at 130 V? (*b*) If the supply voltage is 245 V, over what range of load resistance will the voltage be regulated?

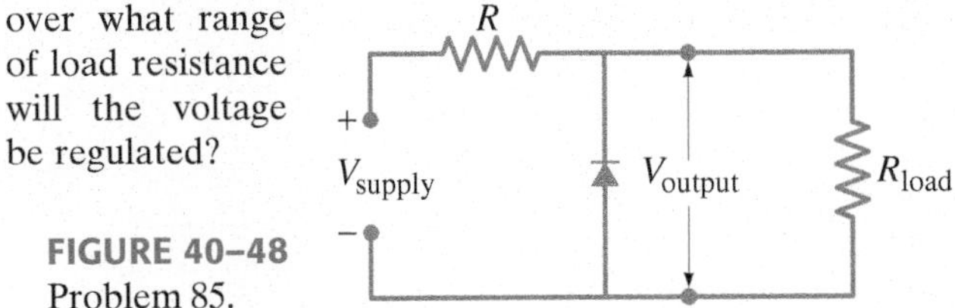

FIGURE 40–48 Problem 85.

86. A full-wave rectifier (Fig. 40–40) uses two diodes to rectify a 95-V rms 60 Hz ac voltage. If $R = 7.8\,\mathrm{k\Omega}$ and $C = 36\,\mu\mathrm{F}$, what will be the approximate percent variation in the output voltage? The variation in output voltage (Fig. 40–40c) is called *ripple voltage*. [*Hint*: See Section 26–5 and assume the discharge of the capacitor is approximately linear.]

* Numerical/Computer

*87. (II) Write a program that will determine the Fermi–Dirac probability function (Eq. 40–14). Make separate plots of this function versus E/E_{F} for copper at (*a*) $T = 500\,\mathrm{K}$; (*b*) $T = 1000\,\mathrm{K}$; (*c*) $T = 5000\,\mathrm{K}$; and (*d*) $T = 10{,}000\,\mathrm{K}$. For copper, $E_{\mathrm{F}} = 7.0\,\mathrm{eV}$. Interpret each plot accordingly.

*88. (III) A simple picture of an H_2 molecule sharing two electrons is shown in Fig. 40–49. We assume the electrons are symmetrically located between the two protons, which are separated by $r_0 = 0.074\,\mathrm{nm}$. (*a*) When the electrons are separated by a distance *d*, write the total potential energy *U* in terms of *d* and r_0. (*b*) Make a graph of *U* in eV as a function of *d* in nm, and state where *U* has a minimum on your graph, and for what range of *d* values *U* is negative. (*c*) Determine analytically the value of *d* that gives minimum *U* (greatest stability).

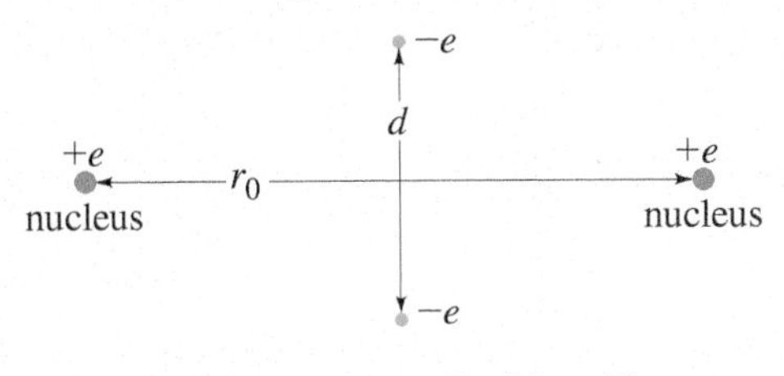

FIGURE 40–49 Problem 88.

*89. (III) Estimate the current produced per $\mathrm{cm^2}$ of area in a flat silicon semiconductor placed perpendicular to sunlight. Assume the sunlight has an intensity of $1000\,\mathrm{W/m^2}$ and that only photons that have more energy than the band gap can create an electron–hole pair in the semiconductor. Assume the Sun is a blackbody emitter at 6000 K, and find the fraction of photons that have energy above the band gap (1.14 eV). See Section 37–1 and integrate the Planck formula numerically.

Answers to Exercises

A: 0, 5.00×10^{-4} eV, 1.50×10^{-3} eV.

B: (*e*).

C: (*a*).

D: (*c*).

This archeologist has unearthed the remains of a sea-turtle within an ancient man-made stone circle. Carbon dating of the remains can tell her when humans inhabited the site.

In this Chapter we begin our discussion of nuclear physics. We study the properties of nuclei, the various forms of radioactivity, and how radioactive decay can be used in a variety of fields to determine the age of old objects, from bones and trees to rocks and other mineral substances, and obtain information on the history of the Earth.

Nuclear Physics and Radioactivity

CONTENTS

CHAPTER-OPENING QUESTION—**Guess now!**

If half of an 80-μg sample of $^{60}_{27}$Co decays in 5.3 years, how much $^{60}_{27}$Co is left in 15.9 years?

(a) 10 μg.
(b) 20 μg.
(c) 30 μg.
(d) 40 μg.
(e) 0 μg.

In the early part of the twentieth century, Rutherford's experiments led to the idea that at the center of an atom there is a tiny but massive nucleus. At the same time that quantum theory was being developed and scientists were attempting to understand the structure of the atom and its electrons, investigations into the nucleus itself had also begun. In this Chapter and the next, we take a brief look at *nuclear physics*.

41–1 Structure and Properties of the Nucleus

An important question for physicists was whether the nucleus had a structure, and what that structure might be. We now understand the nucleus to be a complicated entity that is not fully understood even today. By the early 1930s, a model of the nucleus had been developed that is still useful. According to this model, a nucleus is considered to be an aggregate of two types of particles: protons and neutrons. (These "particles" also have wave properties, but for ease of visualization and language, we often refer to them simply as "particles.") A **proton** is the nucleus of the simplest atom, hydrogen. It has a positive charge ($= +e = +1.60 \times 10^{-19}$ C, the same magnitude as for the electron) and a mass

$$m_{\mathrm{p}} = 1.67262 \times 10^{-27}\,\mathrm{kg}.$$

The **neutron**, whose existence was ascertained in 1932 by the English physicist James Chadwick (1891–1974), is electrically neutral ($q = 0$), as its name implies. Its mass is very slightly larger than that of the proton:

$$m_{\mathrm{n}} = 1.67493 \times 10^{-27}\,\mathrm{kg}.$$

These two constituents of a nucleus, neutrons and protons, are referred to collectively as **nucleons**.

Although the hydrogen nucleus consists of a single proton alone, the nuclei of all other elements consist of both neutrons and protons. The different nuclei are often referred to as **nuclides**. The number of protons in a nucleus (or nuclide) is called the **atomic number** and is designated by the symbol Z. The total number of nucleons, neutrons plus protons, is designated by the symbol A and is called the **atomic mass number**, or sometimes simply **mass number**. This name is used since the mass of a nucleus is very closely A times the mass of one nucleon. A nuclide with 7 protons and 8 neutrons thus has $Z = 7$ and $A = 15$. The **neutron number** N is $N = A - Z$.

To specify a given nuclide, we need give only A and Z. A special symbol is commonly used which takes the form

$${}^{A}_{Z}\mathrm{X},$$

where X is the chemical symbol for the element (see Appendix F, and the Periodic Table inside the back cover), A is the atomic mass number, and Z is the atomic number. For example, ${}^{15}_{7}\mathrm{N}$ means a nitrogen nucleus containing 7 protons and 8 neutrons for a total of 15 nucleons. In a neutral atom, the number of electrons orbiting the nucleus is equal to the atomic number Z (since the charge on an electron has the same magnitude but opposite sign to that of a proton). The main properties of an atom, and how it interacts with other atoms, are largely determined by the number of electrons in the neutral atom. Hence Z determines what kind of atom it is: carbon, oxygen, gold, or whatever. It is redundant to specify both the symbol of a nucleus and its atomic number Z as described above. If the nucleus is nitrogen, for example, we know immediately that $Z = 7$. The subscript Z is thus sometimes dropped and ${}^{15}_{7}\mathrm{N}$ is then written simply ${}^{15}\mathrm{N}$; in words we say "nitrogen fifteen."

For a particular type of atom (say, carbon), nuclei are found to contain different numbers of neutrons, although they all have the same number of protons. For example, carbon nuclei always have 6 protons, but they may have 5, 6, 7, 8, 9, or 10 neutrons. Nuclei that contain the same number of protons but different numbers of neutrons are called **isotopes**. Thus, ${}^{11}_{6}\mathrm{C}$, ${}^{12}_{6}\mathrm{C}$, ${}^{13}_{6}\mathrm{C}$, ${}^{14}_{6}\mathrm{C}$, ${}^{15}_{6}\mathrm{C}$, and ${}^{16}_{6}\mathrm{C}$ are all isotopes of carbon. The isotopes of a given element are not all equally common. For example, 98.9% of naturally occurring carbon (on Earth) is the isotope ${}^{12}_{6}\mathrm{C}$, and about 1.1% is ${}^{13}_{6}\mathrm{C}$. These percentages are referred to as the **natural abundances**.† Even hydrogen has isotopes: 99.99% of natural hydrogen is ${}^{1}_{1}\mathrm{H}$, a simple proton, as the nucleus; there are also ${}^{2}_{1}\mathrm{H}$, called **deuterium**, and ${}^{3}_{1}\mathrm{H}$, **tritium**, which besides the proton contain 1 or 2 neutrons.

†The mass value for each element as given in the Periodic Table (inside back cover) is an average weighted according to the natural abundances of its isotopes.

Many isotopes that do not occur naturally can be produced in the laboratory by means of nuclear reactions (more on this later). Indeed, all elements beyond uranium ($Z > 92$) do not occur naturally on Earth and are only produced artificially (that is, in the laboratory), as are many nuclides with $Z \leq 92$.

The approximate size of nuclei was determined originally by Rutherford from the scattering of charged particles by thin metal foils (Fig. 37–17). We cannot speak about a definite size for nuclei because of the wave–particle duality: their spatial extent must remain somewhat fuzzy. Nonetheless a rough "size" can be measured by scattering high-speed electrons off nuclei. It is found that nuclei have a roughly spherical shape with a radius that increases with A according to the approximate formula

$$r \approx (1.2 \times 10^{-15}\,\mathrm{m})\left(A^{\frac{1}{3}}\right). \tag{41–1}$$

Since the volume of a sphere is $V = \frac{4}{3}\pi r^3$, we see that the volume of a nucleus is approximately proportional to the number of nucleons, $V \propto A$. This is what we would expect if nucleons were like impenetrable billiard balls: if you double the number of balls, you double the total volume. Hence, all nuclei have nearly the same density, and it is enormous (see Example 41–1).

The metric abbreviation for 10^{-15} m is the fermi (after Enrico Fermi) or the femtometer, fm (Table 1–4 or inside the front cover). Thus $1.2 \times 10^{-15}\,\mathrm{m} = 1.2\,\mathrm{fm}$ or 1.2 fermis.

Because nuclear radii vary as $A^{\frac{1}{3}}$, the largest nuclei, such as uranium with $A = 238$, have a radius only about $\sqrt[3]{238} \approx 6$ times that of the smallest, hydrogen ($A = 1$).

EXAMPLE 41–1 ESTIMATE **Nuclear and atomic densities.** Compare the density of nuclear matter to the density of normal solids.

APPROACH The density of normal liquids and solids is on the order of 10^3 to $10^4\,\mathrm{kg/m^3}$ (see Table 13–1), and because the atoms are close packed, atoms have about this density too. We therefore compare the density (mass per volume) of a nucleus to that of its atom as a whole.

SOLUTION The mass of a proton is greater than the mass of an electron by a factor

$$\frac{1.7 \times 10^{-27}\,\mathrm{kg}}{9.1 \times 10^{-31}\,\mathrm{kg}} \approx 2 \times 10^3.$$

Thus, over 99.9% of the mass of an atom is in the nucleus, and for our estimate we can say the mass of the atom equals the mass of the nucleus, $m_{\mathrm{nucl}}/m_{\mathrm{atom}} = 1$. Atoms have a radius of about 10^{-10} m (Chapter 37) and nuclei on the order of 10^{-15} m (Eq. 41–1). Thus the ratio of nuclear density to atomic density is about

$$\frac{\rho_{\mathrm{nucl}}}{\rho_{\mathrm{atom}}} = \frac{(m_{\mathrm{nucl}}/V_{\mathrm{nucl}})}{(m_{\mathrm{atom}}/V_{\mathrm{atom}})} = \left(\frac{m_{\mathrm{nucl}}}{m_{\mathrm{atom}}}\right)\frac{\frac{4}{3}\pi r_{\mathrm{atom}}^3}{\frac{4}{3}\pi r_{\mathrm{nucl}}^3} \approx (1)\frac{(10^{-10})^3}{(10^{-15})^3} = 10^{15}.$$

The nucleus is 10^{15} times more dense than ordinary matter.

The masses of nuclei can be determined from the radius of curvature of fast-moving nuclei (as ions) in a known magnetic field using a mass spectrometer, as discussed in Section 27–9. Indeed the existence of different isotopes of the same element (different number of neutrons) was discovered using this device. Nuclear masses can be specified in **unified atomic mass units** (u). On this scale, a neutral $^{12}_{6}\mathrm{C}$ atom is given the precise value 12.000000 u. A neutron then has a measured mass of 1.008665 u, a proton 1.007276 u, and a neutral hydrogen atom $^{1}_{1}\mathrm{H}$ (proton plus electron) 1.007825 u. The masses of many nuclides are given in Appendix F. It should be noted that the masses in this Table, as is customary, are for the *neutral atom* (including electrons), and not for a bare nucleus.

CAUTION

Masses are for neutral atom (nucleus plus electrons)

Masses are often specified using the electron-volt energy unit. This can be done because mass and energy are related, and the precise relationship is given by Einstein's equation $E = mc^2$ (Chapter 36). Since the mass of a proton is 1.67262×10^{-27} kg, or 1.007276 u, then

$$1.0000\,\mathrm{u} = (1.0000\,\mathrm{u})\left(\frac{1.67262 \times 10^{-27}\,\mathrm{kg}}{1.007276\,\mathrm{u}}\right) = 1.66054 \times 10^{-27}\,\mathrm{kg};$$

this is equivalent to an energy (see Table inside front cover) in MeV $(= 10^6\,\mathrm{eV})$ of

$$E = mc^2 = \frac{(1.66054 \times 10^{-27}\,\mathrm{kg})(2.9979 \times 10^8\,\mathrm{m/s})^2}{(1.6022 \times 10^{-19}\,\mathrm{J/eV})} = 931.5\,\mathrm{MeV}.$$

Thus

$$1\,\mathrm{u} = 1.6605 \times 10^{-27}\,\mathrm{kg} = 931.5\,\mathrm{MeV}/c^2.$$

The masses of some of the basic particles are given in Table 41–1.

TABLE 41–1 Masses in Kilograms, Unified Atomic Mass Units, and MeV/c^2

	Mass		
Object	**kg**	**u**	**MeV/c^2**
Electron	9.1094×10^{-31}	0.00054858	0.51100
Proton	1.67262×10^{-27}	1.007276	938.27
$^1_1$H atom	1.67353×10^{-27}	1.007825	938.78
Neutron	1.67493×10^{-27}	1.008665	939.57

Just as an electron has intrinsic spin and orbital angular momentum quantum numbers, so too do nuclei and their constituents, the proton and neutron. Both the proton and the neutron are spin-$\frac{1}{2}$ particles, just like the electron. A nucleus, made up of protons and neutrons, has a **nuclear spin** quantum number I that is the vector sum of the spins of all its nucleons (plus any orbital angular momentum), and can be either integer or half integer, depending on whether it is made up of an even or an odd number of nucleons. [Orbital angular momentum is integer and doesn't affect half integer or whole for I.] The **total nuclear angular momentum** of a nucleus is given, as might be expected (see Section 39–2 and Eq. 39–15), by $\sqrt{I(I+1)}\,\hbar$.

Nuclear magnetic moments are measured in terms of the **nuclear magneton**

$$\mu_\mathrm{N} = \frac{e\hbar}{2m_\mathrm{p}}, \qquad \textbf{(41–2)}$$

which is defined by analogy with the Bohr magneton for electrons ($\mu_\mathrm{B} = e\hbar/2m_\mathrm{e}$, Section 39–7). Since μ_N contains the proton mass, m_p, instead of the electron mass, it is about 2000 times smaller. The electron spin magnetic moment is about 2 Bohr magnetons. The proton's magnetic moment μ_p has been measured to be

$$\mu_\mathrm{p} = 2.7928\,\mu_\mathrm{N}.$$

There is no satisfactory explanation for this large factor. The neutron has a magnetic moment

$$\mu_\mathrm{n} = -1.9135\,\mu_\mathrm{N},$$

which suggests that, although the neutron carries no net charge, it may have internal structure (quarks, as we discuss later). The minus sign for μ_n indicates that its magnetic moment is opposite to its spin.

Important applications based on nuclear spin are nuclear magnetic resonance (NMR) and magnetic resonance imaging (MRI). They are discussed in the next Chapter (Section 42–10).

41–2 Binding Energy and Nuclear Forces

Binding Energies

The total mass of a stable nucleus is always less than the sum of the masses of its separate protons and neutrons, as the following Example shows.

PROBLEM SOLVING
Keep track of electron masses

EXAMPLE 41–2 $^4_2$He mass compared to its constituents. Compare the mass of a $^4_2$He atom to the total mass of its constituent particles.

APPROACH The $^4_2$He nucleus contains 2 protons and 2 neutrons. Tables normally give the masses of neutral atoms—that is, nucleus plus its Z electrons—since this is how masses are measured. We must therefore be sure to balance out the electrons when we compare masses. Thus we use the mass of $^1_1$H rather than that of a proton alone. We look up the mass of the $^4_2$He atom in Appendix F (it includes the mass of 2 electrons), as well as the mass for the 2 neutrons and 2 hydrogen atoms (= 2 protons + 2 electrons).

SOLUTION The mass of a neutral $^4_2$He atom, from Appendix F, is 4.002603 u. The mass of two neutrons and two H atoms (2 protons including the 2 electrons) is

$$\begin{aligned} 2m_{\mathrm{n}} &= 2(1.008665\,\mathrm{u}) = 2.017330\,\mathrm{u} \\ 2m(^1_1\mathrm{H}) &= 2(1.007825\,\mathrm{u}) = \underline{2.015650\,\mathrm{u}} \\ \mathrm{sum} &= 4.032980\,\mathrm{u}. \end{aligned}$$

Thus the mass of $^4_2$He is measured to be less than the masses of its constituents by an amount $4.032980\,\mathrm{u} - 4.002603\,\mathrm{u} = 0.030377\,\mathrm{u}$.

Where has this lost mass of 0.030377 u disappeared to? It must be $E = mc^2$.

If the four nucleons suddenly came together to form a $^4_2$He nucleus, the mass "loss" would appear as energy of another kind (such as γ radiation, or kinetic energy). The mass (or energy) difference in the case of $^4_2$He, given in energy units, is $(0.030377\,\mathrm{u})(931.5\,\mathrm{MeV/u}) = 28.30\,\mathrm{MeV}$. This difference is referred to as the **total binding energy** of the nucleus. The total binding energy represents the amount of energy that must be put *into* a nucleus in order to break it apart into its constituents. If the mass of, say, a $^4_2$He nucleus were exactly equal to the mass of two neutrons plus two protons, the nucleus could fall apart without any input of energy. To be stable, the mass of a nucleus *must* be less than that of its constituent nucleons, so that energy input *is* needed to break it apart. Note that the binding energy is not something a nucleus has—it is energy it "lacks" relative to the total mass of its separate constituents.

We saw in Chapter 37 that the binding energy of the one electron in the hydrogen atom is 13.6 eV; so the mass of a $^1_1$H atom is less than that of a single proton plus a single electron by $13.6\,\mathrm{eV}/c^2$. Compared to the total mass of the hydrogen atom $(939\,\mathrm{MeV}/c^2)$, this is incredibly small, 1 part in 10^8. The binding energies of nuclei are on the order of MeV, so the eV binding energies of electrons can be ignored. Note that nuclear binding energies, compared to nuclear masses, are on the order of $(28\,\mathrm{MeV}/4000\,\mathrm{MeV}) \approx 1 \times 10^{-2}$, where we used helium's binding energy (see above) and mass $\approx 4 \times 940\,\mathrm{MeV} \approx 4000\,\mathrm{MeV}$.

EXERCISE A Determine how much less the mass of the $^7_3$Li nucleus is compared to that of its constituents.

The **binding energy per nucleon** is defined as the total binding energy of a nucleus divided by A, the total number of nucleons. We calculated above that the binding energy of $^4_2$He is 28.3 MeV, so its binding energy per nucleon is $28.3\,\mathrm{MeV}/4 = 7.1\,\mathrm{MeV}$. Figure 41–1 shows the binding energy per nucleon as a function of A for stable nuclei. The curve rises as A increases and reaches a plateau at about 8.7 MeV per nucleon above $A \approx 40$. Beyond $A \approx 80$, the curve decreases slowly, indicating that larger nuclei are held together a little less tightly than those in the middle of the Periodic Table. We will see later that these characteristics allow the release of nuclear energy in the processes of fission and fusion.

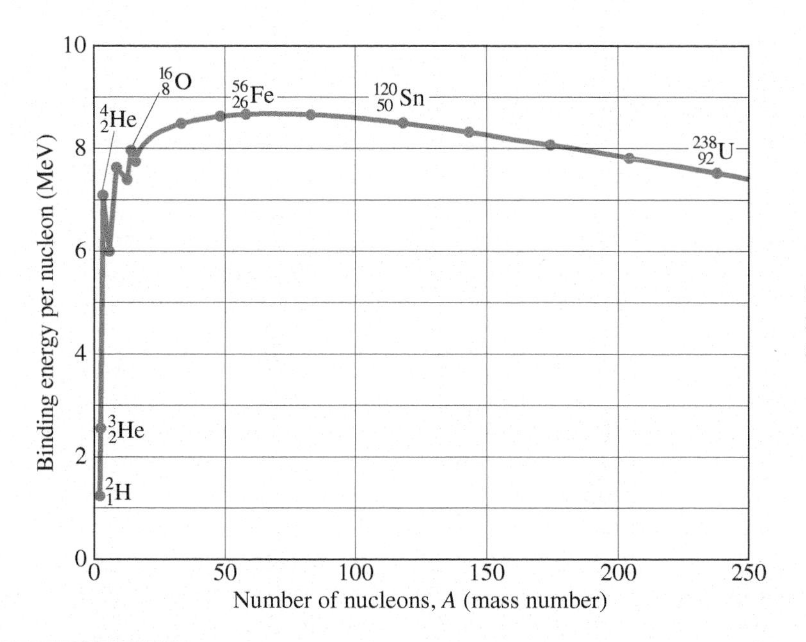

FIGURE 41–1 Binding energy per nucleon for the more stable nuclides as a function of mass number A.

EXAMPLE 41–3 Binding energy for iron. Calculate the total binding energy and the binding energy per nucleon for $^{56}_{26}Fe$, the most common stable isotope of iron.

APPROACH We subtract the mass of an $^{56}_{26}Fe$ atom from the total mass of 26 hydrogen atoms and 30 neutrons, all found in Appendix F. Then we convert mass units to energy units; finally we divide by $A = 56$, the total number of nucleons.

SOLUTION $^{56}_{26}Fe$ has 26 protons and 30 neutrons whose separate masses are

$$\begin{aligned} 26m(^1_1\mathrm{H}) &= (26)(1.007825\,\mathrm{u}) = 26.20345\,\mathrm{u} \text{ (includes 26 electrons)} \\ 30m_\mathrm{n} &= (30)(1.008665\,\mathrm{u}) = 30.25995\,\mathrm{u} \\ \text{sum} &= 56.46340\,\mathrm{u}. \\ \text{Subtract mass of } ^{56}_{26}\mathrm{Fe}: &= -55.93494\,\mathrm{u} \text{ (Appendix F)} \\ \Delta m &= 0.52846\,\mathrm{u}. \end{aligned}$$

The total binding energy is thus

$$(0.52846\,\mathrm{u})(931.5\,\mathrm{MeV/u}) = 492.26\,\mathrm{MeV}$$

and the binding energy per nucleon is

$$\frac{492.26\,\mathrm{MeV}}{56\text{ nucleons}} = 8.79\,\mathrm{MeV}.$$

NOTE The binding energy per nucleon graph (Fig. 41–1) peaks about here, for iron, so the iron nucleus (and its neighbors) is the most stable of nuclei.

EXERCISE B Determine the binding energy per nucleon for $^{16}_{8}O$.

EXAMPLE 41–4 Binding energy of last neutron. What is the binding energy of the last neutron in $^{13}_{6}C$?

APPROACH If $^{13}_{6}C$ lost one neutron, it would be $^{12}_{6}C$. We subtract the mass of $^{13}_{6}C$ from the masses of $^{12}_{6}C$ and a free neutron.

SOLUTION Obtaining the masses from Appendix F, we have

$$\begin{aligned} \text{Mass } ^{12}_{6}\mathrm{C} &= 12.000000\,\mathrm{u} \\ \text{Mass } ^{1}_{0}\mathrm{n} &= 1.008665\,\mathrm{u} \\ \text{Total} &= 13.008665\,\mathrm{u}. \\ \text{Subtract mass of } ^{13}_{6}\mathrm{C}: &\quad -13.003355\,\mathrm{u} \\ \Delta m &= 0.005310\,\mathrm{u} \end{aligned}$$

which in energy is $(931.5\,\mathrm{MeV/u})(0.005310\,\mathrm{u}) = 4.95\,\mathrm{MeV}$. That is, it would require 4.95 MeV input of energy to remove one neutron from $^{13}_{6}C$.

Nuclear Forces

We can analyze nuclei not only from the point of view of energy, but also from the point of view of the forces that hold them together. We would not expect a collection of protons and neutrons to come together spontaneously, since protons are all positively charged and thus exert repulsive electric forces on each other. Since stable nuclei *do* stay together, it is clear that another force must be acting. Because this new force is stronger than the electric force, it is called the **strong nuclear force**. The strong nuclear force is an attractive force that acts between all nucleons—protons and neutrons alike. Thus protons attract each other via the strong nuclear force at the same time they repel each other via the electric force. Neutrons, since they are electrically neutral, only attract other neutrons or protons via the strong nuclear force.

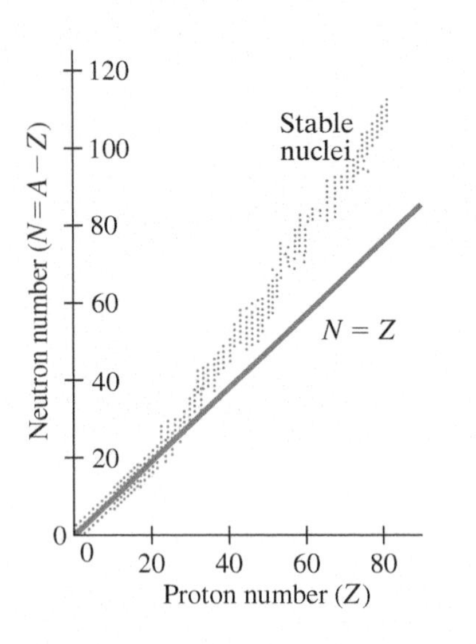

FIGURE 41–2 Number of neutrons versus number of protons for stable nuclides, which are represented by dots. The straight line represents $N = Z$.

The strong nuclear force turns out to be far more complicated than the gravitational and electromagnetic forces. One important aspect of the strong nuclear force is that it is a **short-range** force: it acts only over a very short distance. It is very strong between two nucleons if they are less than about 10^{-15} m apart, but it is essentially zero if they are separated by a distance greater than this. Compare this to electric and gravitational forces, which decrease as $1/r^2$ but continue acting over any distances and are therefore called **long-range** forces.

The strong nuclear force has some strange features. For example, if a nuclide contains too many or too few neutrons relative to the number of protons, the binding of the nucleons is reduced; nuclides that are too unbalanced in this regard are unstable. As shown in Fig. 41–2, stable nuclei tend to have the same number of protons as neutrons $(N = Z)$ up to about $A = 30$. Beyond this, stable nuclei contain more neutrons than protons. This makes sense since, as Z increases, the electrical repulsion increases, so a greater number of neutrons—which exert only the attractive strong nuclear force—are required to maintain stability. For very large Z, no number of neutrons can overcome the greatly increased electric repulsion. Indeed, there are no completely stable nuclides above $Z = 82$.

What we mean by a *stable nucleus* is one that stays together indefinitely. What then is an *unstable nucleus*? It is one that comes apart; and this results in radioactive decay. Before we discuss the important subject of radioactivity (next Section), we note that there is a second type of nuclear force that is much weaker than the strong nuclear force. It is called the **weak nuclear force**, and we are aware of its existence only because it shows itself in certain types of radioactive decay. These two nuclear forces, the strong and the weak, together with the gravitational and electromagnetic forces, comprise the four basic types of force in nature.

41–3 Radioactivity

Nuclear physics had its beginnings in 1896. In that year, Henri Becquerel (1852–1908) made an important discovery: in his studies of phosphorescence, he found that a certain mineral (which happened to contain uranium) would darken a photographic plate even when the plate was wrapped to exclude light. It was clear that the mineral emitted some new kind of radiation that, unlike X-rays, occurred without any external stimulus. This new phenomenon eventually came to be called **radioactivity**.

FIGURE 41–3 Marie and Pierre Curie in their laboratory (about 1906) where radium was discovered.

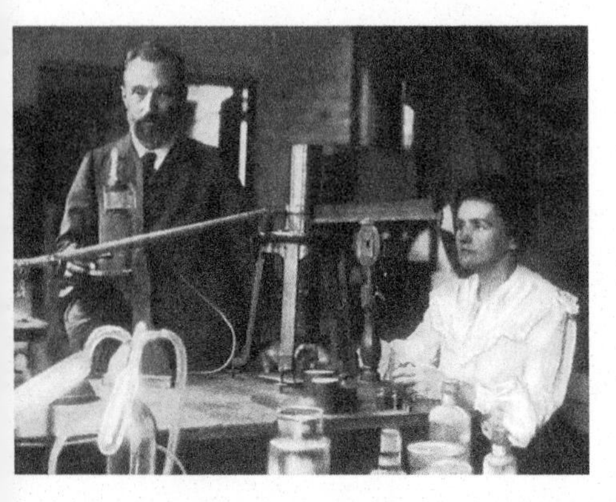

Soon after Becquerel's discovery, Marie Curie (1867–1934) and her husband, Pierre Curie (1859–1906), isolated two previously unknown elements that were very highly radioactive (Fig. 41–3). These were named polonium and radium. Other radioactive elements were soon discovered as well. The radioactivity was found in every case to be unaffected by the strongest physical and chemical treatments, including strong heating or cooling or the action of strong chemical reagents. It was suspected that the source of radioactivity must be deep within the atom, emanating from the nucleus. It became apparent that radioactivity is the result of the **disintegration** or **decay** of an unstable nucleus. Certain isotopes are not stable, and they decay with the emission of some type of radiation or "rays."

Many unstable isotopes occur in nature, and such radioactivity is called "natural radioactivity." Other unstable isotopes can be produced in the laboratory by nuclear reactions (Section 42–1); these are said to be produced "artificially" and to have "artificial radioactivity." Radioactive isotopes are sometimes referred to as **radioisotopes** or **radionuclides**.

Rutherford and others began studying the nature of the rays emitted in radioactivity about 1898. They classified the rays into three distinct types according to their penetrating power. One type of radiation could barely penetrate a piece of paper. The second type could pass through as much as 3 mm of aluminum. The third was extremely penetrating: it could pass through several centimeters of lead and still be detected on the other side. They named these three types of radiation alpha (α), beta (β), and gamma (γ), respectively, after the first three letters of the Greek alphabet.

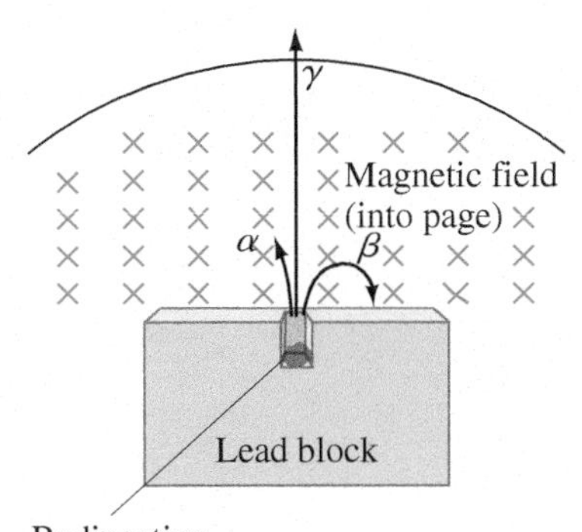

FIGURE 41–4 Alpha and beta rays are bent in opposite directions by a magnetic field, whereas gamma rays are not bent at all.

Each type of ray was found to have a different charge and hence is bent differently in a magnetic field, Fig. 41–4; α rays are positively charged, β rays are negatively charged, and γ rays are neutral. It was soon found that all three types of radiation consisted of familiar kinds of particles. Gamma rays are very high-energy photons whose energy is even higher than that of X-rays. Beta rays are electrons, identical to those that orbit the nucleus, but they are created within the nucleus itself. Alpha rays (or α particles) are simply the nuclei of helium atoms, ${}^4_2\text{He}$; that is, an α ray consists of two protons and two neutrons bound together.

We now discuss each of these three types of radioactivity, or decay, in more detail.

41–4 Alpha Decay

Experiments show that when nuclei decay, the number of nucleons (= mass number A) is conserved, as well as electric charge (= Ze). When a nucleus emits an α particle (${}^4_2\text{He}$), the remaining nucleus will be different from the original: it has lost two protons and two neutrons. Radium 226 (${}^{226}_{88}\text{Ra}$), for example, is an α emitter. It decays to a nucleus with $Z = 88 - 2 = 86$ and $A = 226 - 4 = 222$. The nucleus with $Z = 86$ is radon (Rn)—see Appendix F or the Periodic Table. Thus the radium decays to radon with the emission of an α particle. This is written

$$ {}^{226}_{88}\text{Ra} \rightarrow {}^{222}_{86}\text{Rn} + {}^4_2\text{He}. $$

See Fig. 41–5.

FIGURE 41–5 Radioactive decay of radium to radon with emission of an alpha particle.

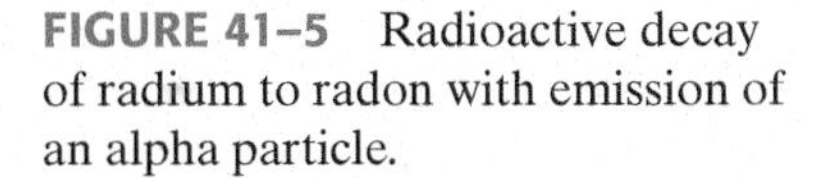

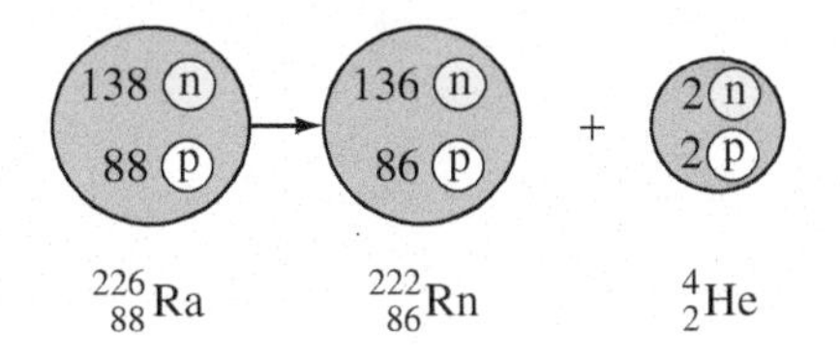

When α decay occurs, a different element is formed. The **daughter** nucleus (${}^{222}_{86}\text{Rn}$ in this case) is different from the **parent** nucleus (${}^{226}_{88}\text{Ra}$ in this case). This changing of one element into another is called **transmutation**.

Alpha decay can be written in general as

$$ {}^A_Z\text{N} \rightarrow {}^{A-4}_{Z-2}\text{N}' + {}^4_2\text{He} \qquad [\alpha \text{ decay}] $$

where N is the parent, N' the daughter, and Z and A are the atomic number and atomic mass number, respectively, of the parent.

EXERCISE C ${}^{154}_{66}\text{Dy}$ decays by α emission to what element? (*a*) Pb, (*b*) Gd, (*c*) Sm, (*d*) Er, (*e*) Yb.

Alpha decay occurs because the strong nuclear force is unable to hold very large nuclei together. The nuclear force is a short-range force: it acts only between neighboring nucleons. But the electric force acts all the way across a large nucleus. For very large nuclei, the large Z means the repulsive electric force becomes so large (Coulomb's law) that the strong nuclear force is unable to hold the nucleus together.

We can express the instability of the parent nucleus in terms of energy (or mass): the mass of the parent nucleus is greater than the mass of the daughter nucleus plus the mass of the α particle. The mass difference appears as kinetic energy, which is carried away by the α particle and the recoiling daughter nucleus. The total energy released is called the **disintegration energy**, Q, or the ***Q*-value** of the decay. From conservation of energy,

$$M_{\mathrm{P}}c^2 = M_{\mathrm{D}}c^2 + m_\alpha c^2 + Q,$$

where Q equals the kinetic energy of the daughter and α particle, and M_{P}, M_{D}, and m_α are the masses of the parent, daughter, and α particle, respectively. Thus

$$Q = M_{\mathrm{P}}c^2 - (M_{\mathrm{D}} + m_\alpha)c^2. \qquad \textbf{(41–3)}$$

If the parent had *less* mass than the daughter plus the α particle (so $Q < 0$), the decay could not occur spontaneously, for the conservation of energy law would be violated.

EXAMPLE 41–5 **Uranium decay energy release.** Calculate the disintegration energy when ${}^{232}_{92}\mathrm{U}$ (mass = 232.037156 u) decays to ${}^{228}_{90}\mathrm{Th}$ (228.028741 u) with the emission of an α particle. (As always, masses given are for neutral atoms.)

APPROACH We use conservation of energy as expressed in Eq. 41–3. ${}^{232}_{92}\mathrm{U}$ is the parent, ${}^{228}_{90}\mathrm{Th}$ is the daughter.

SOLUTION Since the mass of the ${}^{4}_{2}\mathrm{He}$ is 4.002603 u (Appendix F), the total mass in the final state is

$$228.028741\ \mathrm{u} + 4.002603\ \mathrm{u} = 232.031344\ \mathrm{u}.$$

The mass lost when the ${}^{232}_{92}\mathrm{U}$ decays is

$$232.037156\ \mathrm{u} - 232.031344\ \mathrm{u} = 0.005812\ \mathrm{u}.$$

Since $1\ \mathrm{u} = 931.5\ \mathrm{MeV}$, the energy Q released is

$$Q = (0.005812\ \mathrm{u})(931.5\ \mathrm{MeV/u}) = 5.4\ \mathrm{MeV}$$

and this energy appears as kinetic energy of the α particle and the daughter nucleus.

EXAMPLE 41–6 **Kinetic energy of the α in ${}^{232}_{92}\mathrm{U}$ decay.** For the ${}^{232}_{92}\mathrm{U}$ decay of Example 41–5, how much of the 5.4-MeV disintegration energy will be carried off by the α particle?

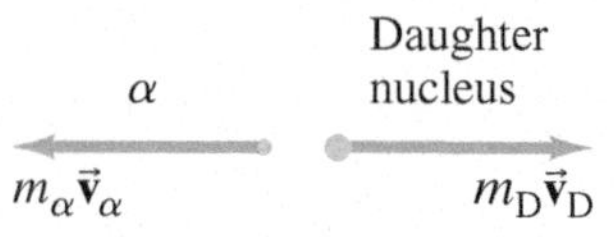

APPROACH In any reaction, momentum must be conserved as well as energy.

SOLUTION Before disintegration, the nucleus can be assumed to be at rest, so the total momentum was zero. After disintegration, the total vector momentum must still be zero so the magnitude of the α particle's momentum must equal the magnitude of the daughter's momentum (Fig. 41–6):

FIGURE 41–6 Momentum conservation in Example 41–6.

$$m_\alpha v_\alpha = m_{\mathrm{D}} v_{\mathrm{D}}.$$

Thus $v_\alpha = m_{\mathrm{D}} v_{\mathrm{D}}/m_\alpha$ and the α's kinetic energy is

$$K_\alpha = \tfrac{1}{2}m_\alpha v_\alpha^2 = \tfrac{1}{2}m_\alpha\left(\frac{m_{\mathrm{D}}v_{\mathrm{D}}}{m_\alpha}\right)^2 = \tfrac{1}{2}m_{\mathrm{D}}v_{\mathrm{D}}^2\left(\frac{m_{\mathrm{D}}}{m_\alpha}\right) = \left(\frac{m_{\mathrm{D}}}{m_\alpha}\right)K_{\mathrm{D}}$$

$$= \left(\frac{228.028741\ \mathrm{u}}{4.002603\ \mathrm{u}}\right)K_{\mathrm{D}} = 57K_{\mathrm{D}}.$$

The total disintegration energy is $Q = K_\alpha + K_{\mathrm{D}} = 57K_{\mathrm{D}} + K_{\mathrm{D}} = 58K_{\mathrm{D}}$. Hence

$$K_\alpha = \frac{57}{58}Q = 5.3\ \mathrm{MeV}.$$

The lighter α particle carries off (57/58) or 98% of the total kinetic energy.

α-Decay Theory—Tunneling

If the mass of the daughter nucleus plus the mass of the α particle is less than the mass of the parent nucleus (so the parent is energetically allowed to decay), why are there any parent nuclei at all? That is, why haven't radioactive nuclei all decayed long ago, right after they were formed (in supernovae)? We can understand decay using a model of a nucleus that has an alpha particle trapped inside it. The potential energy "seen" by the α particle would have a shape something like that shown in Fig. 41–7. The potential energy well (approximately square) between $r = 0$ and $r = R_0$ represents the short-range attractive nuclear force.

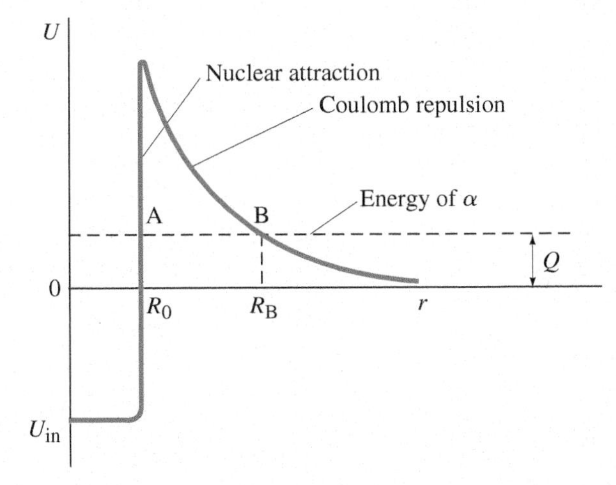

FIGURE 41–7 Potential energy for alpha particle and (daughter) nucleus, showing the Coulomb barrier through which the α particle must tunnel to escape. The Q-value of the reaction is also shown. This plot assumes spherical symmetry, so the central well has diameter $2R_0$.

Beyond the nuclear radius, R_0, the Coulomb repulsion dominates (since the nuclear force drops to zero) and we see the characteristic $1/r$ dependence of the Coulomb potential. The α particle, trapped within the nucleus, can be thought of as moving back and forth between the potential walls. Since the potential energy just beyond $r = R_0$ is greater than the energy of the α particle (dashed line), the α particle could not escape the nucleus if it were governed by classical physics. But according to quantum mechanics, there is a nonzero probability that the α particle can **tunnel** through the Coulomb barrier, from point A to point B in Fig. 41–7, as we discussed in Section 38–10. The height and width of the barrier affect the rate at which nuclei decay (Section 41–8). Because of this barrier, the lifetimes of α-unstable nuclei can be quite long, from a fraction of a microsecond to over 10^{10} years. Note in Fig. 41–7 that the Q-value represents the total kinetic energy when the α particle is far from the nucleus.

A simple way to look at tunneling is via the uncertainty principle which tells us that energy conservation can be violated by an amount ΔE for a length of time Δt given by

$$(\Delta E)(\Delta t) \approx \frac{h}{2\pi}.$$

Thus quantum mechanics allows conservation of energy to be violated for brief periods that may be long enough for an α particle to "tunnel" through the barrier. ΔE would represent the energy difference between the average barrier height and the particle's energy, and Δt the time to pass through the barrier. The higher and wider the barrier, the less time the α particle has to escape and the less likely it is to do so. It is therefore the height and width of this barrier that controls the rate of decay and half-life of an isotope.

Why α Particles?

Why, you may wonder, do nuclei emit this combination of four nucleons called an α particle? Why not just four separate nucleons, or even one? The answer is that the α particle is very strongly bound, so that its mass is significantly less than that of four separate nucleons. As we saw in Example 41–2, two protons and two neutrons separately have a total mass of about 4.032980 u (electrons included). For the decay of ${}^{232}_{92}\text{U}$ discussed in Example 41–5, the total mass of the daughter ${}^{228}_{90}\text{Th}$ plus four separate nucleons is $228.028741 + 4.032980 = 232.061721$ u, which is greater than the mass of the ${}^{232}_{92}\text{U}$ parent (232.037156). Such a decay could not occur because it would violate the conservation of energy. Indeed, we have never seen ${}^{232}_{92}\text{U} \rightarrow {}^{228}_{90}\text{Th} + 2\text{n} + 2\text{p}$. Similarly, it is almost always true that the emission of a single nucleon is energetically not possible.

Smoke Detectors—An Application

PHYSICS APPLIED
Smoke detector

One widespread application of nuclear physics is present in nearly every home in the form of an ordinary **smoke detector**. The most common type of detector contains about 0.2 mg of the radioactive americium isotope, ${}^{241}_{95}\text{Am}$, in the form of AmO_2. The radiation continually ionizes the nitrogen and oxygen molecules in the air space between two oppositely charged plates. The resulting conductivity allows a small steady electric current. If smoke enters, the radiation is absorbed by the smoke particles rather than by the air molecules, thus reducing the current. The current drop is detected by the device's electronics and sets off the alarm. The radiation dose that escapes from an intact americium smoke detector is much less than the natural radioactive background, and so can be considered relatively harmless. There is no question that smoke detectors save lives and reduce property damage.

41–5 Beta Decay

β^- Decay

Transmutation of elements also occurs when a nucleus decays by β decay—that is, with the emission of an electron or β^- particle. The nucleus ${}^{14}_{6}\text{C}$, for example, emits an electron when it decays:

$$ {}^{14}_{6}\text{C} \rightarrow {}^{14}_{7}\text{N} + \text{e}^- + \text{neutrino}, $$

where e^- is the symbol for the electron. The particle known as the neutrino, whose charge $q = 0$ and whose mass is very small or zero, was not initially detected and was only later hypothesized to exist, as we shall discuss later in this Section. No nucleons are lost when an electron is emitted, and the total number of nucleons, A, is the same in the daughter nucleus as in the parent. But because an electron has been emitted from the nucleus itself, the charge on the daughter nucleus is $+1e$ greater than that on the parent. The parent nucleus in the decay written above had $Z = +6$, so from charge conservation the nucleus remaining behind must have a charge of $+7e$. So the daughter nucleus has $Z = 7$, which is nitrogen.

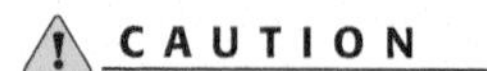

CAUTION
β-decay e^- comes from nucleus (it is not an orbital electron)

It must be carefully noted that the electron emitted in β decay is *not* an orbital electron. Instead, the electron is created *within the nucleus itself.* What happens is that one of the neutrons changes to a proton and in the process (to conserve charge) emits an electron. Indeed, free neutrons actually do decay in this fashion:

$$ \text{n} \rightarrow \text{p} + \text{e}^- + \text{neutrino}. $$

To remind us of their origin in the nucleus, the electrons emitted in β decay are often referred to as "β particles." They are, nonetheless, indistinguishable from orbital electrons.

EXAMPLE 41–7 Energy release in $^{14}_{6}$C decay. How much energy is released when $^{14}_{6}$C decays to $^{14}_{7}$N by β emission?

APPROACH We find the mass difference before and after decay, Δm. The energy released is $E = (\Delta m)c^2$. The masses given in Appendix F are those of the neutral atom, and we have to keep track of the electrons involved. Assume the parent nucleus has six orbiting electrons so it is neutral; its mass is 14.003242 u. The daughter in this decay, $^{14}_{7}$N, is not neutral since it has the same six orbital electrons circling it but the nucleus has a charge of $+7e$. However, the mass of this daughter with its six electrons, plus the mass of the emitted electron (which makes a total of seven electrons), is just the mass of a neutral nitrogen atom.

SOLUTION The total mass in the final state is

$$(\text{mass of } ^{14}_{7}\text{N nucleus} + 6 \text{ electrons}) + (\text{mass of 1 electron}),$$

and this is equal to

$$\text{mass of neutral } ^{14}_{7}\text{N (includes 7 electrons)},$$

which from Appendix F is a mass of 14.003074 u. So the mass difference is $14.003242\text{ u} - 14.003074\text{ u} = 0.000168\text{ u}$, which is equivalent to an energy change $\Delta m\, c^2 = (0.000168\text{ u})(931.5\text{ MeV/u}) = 0.156\text{ MeV}$ or 156 keV.

NOTE The neutrino doesn't contribute to either the mass or charge balance since it has $q = 0$ and $m \approx 0$.

⚠ CAUTION

Be careful with atomic and electron masses in β decay

According to Example 41–7, we would expect the emitted electron to have a kinetic energy of 156 keV. (The daughter nucleus, because its mass is very much larger than that of the electron, recoils with very low velocity and hence gets very little of the kinetic energy—see Example 41–6.) Indeed, very careful measurements indicate that a few emitted β particles do have kinetic energy close to this calculated value. But the vast majority of emitted electrons have somewhat less energy. In fact, the energy of the emitted electron can be anywhere from zero up to the maximum value as calculated above. This range of electron kinetic energy was found for any β decay. It was as if the law of conservation of energy was being violated, and indeed Bohr actually considered this possibility. Careful experiments indicated that linear momentum and angular momentum also did not seem to be conserved. Physicists were troubled at the prospect of giving up these laws, which had worked so well in all previous situations.

In 1930, Wolfgang Pauli proposed an alternate solution: perhaps a new particle that was very difficult to detect was emitted during β decay in addition to the electron. This hypothesized particle could be carrying off the energy, momentum, and angular momentum required to maintain the conservation laws. This new particle was named the **neutrino**—meaning "little neutral one"—by the great Italian physicist Enrico Fermi (1901–1954; Fig. 41–8), who in 1934 worked out a detailed theory of β decay. (It was Fermi who, in this theory, postulated the existence of the fourth force in nature which we call the *weak nuclear force*.) The neutrino has zero charge, spin of $\frac{1}{2}\hbar$, and was long thought to have zero mass, although today we are quite sure that it has a very tiny mass ($< 0.14\text{ eV}/c^2$). If its mass were zero, it would be much like a photon in that it is neutral and would travel at the speed of light. But the neutrino is very difficult to detect. In 1956, complex experiments produced further evidence for the existence of the neutrino; but by then, most physicists had already accepted its existence.

USA 34 Enrico Fermi

2001

FIGURE 41–8 Enrico Fermi, as portrayed on a US postage stamp. Fermi contributed significantly to both theoretical and experimental physics, a feat almost unique in modern times.

The symbol for the neutrino is the Greek letter nu (ν). The correct way of writing the decay of $^{14}_{6}$C is then

$$^{14}_{6}\text{C} \rightarrow {}^{14}_{7}\text{N} + \text{e}^- + \bar{\nu}.$$

The bar ($^{-}$) over the neutrino symbol is to indicate that it is an "antineutrino." (Why this is called an antineutrino rather than simply a neutrino need not concern us now; it is discussed in Chapter 43.)

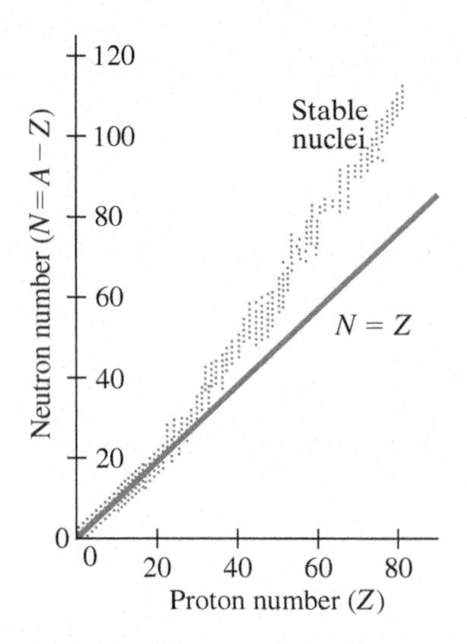

FIGURE 41–2 (Repeated.) Number of neutrons versus number of protons for stable nuclides, which are represented by dots. The straight line represents $N = Z$.

β^+ Decay

Many isotopes decay by electron emission. They are always isotopes that have too many neutrons compared to the number of protons. That is, they are isotopes that lie above the stable isotopes plotted in Fig. 41–2. But what about unstable isotopes that have too few neutrons compared to their number of protons—those that fall below the stable isotopes of Fig. 41–2? These, it turns out, decay by emitting a **positron** instead of an electron. A positron (sometimes called an e^+ or β^+ particle) has the same mass as the electron, but it has a positive charge of $+1e$. Because it is so like an electron, except for its charge, the positron is called the **antiparticle**† to the electron. An example of a β^+ decay is that of ${}^{19}_{10}\text{Ne}$:

$${}^{19}_{10}\text{Ne} \rightarrow {}^{19}_{9}\text{F} + e^+ + \nu,$$

where e^+ stands for a positron. Note that the ν emitted here is a neutrino, whereas that emitted in β^- decay is called an antineutrino. Thus an antielectron (= positron) is emitted with a neutrino, whereas an antineutrino is emitted with an electron; this gives a certain balance as discussed in Chapter 43.

We can write β^- and β^+ decay, in general, as follows:

$${}^{A}_{Z}\text{N} \rightarrow {}^{A}_{Z+1}\text{N}' + e^- + \bar{\nu} \qquad [\beta^- \text{ decay}]$$
$${}^{A}_{Z}\text{N} \rightarrow {}^{A}_{Z-1}\text{N}' + e^+ + \nu, \qquad [\beta^+ \text{ decay}]$$

where N is the parent nucleus and N′ is the daughter.

Electron Capture

Besides β^- and β^+ emission, there is a third related process. This is **electron capture** (abbreviated EC in Appendix F) and occurs when a nucleus absorbs one of its orbiting electrons. An example is ${}^{7}_{4}\text{Be}$, which as a result becomes ${}^{7}_{3}\text{Li}$. The process is written

$${}^{7}_{4}\text{Be} + e^- \rightarrow {}^{7}_{3}\text{Li} + \nu,$$

or, in general,

$${}^{A}_{Z}\text{N} + e^- \rightarrow {}^{A}_{Z-1}\text{N}' + \nu. \qquad [\text{electron capture}]$$

Usually it is an electron in the innermost (K) shell that is captured, in which case the process is called **K-capture**. The electron disappears in the process, and a proton in the nucleus becomes a neutron; a neutrino is emitted as a result. This process is inferred experimentally by detection of emitted X-rays (due to other electrons jumping down to fill the empty state) of just the proper energy.

In β decay, it is the weak nuclear force that plays the crucial role. The neutrino is unique in that it interacts with matter only via the weak force, which is why it is so hard to detect.

FIGURE 41–9 Energy-level diagram showing how ${}^{12}_{5}\text{B}$ can decay to the ground state of ${}^{12}_{6}\text{C}$ by β decay (total energy released = 13.4 MeV), or can instead β decay to an excited state of ${}^{12}_{6}\text{C}$ (indicated by *), which subsequently decays to its ground state by emitting a 4.4-MeV γ ray.

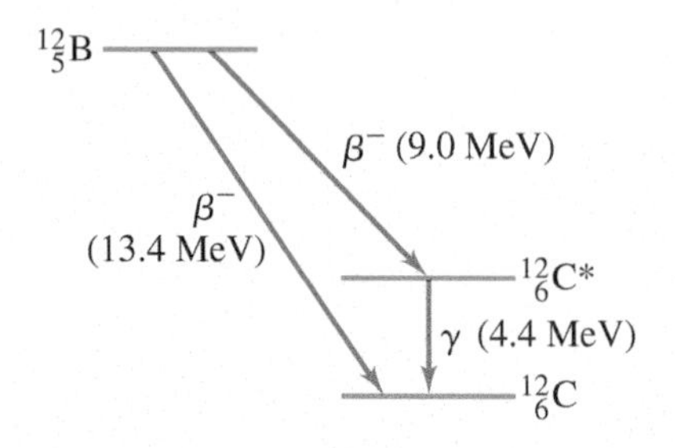

41–6 Gamma Decay

Gamma rays are photons having very high energy. They have their origin in the decay of a nucleus, much like emission of photons by excited atoms. Like an atom, a nucleus itself can be in an excited state. When it jumps down to a lower energy state, or to the ground state, it emits a photon which we call a γ ray. The possible energy levels of a nucleus are much farther apart than those of an atom: on the order of keV or MeV, as compared to a few eV for electrons in an atom. Hence, the emitted photons have energies that can range from a few keV to several MeV. For a given decay, the γ ray always has the same energy. Since a γ ray carries no charge, there is no change in the element as a result of a γ decay.

How does a nucleus get into an excited state? It may occur because of a violent collision with another particle. More commonly, the nucleus remaining after a previous radioactive decay may be in an excited state. A typical example is shown in the energy-level diagram of Fig. 41–9. ${}^{12}_{5}\text{B}$ can decay by β decay directly

†Discussed in Chapter 43. Briefly, an antiparticle has the same mass as its corresponding particle, but opposite charge. A particle and its antiparticle can quickly annihilate each other, releasing energy (γ rays).

to the ground state of $^{12}_{6}C$; or it can go by β decay to an excited state of $^{12}_{6}C$, which then decays by emission of a 4.4-MeV γ ray to the ground state.

We can write γ decay as

$$^{A}_{Z}N^* \rightarrow {}^{A}_{Z}N + \gamma, \qquad [\gamma \text{ decay}]$$

where the asterisk means "excited state" of that nucleus.

What, you may wonder, is the difference between a γ ray and an X-ray? They both are electromagnetic radiation (photons) and, though γ rays usually have higher energy than X-rays, their range of energies overlap to some extent. The difference is not intrinsic. We use the term X-ray if the photon is produced by an electron–atom interaction, and γ ray if the photon is produced in a nuclear process.

Isomers; Internal Conversion

In some cases, a nucleus may remain in an excited state for some time before it emits a γ ray. The nucleus is then said to be in a **metastable state** and is called an **isomer**.

An excited nucleus can sometimes return to the ground state by another process known as **internal conversion** with no γ ray emitted. In this process, the excited nucleus interacts with one of the orbital electrons and ejects this electron from the atom with the same kinetic energy (minus the binding energy of the electron) that an emitted γ ray would have had.

41–7 Conservation of Nucleon Number and Other Conservation Laws

In all three types of radioactive decay, the classical conservation laws hold. Energy, linear momentum, angular momentum, and electric charge are all conserved. These quantities are the same before the decay as after. But a new conservation law is also revealed, the **law of conservation of nucleon number**. According to this law, the total number of nucleons (A) remains constant in any process, although one type can change into the other type (protons into neutrons or vice versa). This law holds in all three types of decay. Table 41–2 gives a summary of α, β, and γ decay. [In Chapter 43 we will generalize this and call it conservation of baryon number.]

TABLE 41–2 The Three Types of Radioactive Decay

α decay:
$^{A}_{Z}N \rightarrow {}^{A-4}_{Z-2}N' + {}^{4}_{2}He$
β decay:
$^{A}_{Z}N \rightarrow {}^{A}_{Z+1}N' + e^- + \bar{\nu}$
$^{A}_{Z}N \rightarrow {}^{A}_{Z-1}N' + e^+ + \nu$
$^{A}_{Z}N + e^- \rightarrow {}^{A}_{Z-1}N' + \nu$ [EC]†
γ decay:
$^{A}_{Z}N^* \rightarrow {}^{A}_{Z}N + \gamma$

† Electron capture.
*Indicates the excited state of a nucleus.

41–8 Half-Life and Rate of Decay

A macroscopic sample of any radioactive isotope consists of a vast number of radioactive nuclei. These nuclei do not all decay at one time. Rather, they decay one by one over a period of time. This is a random process: we can not predict exactly when a given nucleus will decay. But we can determine, on a probabilistic basis, approximately how many nuclei in a sample will decay over a given time period, by assuming that each nucleus has the same probability of decaying in each second that it exists.

The number of decays ΔN that occur in a very short time interval Δt is then proportional to Δt and to the total number N of radioactive nuclei present:

$$\Delta N = -\lambda N \, \Delta t \qquad \textbf{(41–4a)}$$

where the minus sign means N is decreasing. We rewrite this to get the rate of decay:

$$\frac{\Delta N}{\Delta t} = -\lambda N. \qquad \textbf{(41–4b)}$$

In these equations, λ is a constant of proportionality called the **decay constant**, which is different for different isotopes. The greater λ is, the greater the rate of decay and the more "radioactive" that isotope is said to be. The number of decays that occur in the short time interval Δt is designated ΔN because each decay that

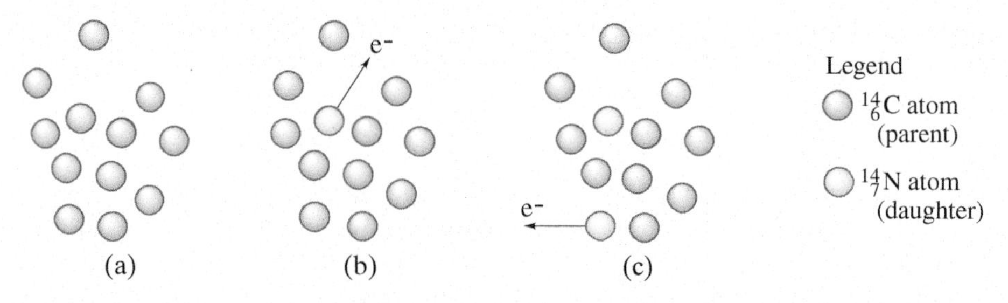

FIGURE 41–10 Radioactive nuclei decay one by one. Hence, the number of parent nuclei in a sample is continually decreasing. When a $^{14}_{6}$C nucleus emits an electron, the nucleus becomes a $^{14}_{7}$N nucleus.

occurs corresponds to a decrease by one in the number N of nuclei present. That is, radioactive decay is a "one-shot" process, Fig. 41–10. Once a particular parent nucleus decays into its daughter, it cannot do it again.

If we take the limit $\Delta t \rightarrow 0$ in Eq. 41–4, ΔN will be small compared to N, and we can write the equation in infinitesimal form as

$$dN = -\lambda N\,dt. \tag{41–5}$$

We can determine N as a function of t by rearranging this equation to

$$\frac{dN}{N} = -\lambda\,dt$$

and then integrating from $t = 0$ to $t = t$:

$$\int_{N_0}^{N} \frac{dN}{N} = -\int_0^t \lambda\,dt,$$

where N_0 is the number of parent nuclei present at $t = 0$ and N is the number remaining at time t. The integration gives

$$\ln\frac{N}{N_0} = -\lambda t$$

or

$$N = N_0 e^{-\lambda t}. \tag{41–6}$$

Equation 41–6 is called the **radioactive decay law**. It tells us that the number of radioactive nuclei in a given sample decreases exponentially in time. This is shown in Fig. 41–11a for the case of $^{14}_{6}$C whose decay constant is $\lambda = 3.83 \times 10^{-12}\,\mathrm{s}^{-1}$.

The rate of decay in a pure sample, or number of decays per second, is

$$\left|\frac{dN}{dt}\right|,$$

also called the **activity** of the sample. We use absolute value signs to make activity a positive number (dN/dt is negative because the number of parent nuclei N is decreasing). The symbol R is also used for activity, $R = |dN/dt|$.

FIGURE 41–11 (a) The number N of parent nuclei in a given sample of $^{14}_{6}$C decreases exponentially. We set $N_0 = 1.00 \times 10^{22}$ here, as we do in the text shortly. (b) The number of decays per second also decreases exponentially. The half-life (Eq. 41–8) of $^{14}_{6}$C is about 5730 yr, which means that the number of parent nuclei, N, and the rate of decay, $|dN/dt|$, decrease by half every 5730 yr.

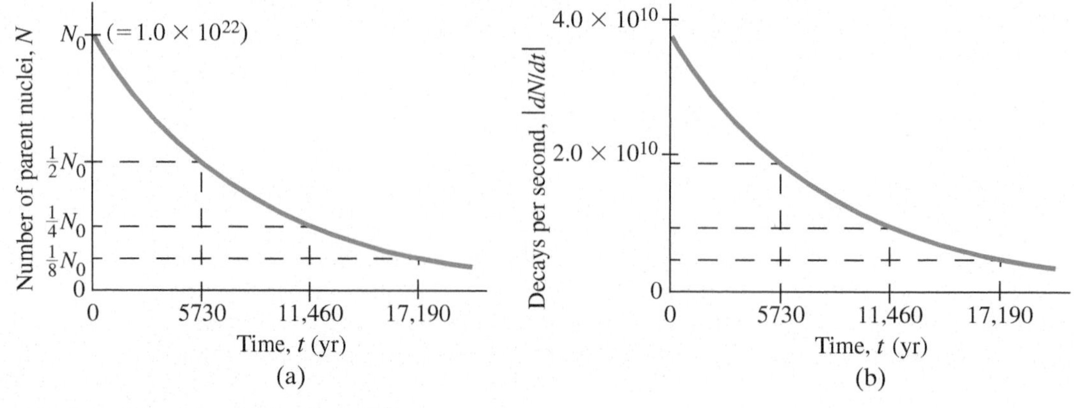

From Eqs. 41–5 and 41–6,

$$\left|\frac{dN}{dt}\right| = \lambda N = \lambda N_0 e^{-\lambda t}. \tag{41–7a}$$

At $t = 0$, the activity is

$$\left|\frac{dN}{dt}\right|_0 = \lambda N_0. \tag{41–7b}$$

Hence, at any other time t the activity is

$$\left|\frac{dN}{dt}\right| = \left|\frac{dN}{dt}\right|_0 e^{-\lambda t}, \tag{41–7c}$$

so the activity decreases exponentially in time at the same rate as for N (Fig. 41–11b). Equation 41–7c is sometimes referred to as the **radioactive decay law** (so is Eq. 41–6), and can be written using R to represent activity, $R = |dN/dt|$, as

$$R = R_0 e^{-\lambda t}. \tag{41–7d}$$

The rate of decay of any isotope is often specified by giving its half-life rather than the decay constant λ. The **half-life** of an isotope is defined as the time it takes for half the original amount of isotope in a given sample to decay. For example, the half-life of $^{14}_{6}C$ is about 5730 years. If at some time a piece of petrified wood contains, say, 1.00×10^{22} $^{14}_{6}C$ nuclei, then 5730 yr later it will contain only 0.50×10^{22} nuclei. After another 5730 yr it will contain 0.25×10^{22} nuclei, and so on. This is characteristic of the exponential function, and is shown in Fig. 41–11a. Since the rate of decay $|dN/dt|$ is proportional to N, it too decreases by a factor of 2 every half-life, Fig. 41–11b.

EXERCISE D The half-life of $^{22}_{11}Na$ is 2.6 years. How much will be left of a 1.0-μg sample of $^{22}_{11}Na$ after 5.2 yr? (*a*) None. (*b*) $\frac{1}{8}\,\mu$g. (*c*) $\frac{1}{4}\,\mu$g. (*d*) $\frac{1}{2}\,\mu$g. (*e*) 0.693 μg.

EXERCISE E Return to the Chapter-Opening Question, page 1104, and answer it again now. Try to explain why you may have answered differently the first time.

The half-lives of known radioactive isotopes vary from as short as 10^{-22} s to about 10^{28} s (about 10^{21} yr). The half-lives of many isotopes are given in Appendix F. It should be clear that the half-life (which we designate $T_{\frac{1}{2}}$) bears an inverse relationship to the decay constant. The longer the half-life of an isotope, the more slowly it decays, and hence λ is smaller. The precise relation is obtained from Eq. 41–6 by setting $N = N_0/2$ at $t = T_{\frac{1}{2}}$:

$$\frac{N_0}{2} = N_0 e^{-\lambda T_{\frac{1}{2}}} \qquad \text{or} \qquad e^{\lambda T_{\frac{1}{2}}} = 2.$$

We take natural logs of both sides ("ln" and "e" are inverse operations, meaning $\ln(e^x) = x$) and find

$$\ln\left(e^{\lambda T_{\frac{1}{2}}}\right) = \ln 2, \qquad \text{so} \qquad \lambda T_{\frac{1}{2}} = \ln 2$$

and

$$T_{\frac{1}{2}} = \frac{\ln 2}{\lambda} = \frac{0.693}{\lambda}. \tag{41–8}$$

We can then write Eq. 41–6 as

$$N = N_0 e^{-0.693t/T_{\frac{1}{2}}}.$$

Sometimes the **mean life** τ of an isotope is quoted, which is defined as $\tau = 1/\lambda$ (see also Problem 80), so then Eq. 41–6 can be written

$$N = N_0 e^{-t/\tau}$$

just as for *RC* and *LR* circuits (Chapters 26 and 30, where τ is called the time constant). Since

$$\tau = \frac{1}{\lambda} = \frac{T_{\frac{1}{2}}}{0.693} \tag{41–9a}$$

the mean life and half-life differ by a factor 0.693; confusing them can cause serious error. The *radioactive decay law*, Eq. 41–7d, can be written simply as

$$R = R_0 e^{-t/\tau}. \tag{41–9b}$$

EXAMPLE 41–8 **Sample activity.** The isotope $^{14}_{6}C$ has a half-life of 5730 yr. If a sample contains 1.00×10^{22} carbon-14 nuclei, what is the activity of the sample?

APPROACH We first use the half-life to find the decay constant (Eq. 41–8), and use that to find the activity, Eq. 41–7b or 41–5. The number of seconds in a year is $(60\,\text{s/min})(60\,\text{min/h})(24\,\text{h/d})(365\tfrac{1}{4}\,\text{d/yr}) = 3.156 \times 10^7\,\text{s}$.

SOLUTION The decay constant λ from Eq. 41–8 is

$$\lambda = \frac{0.693}{T_{\frac{1}{2}}} = \frac{0.693}{(5730\,\text{yr})(3.156 \times 10^7\,\text{s/yr})} = 3.83 \times 10^{-12}\,\text{s}^{-1}.$$

From Eq. 41–7b, the activity or rate of decay is

$$\left|\frac{dN}{dt}\right|_0 = \lambda N_0 = (3.83 \times 10^{-12}\,\text{s}^{-1})(1.00 \times 10^{22}) = 3.83 \times 10^{10}\,\text{decays/s}.$$

Notice that the graph of Fig. 41–11b starts at this value, corresponding to the original value of $N_0 = 1.0 \times 10^{22}$ nuclei in Fig. 41–11a.

NOTE The unit "decays/s" is often written simply as s^{-1} since "decays" is not a unit but refers only to the number. This simple unit of activity is called the becquerel: $1\,\text{Bq} = 1\,\text{decay/s}$, as discussed in Chapter 42.

EXERCISE F Determine the decay constant for radium $(T_{\frac{1}{2}} = 1600\,\text{yr})$.

CONCEPTUAL EXAMPLE 41–9 **Safety: Activity versus half-life.** One might think that a short half-life material is safer than a long half-life material because it will not last as long. Is that true?

RESPONSE No. A shorter half-life means the activity is higher and thus more "radioactive" and can cause more biological damage. On the other hand, a longer half-life for the same sample size N means a lower activity but we have to worry about it for longer and find safe storage until it reaches a safe (low) level of activity.

EXAMPLE 41–10 **A sample of radioactive $^{13}_{7}N$.** A laboratory has 1.49 μg of pure $^{13}_{7}N$, which has a half-life of 10.0 min (600 s). (*a*) How many nuclei are present initially? (*b*) What is the activity initially? (*c*) What is the activity after 1.00 h? (*d*) After approximately how long will the activity drop to less than one per second $(= 1\,\text{s}^{-1})$?

APPROACH We use the definition of the mole and Avogadro's number (Sections 17–7 and 17–9) to find the number of nuclei. For (*b*) we get λ from the given half-life and use Eq. 41–7b for the activity. For (*c*) and (*d*) we use Eq. 41–7c.

SOLUTION (*a*) The atomic mass is 13.0, so 13.0 g will contain 6.02×10^{23} nuclei (Avogadro's number). We have only $1.49 \times 10^{-6}\,\text{g}$, so the number of nuclei N_0 that we have initially is given by the ratio

$$\frac{N_0}{6.02 \times 10^{23}} = \frac{1.49 \times 10^{-6}\,\text{g}}{13.0\,\text{g}}.$$

Solving, we find $N_0 = 6.90 \times 10^{16}$ nuclei.

(*b*) From Eq. 41–8,

$$\lambda = 0.693/T_{\frac{1}{2}} = (0.693)/(600\,\text{s}) = 1.155 \times 10^{-3}\,\text{s}^{-1}.$$

Then, at $t = 0$ (Eq. 41–7b),

$$\left|\frac{dN}{dt}\right|_0 = \lambda N_0 = (1.155 \times 10^{-3}\,\text{s}^{-1})(6.90 \times 10^{16}) = 7.97 \times 10^{13}\,\text{decays/s}.$$

(*c*) After $1.00\,\text{h} = 3600\,\text{s}$, the magnitude of the activity will be (Eq. 41–7c)

$$\left|\frac{dN}{dt}\right| = \left|\frac{dN}{dt}\right|_0 e^{-\lambda t} = (7.97 \times 10^{13}\,\text{s}^{-1})e^{-(1.155\times10^{-3}\,\text{s}^{-1})(3600\,\text{s})} = 1.25 \times 10^{12}\,\text{s}^{-1}.$$

(*d*) We want to determine the time t when $|dN/dt| = 1.00\,\text{s}^{-1}$. From Eq. 41–7c, we have

$$e^{-\lambda t} = \frac{|dN/dt|}{|dN/dt|_0} = \frac{1.00\,\text{s}^{-1}}{7.97 \times 10^{13}\,\text{s}^{-1}} = 1.25 \times 10^{-14}.$$

We take the natural log (ln) of both sides $(\ln e^{-\lambda t} = -\lambda t)$ and divide by λ to find

$$t = -\frac{\ln(1.25 \times 10^{-14})}{\lambda} = 2.77 \times 10^4\,\mathrm{s} = 7.70\,\mathrm{h}.$$

Easy Alternate Solution to (c) 1.00 h = 60.0 minutes is 6 half-lives, so the activity will decrease to $(\frac{1}{2})(\frac{1}{2})(\frac{1}{2})(\frac{1}{2})(\frac{1}{2})(\frac{1}{2}) = (\frac{1}{2})^6 = \frac{1}{64}$ of its original value, or $(7.97 \times 10^{13})/(64) = 1.25 \times 10^{12}$ per second.

EXERCISE G Technicium $^{98}_{43}\mathrm{Tc}$ has a half-life of 4.2×10^6 yr. Strontium $^{90}_{38}\mathrm{Sr}$ has a half-life of 28.79 yr. Which statements are true?

(*a*) The decay constant of Sr is greater than the decay constant of Tc.
(*b*) The activity of 100 g of Sr is less than the activity of 100 g of Tc.
(*c*) The long half-life of Tc means that it decays by alpha decay.
(*d*) A Tc atom has a higher probability of decaying in 1 yr than a Sr atom.
(*e*) 28.79 g of Sr has the same activity as 4.2×10^6 g of Tc.

41–9 Decay Series

It is often the case that one radioactive isotope decays to another isotope that is also radioactive. Sometimes this daughter decays to yet a third isotope which also is radioactive. Such successive decays are said to form a **decay series**. An important example is illustrated in Fig. 41–12. As can be seen, $^{238}_{92}\mathrm{U}$ decays by α emission to $^{234}_{90}\mathrm{Th}$, which in turn decays by β decay to $^{234}_{91}\mathrm{Pa}$. The series continues as shown, with several possible branches near the bottom, ending at the stable lead isotope, $^{206}_{82}\mathrm{Pb}$. The two last decays can be

$$^{206}_{81}\mathrm{Tl} \rightarrow {}^{206}_{82}\mathrm{Pb} + e^- + \bar{\nu}, \qquad (T_{\frac{1}{2}} = 4.2\,\mathrm{min})$$

or

$$^{210}_{84}\mathrm{Po} \rightarrow {}^{206}_{82}\mathrm{Pb} + \alpha. \qquad (T_{\frac{1}{2}} = 138\,\mathrm{days})$$

Other radioactive series also exist.

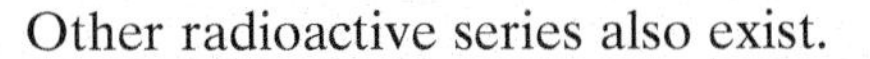

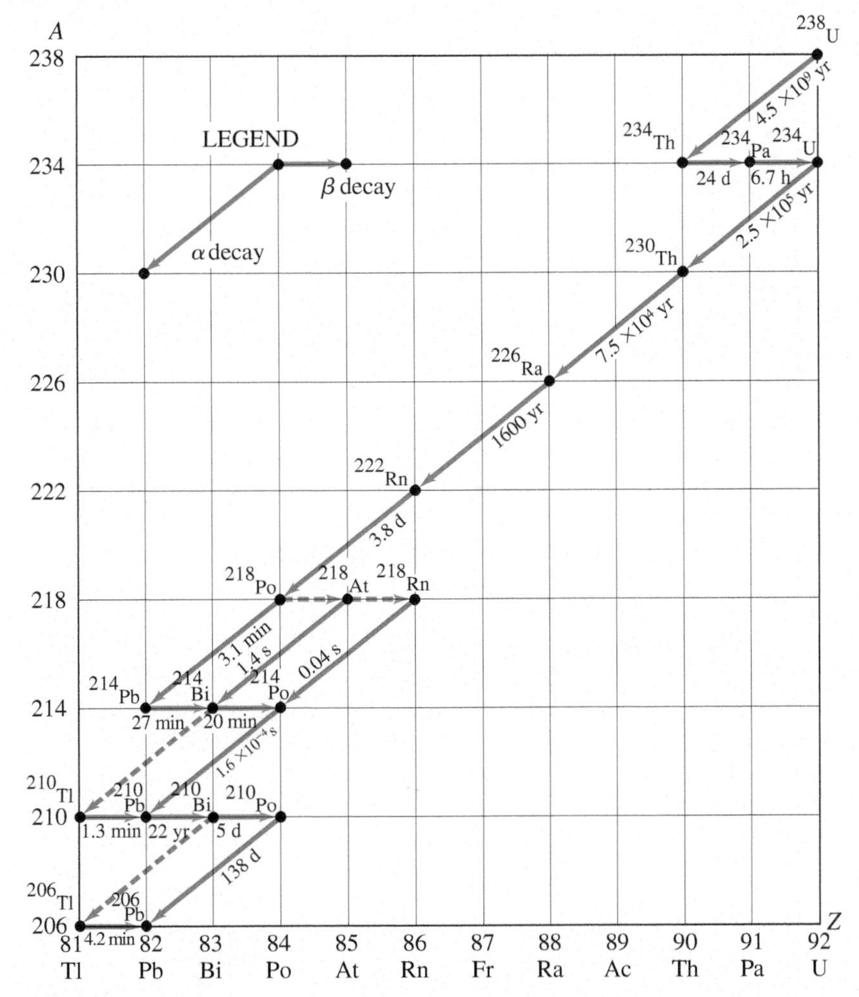

FIGURE 41–12 Decay series beginning with $^{238}_{92}\mathrm{U}$. Nuclei in the series are specified by a dot representing A and Z values. Half-lives are given in seconds (s), minutes (min), hours (h), days (d), or years (yr). Note that a horizontal arrow represents β decay (A does not change), whereas a diagonal line represents α decay (A changes by 4, Z changes by 2). For the four nuclides shown that can decay by both α and β decay, the more prominent decay (in these four cases, $>99.9\%$) is shown as a solid arrow and the less common decay ($<0.1\%$) as a dashed arrow.

Because of such decay series, certain radioactive elements are found in nature that otherwise would not be. When the solar system (including Earth) was formed about 5 billion years ago, it is believed that nearly all nuclides were present, having been formed (by fusion and neutron capture, Sections 42–4 and 44–2) in a nearby supernova explosion (Section 44–2). Many isotopes with short half-lives decayed quickly and no longer are detected in nature today. But long-lived isotopes, such as $^{238}_{92}U$ with a half-life of 4.5×10^9 yr, still do exist in nature today. Indeed, about half of the original $^{238}_{92}U$ still remains. We might expect, however, that radium ($^{226}_{88}Ra$), with a half-life of 1600 yr, would long since have disappeared from the Earth. Indeed, the original $^{226}_{88}Ra$ nuclei must by now have all decayed. However, because $^{238}_{92}U$ decays (in several steps) to $^{226}_{88}Ra$, the supply of $^{226}_{88}Ra$ is continually replenished, which is why it is still found on Earth today. The same can be said for many other radioactive nuclides.

CONCEPTUAL EXAMPLE 41–11 **Decay chain.** In the decay chain of Fig. 41–12, if we start looking below $^{234}_{92}U$, we see four successive nuclides with half-lives of 250,000 yr, 75,000 yr, 1600 yr, and a little under 4 days. Each decay in the chain has an alpha particle of a characteristic energy, and so we can monitor the radioactive decay rate of each nuclide. Given a sample that was pure $^{234}_{92}U$ a million years ago, which alpha decay would you expect to have the highest activity in the sample?

RESPONSE The first instinct is to say that the process with the shortest half-life would show the highest activity. Surprisingly, perhaps, the activities of the four nuclides in this sample are all the same. The reason is that in each case the decay of the parent acts as a bottleneck to the decay of the daughter. Compared to the 1600-yr half-life of $^{226}_{88}Ra$, for example, its daughter $^{222}_{86}Rn$ decays almost immediately, but it cannot decay until it is made. (This is like an automobile assembly line: if worker A takes 20 minutes to do a task and then worker B takes only 1 minute to do the next task, worker B still does only one car every 20 minutes.)

41–10 Radioactive Dating

Radioactive decay has many interesting applications. One is the technique of *radioactive dating* by which the age of ancient materials can be determined.

PHYSICS APPLIED
Carbon-14 dating

The age of any object made from once-living matter, such as wood, can be determined using the natural radioactivity of $^{14}_{6}C$. All living plants absorb carbon dioxide (CO_2) from the air and use it to synthesize organic molecules. The vast majority of these carbon atoms are $^{12}_{6}C$, but a small fraction, about 1.3×10^{-12}, is the radioactive isotope $^{14}_{6}C$. The ratio of $^{14}_{6}C$ to $^{12}_{6}C$ in the atmosphere has remained roughly constant over many thousands of years, in spite of the fact that $^{14}_{6}C$ decays with a half-life of about 5730 yr. This is because energetic nuclei in the cosmic radiation, which impinges on the Earth from outer space, strike nuclei of atoms in the atmosphere and break those nuclei into pieces, releasing free neutrons. Those neutrons can collide with nitrogen nuclei in the atmosphere to produce the nuclear transformation $n + ^{14}_{7}N \rightarrow ^{14}_{6}C + p$. That is, a neutron strikes and is absorbed by a $^{14}_{7}N$ nucleus, and a proton is knocked out in the process. The remaining nucleus is $^{14}_{6}C$. This continual production of $^{14}_{6}C$ in the atmosphere roughly balances the loss of $^{14}_{6}C$ by radioactive decay.

As long as a plant or tree is alive, it continually uses the carbon from carbon dioxide in the air to build new tissue and to replace old. Animals eat plants, so they too are continually receiving a fresh supply of carbon for their tissues. Organisms cannot distinguish[†] $^{14}_{6}C$ from $^{12}_{6}C$, and since the ratio of $^{14}_{6}C$ to $^{12}_{6}C$ in the atmosphere remains nearly constant, the ratio of the two isotopes within the living organism remains nearly constant as well. When an organism dies, carbon dioxide is no longer absorbed and utilized. Because the $^{14}_{6}C$ decays radioactively, the ratio of $^{14}_{6}C$ to $^{12}_{6}C$ in a dead organism decreases over time. Since the half-life of $^{14}_{6}C$ is about 5730 yr, the $^{14}_{6}C/^{12}_{6}C$ ratio decreases by half every 5730 yr. If, for example,

[†]Organisms operate almost exclusively via chemical reactions—which involve only the outer orbital electrons of the atom; extra neutrons in the nucleus have essentially no effect.

the $^{14}_{6}C/^{12}_{6}C$ ratio of an ancient wooden tool is half of what it is in living trees, then the object must have been made from a tree that was felled about 5730 years ago.

Actually, corrections must be made for the fact that the $^{14}_{6}C/^{12}_{6}C$ ratio in the atmosphere has not remained precisely constant over time. The determination of what this ratio has been over the centuries has required techniques such as comparing the expected ratio to the actual ratio for objects whose age is known, such as very old trees whose annual rings can be counted reasonably accurately.

EXAMPLE 41–12 **An ancient animal.** The mass of carbon in an animal bone fragment found in an archeological site is 200 g. If the bone registers an activity of 16 decays/s, what is its age?

PHYSICS APPLIED
Archeological dating

APPROACH First we determine how many $^{14}_{6}C$ atoms there were in our 200-g sample when the animal was alive, given the known fraction of $^{14}_{6}C$, 1.3×10^{-12}. Then we use Eq. 41–7b to find the activity back then, and Eq. 41–7c to find out how long ago that was by solving for the time t.

SOLUTION The 200 g of carbon is nearly all $^{12}_{6}C$; 12.0 g of $^{12}_{6}C$ contains 6.02×10^{23} atoms, so 200 g contains

$$\left(\frac{6.02 \times 10^{23}\ \text{atoms/mol}}{12\ \text{g/mol}}\right)(200\ \text{g}) = 1.00 \times 10^{25}\ \text{atoms.}$$

When the animal was alive, the ratio of $^{14}_{6}C$ to $^{12}_{6}C$ in the bone was 1.3×10^{-12}. The number of $^{14}_{6}C$ nuclei at that time was

$$N_0 = (1.00 \times 10^{25}\ \text{atoms})(1.3 \times 10^{-12}) = 1.3 \times 10^{13}\ \text{atoms.}$$

From Eq. 41–7b, with $\lambda = 3.83 \times 10^{-12}\ \text{s}^{-1}$ for $^{14}_{6}C$ (Example 41–8), the magnitude of the activity when the animal was still alive ($t = 0$) was

$$\left|\frac{dN}{dt}\right|_0 = \lambda N_0 = (3.83 \times 10^{-12}\ \text{s}^{-1})(1.3 \times 10^{13}) = 50\ \text{s}^{-1}.$$

From Eq. 41–7c,

$$\left|\frac{dN}{dt}\right| = \left|\frac{dN}{dt}\right|_0 e^{-\lambda t},$$

where $|dN/dt|$ is given as $16\ \text{s}^{-1}$. Then

$$e^{\lambda t} = \frac{|dN/dt|_0}{|dN/dt|} = \frac{50\ \text{s}^{-1}}{16\ \text{s}^{-1}}.$$

We take the natural log (ln) of both sides (and divide by λ) to get

$$t = \frac{1}{\lambda}\ln\left[\frac{|dN/dt|_0}{|dN/dt|}\right] = \frac{1}{3.83 \times 10^{-12}\ \text{s}^{-1}}\ln\left[\frac{50\ \text{s}^{-1}}{16\ \text{s}^{-1}}\right]$$
$$= 2.98 \times 10^{11}\ \text{s} = 9400\ \text{yr},$$

which is the time elapsed since the death of the animal.

Geological Time Scale Dating

Carbon dating is useful only for determining the age of objects less than about 60,000 years old. The amount of $^{14}_{6}C$ remaining in objects older than that is usually too small to measure accurately, although new techniques are allowing detection of even smaller amounts of $^{14}_{6}C$, pushing the time frame further back. On the other hand, radioactive isotopes with longer half-lives can be used in certain circumstances to obtain the age of older objects. For example, the decay of $^{238}_{92}U$, because of its long half-life of 4.5×10^{9} years, is useful in determining the ages of rocks on a geologic time scale. When molten material on Earth long ago solidified into rock as the temperature dropped, different compounds solidified according to the melting points, and thus different compounds separated to some extent. Uranium

PHYSICS APPLIED
Geological dating

present in a material became fixed in position and the daughter nuclei that result from the decay of uranium were also fixed in that position. Thus, by measuring the amount of ${}^{238}_{92}\mathrm{U}$ remaining in the material relative to the amount of daughter nuclei, the time when the rock solidified can be determined.

Radioactive dating methods using ${}^{238}_{92}\mathrm{U}$ and other isotopes have shown the age of the oldest Earth rocks to be about 4×10^9 yr. The age of rocks in which the oldest fossilized organisms are embedded indicates that life appeared more than $3\frac{1}{2}$ billion years ago. The earliest fossilized remains of mammals are found in rocks 200 million years old, and humanlike creatures seem to have appeared about 2 million years ago. Radioactive dating has been indispensable for the reconstruction of Earth's history.

41–11 Detection of Radiation

Individual particles such as electrons, protons, α particles, neutrons, and γ rays are not detected directly by our senses. Consequently, a variety of instruments have been developed to detect them.

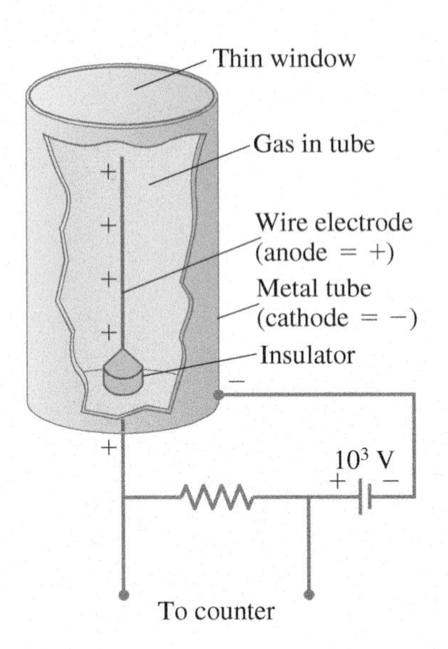

FIGURE 41–13 Diagram of a Geiger counter.

Counters

One of the most common detectors is the **Geiger counter**. As shown in Fig. 41–13, it consists of a cylindrical metal tube filled with a certain type of gas. A long wire runs down the center and is kept at a high positive voltage ($\approx 10^3$ V) with respect to the outer cylinder. The voltage is just slightly less than that required to ionize the gas atoms. When a charged particle enters through the thin "window" at one end of the tube, it ionizes a few atoms of the gas. The freed electrons are attracted toward the positive wire, and as they are accelerated they strike and ionize additional atoms. An "avalanche" of electrons is quickly produced, and when it reaches the wire anode, it produces a voltage pulse. The pulse, after being amplified, can be sent to an electronic counter, which counts how many particles have been detected. Or the pulses can be sent to a loudspeaker and each detection of a particle is heard as a "click." Only a fraction of the radiation emitted by a sample is detected by any detector.

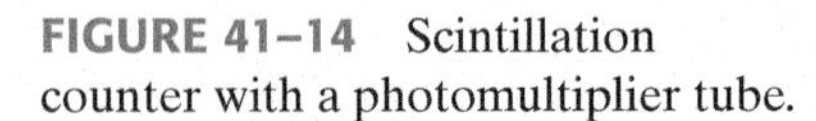

FIGURE 41–14 Scintillation counter with a photomultiplier tube.

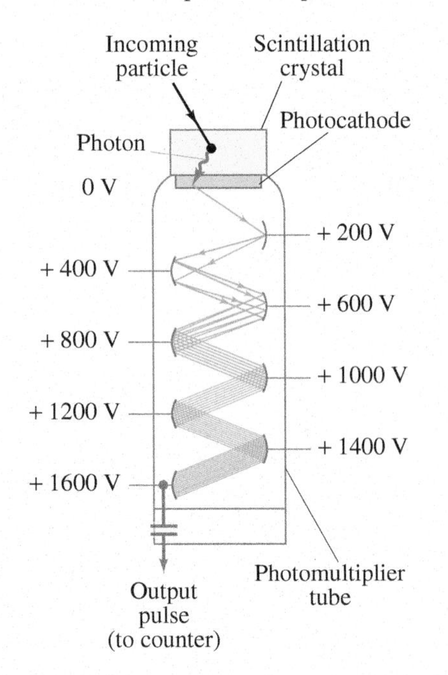

A **scintillation counter** makes use of a solid, liquid, or gas known as a **scintillator** or **phosphor**. The atoms of a scintillator are easily excited when struck by an incoming particle and emit visible light when they return to their ground states. Typical scintillators are crystals of NaI and certain plastics. One face of a solid scintillator is cemented to a photomultiplier tube, and the whole is wrapped with opaque material to keep it light-tight (in the dark) or is placed within a light-tight container. The **photomultiplier** (PM) **tube** converts the energy of the scintillator-emitted photon(s) into an electric signal. A PM tube is a vacuum tube containing several electrodes (typically 8 to 14), called *dynodes,* which are maintained at successively higher voltages as shown in Fig. 41–14. At its top surface is a photoelectric surface, called the *photocathode*, whose work function (Section 37–2) is low enough that an electron is easily released when struck by a photon from the scintillator. Such an electron is accelerated toward the positive voltage of the first dynode. When it strikes the first dynode, the electron has acquired sufficient kinetic energy so that it can eject two to five more electrons. These, in turn, are accelerated toward the higher voltage second dynode, and a multiplication process begins. The number of electrons striking the last dynode may be 10^6 or more. Thus the passage of a particle through the scintillator results in an electric signal at the output of the PM tube that can be sent to an electronic counter just as for a Geiger tube. Solid scintillators are much more dense than the gas of a Geiger counter, and so are much more efficient detectors—especially for γ rays, which interact less with matter than do α or β particles. Scintillators that can measure the total energy deposited are much used today and are called **calorimeters**.

In tracer work (Section 42–8), **liquid scintillators** are often used. Radioactive samples taken at different times or from different parts of an organism are placed directly in small bottles containing the liquid scintillator. This is particularly convenient for detection of β rays from $^{3}_{1}\mathrm{H}$ and $^{14}_{6}\mathrm{C}$, which have very low energies and have difficulty passing through the outer covering of a crystal scintillator or Geiger tube. A PM tube is still used to produce the electric signal from the liquid scintillator.

A **semiconductor detector** consists of a reverse-biased *pn* junction diode (Sections 40–8 and 40–9). A particle passing through the junction can excite electrons into the conduction band, leaving holes in the valence band. The freed charges produce a short electrical pulse that can be counted just as for Geiger and scintillation counters.

Hospital workers and others who work around radiation may carry *film badges* which detect the accumulation of radiation exposure. The film inside is periodically replaced and developed, the darkness being related to total exposure (see Section 42–6).

Visualization

The devices discussed so far are used for counting the number of particles (or decays of a radioactive isotope). There are also devices that allow the track of charged particles to be *seen*. Very important are semiconductor detectors. **Silicon wafer semiconductors** have their surface etched into separate tiny pixels, each providing particle position information. They are much used in elementary particle physics (Chapter 43) to track the positions of particles produced and to determine their point of origin and/or their momentum (with the help of a magnetic field). The pixel arrangement can be CCD or CMOS (Section 33–5), the latter able to incorporate electronics inside, allowing fast readout.

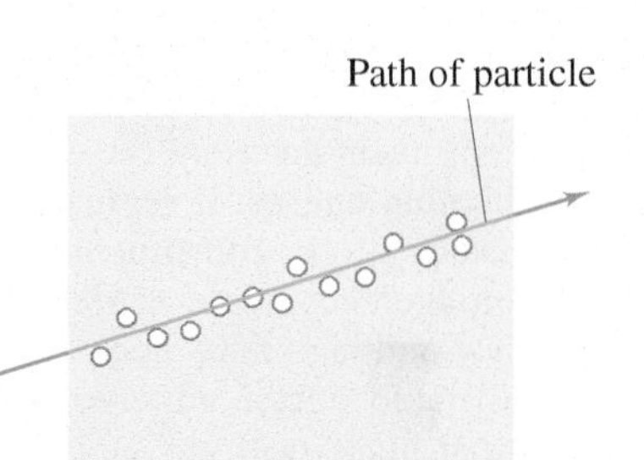

FIGURE 41–15 In a cloud chamber or bubble chamber, droplets or bubbles are formed around ions produced by the passage of a charged particle.

One of the oldest tracking devices is the **photographic emulsion**, which can be small and portable, used now particularly for cosmic-ray studies from balloons. A charged particle passing through an emulsion ionizes the atoms along its path. These points undergo a chemical change, and when the emulsion is developed (like film) the particle's path is revealed.

In a **cloud chamber**, used in the early days of nuclear physics, a gas is cooled to a temperature slightly below its usual condensation point ("supercooled"), and gas molecules condense on any ionized molecules present. Tiny droplets form around ions produced when a charged particle passes through (Fig. 41–15). Light scattering from these droplets reveals the track of the particle.

The **bubble chamber**, invented in 1952 by D. A. Glaser (1926–), makes use of a superheated liquid kept close to its normal boiling point. Bubbles characteristic of boiling form around ions produced by the passage of a charged particle, revealing paths of particles that recently passed through. Because a bubble chamber uses a liquid, often liquid hydrogen, many more interactions can occur than in a cloud chamber. A magnetic field is usually applied across the chamber so the momentum of the moving particles can be determined from the radius of curvature of their paths.

FIGURE 41–16 Wire-drift chamber inside the Collider Detector at Fermilab (CDF). The photo at the start of Chapter 43 (page 1164) was done with this detector.

A **wire drift chamber** consists of a set of closely spaced fine wires immersed in a gas (Fig. 41–16). Many wires are grounded, and the others between are kept at very high voltage. A charged particle passing through produces ions in the gas. Freed electrons drift toward the nearest high voltage wire, creating an "avalanche" of many more ions, and producing an electric pulse or signal at that wire. The positions of the particles are determined electronically by the position of the wire and by the time it takes the pulses to reach "readout" electronics at the ends of the wires. The paths of the particles are reconstructed electronically by computers which can "draw" a picture of the tracks, as shown in the photo at the start of Chapter 43. An external magnetic field curves the paths, allowing the momentum of the particles to be measured.

In many detectors, the energy of the particles can be measured by the strength of the electronic signal; such detectors are referred to as **calorimeters**, as already mentioned.

Summary

Nuclear physics is the study of atomic nuclei. Nuclei contain **protons** and **neutrons**, which are collectively known as **nucleons**. The total number of nucleons, A, is the nucleus's **atomic mass number**. The number of protons, Z, is the **atomic number**. The number of neutrons equals $A - Z$. **Isotopes** are nuclei with the same Z, but with different numbers of neutrons. For an element X, an isotope of given Z and A is represented by

$$^{A}_{Z}\mathrm{X}.$$

The nuclear radius is approximately proportional to $A^{\frac{1}{3}}$, indicating that all nuclei have about the same density. Nuclear masses are specified in **unified atomic mass units** (u), where the mass of $^{12}_{6}\mathrm{C}$ (including its 6 electrons) is defined as exactly 12.000000 u. In terms of the energy equivalent (because $E = mc^2$),

$$1\,\mathrm{u} = 931.5\,\mathrm{MeV}/c^2 = 1.66 \times 10^{-27}\,\mathrm{kg}.$$

The mass of a stable nucleus is less than the sum of the masses of its constituent nucleons. The difference in mass (times c^2) is the **total binding energy**. It represents the energy needed to break the nucleus into its constituent nucleons. The **binding energy per nucleon** averages about 8 MeV per nucleon, and is lowest for low mass and high mass nuclei.

Unstable nuclei undergo **radioactive decay**; they change into other nuclei with the emission of an α, β, or γ particle. An α particle is a $^{4}_{2}\mathrm{He}$ nucleus; a β particle is an electron or positron; and a γ ray is a high-energy photon. In β decay, a **neutrino** is also emitted. The transformation of the **parent** into the **daughter** nucleus is called **transmutation** of the elements. Radioactive decay occurs spontaneously only when the mass of the products is less than the mass of the parent nucleus. The loss in mass appears as kinetic energy of the products.

Alpha decay occurs via the purely quantum mechanical process of **tunneling** through a barrier.

Nuclei are held together by the **strong nuclear force**. The **weak nuclear force** makes itself apparent in β decay. These two forces, plus the gravitational and electromagnetic forces, are the four known types of force.

Electric charge, linear and angular momentum, mass–energy, and **nucleon number** are **conserved** in all decays.

Radioactive decay is a statistical process. For a given type of radioactive nucleus, the number of nuclei that decay (ΔN) in a time Δt is proportional to the number N of parent nuclei present:

$$\Delta N = -\lambda N\,\Delta t; \qquad \textbf{(41–4a)}$$

the minus sign means N *decreases* in time.

The proportionality constant λ is called the **decay constant** and is characteristic of the given nucleus. The number N of nuclei remaining after a time t decreases exponentially

$$N = N_0 e^{-\lambda t}, \qquad \textbf{(41–6)}$$

as does the **activity**, $R = |dN/dt|$:

$$\left|\frac{dN}{dt}\right| = \left|\frac{dN}{dt}\right|_0 e^{-\lambda t}. \qquad \textbf{(41–7c)}$$

The **half-life**, $T_{\frac{1}{2}}$, is the time required for half the nuclei of a radioactive sample to decay. It is related to the decay constant by

$$T_{\frac{1}{2}} = \frac{0.693}{\lambda}. \qquad \textbf{(41–8)}$$

Radioactive decay can be used to determine the age of certain objects, such as once-living biological material ($^{14}_{6}\mathrm{C}$) or geological formations ($^{238}_{92}\mathrm{U}$).

Particle **detectors** include **Geiger counters**, **scintillators** with attached **photomultiplier tubes**, and **semiconductor detectors**. Detectors that can image particle tracks include **semiconductors**, photographic **emulsions**, **bubble chambers**, and **wire drift chambers**.

Questions

1. What do different isotopes of a given element have in common? How are they different?
2. What are the elements represented by the X in the following: (*a*) $^{232}_{92}\mathrm{X}$; (*b*) $^{18}_{7}\mathrm{X}$; (*c*) $^{1}_{1}\mathrm{X}$; (*d*) $^{82}_{38}\mathrm{X}$; (*e*) $^{247}_{97}\mathrm{X}$?
3. How many protons and how many neutrons do each of the isotopes in Question 2 have?
4. Identify the element that has 88 nucleons and 50 neutrons.
5. Why are the atomic masses of many elements (see the Periodic Table) not close to whole numbers?
6. How do we know there is such a thing as the strong nuclear force?
7. What are the similarities and the differences between the strong nuclear force and the electric force?
8. What is the experimental evidence in favor of radioactivity being a nuclear process?
9. The isotope $^{64}_{29}\mathrm{Cu}$ is unusual in that it can decay by γ, β^-, and β^+ emission. What is the resulting nuclide for each case?
10. A $^{238}_{92}\mathrm{U}$ nucleus decays via α decay to a nucleus containing how many neutrons?
11. Describe, in as many ways as you can, the difference between α, β, and γ rays.
12. What element is formed by the radioactive decay of (*a*) $^{24}_{11}\mathrm{Na}\ (\beta^-)$; (*b*) $^{22}_{11}\mathrm{Na}\ (\beta^+)$; (*c*) $^{210}_{84}\mathrm{Po}\ (\alpha)$?
13. What element is formed by the decay of (*a*) $^{32}_{15}\mathrm{P}\ (\beta^-)$; (*b*) $^{35}_{16}\mathrm{S}\ (\beta^-)$; (*c*) $^{211}_{83}\mathrm{Bi}\ (\alpha)$?
14. Fill in the missing particle or nucleus:
    (*a*) $^{45}_{20}\mathrm{Ca} \rightarrow\ ? + e^- + \bar{\nu}$
    (*b*) $^{58}_{29}\mathrm{Cu}^* \rightarrow\ ? + \gamma$
    (*c*) $^{46}_{24}\mathrm{Cr} \rightarrow {}^{46}_{23}\mathrm{V} + ?$
    (*d*) $^{234}_{94}\mathrm{Pu} \rightarrow\ ? + \alpha$
    (*e*) $^{239}_{93}\mathrm{Np} \rightarrow {}^{239}_{94}\mathrm{Pu} + ?$
15. Immediately after a $^{238}_{92}\mathrm{U}$ nucleus decays to $^{234}_{90}\mathrm{Th} + {}^{4}_{2}\mathrm{He}$, the daughter thorium nucleus may still have 92 electrons circling it. Since thorium normally holds only 90 electrons, what do you suppose happens to the two extra ones?
16. When a nucleus undergoes either β^- or β^+ decay, what happens to the energy levels of the atomic electrons? What is likely to happen to these electrons following the decay?

17. The alpha particles from a given alpha-emitting nuclide are generally monoenergetic; that is, they all have the same kinetic energy. But the beta particles from a beta-emitting nuclide have a spectrum of energies. Explain the difference between these two cases.

18. Do isotopes that undergo electron capture generally lie above or below the stable nuclides in Fig. 41–2?

19. Can hydrogen or deuterium emit an α particle? Explain.

20. Why are many artificially produced radioactive isotopes rare in nature?

21. An isotope has a half-life of one month. After two months, will a given sample of this isotope have completely decayed? If not, how much remains?

22. Why are none of the elements with $Z > 92$ stable?

23. A proton strikes a $^{6}_{3}\mathrm{Li}$ nucleus. As a result, an α particle and another particle are released. What is the other particle?

24. Can $^{14}_{6}\mathrm{C}$ dating be used to measure the age of stone walls and tablets of ancient civilizations? Explain.

25. In both internal conversion and β decay, an electron is emitted. How could you determine which decay process occurred?

26. Describe how the potential energy curve for an α particle in an α-emitting nucleus differs from that for a stable nucleus.

27. Explain the absence of β^+ emitters in the radioactive decay series of Fig. 41–12.

28. As $^{222}_{86}\mathrm{Rn}$ decays into $^{206}_{82}\mathrm{Pb}$, how many alpha and beta particles are emitted? Does it matter which path in the decay series is chosen? Why or why not?

Problems

41–1 Nuclear Properties

1. (I) A pi meson has a mass of 139 MeV/c^2. What is this in atomic mass units?

2. (I) What is the approximate radius of an alpha particle ($^{4}_{2}\mathrm{He}$)?

3. (I) By what % is the radius of $^{238}_{92}\mathrm{U}$ greater than the radius of $^{232}_{92}\mathrm{U}$?

4. (II) (*a*) What is the approximate radius of a $^{112}_{48}\mathrm{Cd}$ nucleus? (*b*) Approximately what is the value of A for a nucleus whose radius is 3.7×10^{-15} m?

5. (II) What is the mass of a bare α particle (without electrons) in MeV/c^2?

6. (II) Suppose two alpha particles were held together so they were just touching. Estimate the electrostatic repulsive force each would exert on the other. What would be the acceleration of an alpha particle subjected to this force?

7. (II) (*a*) Show that the density of nuclear matter is essentially the same for all nuclei. (*b*) What would be the radius of the Earth if it had its actual mass but had the density of nuclei? (*c*) What would be the radius of a $^{238}_{92}\mathrm{U}$ nucleus if it had the density of the Earth?

8. (II) What stable nucleus has approximately half the radius of a uranium nucleus? [*Hint*: Find A and use Appendix F to get Z.]

9. (II) If an alpha particle were released from rest near the surface of a $^{257}_{100}\mathrm{Fm}$ nucleus, what would its kinetic energy be when far away?

10. (II) (*a*) What is the fraction of the hydrogen atom's mass that is in the nucleus? (*b*) What is the fraction of the hydrogen atom's volume that is occupied by the nucleus?

11. (II) Approximately how many nucleons are there in a 1.0-kg object? Does it matter what the object is made of? Why or why not?

12. (II) How much kinetic energy must an α particle have to just "touch" the surface of a $^{238}_{92}\mathrm{U}$ nucleus?

41–2 Binding Energy

13. (I) Estimate the total binding energy for $^{63}_{29}\mathrm{Cu}$, using Fig. 41–1.

14. (II) Use Appendix F to calculate the binding energy of $^{2}_{1}\mathrm{H}$ (deuterium).

15. (II) Determine the binding energy of the last neutron in a $^{32}_{15}\mathrm{P}$ nucleus.

16. (II) Calculate the total binding energy, and the binding energy per nucleon, for (*a*) $^{7}_{3}\mathrm{Li}$, (*b*) $^{197}_{79}\mathrm{Au}$. Use Appendix F.

17. (II) Compare the average binding energy of a nucleon in $^{23}_{11}\mathrm{Na}$ to that in $^{24}_{11}\mathrm{Na}$.

18. (III) How much energy is required to remove (*a*) a proton, (*b*) a neutron, from $^{15}_{7}\mathrm{N}$? Explain the difference in your answers.

19. (III) (*a*) Show that the nucleus $^{8}_{4}\mathrm{Be}$ (mass = 8.005305 u) is unstable and will decay into two α particles. (*b*) Is $^{12}_{6}\mathrm{C}$ stable against decay into three α particles? Show why or why not.

41–3 to 41–7 Radioactive Decay

20. (I) How much energy is released when tritium, $^{3}_{1}\mathrm{H}$, decays by β^- emission?

21. (I) What is the maximum kinetic energy of an electron emitted in the β decay of a free neutron?

22. (I) Show that the decay $^{11}_{6}\mathrm{C} \rightarrow {}^{10}_{5}\mathrm{B} + \mathrm{p}$ is not possible because energy would not be conserved.

23. (I) The $^{7}_{3}\mathrm{Li}$ nucleus has an excited state 0.48 MeV above the ground state. What wavelength gamma photon is emitted when the nucleus decays from the excited state to the ground state?

24. (II) Give the result of a calculation that shows whether or not the following decays are possible:
(*a*) $^{233}_{92}\mathrm{U} \rightarrow {}^{232}_{92}\mathrm{U} + \mathrm{n}$;
(*b*) $^{14}_{7}\mathrm{N} \rightarrow {}^{13}_{7}\mathrm{N} + \mathrm{n}$;
(*c*) $^{40}_{19}\mathrm{K} \rightarrow {}^{39}_{19}\mathrm{K} + \mathrm{n}$.

25. (II) $^{24}_{11}\mathrm{Na}$ is radioactive. (*a*) Is it a β^- or β^+ emitter? (*b*) Write down the decay reaction, and estimate the maximum kinetic energy of the emitted β.

26. (II) When $^{23}_{10}\mathrm{Ne}$ (mass = 22.9945 u) decays to $^{23}_{11}\mathrm{Na}$ (mass = 22.9898 u), what is the maximum kinetic energy of the emitted electron? What is its minimum energy? What is the energy of the neutrino in each case? Ignore recoil of the daughter nucleus.

27. (II) A $^{238}_{92}\mathrm{U}$ nucleus emits an α particle with kinetic energy = 4.20 MeV. (*a*) What is the daughter nucleus, and (*b*) what is the approximate atomic mass (in u) of the daughter atom? Ignore recoil of the daughter nucleus.

28. (II) What is the maximum kinetic energy of the emitted β particle during the decay of $^{60}_{27}\mathrm{Co}$?

29. (II) A nucleus of mass 256 u, initially at rest, emits an α particle with a kinetic energy of 5.0 MeV. What is the kinetic energy of the recoiling daughter nucleus?

30. (II) The isotope ${}^{218}_{84}\mathrm{Po}$ can decay by either α or β^- emission. What is the energy release in each case? The mass of ${}^{218}_{84}\mathrm{Po}$ is 218.008965 u.

31. (II) The nuclide ${}^{32}_{15}\mathrm{P}$ decays by emitting an electron whose maximum kinetic energy can be 1.71 MeV. (*a*) What is the daughter nucleus? (*b*) Calculate the daughter's atomic mass (in u).

32. (II) A photon with a wavelength of 1.00×10^{-13} m is ejected from an atom. Calculate its energy and explain why it is a γ ray from the nucleus or a photon from the atom.

33. (II) How much energy is released in electron capture by beryllium: ${}^{7}_{4}\mathrm{Be} + e^- \rightarrow {}^{7}_{3}\mathrm{Li} + \nu$?

34. (II) How much recoil energy does a ${}^{40}_{19}\mathrm{K}$ nucleus get when it emits a 1.46-MeV gamma ray?

35. (II) Determine the maximum kinetic energy of β^+ particles released when ${}^{11}_{6}\mathrm{C}$ decays to ${}^{11}_{5}\mathrm{B}$. What is the maximum energy the neutrino can have? What is the minimum energy of each?

36. (III) The α particle emitted when ${}^{238}_{92}\mathrm{U}$ decays has 4.20 MeV of kinetic energy. Calculate the recoil kinetic energy of the daughter nucleus and the Q-value of the decay.

37. (III) What is the energy of the α particle emitted in the decay ${}^{210}_{84}\mathrm{Po} \rightarrow {}^{206}_{82}\mathrm{Pb} + \alpha$? Take into account the recoil of the daughter nucleus.

38. (III) Show that when a nucleus decays by β^+ decay, the total energy released is equal to

$$(M_P - M_D - 2m_e)c^2,$$

where M_P and M_D are the masses of the parent and daughter atoms (neutral), and m_e is the mass of an electron or positron.

41–8 to 41–10 Half-Life, Decay Rates, Decay Series, Dating

39. (I) (*a*) What is the decay constant of ${}^{238}_{92}\mathrm{U}$ whose half-life is 4.5×10^9 yr? (*b*) The decay constant of a given nucleus is $3.2 \times 10^{-5}\,\mathrm{s}^{-1}$. What is its half-life?

40. (I) A radioactive material produces 1280 decays per minute at one time, and 3.6 h later produces 320 decays per minute. What is its half-life?

41. (I) What fraction of a sample of ${}^{68}_{32}\mathrm{Ge}$, whose half-life is about 9 months, will remain after 2.0 yr?

42. (I) What is the activity of a sample of ${}^{14}_{6}\mathrm{C}$ that contains 8.1×10^{20} nuclei?

43. (I) What fraction of a sample is left after exactly 6 half-lives?

44. (II) A sample of ${}^{60}_{27}\mathrm{Co}$ and a sample of ${}^{131}_{53}\mathrm{I}$ both have N_0 atoms at $t = 0$. How long will it take until both have the same activity? (Use Appendix F for half-life data.)

45. (II) How many nuclei of ${}^{238}_{92}\mathrm{U}$ remain in a rock if the activity registers 340 decays per second?

46. (II) In a series of decays, the nuclide ${}^{235}_{92}\mathrm{U}$ becomes ${}^{207}_{82}\mathrm{Pb}$. How many α and β^- particles are emitted in this series?

47. (II) The iodine isotope ${}^{131}_{53}\mathrm{I}$ is used in hospitals for diagnosis of thyroid function. If 782 μg are ingested by a patient, determine the activity (*a*) immediately, (*b*) 1.00 h later when the thyroid is being tested, and (*c*) 4.0 months later. Use Appendix F.

48. (II) ${}^{124}_{55}\mathrm{Cs}$ has a half-life of 30.8 s. (*a*) If we have 7.8 μg initially, how many Cs nuclei are present? (*b*) How many are present 2.6 min later? (*c*) What is the activity at this time? (*d*) After how much time will the activity drop to less than about 1 per second?

49. (II) Calculate the mass of a sample of pure ${}^{40}_{19}\mathrm{K}$ with an initial decay rate of $2.0 \times 10^5\,\mathrm{s}^{-1}$. The half-life of ${}^{40}_{19}\mathrm{K}$ is 1.265×10^9 yr.

50. (II) Calculate the activity of a pure 8.7-μg sample of ${}^{32}_{15}\mathrm{P}$ $(T_{\frac{1}{2}} = 1.23 \times 10^6\,\mathrm{s})$.

51. (II) The activity of a sample of ${}^{35}_{16}\mathrm{S}$ $(T_{\frac{1}{2}} = 87.32\,\text{days})$ is 3.65×10^4 decays per second. What is the mass of the sample?

52. (II) A sample of ${}^{233}_{92}\mathrm{U}$ $(T_{\frac{1}{2}} = 1.59 \times 10^5\,\mathrm{yr})$ contains 5.50×10^{18} nuclei. (*a*) What is the decay constant? (*b*) Approximately how many disintegrations will occur per minute?

53. (II) The activity of a sample drops by a factor of 4.0 in 8.6 minutes. What is its half-life?

54. (II) A 385-g sample of pure carbon contains 1.3 parts in 10^{12} (atoms) of ${}^{14}_{6}\mathrm{C}$. How many disintegrations occur per second?

55. (II) A sample of ${}^{238}_{92}\mathrm{U}$ is decaying at a rate of 3.70×10^2 decays/s. What is the mass of the sample?

56. (II) **Rubidium–strontium dating.** The rubidium isotope ${}^{87}_{37}\mathrm{Rb}$, a β emitter with a half-life of 4.75×10^{10} yr, is used to determine the age of rocks and fossils. Rocks containing fossils of ancient animals contain a ratio of ${}^{87}_{38}\mathrm{Sr}$ to ${}^{87}_{37}\mathrm{Rb}$ of 0.0260. Assuming that there was no ${}^{87}_{38}\mathrm{Sr}$ present when the rocks were formed, estimate the age of these fossils.

57. (II) The activity of a radioactive source decreases by 2.5% in 31.0 hours. What is the half-life of this source?

58. (II) ${}^{7}_{4}\mathrm{Be}$ decays with a half-life of about 53 d. It is produced in the upper atmosphere, and filters down onto the Earth's surface. If a plant leaf is detected to have 350 decays/s of ${}^{7}_{4}\mathrm{Be}$, (*a*) how long do we have to wait for the decay rate to drop to 15 per second? (*b*) Estimate the initial mass of ${}^{7}_{4}\mathrm{Be}$ on the leaf.

59. (II) Two of the naturally occurring radioactive decay sequences start with ${}^{232}_{90}\mathrm{Th}$ and with ${}^{235}_{92}\mathrm{U}$. The first five decays of these two sequences are:

$$\alpha, \beta, \beta, \alpha, \alpha$$

and

$$\alpha, \beta, \alpha, \beta, \alpha.$$

Determine the resulting intermediate daughter nuclei in each case.

60. (II) An ancient wooden club is found that contains 85 g of carbon and has an activity of 7.0 decays per second. Determine its age assuming that in living trees the ratio of ${}^{14}\mathrm{C}/{}^{12}\mathrm{C}$ atoms is about 1.3×10^{-12}.

61. (III) At $t = 0$, a pure sample of radioactive nuclei contains N_0 nuclei whose decay constant is λ. Determine a formula for the number of daughter nuclei, N_D, as a function of time; assume the daughter is stable and that $N_D = 0$ at $t = 0$.

General Problems

62. Which radioactive isotope of lead is being produced if the measured activity of a sample drops to 1.050% of its original activity in 4.00 h?

63. An old wooden tool is found to contain only 6.0% of the ${}^{14}_{6}C$ that an equal mass of fresh wood would. How old is the tool?

64. A neutron star consists of neutrons at approximately nuclear density. Estimate, for a 10-km-diameter neutron star, (*a*) its mass number, (*b*) its mass (kg), and (*c*) the acceleration of gravity at its surface.

65. **Tritium dating**. The ${}^{3}_{1}H$ isotope of hydrogen, which is called *tritium* (because it contains three nucleons), has a half-life of 12.3 yr. It can be used to measure the age of objects up to about 100 yr. It is produced in the upper atmosphere by cosmic rays and brought to Earth by rain. As an application, determine approximately the age of a bottle of wine whose ${}^{3}_{1}H$ radiation is about $\frac{1}{10}$ that present in new wine.

66. Some elementary particle theories (Section 43–11) suggest that the proton may be unstable, with a half-life $\geq 10^{33}$ yr. How long would you expect to wait for one proton in your body to decay (approximate your body as all water)?

67. Show, using the decays given in Section 41–5, that the neutrino has either spin $\frac{1}{2}$ or $\frac{3}{2}$.

68. The original experiments which established that an atom has a heavy, positive nucleus were done by shooting alpha particles through gold foil. The alpha particles used had a kinetic energy of 7.7 MeV. What is the closest they could get to a gold nucleus? How does this compare with the size of the nucleus?

69. How long must you wait (in half-lives) for a radioactive sample to drop to 1.00% of its original activity?

70. If the potassium isotope ${}^{40}_{19}K$ gives 45 decays/s in a liter of milk, estimate how much ${}^{40}_{19}K$ and regular ${}^{39}_{19}K$ are in a liter of milk. Use Appendix F.

71. (*a*) In α decay of, say, a ${}^{226}_{88}Ra$ nucleus, show that the nucleus carries away a fraction $1/(1 + \frac{1}{4}A_D)$ of the total energy available, where A_D is the mass number of the daughter nucleus. [*Hint*: Use conservation of momentum as well as conservation of energy.] (*b*) Approximately what percentage of the energy available is thus carried off by the α particle when ${}^{226}_{88}Ra$ decays?

72. Strontium-90 is produced as a nuclear fission product of uranium in both reactors and atomic bombs. Look at its location in the Periodic Table to see what other elements it might be similar to chemically, and tell why you think it might be dangerous to ingest. It has too many neutrons, and it decays with a half-life of about 29 yr. How long will we have to wait for the amount of ${}^{90}_{38}Sr$ on the Earth's surface to reach 1% of its current level, assuming no new material is scattered about? Write down the decay reaction, including the daughter nucleus. The daughter is radioactive: write down its decay.

73. Using the uncertainty principle and the radius of a nucleus, estimate the minimum possible kinetic energy of a nucleon in, say, iron. Ignore relativistic corrections. [*Hint*: A particle can have a momentum at least as large as its momentum uncertainty.]

74. (*a*) Calculate the kinetic energy of the α particle emitted when ${}^{236}_{92}U$ decays. (*b*) Use Eq. 41–1 to estimate the radius of an α particle and a ${}^{232}_{90}Th$ nucleus. Use this to estimate (*c*) the maximum height of the Coulomb barrier, and (*d*) its width AB in Fig. 41–7.

75. The nuclide ${}^{191}_{76}Os$ decays with β^- energy of 0.14 MeV accompanied by γ rays of energy 0.042 MeV and 0.129 MeV. (*a*) What is the daughter nucleus? (*b*) Draw an energy-level diagram showing the ground states of the parent and daughter and excited states of the daughter. (*c*) To which of the daughter states does β^- decay of ${}^{191}_{76}Os$ occur?

76. Determine the activities of (*a*) 1.0 g of ${}^{131}_{53}I$ ($T_{\frac{1}{2}} = 8.02$ days) and (*b*) 1.0 g of ${}^{238}_{92}U$ ($T_{\frac{1}{2}} = 4.47 \times 10^9$ yr).

77. Use Fig. 41–1 to estimate the total binding energy for copper and then estimate the energy, in joules, needed to break a 3.0-g copper penny into its constituent nucleons.

78. Instead of giving atomic masses for nuclides as in Appendix F, some Tables give the **mass excess**, Δ, defined as $\Delta = M - A$, where A is the atomic mass number and M is the mass in u. Determine the mass excess, in u and in MeV/c^2, for: (*a*) ${}^{4}_{2}He$; (*b*) ${}^{12}_{6}C$; (*c*) ${}^{86}_{38}Sr$; (*d*) ${}^{235}_{92}U$. (*e*) From a glance at Appendix F, can you make a generalization about the sign of Δ as a function of Z or A?

79. When water is placed near an intense neutron source, the neutrons can be slowed down by collisions with the water molecules and eventually captured by a hydrogen nucleus to form the stable isotope called deuterium, ${}^{2}_{1}H$, giving off a gamma ray. What is the energy of the gamma ray?

80. (*a*) Show that the **mean life** of a radioactive nuclide, defined as

$$\tau = \frac{\int_0^\infty t\,N(t)\,dt}{\int_0^\infty N(t)\,dt},$$

is $\tau = 1/\lambda$. (*b*) What fraction of the original number of nuclei remains after one mean life?

81. (*a*) A 72-gram sample of natural carbon contains the usual fraction of ${}^{14}_{6}C$. Estimate how long it will take before there is only one ${}^{14}_{6}C$ nucleus left. (*b*) How does the answer in (*a*) change if the sample is 270 grams? What does this tell you about the limits of carbon dating?

82. If the mass of the proton were just a little closer to the mass of the neutron, the following reaction would be possible even at low collision energies:

$$e^- + p \rightarrow n + \nu.$$

(*a*) Why would this situation be catastrophic? (*b*) By what percentage would the proton's mass have to be increased to make this reaction possible?

83. What is the ratio of the kinetic energies for an alpha particle and a beta particle if both make tracks with the same radius of curvature in a magnetic field, oriented perpendicular to the paths of the particles?

84. A 1.00-g sample of natural samarium emits α particles at a rate of $120\,\mathrm{s}^{-1}$ due to the presence of $^{147}_{62}\mathrm{Sm}$. The natural abundance of $^{147}_{62}\mathrm{Sm}$ is 15%. Calculate the half-life for this decay process.

85. Almost all of naturally occurring uranium is $^{238}_{92}\mathrm{U}$ with a half-life of 4.468×10^9 yr. Most of the rest of natural uranium is $^{235}_{92}\mathrm{U}$ with a half-life of 7.04×10^8 yr. Today a sample contains 0.720% $^{235}_{92}\mathrm{U}$. (*a*) What was this percentage 1.0 billion years ago? (*b*) What percentage of $^{235}_{92}\mathrm{U}$ will remain 100 million years from now?

86. A typical banana contains 400 mg of potassium, of which a small fraction is the radioactive isotope $^{40}_{19}\mathrm{K}$ (see Appendix F). Estimate the activity of an average banana due to $^{40}_{19}\mathrm{K}$.

87. Some radioactive isotopes have half-lives that are larger than the age of the universe (like gadolinium or samarium). The only way to determine these half-lives is to monitor the decay rate of a sample that contains these isotopes. For example, suppose we find an asteroid that currently contains about 15,000 kg of $^{152}_{64}\mathrm{Gd}$ (gadolinium) and we detect an activity of 1 decay/s. What is the half-life of gadolinium (in years)?

88. Decay series, such as that shown in Fig. 41–12, can be classified into four families, depending on whether the mass numbers have the form $4n$, $4n + 1$, $4n + 2$, or $4n + 3$, where n is an integer. Justify this statement and show that for a nuclide in any family, all its daughters will be in the same family.

* Numerical/Computer

*89. (I) A laboratory has a 1.80-μg sample of radioactive $^{13}_{7}\mathrm{N}$ whose decay constant $\lambda = 1.16 \times 10^{-3}\,\mathrm{s}^{-1}$. Calculate the initial number of nuclei, N_0, present in the sample. Use the radioactive decay law, $N = N_0 e^{-\lambda t}$, to determine the number of nuclei N present at time t for $t = 0$ to 30 minutes (1800 s) in steps of 0.5 min (30 s). Make a graph of N versus t and from the graph determine the half-life of the sample.

*90. (II) Construct a spreadsheet (or other numerical tool) that will reproduce Fig. 41–1, the graph of binding energy per nucleon (in MeV) vs. the mass number A. Using Appendix F, calculate the binding energy per nucleon for the most stable isotope of each possible mass number $A \geq 2$. [The first few values will be for $^2_1\mathrm{H}$, $^3_2\mathrm{He}$ (it is more stable than $^3_1\mathrm{H}$), $^4_2\mathrm{He}$, $^6_3\mathrm{Li}$, and $^7_3\mathrm{Li}$ (since it is more stable than $^7_4\mathrm{Be}$).] To reduce the amount of data, for $A \geq 20$ plot only points for even values of A, and plot to a maximum of $A = 142$.

Answers to Exercises

A: 0.042130 u.

B: 7.98 MeV/nucleon.

C: (*b*).

D: (*c*).

E: (*a*).

F: $1.37 \times 10^{-11}\,\mathrm{s}^{-1}$.

G: (*a*).

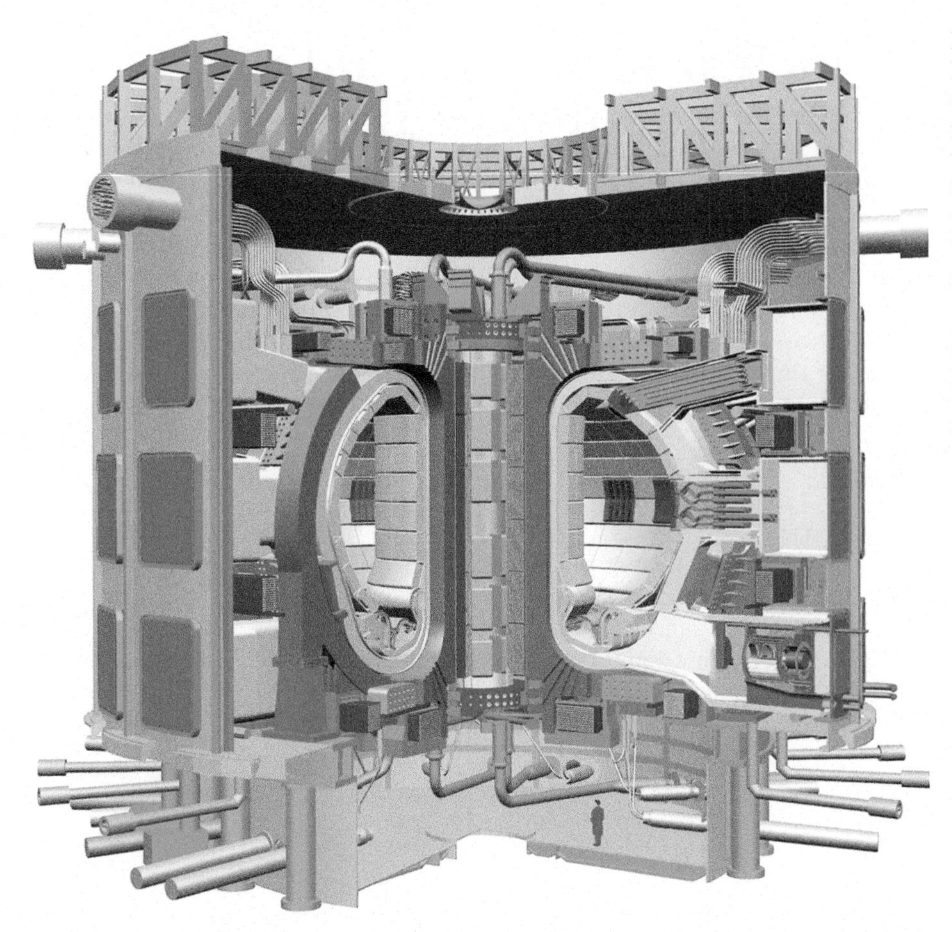

Diagram of ITER (International Thermonuclear Experimental Reactor), which will hopefully begin operation in the 2020s. Inside its cavity, over 12 m in diameter, a plasma of electrons and light nuclei will be heated to high temperatures that rival the Sun. Confining a plasma by magnetic fields has proved difficult, and intense research is needed if the fusion of smaller nuclei is to fulfill its promise as a source of abundant and relatively clean power.

This Chapter covers the basic physics topics of nuclear reactions, nuclear fission, nuclear fusion, and how we obtain nuclear energy. We also examine the health aspects of radiation dosimetry, therapy, and imaging by CAT, PET, SPET, and MRI.

Nuclear Energy; Effects and Uses of Radiation

CHAPTER-OPENING QUESTIONS—**Guess now!**

1. The Sun is powered by
   (a) nuclear alpha decay.
   (b) nuclear beta decay.
   (c) nuclear gamma decay.
   (d) nuclear fission.
   (e) nuclear fusion.

2. Which radiation induces the most biological damage for a given amount of energy deposited in tissue?
   (a) Alpha particles.
   (b) Gamma radiation.
   (c) Beta radiation.
   (d) They all do the same damage for the same deposited energy.
   (e) It depends on the type of tissue.

We continue our study of nuclear physics in this Chapter. We begin with a discussion of nuclear reactions, and then we examine the important large energy-releasing processes of fission and fusion. This Chapter also deals with the effects of nuclear radiation passing through matter, particularly biological matter, and how radiation is used medically for therapy, diagnosis, and imaging techniques.

CONTENTS

42–1 Nuclear Reactions and the Transmutation of Elements

When a nucleus undergoes α or β decay, the daughter nucleus is a different element from the parent. The transformation of one element into another, called **transmutation**, also occurs by means of nuclear reactions. A **nuclear reaction** is said to occur when a given nucleus is struck by another nucleus, or by a simpler particle such as a γ ray or neutron, and an interaction takes place. Ernest Rutherford was the first to report seeing a nuclear reaction. In 1919 he observed that some of the α particles passing through nitrogen gas were absorbed and protons emitted. He concluded that nitrogen nuclei had been transformed into oxygen nuclei via the reaction

$${}^{4}_{2}\mathrm{He} + {}^{14}_{7}\mathrm{N} \rightarrow {}^{17}_{8}\mathrm{O} + {}^{1}_{1}\mathrm{H},$$

where ${}^{4}_{2}\mathrm{He}$ is an α particle, and ${}^{1}_{1}\mathrm{H}$ is a proton.

Since then, a great many nuclear reactions have been observed. Indeed, many of the radioactive isotopes used in the laboratory are made by means of nuclear reactions. Nuclear reactions can be made to occur in the laboratory, but they also occur regularly in nature. In Chapter 41 we saw an example: ${}^{14}_{6}\mathrm{C}$ is continually being made in the atmosphere via the reaction $\mathrm{n} + {}^{14}_{7}\mathrm{N} \rightarrow {}^{14}_{6}\mathrm{C} + \mathrm{p}$.

Nuclear reactions are sometimes written in a shortened form: for example, the reaction

$$\mathrm{n} + {}^{14}_{7}\mathrm{N} \rightarrow {}^{14}_{6}\mathrm{C} + \mathrm{p}$$

can be written

$${}^{14}_{7}\mathrm{N}\ (\mathrm{n}, \mathrm{p})\ {}^{14}_{6}\mathrm{C}.$$

The symbols outside the parentheses on the left and right represent the initial and final nuclei, respectively. The symbols inside the parentheses represent the bombarding particle (first) and the emitted small particle (second).

In any nuclear reaction, both electric charge and nucleon number are conserved. These conservation laws are often useful, as the following Example shows.

CONCEPTUAL EXAMPLE 42–1 **Deuterium reaction.** A neutron is observed to strike an ${}^{16}_{8}\mathrm{O}$ nucleus, and a deuteron is given off. (A **deuteron**, or **deuterium**, is the isotope of hydrogen containing one proton and one neutron, ${}^{2}_{1}\mathrm{H}$; it is sometimes given the symbol d or D.) What is the nucleus that results?

RESPONSE We have the reaction $\mathrm{n} + {}^{16}_{8}\mathrm{O} \rightarrow \, ? + {}^{2}_{1}\mathrm{H}$. The total number of nucleons initially is $1 + 16 = 17$, and the total charge is $0 + 8 = 8$. The same totals apply after the reaction. Hence the product nucleus must have $Z = 7$ and $A = 15$. From the Periodic Table, we find that it is nitrogen that has $Z = 7$, so the nucleus produced is ${}^{15}_{7}\mathrm{N}$.

EXERCISE A Determine the resulting nucleus in the reaction $\mathrm{n} + {}^{137}_{56}\mathrm{Ba} \rightarrow \, ? + \gamma$.

Energy and momentum are also conserved in nuclear reactions, and can be used to determine whether or not a given reaction can occur. For example, if the total mass of the final products is less than the total mass of the initial particles, this decrease in mass (recall $\Delta E = \Delta m\, c^2$) is converted to kinetic energy (K) of the outgoing particles. But if the total mass of the products is greater than the total mass of the initial reactants, the reaction requires energy. The reaction will then not occur unless the bombarding particle has sufficient kinetic energy. Consider a nuclear reaction of the general form

$$\mathrm{a} + \mathrm{X} \rightarrow \mathrm{Y} + \mathrm{b}, \qquad \textbf{(42–1)}$$

where particle "a" is a moving projectile (or small nucleus) that strikes nucleus X, producing nucleus Y and particle b (typically, p, n, α, γ). We define the **reaction energy**, or ***Q*-value**, in terms of the masses involved, as

$$Q = (M_a + M_X - M_b - M_Y)c^2. \qquad \textbf{(42–2a)}$$

For a γ ray, $M = 0$.

Because energy is conserved, Q has to be equal to the change in kinetic energy (final minus initial):

$$Q = K_b + K_Y - K_a - K_X. \qquad \textbf{(42–2b)}$$

If X is a target nucleus at rest (or nearly so) struck by incoming particle a, then $K_X = 0$. For $Q > 0$, the reaction is said to be *exothermic* or *exoergic*; energy is released in the reaction, so the total kinetic energy is greater after the reaction than before. If Q is negative ($Q < 0$), the reaction is said to be *endothermic* or *endoergic*: the final total kinetic energy is less than the initial kinetic energy, and an energy input is required to make the reaction happen. The energy input comes from the kinetic energy of the initial colliding particles (a and X).

EXAMPLE 42–2 **A slow-neutron reaction.** The nuclear reaction

$$n + {}^{10}_{5}B \rightarrow {}^{7}_{3}Li + {}^{4}_{2}He$$

is observed to occur even when very slow-moving neutrons (mass $M_n = 1.0087$ u) strike boron atoms at rest. For a particular reaction in which $K_n \approx 0$, the outgoing helium ($M_{He} = 4.0026$ u) is observed to have a speed of 9.30×10^6 m/s. Determine (*a*) the kinetic energy of the lithium ($M_{Li} = 7.0160$ u), and (*b*) the Q-value of the reaction.

APPROACH Since the neutron and boron are both essentially at rest, the total momentum before the reaction is zero; momentum is conserved and so must be zero afterward as well. Thus,

$$M_{Li} v_{Li} = M_{He} v_{He}.$$

We solve this for v_{Li} and substitute it into the equation for kinetic energy. In (*b*) we use Eq. 42–2b.

SOLUTION (*a*) We can use classical kinetic energy with little error, rather than relativistic formulas, because $v_{He} = 9.30 \times 10^6$ m/s is not close to the speed of light c, and v_{Li} will be even less since $M_{Li} > M_{He}$. Thus we can write:

$$K_{Li} = \frac{1}{2} M_{Li} v_{Li}^2 = \frac{1}{2} M_{Li}\left(\frac{M_{He} v_{He}}{M_{Li}}\right)^2 = \frac{M_{He}^2 v_{He}^2}{2M_{Li}}.$$

We put in numbers, changing the mass in u to kg and recall that 1.60×10^{-13} J = 1 MeV:

$$K_{Li} = \frac{(4.0026\,\text{u})^2(1.66 \times 10^{-27}\,\text{kg/u})^2(9.30 \times 10^6\,\text{m/s})^2}{2(7.0160\,\text{u})(1.66 \times 10^{-27}\,\text{kg/u})} = 1.64 \times 10^{-13}\,\text{J} = 1.02\,\text{MeV}.$$

(*b*) We are given the data $K_a = K_X = 0$ in Eq. 42–2b, so $Q = K_{Li} + K_{He}$, where

$$K_{He} = \tfrac{1}{2} M_{He} v_{He}^2 = \tfrac{1}{2}(4.0026\,\text{u})(1.66 \times 10^{-27}\,\text{kg/u})(9.30 \times 10^6\,\text{m/s})^2 = 2.87 \times 10^{-13}\,\text{J} = 1.80\,\text{MeV}.$$

Hence, $Q = 1.02\,\text{MeV} + 1.80\,\text{MeV} = 2.82\,\text{MeV}$.

EXAMPLE 42–3 **Will the reaction "go"?** Can the reaction

$$p + {}^{13}_{6}C \rightarrow {}^{13}_{7}N + n$$

occur when ${}^{13}_{6}C$ is bombarded by 2.0-MeV protons?

APPROACH The reaction will "go" if the reaction is exothermic ($Q > 0$) and even if $Q < 0$ if the input momentum and kinetic energy are sufficient. First we calculate Q from the difference between final and initial masses using Eq. 42–2a, and look up the masses in Appendix F.

SOLUTION The total masses before and after the reaction are:

Before		After	
$M({}^{13}_{6}C) =$	13.003355	$M({}^{13}_{7}N) =$	13.005739
$M({}^{1}_{1}H) =$	1.007825	$M(n) =$	1.008665.
	14.011180		14.014404

(We must use the mass of the ${}^{1}_{1}H$ atom rather than that of the bare proton because the masses of ${}^{13}_{6}C$ and ${}^{13}_{7}N$ include the electrons, and we must include an equal number of electron masses on each side of the equation since none are created or destroyed.) The products have an excess mass of

$$(14.014404 - 14.011180)\,u = 0.003224\,u \times 931.5\,\text{MeV}/u = 3.00\,\text{MeV}.$$

Thus $Q = -3.00\,\text{MeV}$, and the reaction is endothermic. This reaction requires energy, and the 2.0-MeV protons do not have enough to make it go.

NOTE The incoming proton in this Example would have to have somewhat more than 3.00 MeV of kinetic energy to make this reaction go; 3.00 MeV would be enough to conserve energy, but a proton of this energy would produce the ${}^{13}_{7}N$ and n with no kinetic energy and hence no momentum. Since an incident 3.0-MeV proton has momentum, conservation of momentum would be violated. A calculation using conservation of energy *and* of momentum, as we did in Examples 41–6 and 42–2, shows that the minimum proton energy, called the **threshold energy**, is 3.23 MeV in this case (= Problem 16).

Neutron captured by ${}^{238}_{92}U$.

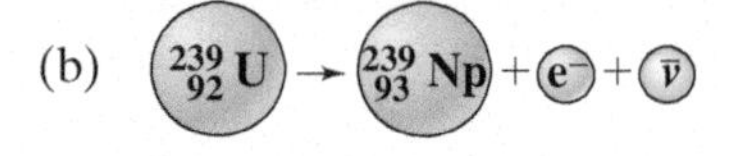

${}^{239}_{92}U$ decays by β decay to neptunium-239.

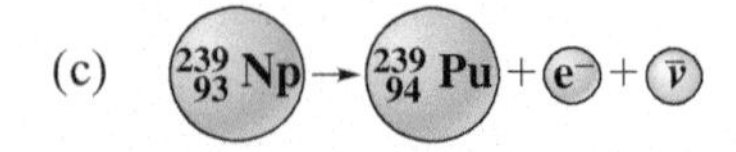

${}^{239}_{93}Np$ itself decays by β decay to produce plutonium-239.

FIGURE 42–1 Neptunium and plutonium are produced in this series of reactions, after bombardment of ${}^{238}_{92}U$ by neutrons.

Neutron Physics

The artificial transmutation of elements took a great leap forward in the 1930s when Enrico Fermi realized that neutrons would be the most effective projectiles for causing nuclear reactions and in particular for producing new elements. Because neutrons have no net electric charge, they are not repelled by positively charged nuclei as are protons or alpha particles. Hence the probability of a neutron reaching the nucleus and causing a reaction is much greater than for charged projectiles,† particularly at low energies. Between 1934 and 1936, Fermi and his co-workers in Rome produced many previously unknown isotopes by bombarding different elements with neutrons. Fermi realized that if the heaviest known element, uranium, is bombarded with neutrons, it might be possible to produce new elements with atomic numbers greater than that of uranium. After several years of hard work, it was suspected that two new elements had been produced, neptunium ($Z = 93$) and plutonium ($Z = 94$). The full confirmation that such "transuranic" elements could be produced came several years later at the University of California, Berkeley. The reactions are shown in Fig. 42–1.

It was soon shown that what Fermi had actually observed when he bombarded uranium was an even stranger process—one that was destined to play an extraordinary role in the world at large. We discuss it in Section 42–3.

†That is, positively charged particles. Electrons rarely cause nuclear reactions because they do not interact via the strong nuclear force.

42–2 Cross Section

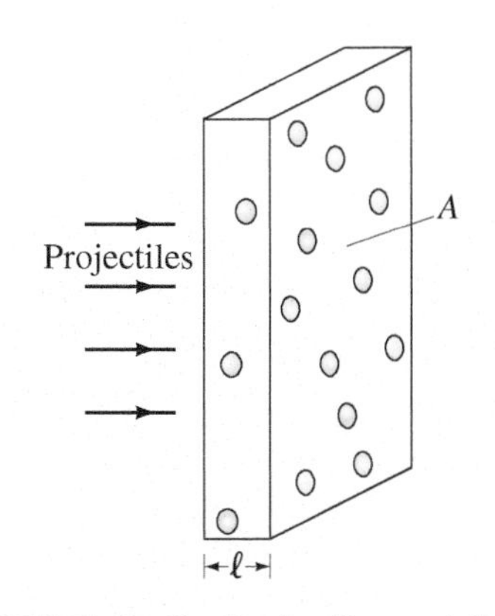

FIGURE 42–2 Projectile particles strike a target of area A and thickness ℓ made up of n nuclei per unit volume.

Some reactions have a higher probability of occurring than others. The reaction probability is specified by a quantity called the **cross section**. Although the size of a nucleus, like that of an atom, is not a clearly defined quantity since the edges are not distinct like those of a tennis ball or baseball, we can nonetheless define a *cross section* for nuclei undergoing collisions by using an analogy. Suppose that projectile particles strike a stationary target of total area A and thickness ℓ, as shown in Fig. 42–2. Assume also that the target is made up of identical objects (such as marbles or nuclei), each of which has a cross-sectional area σ, and we assume the incoming projectiles are small by comparison. We assume that the target objects are fairly far apart and the thickness ℓ is so small that we don't have to worry about overlapping. This is often a reasonable assumption because nuclei have diameters on the order of 10^{-14} m but are at least 10^{-10} m (atomic size) apart even in solids. If there are n nuclei per unit volume, the total cross-sectional area of all these tiny targets is

$$A' = nA\ell\sigma$$

since $nA\ell = (n)(\text{volume})$ is the total number of targets and σ is the cross-sectional area of each. If $A' \ll A$, most of the incident projectile particles will pass through the target without colliding. If R_0 is the rate at which the projectile particles strike the target (number/second), the rate at which collisions occur, R, is

$$R = R_0 \frac{A'}{A} = R_0 \frac{nA\ell\sigma}{A}$$

so

$$R = R_0 n\ell\sigma.$$

Thus, by measuring the collision rate, R, we can determine σ:

$$\sigma = \frac{R}{R_0 n\ell}. \qquad \textbf{(42–3)}$$

If nuclei were simple billiard balls, and R the number of particles that are deflected per second, σ would represent the real cross-sectional area of each ball. But nuclei are complicated objects that cannot be considered to have distinct boundaries. Furthermore, collisions can be either elastic or inelastic, and reactions can occur in which the nature of the particles can change. By measuring R for each possible process, we can determine an **effective cross section**, σ, for each process. None of these cross sections is necessarily related to a geometric cross-sectional area. Rather, σ is an "effective" target area. It is a *measure of the probability of a collision or of a particular reaction occurring* per target nucleus, independent of the dimensions of the entire target. The concept of cross section is useful because σ depends only on the properties of the interacting particles, whereas R depends on the thickness and area of the physical (macroscopic) target, on the number of particles in the incident beam, and so on.

When a given pair of particles interact, we define their **elastic cross section** σ_{el} using Eq. 42–3, where R for a given experimental setup is the rate of elastic collisions (or **elastic scattering**), by which we mean collisions for which the final particles are the same as the initial particles ($\text{a} = \text{b}$, $\text{X} = \text{Y}$ in Eq. 42–1) and $Q = 0$. Similarly, the inelastic cross section, σ_{inel}, is related to the rate of inelastic collisions, or **inelastic scattering**, which involves the same final and initial particles but $Q \neq 0$, usually because excited states are involved. For each reaction in which the final particles are different than the initial particles, there is a particular cross section. For protons (p) incident on ${}^{13}_{6}\text{C}$, for example, we could have various reactions, such as $\text{p} + {}^{13}_{6}\text{C} \rightarrow {}^{13}_{7}\text{N} + \text{n}$ or $\text{p} + {}^{13}_{6}\text{C} \rightarrow {}^{10}_{5}\text{B} + {}^{4}_{2}\text{He}$, and so on. The sum of all the separate reaction cross sections (for a given pair of initial particles) is called the **total reaction cross section**, σ_{R}. The **total cross section**, σ_{T}, is

$$\sigma_{\text{T}} = \sigma_{\text{el}} + \sigma_{\text{inel}} + \sigma_{\text{R}}$$

and is a measure of all possible interactions or collisions starting with given initial particles. Said another way, σ_{T} is a measure of how many of the incident particles interact in some way and hence are eliminated from the incident beam.

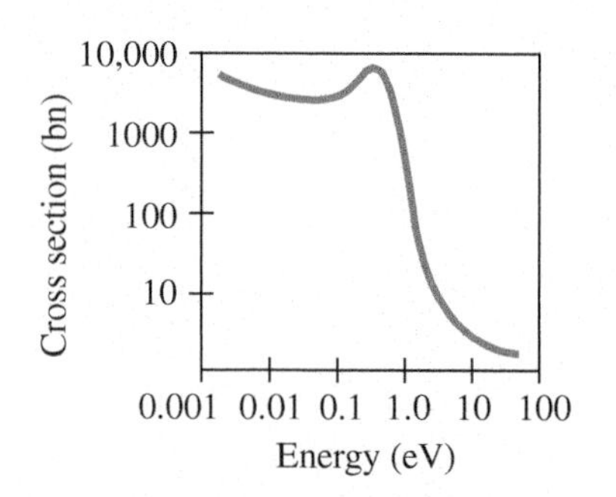

FIGURE 42–3 The neutron cross section for cadmium-114 as a function of incoming neutron kinetic energy. It is extraordinarily large for $K \lesssim 1\,\text{eV}$. Note that both scales are logarithmic.

We can also define **differential cross sections**, which represent the probability of the deflected (or emitted) particles leaving at particular angles.

It is said that when one of the first nuclear cross sections was measured, a physicist, surprised that it was as large as it was ($\approx 10^{-28}\,\text{m}^2$), remarked, "it's as big as a barn." Ever since then nuclear cross sections have been measured in "barns," where $1\text{ barn (bn)} = 10^{-28}\,\text{m}^2$.

The value of σ for a given reaction depends on, among other things, the incident kinetic energy. Typical nuclear cross sections are on the order of barns, but they can vary from millibarns to kilobarns or more. Figure 42–3 shows the cross section for neutron capture in cadmium ($\text{n} + {}^{114}_{48}\text{Cd} \rightarrow {}^{115}_{48}\text{Cd} + \gamma$) as a function of neutron kinetic energy. Neutron cross sections for most materials are greater at low energies, as in Fig. 42–3. To produce nuclear reactions at a high rate it is therefore desirable that the bombarding neutrons have low energy. Neutrons that have been slowed down and have reached equilibrium with matter at room temperature $\left(\frac{3}{2}kT \approx 0.04\,\text{eV at } T = 300\,\text{K}\right)$ are called **thermal neutrons**.

EXAMPLE 42–4 Using cross section. The reaction

$$\text{p} + {}^{56}_{26}\text{Fe} \rightarrow {}^{56}_{27}\text{Co} + \text{n}$$

has a cross section of 0.65 bn for a particular incident proton energy. Suppose the iron target has an area of $1.5\,\text{cm}^2$, and is $2.0\,\mu\text{m}$ thick. The density of iron is $7.8 \times 10^3\,\text{kg/m}^3$. If the protons are incident at a rate of 2.0×10^{13} particles/s, calculate the rate at which neutrons are produced.

APPROACH We use Eq. 42–3 in the form: $R = R_0 n \ell \sigma$.

SOLUTION We are given $R_0 = 2.0 \times 10^{13}\,\text{particles/s}$, $\ell = 2.0 \times 10^{-6}\,\text{m}$, and $\sigma = 0.65\,\text{bn}$. Recalling from Chapter 17 that one mole (mass = 56 g for iron) contains 6.02×10^{23} atoms, then the number of iron atoms per unit volume is

$$n = (6.02 \times 10^{23}\,\text{atoms/mole})\frac{(7.8 \times 10^3\,\text{kg/m}^3)}{(56 \times 10^{-3}\,\text{kg/mole})} = 8.4 \times 10^{28}\,\text{atoms/m}^3.$$

Then we find that the rate at which neutrons are produced is

$$\begin{aligned} R &= R_0 n \ell \sigma \\ &= (2.0 \times 10^{13}\,\text{particles/s})(8.4 \times 10^{28}\,\text{atoms/m}^3)(2.0 \times 10^{-6}\,\text{m})(0.65 \times 10^{-28}\,\text{m}^2) \\ &= 2.2 \times 10^8\,\text{particles/s}. \end{aligned}$$

42–3 Nuclear Fission; Nuclear Reactors

In 1938, the German scientists Otto Hahn and Fritz Strassmann made an amazing discovery. Following up on Fermi's work, they found that uranium bombarded by neutrons sometimes produced smaller nuclei that were roughly half the size of the original uranium nucleus. Lise Meitner and Otto Frisch quickly realized what had happened: the uranium nucleus, after absorbing a neutron, actually had split into two roughly equal pieces. This was startling, for until then the known nuclear reactions involved knocking out only a tiny fragment (for example, n, p, or α) from a nucleus.

Nuclear Fission and Chain Reactions

This new phenomenon was named **nuclear fission** because of its resemblance to biological fission (cell division). It occurs much more readily for ${}^{235}_{92}\text{U}$ than for the more common ${}^{238}_{92}\text{U}$. The process can be visualized by imagining the uranium nucleus to be like a liquid drop. According to this **liquid-drop model**, the neutron absorbed by the ${}^{235}_{92}\text{U}$ nucleus gives the nucleus extra internal energy (like heating a drop of water). This intermediate state, or **compound nucleus**, is ${}^{236}_{92}\text{U}$ (because of the absorbed neutron). The extra energy of this nucleus—it is in an excited state—appears as increased motion of the individual nucleons, which causes the nucleus to take on abnormal elongated shapes, Fig. 42–4. When the nucleus elongates (in this model) into the shape shown in Fig. 42–4c, the attraction of the two ends via the short-range nuclear force is greatly weakened by the increased separation distance, and the electric repulsive force becomes dominant, and the nucleus splits in two (Fig. 42–4d). The two resulting nuclei, X_1 and X_2, are called **fission fragments**, and in the process a number

FIGURE 42–4 Fission of a ${}^{235}_{92}\text{U}$ nucleus after capture of a neutron, according to the liquid-drop model.

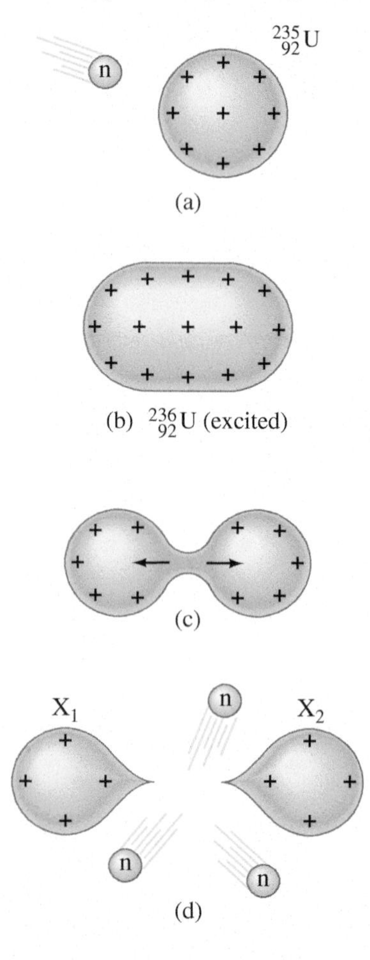

of neutrons (typically two or three) are also given off. The reaction can be written

$$\mathrm{n} + {}^{235}_{92}\mathrm{U} \rightarrow {}^{236}_{92}\mathrm{U} \rightarrow \mathrm{X}_1 + \mathrm{X}_2 + \text{neutrons}. \quad \textbf{(42–4)}$$

The compound nucleus, ${}^{236}_{92}\mathrm{U}$, exists for less than 10^{-12} s, so the process occurs very quickly. The two fission fragments, X_1 and X_2, more often split the original uranium mass as about 40%–60% rather than precisely half and half. A typical fission reaction is

$$\mathrm{n} + {}^{235}_{92}\mathrm{U} \rightarrow {}^{141}_{56}\mathrm{Ba} + {}^{92}_{36}\mathrm{Kr} + 3\mathrm{n}, \quad \textbf{(42–5)}$$

although many others also occur.

CONCEPTUAL EXAMPLE 42–5 **Counting nucleons.** Identify the element X in the fission reaction $\mathrm{n} + {}^{235}_{92}\mathrm{U} \rightarrow {}^{A}_{Z}\mathrm{X} + {}^{93}_{38}\mathrm{Sr} + 2\mathrm{n}$.

RESPONSE The number of nucleons is conserved (Section 41–7). The uranium nucleus with 235 nucleons plus the incoming neutron make $235 + 1 = 236$ nucleons. So there must be 236 nucleons after the reaction. The Sr has 93 nucleons, and the two neutrons make 95 nucleons, so X has $A = 236 - 95 = 141$. Electric charge is also conserved: before the reaction, the total charge is $92e$. After the reaction the total charge is $(Z + 38)e$ and must equal $92e$. Thus $Z = 92 - 38 = 54$. The element with $Z = 54$ is xenon (see Appendix F or the Periodic Table inside the back cover), so the isotope is ${}^{141}_{54}\mathrm{Xe}$.

Figure 42–5 shows the distribution of ${}^{235}_{92}\mathrm{U}$ fission fragments according to mass. Only rarely (about 1 in 10^4) does a fission result in equal mass fragments (arrow in Fig. 42–5).

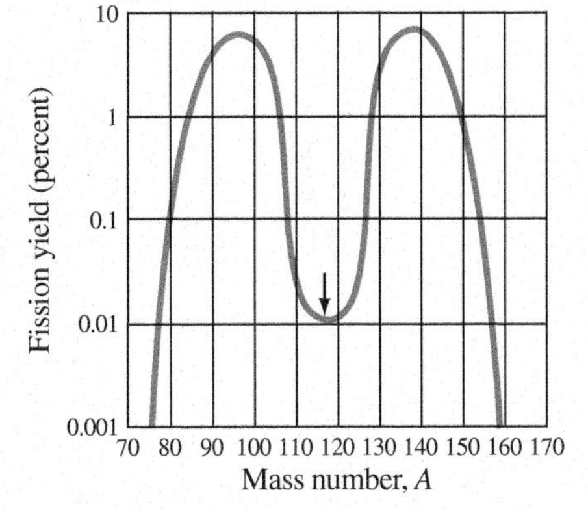

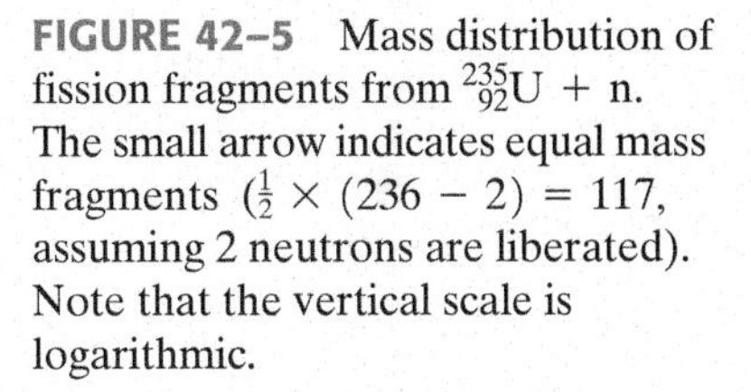
FIGURE 42–5 Mass distribution of fission fragments from ${}^{235}_{92}\mathrm{U}$ + n. The small arrow indicates equal mass fragments ($\frac{1}{2} \times (236 - 2) = 117$, assuming 2 neutrons are liberated). Note that the vertical scale is logarithmic.

A tremendous amount of energy is released in a fission reaction because the mass of ${}^{235}_{92}\mathrm{U}$ is considerably greater than the total mass of the fission fragments plus released neutrons. This can be seen from the binding-energy-per-nucleon curve of Fig. 41–1; the binding energy per nucleon for uranium is about 7.6 MeV/nucleon, but for fission fragments that have intermediate mass (in the center portion of the graph, $A \approx 100$), the average binding energy per nucleon is about 8.5 MeV/nucleon. Since the fission fragments are more tightly bound, the sum of their masses is less than the mass of the uranium. The difference in mass, or energy, between the original uranium nucleus and the fission fragments is about $8.5 - 7.6 = 0.9\,\mathrm{MeV}$ per nucleon. Since there are 236 nucleons involved in each fission, the total energy released per fission is

$$(0.9\,\mathrm{MeV/nucleon})(236\,\text{nucleons}) \approx 200\,\mathrm{MeV}. \quad \textbf{(42–6)}$$

This is an enormous amount of energy for one single nuclear event. At a practical level, the energy from one fission is tiny. But if many such fissions could occur in a short time, an enormous amount of energy at the macroscopic level would be available. A number of physicists, including Fermi, recognized that the neutrons released in each fission (Eqs. 42–4 and 42–5) could be used to create a **chain reaction**. That is, one neutron initially causes one fission of a uranium nucleus; the two or three neutrons released can go on to cause additional fissions, so the process multiplies as shown schematically in Fig. 42–6.

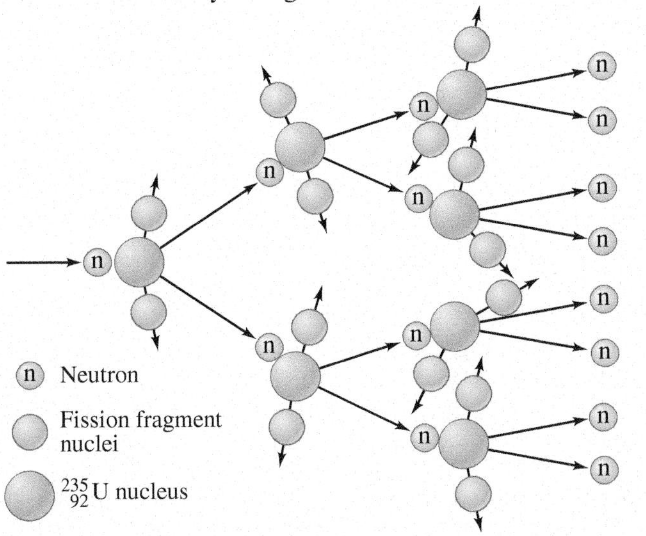

FIGURE 42–6 Chain reaction.

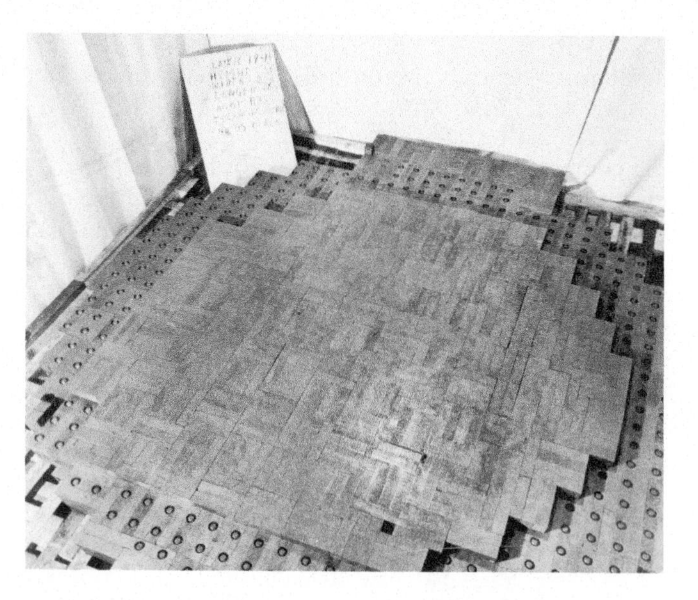

FIGURE 42–7 This is the only photograph of the first nuclear reactor, built by Fermi under the grandstand of Stagg Field at the University of Chicago. It is shown here under construction as a layer of graphite (used as moderator) was being placed over a layer of natural uranium. On December 2, 1942, Fermi slowly withdrew the cadmium control rods and the reactor went critical. This first self-sustaining chain reaction was announced to Washington, via telephone, by Arthur Compton who witnessed the event and reported: "The Italian navigator has just landed in the new world."

If a **self-sustaining chain reaction** was actually possible in practice, the enormous energy available in fission could be released on a larger scale. Fermi and his co-workers (at the University of Chicago) showed it was possible by constructing the first **nuclear reactor** in 1942 (Fig. 42–7).

Nuclear Reactors

Several problems have to be overcome to make any nuclear reactor function. First, the probability that a $^{235}_{92}U$ nucleus will absorb a neutron is large only for slow neutrons, but the neutrons emitted during a fission (which are needed to sustain a chain reaction) are moving very fast. A substance known as a **moderator** must be used to slow down the neutrons. The most effective moderator will consist of atoms whose mass is as close as possible to that of the neutrons. (To see why this is true, recall from Chapter 9 that a billiard ball striking an equal mass ball at rest can itself be stopped in one collision; but a billiard ball striking a heavy object bounces off with nearly unchanged speed.) The best moderator would thus contain $^{1}_{1}H$ atoms. Unfortunately, $^{1}_{1}H$ tends to absorb neutrons. But the isotope of hydrogen called *deuterium*, $^{2}_{1}H$, does not absorb many neutrons and is thus an almost ideal moderator. Either $^{1}_{1}H$ or $^{2}_{1}H$ can be used in the form of water. In the latter case, it is **heavy water**, in which the hydrogen atoms have been replaced by deuterium. Another common moderator is *graphite*, which consists of $^{12}_{6}C$ atoms.

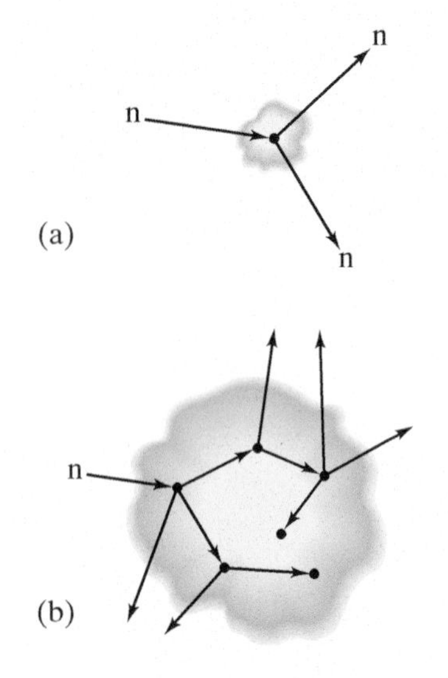

FIGURE 42–8 If the amount of uranium exceeds the critical mass, as in (b), a sustained chain reaction is possible. If the mass is less than critical, as in (a), too many neutrons escape before additional fissions occur, and the chain reaction is not sustained.

A second problem is that the neutrons produced in one fission may be absorbed and produce other nuclear reactions with other nuclei in the reactor, rather than produce further fissions. In a "light-water" reactor, the $^{1}_{1}H$ nuclei absorb neutrons, as does $^{238}_{92}U$ to form $^{239}_{92}U$ in the reaction $n + ^{238}_{92}U \rightarrow ^{239}_{92}U + \gamma$. Naturally occurring uranium† contains 99.3% $^{238}_{92}U$ and only 0.7% fissionable $^{235}_{92}U$. To increase the probability of fission of $^{235}_{92}U$ nuclei, natural uranium can be **enriched** to increase the percentage of $^{235}_{92}U$ by using processes such as diffusion or centrifugation. Enrichment is not usually necessary for reactors using heavy water as moderator since heavy water doesn't absorb neutrons.

The third problem is that some neutrons will escape through the surface of the reactor core before they can cause further fissions (Fig. 42–8). Thus the mass of fuel must be sufficiently large for a self-sustaining chain reaction to take place. The minimum mass of uranium needed is called the **critical mass**. The value of the critical mass depends on the moderator, the fuel ($^{239}_{94}Pu$ may be used instead of $^{235}_{92}U$), and how much the fuel is enriched, if at all. Typical values are on the order of a few kilograms (that is, neither grams nor thousands of kilograms).

To have a self-sustaining chain reaction, on average at least one neutron produced in each fission must go on to produce another fission. The average number of neutrons per fission that do go on to produce further fissions is called the **multiplication factor**, f. For a self-sustaining chain reaction, we must have $f \geq 1$.

† $^{238}_{92}U$ will fission, but only with fast neutrons ($^{238}_{92}U$ is more stable than $^{235}_{92}U$). The probability of absorbing a fast neutron and producing a fission is too low to produce a self-sustaining chain reaction.

If $f < 1$, the reactor is "subcritical." If $f > 1$, it is "supercritical" (and could become dangerously explosive). Reactors are equipped with movable **control rods** (good neutron absorbers like cadmium or boron), whose function is to absorb neutrons and maintain the reactor at just barely "critical," $f = 1$. The release of neutrons and subsequent fissions occur so quickly that manipulation of the control rods to maintain $f = 1$ would not be possible if it weren't for the small percentage ($\approx 1\%$) of so-called **delayed neutrons**. They come from the decay of neutron-rich fission fragments (or their daughters) having lifetimes on the order of seconds—sufficient to allow enough reaction time to operate the control rods and maintain $f = 1$.

Nuclear reactors have been built for use in research and to produce electric power. Fission produces many neutrons and a "research reactor" is basically an intense source of neutrons. These neutrons can be used as projectiles in nuclear reactions to produce nuclides not found in nature, including isotopes used as tracers and for therapy. A "power reactor" is used to produce electric power. The energy released in the fission process appears as heat, which is used to boil water and produce steam to drive a turbine connected to an electric generator (Fig. 42–9). The **core** of a nuclear reactor consists of the fuel and a moderator (water in most U.S. commercial reactors). The fuel is usually uranium enriched so that it contains 2 to 4 percent $^{235}_{92}U$. Water at high pressure or other liquid (such as liquid sodium) is allowed to flow through the core. The thermal energy it absorbs is used to produce steam in the heat exchanger, so the fissionable fuel acts as the heat input for a heat engine (Chapter 20).

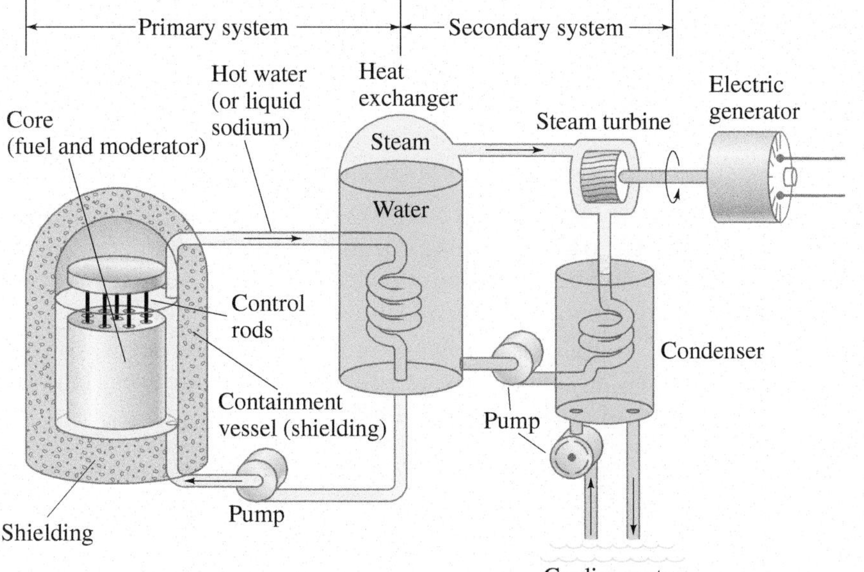

FIGURE 42–9 A nuclear reactor. The heat generated by the fission process in the fuel rods is carried off by hot water or liquid sodium and is used to boil water to steam in the heat exchanger. The steam drives a turbine to generate electricity and is then cooled in the condenser.

There are problems associated with nuclear power plants. Besides the usual thermal pollution associated with any heat engine (Section 20–11), there is the serious problem of disposal of the radioactive fission fragments produced in the reactor, plus radioactive nuclides produced by neutrons interacting with the structural parts of the reactor. Fission fragments, like their uranium or plutonium parents, have about 50% more neutrons than protons. Nuclei with atomic number in the typical range for fission fragments ($Z \approx 30$ to 60) are stable only if they have more nearly equal numbers of protons and neutrons (see Fig. 41–2). Hence the highly neutron-rich fission fragments are very unstable and decay radioactively. The accidental release of highly radioactive fission fragments into the atmosphere poses a serious threat to human health (Section 42–5), as does possible leakage of the radioactive wastes when they are disposed of. The accidents at Three Mile Island, Pennsylvania (1979), and at Chernobyl, Russia (1986), have illustrated some of the dangers and have shown that nuclear plants must be constructed, maintained, and operated with great care and precision (Fig. 42–10).

FIGURE 42–10 Devastation around Chernobyl in Russia, after the nuclear power plant meltdown in 1986.

Finally, the lifetime of nuclear power plants is limited to 30-some years, due to buildup of radioactivity and the fact that the structural materials themselves are weakened by the intense conditions inside. "Decommissioning" of a power plant could take a number of forms, but the cost of any method of decommissioning a large plant is very great.

So-called **breeder reactors** were proposed as a solution to the problem of limited supplies of fissionable uranium, ${}^{235}_{92}U$. A breeder reactor is one in which some of the neutrons produced in the fission of ${}^{235}_{92}U$ are absorbed by ${}^{238}_{92}U$, and ${}^{239}_{94}Pu$ is produced via the set of reactions shown in Fig. 42–1. ${}^{239}_{94}Pu$ is fissionable with slow neutrons, so after separation it can be used as a fuel in a nuclear reactor. Thus a breeder reactor "breeds" new fuel $({}^{239}_{94}Pu)$ from otherwise useless ${}^{238}_{92}U$. Since natural uranium is 99.3% ${}^{238}_{92}U$, this means that the supply of fissionable fuel could be increased by more than a factor of 100. But breeder reactors have the same problems as other reactors, plus other serious problems. Not only is plutonium considered to be a serious health hazard in itself (radioactive with a half-life of 24,000 years), but plutonium produced in a reactor can readily be used in a bomb, increasing the danger of nuclear proliferation and theft of fuel by terrorists to produce a bomb.

Nuclear power presents risks. Other large-scale energy-conversion methods, such as conventional oil and coal-burning steam plants, also present health and environmental hazards; some of them were discussed in Section 20–11, and include air pollution, oil spills, and the release of CO_2 gas which can trap heat as in a greenhouse to raise the Earth's temperature. The solution to the world's needs for energy is not only technological, but also economic and political. A major factor surely is to "conserve"—to minimize our use of energy. "Reduce, reuse, recycle."

EXAMPLE 42–6 **Uranium fuel amount.** Estimate the minimum amount of ${}^{235}_{92}U$ that needs to undergo fission in order to run a 1000-MW power reactor per year of continuous operation. Assume an efficiency (Chapter 20) of about 33%.

APPROACH At 33% efficiency, we need $3 \times 1000\,\text{MW} = 3000 \times 10^6\,\text{J/s}$ input. Each fission releases about 200 MeV (Eq. 42–6), so we divide the energy for a year by 200 MeV to get the number of fissions needed per year. Then we multiply by the mass of one uranium atom.

SOLUTION For 1000 MW output, the total power generation needs to be 3000 MW, of which 2000 MW is dumped as "waste" heat. Thus the total energy release in 1 yr $(3 \times 10^7\,\text{s})$ from fission needs to be about

$$(3 \times 10^9\,\text{J/s})(3 \times 10^7\,\text{s}) \approx 10^{17}\,\text{J}.$$

If each fission releases 200 MeV of energy, the number of fissions required for a year is

$$\frac{(10^{17}\,\text{J})}{(2 \times 10^8\,\text{eV/fission})(1.6 \times 10^{-19}\,\text{J/eV})} \approx 3 \times 10^{27}\,\text{fissions.}$$

The mass of a single uranium atom is about $(235\,\text{u})(1.66 \times 10^{-27}\,\text{kg/u}) \approx 4 \times 10^{-25}\,\text{kg}$, so the total uranium mass needed is

$$(4 \times 10^{-25}\,\text{kg/fission})(3 \times 10^{27}\,\text{fissions}) \approx 1000\,\text{kg},$$

or about a ton of ${}^{235}_{92}U$.

PHYSICS APPLIED
Energy in coal vs. uranium

NOTE Since ${}^{235}_{92}U$ makes up only 0.7% of natural uranium, the yearly requirement for uranium is on the order of a hundred tons. This is orders of magnitude less than coal, both in mass and volume. Coal releases $2.8 \times 10^7\,\text{J/kg}$, whereas ${}^{235}_{92}U$ can release $10^{17}\,\text{J}/10^3\,\text{kg} = 10^{14}\,\text{J/kg}$. For natural uranium, the figure is 100 times less, $10^{12}\,\text{J/kg}$.

EXERCISE B A nuclear-powered submarine needs 6000-kW input power. How many ${}^{235}_{92}\mathrm{U}$ fissions is this per second?

Atom Bomb

The first use of fission, however, was not to produce electric power. Instead, it was first used as a fission bomb (called the "atomic bomb"). In early 1940, with Europe already at war, Germany's leader, Adolf Hitler, banned the sale of uranium from the Czech mines he had recently taken over. Research into the fission process suddenly was enshrouded in secrecy. Physicists in the United States were alarmed. A group of them approached Einstein—a man whose name was a household word—to send a letter to President Franklin Roosevelt about the possibilities of using nuclear fission for a bomb far more powerful than any previously known, and inform him that Germany might already have begun development of such a bomb. Roosevelt responded by authorizing the program known as the Manhattan Project, to see if a bomb could be built. Work began in earnest after Fermi's demonstration in 1942 that a sustained chain reaction was possible. A new secret laboratory was developed on an isolated mesa in New Mexico known as Los Alamos. Under the direction of J. Robert Oppenheimer (1904–1967; Fig. 42–11), it became the home of famous scientists from all over Europe and the United States.

FIGURE 42–11 J. Robert Oppenheimer, on the left, with General Leslie Groves, who was the administrative head of Los Alamos during World War II. The photograph was taken at the Trinity site in the New Mexico desert, where the first atomic bomb was exploded.

To build a bomb that was subcritical during transport but that could be made supercritical (to produce a chain reaction) at just the right moment, two pieces of uranium were used, each less than the critical mass but together greater than the critical mass. The two masses, kept separate until the moment of detonation, were then forced together quickly by a kind of gun, and a chain reaction of explosive proportions occurred. An alternate bomb detonated conventional explosives (TNT) surrounding a plutonium sphere to compress it by implosion to double its density, making it more than critical and causing a nuclear explosion. The first fission bomb was tested in the New Mexico desert in July 1945. It was successful. In early August, a fission bomb using uranium was dropped on Hiroshima and a second, using plutonium, was dropped on Nagasaki (Fig. 42–12), both in Japan. World War II ended shortly thereafter.

FIGURE 42–12 Photo taken a month after the bomb was dropped on Nagasaki. The shacks were constructed afterwards from debris in the ruins.

Besides its great destructive power, a fission bomb produces many highly radioactive fission fragments, as does a nuclear reactor. When a fission bomb explodes, these radioactive isotopes are released into the atmosphere and are known as **radioactive fallout**.

Testing of nuclear bombs in the atmosphere after World War II was a cause of concern, for the movement of air masses spread the fallout all over the globe. Radioactive fallout eventually settles to the Earth, particularly in rainfall, and is absorbed by plants and grasses and enters the food chain. This is a far more serious problem than the same radioactivity on the exterior of our bodies, since α and β particles are largely absorbed by clothing and the outer (dead) layer of skin. But inside our bodies via food, the isotopes are in direct contact with living cells. One particularly dangerous radioactive isotope is ${}^{90}_{38}\mathrm{Sr}$, which is chemically much like calcium and becomes concentrated in bone, where it causes bone cancer and destruction of bone marrow. The 1963 treaty signed by over 100 nations that bans nuclear weapons testing in the atmosphere was motivated because of the hazards of fallout.

42–4 Nuclear Fusion

The mass of every stable nucleus is less than the sum of the masses of its constituent protons and neutrons. For example, the mass of the helium isotope ${}^{4}_{2}\mathrm{He}$ is less than the mass of two protons plus the mass of two neutrons, as we saw in Example 41–2. Thus, if two protons and two neutrons were to come together to form a helium nucleus, there would be a loss of mass. This mass loss is manifested in the release of a large amount of energy.

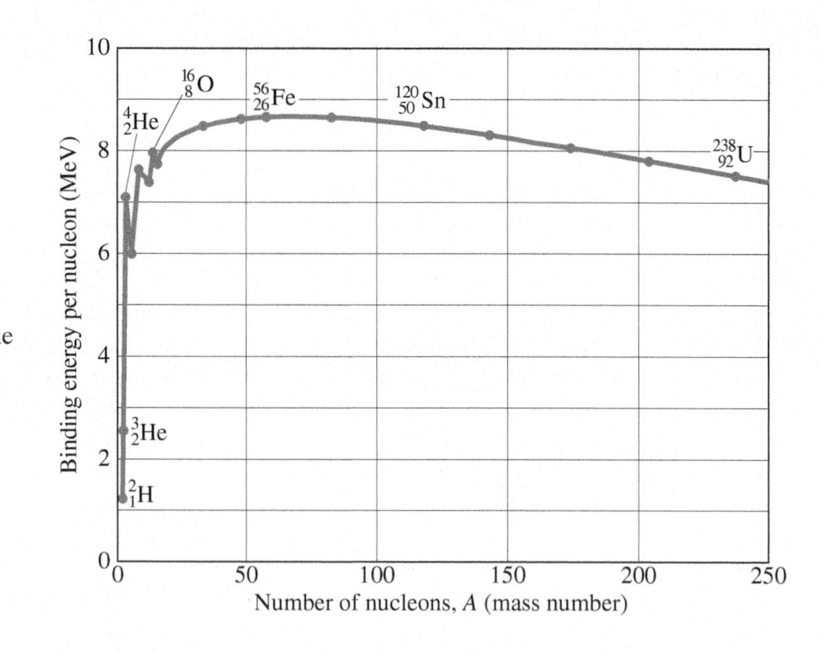

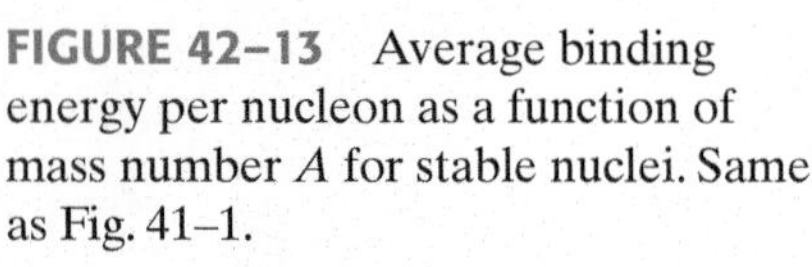

FIGURE 42–13 Average binding energy per nucleon as a function of mass number A for stable nuclei. Same as Fig. 41–1.

Nuclear Fusion; Stars

The process of building up nuclei by bringing together individual protons and neutrons, or building larger nuclei by combining small nuclei, is called **nuclear fusion**. A glance at Fig. 42–13 (same as Fig. 41–1) shows why small nuclei can combine to form larger ones with the release of energy: it is because the binding energy per nucleon is smaller for light nuclei than it is for those of increasing mass (up to about $A \approx 60$). For two positively charged nuclei to get close enough to fuse, they must have very high kinetic energy to overcome the electric repulsion. Many of the elements in the universe were originally formed through the process of nuclear fusion in stars (see Chapter 44), where the temperature is extremely high, corresponding to high kinetic energy (Eq. 18–4). Today, fusion is still producing the prodigious amount of light (EM waves) stars emit, including our Sun.

EXAMPLE 42–7 **Fusion energy release.** One of the simplest fusion reactions involves the production of deuterium, $^{2}_{1}\mathrm{H}$, from a neutron and a proton: $^{1}_{1}\mathrm{H} + \mathrm{n} \rightarrow {}^{2}_{1}\mathrm{H} + \gamma$. How much energy is released in this reaction?

APPROACH The energy released equals the difference in mass (times c^2) between the initial and final masses.

SOLUTION From Appendix F, the initial mass is

$$1.007825\,\mathrm{u} + 1.008665\,\mathrm{u} = 2.016490\,\mathrm{u},$$

and after the reaction the mass is that of the $^{2}_{1}\mathrm{H}$, namely 2.014102 u (the γ is massless). The mass difference is

$$2.016490\,\mathrm{u} - 2.014102\,\mathrm{u} = 0.002388\,\mathrm{u},$$

so the energy released is

$$(\Delta m)c^2 = (0.002388\,\mathrm{u})(931.5\,\mathrm{MeV/u}) = 2.24\,\mathrm{MeV},$$

and it is carried off as kinetic energy of the $^{2}_{1}\mathrm{H}$ nucleus and the γ ray.

The energy output of our Sun is believed to be due principally to the following sequence of fusion reactions:

$$^{1}_{1}\mathrm{H} + {}^{1}_{1}\mathrm{H} \rightarrow {}^{2}_{1}\mathrm{H} + e^{+} + \nu \qquad (0.42\,\mathrm{MeV}) \qquad \textbf{(42–7a)}$$

$$^{1}_{1}\mathrm{H} + {}^{2}_{1}\mathrm{H} \rightarrow {}^{3}_{2}\mathrm{He} + \gamma \qquad (5.49\,\mathrm{MeV}) \qquad \textbf{(42–7b)}$$

$$^{3}_{2}\mathrm{He} + {}^{3}_{2}\mathrm{He} \rightarrow {}^{4}_{2}\mathrm{He} + {}^{1}_{1}\mathrm{H} + {}^{1}_{1}\mathrm{H} \qquad (12.86\,\mathrm{MeV}) \qquad \textbf{(42–7c)}$$

where the energy released (Q-value) for each reaction is given in parentheses, keeping track of atomic electrons as for β-decay (Section 41–5). The net effect of this sequence, which is called the **proton–proton chain**, is for four protons to combine to form one ${}^4_2\text{He}$ nucleus plus two positrons, two neutrinos, and two gamma rays:

$$4\,{}^1_1\text{H} \rightarrow {}^4_2\text{He} + 2\text{e}^+ + 2\nu + 2\gamma. \qquad \textbf{(42–8)}$$

Note that it takes two of each of the first two reactions (Eqs. 42–7a and b) to produce the two ${}^3_2\text{He}$ for the third reaction. Also, each of the two e^+ formed (Eq. 42–7a) quickly annihilates with an electron to produce 2γ rays (Section 37–5) with total energy $2m_\text{e}c^2 = 1.02\,\text{MeV}$. So the total energy release for the net reaction, Eq. 42–8, is

$$2(0.42\,\text{MeV}) + 2(1.02\,\text{MeV}) + 2(5.49\,\text{MeV}) + 12.86\,\text{MeV} = 26.7\,\text{MeV}.$$

The first reaction, the formation of deuterium from two protons (Eq. 42–7a), has a very low probability and so limits the rate at which the Sun produces energy. (Thank goodness; this is why the Sun and other stars have long lifetimes and are still shining brightly.)

EXERCISE C Return to the first Chapter-Opening Question, page 1131, and answer it again now. Try to explain why you may have answered it differently the first time.

EXERCISE D If the Sun is generating a constant amount of energy via fusion, the mass of the Sun must be (*a*) increasing, (*b*) decreasing, (*c*) constant, (*d*) irregular.

EXAMPLE 42–8 ESTIMATE **Estimating fusion energy.** Estimate the energy released if the following reaction occurred:

$${}^2_1\text{H} + {}^2_1\text{H} \rightarrow {}^4_2\text{He}.$$

APPROACH We use Fig. 42–13 for a quick estimate.

SOLUTION We see in Fig. 42–13 that each ${}^2_1\text{H}$ has a binding energy of about $1\frac{1}{4}\,\text{MeV/nucleon}$, which for 2 nuclei of mass 2 is $4 \times \left(1\frac{1}{4}\right) \approx 5\,\text{MeV}$. The ${}^4_2\text{He}$ has a binding energy per nucleon of about 7 MeV for a total of $4 \times 7\,\text{MeV} = 28\,\text{MeV}$. Hence the energy release is about $28\,\text{MeV} - 5\,\text{MeV} = 23\,\text{MeV}$.

In stars hotter than the Sun, it is more likely that the energy output comes principally from the **carbon** (or **CNO**) **cycle**, which comprises the following sequence of reactions:

$$\begin{aligned}
{}^{12}_6\text{C} + {}^1_1\text{H} &\rightarrow {}^{13}_7\text{N} + \gamma \\
{}^{13}_7\text{N} &\rightarrow {}^{13}_6\text{C} + \text{e}^+ + \nu \\
{}^{13}_6\text{C} + {}^1_1\text{H} &\rightarrow {}^{14}_7\text{N} + \gamma \\
{}^{14}_7\text{N} + {}^1_1\text{H} &\rightarrow {}^{15}_8\text{O} + \gamma \\
{}^{15}_8\text{O} &\rightarrow {}^{15}_7\text{N} + \text{e}^+ + \nu \\
{}^{15}_7\text{N} + {}^1_1\text{H} &\rightarrow {}^{12}_6\text{C} + {}^4_2\text{He}.
\end{aligned}$$

No carbon is consumed in this cycle and the net effect is the same as the proton–proton chain, Eq. 42–8 (plus one extra γ). The theory of the proton–proton chain and of the carbon cycle as the source of energy for the Sun and stars was first worked out by Hans Bethe (1906–2005) in 1939.

CONCEPTUAL EXAMPLE 42–9 **Stellar fusion.** What is the heaviest element likely to be produced in fusion processes in stars?

RESPONSE Fusion is possible if the final products have more binding energy (less mass) than the reactants, because then there is a net release of energy. Since the binding energy curve in Fig. 42–13 (or Fig. 41–1) peaks near $A \approx 56$ to 58 which corresponds to iron or nickel, it would not be energetically favorable to produce elements heavier than that. Nevertheless, in the center of massive stars or in supernova explosions, there is enough initial kinetic energy available to drive endothermic reactions that produce heavier elements, as well.

Possible Fusion Reactors

PHYSICS APPLIED

Fusion energy reactors

The possibility of utilizing the energy released in fusion to make a power reactor is very attractive. The fusion reactions most likely to succeed in a reactor involve the isotopes of hydrogen, ${}^2_1\text{H}$ (deuterium) and ${}^3_1\text{H}$ (tritium), and are as follows, with the energy released given in parentheses:

$$ {}^2_1\text{H} + {}^2_1\text{H} \rightarrow {}^3_1\text{H} + {}^1_1\text{H} \qquad (4.03\,\text{MeV}) \qquad \textbf{(42–9a)} $$

$$ {}^2_1\text{H} + {}^2_1\text{H} \rightarrow {}^3_2\text{He} + \text{n} \qquad (3.27\,\text{MeV}) \qquad \textbf{(42–9b)} $$

$$ {}^2_1\text{H} + {}^3_1\text{H} \rightarrow {}^4_2\text{He} + \text{n}. \qquad (17.59\,\text{MeV}) \qquad \textbf{(42–9c)} $$

Comparing these energy yields with that for the fission of ${}^{235}_{92}\text{U}$, we can see that the energy released in fusion reactions can be greater for a given mass of fuel than in fission. Furthermore, as fuel, a fusion reactor could use deuterium, which is very plentiful in the water of the oceans (the natural abundance of ${}^2_1\text{H}$ is 0.0115% on average, or about 1 g of deuterium per 80 L of water). The simple proton–proton reaction of Eq. 42–7a, which could use a much more plentiful source of fuel, ${}^1_1\text{H}$, has such a small probability of occurring that it cannot be considered a possibility on Earth.

Although a useful fusion reactor has not yet been achieved, considerable progress has been made in overcoming the inherent difficulties. The problems are associated with the fact that all nuclei have a positive charge and repel each other. However, if they can be brought close enough together so that the short-range attractive strong nuclear force can come into play, it can pull the nuclei together and fusion will occur. For the nuclei to get close enough together, they must have large kinetic energy to overcome the electric repulsion. High kinetic energies are easily attainable with particle accelerators (Chapter 43), but the number of particles involved is too small. To produce realistic amounts of energy, we must deal with matter in bulk, for which high kinetic energy means higher temperatures. Indeed, very high temperatures are required for fusion to occur, and fusion devices are often referred to as **thermonuclear devices**. The interiors of the Sun and other stars are very hot, many millions of degrees, so the nuclei are moving fast enough for fusion to take place, and the energy released keeps the temperature high so that further fusion reactions can occur. The Sun and the stars represent huge self-sustaining thermonuclear reactors that stay together because of their great gravitational mass; but on Earth, containment of the fast-moving nuclei at the high temperatures and densities required has proven difficult.

It was realized after World War II that the temperature produced within a fission (or "atomic") bomb was close to 10^8 K. This suggested that a fission bomb could be used to ignite a fusion bomb (popularly known as a thermonuclear or hydrogen bomb) to release the vast energy of fusion. The uncontrollable release of fusion energy in an H-bomb (in 1952) was relatively easy to obtain. But to realize usable energy from fusion at a slow and controlled rate has turned out to be a serious challenge.

EXAMPLE 42–10 ESTIMATE **Temperature needed for d–t fusion.** Estimate the temperature required for deuterium–tritium fusion (d–t) to occur.

APPROACH We assume the nuclei approach head-on, each with kinetic energy K, and that the nuclear force comes into play when the distance between their centers equals the sum of their nuclear radii. The electrostatic potential energy (Chapter 23) of the two particles at this distance equals the minimum total kinetic energy of the two particles when far apart. The average kinetic energy is related to Kelvin temperature by Eq. 18–4.

SOLUTION The radii of the two nuclei ($A_\text{d} = 2$ and $A_\text{t} = 3$) are given by Eq. 41–1: $r_\text{d} \approx 1.5$ fm, $r_\text{t} \approx 1.7$ fm, so $r_\text{d} + r_\text{t} = 3.2 \times 10^{-15}$ m. We equate the kinetic energy of the two initial particles to the potential energy when at this distance:

$$ 2K \approx \frac{1}{4\pi\epsilon_0}\frac{e^2}{(r_\text{d} + r_\text{t})} $$

$$ \approx \left(9.0 \times 10^9\,\frac{\text{N}\cdot\text{m}^2}{\text{C}^2}\right)\frac{(1.6 \times 10^{-19}\,\text{C})^2}{(3.2 \times 10^{-15}\,\text{m})(1.6 \times 10^{-19}\,\text{J/eV})} \approx 0.45\,\text{MeV}. $$

Thus, $K \approx 0.22$ MeV, and if we ask that the average kinetic energy be this high, then from Eq. 18–4, $\frac{3}{2}kT = \overline{K}$, we have a temperature of

$$T = \frac{2\overline{K}}{3k} = \frac{2(0.22\,\text{MeV})(1.6 \times 10^{-13}\,\text{J/MeV})}{3(1.38 \times 10^{-23}\,\text{J/K})} \approx 2 \times 10^{9}\,\text{K}.$$

NOTE More careful calculations show that the temperature required for fusion is actually about an order of magnitude less than this rough estimate, partly because it is not necessary that the *average* kinetic energy be 0.22 MeV—a small percentage of nuclei with this much energy (in the high-energy tail of the Maxwell distribution, Fig. 18–3) would be sufficient. Reasonable estimates for a usable fusion reactor are in the range $T \gtrsim 1$ to $4 \times 10^{8}\,\text{K}$.

It is not only a high temperature that is required for a fusion reactor. But there must also be a high density of nuclei to ensure a sufficiently high collision rate. A real difficulty with controlled fusion is to contain nuclei long enough and at a high enough density for sufficient reactions to occur that a usable amount of energy is obtained. At the temperatures needed for fusion, the atoms are ionized, and the resulting collection of nuclei and electrons is referred to as a **plasma**. Ordinary materials vaporize at a few thousand degrees at most, and hence cannot be used to contain a high-temperature plasma. Two major containment techniques are *magnetic confinement* and *inertial confinement*.

In **magnetic confinement**, magnetic fields are used to try to contain the hot plasma. A simple approach is the "magnetic bottle" shown in Fig. 42–14. The paths of the charged particles in the plasma are bent by the magnetic field; where magnetic field lines are close together, the force on the particles reflects them back toward the center. Unfortunately, magnetic bottles develop "leaks" and the charged particles slip out before sufficient fusion takes place. The most promising design today is the **tokamak**, first developed in Russia. A tokamak (Fig. 42–15) is toroid-shaped and involves complicated magnetic fields: current-carrying conductors produce a magnetic field directed along the axis of the toroid ("toroidal" field); an additional field is produced by currents within the plasma itself ("poloidal" field). The combination produces a helical field as shown in Fig. 42–15, confining the plasma, at least briefly, so it doesn't touch the vacuum chamber's metal walls.

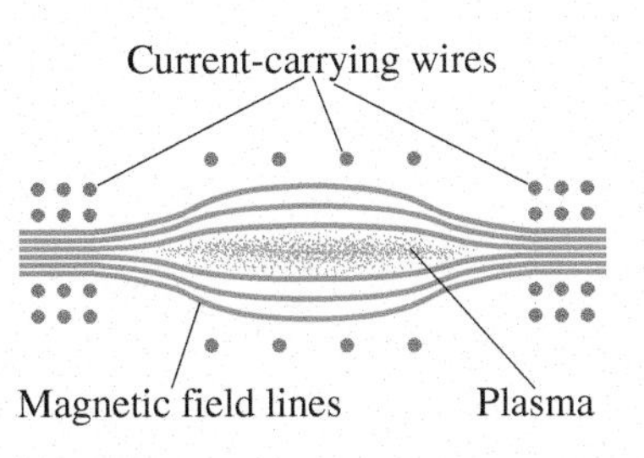

FIGURE 42–14 "Magnetic bottle" used to confine a plasma.

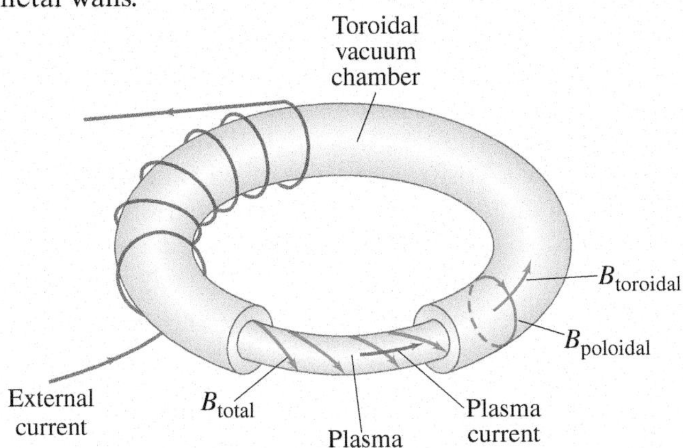

FIGURE 42–15 Tokamak configuration, showing the total $\vec{\mathbf{B}}$ field due to external current plus current in the plasma itself.

In 1957, J. D. Lawson showed that the product of ion density n and confinement time τ must exceed a minimum value of approximately

$$n\tau \gtrsim 3 \times 10^{20}\,\text{s/m}^3.$$

This **Lawson criterion** must be reached to produce **ignition**, meaning fusion that continues after all external heating is turned off. Practically, it is expected to be achieved with $n \approx 1$ to $3 \times 10^{20}\,\text{m}^{-3}$ and $\tau \approx 1$–3 s. To reach **break-even**, the point at which the energy output due to fusion is equal to the energy input to heat the plasma, requires an $n\tau$ about an order of magnitude less. The break-even point was very closely approached in the 1990s at the Tokamak Fusion Test Reactor (TFTR) at Princeton, and the very high temperature needed for ignition $(4 \times 10^{8}\,\text{K})$ was exceeded—although not both of these at the same time.

Magnetic confinement fusion research continues throughout the world. This research will help us in developing the huge multinational test device (European Union, India, Japan, South Korea, Russia, China, and the U.S.), called ITER (International Thermonuclear Experimental Reactor), situated in France. It is expected that ITER will produce temperatures above 10^8 K for extended periods (minutes or hours) and to begin running by the 2020s, with an expected power output of about 500 MW, 10 times the input energy. ITER (see Chapter-Opening Photograph on page 1131) is planned to be the final research step before building a commercial reactor.

The second method for containing the fuel for fusion is **inertial confinement fusion** (ICF): a small pellet or capsule of deuterium and tritium is struck simultaneously from hundreds of directions by very intense laser beams. The intense influx of energy heats and ionizes the pellet into a plasma, compressing it and heating it to temperatures at which fusion can occur ($> 10^8$ K). The confinement time is on the order of 10^{-11} to 10^{-9} s, during which time the ions do not move appreciably because of their own inertia, and fusion can take place.

42–5 Passage of Radiation Through Matter; Radiation Damage

When we speak of *radiation*, we include α, β, γ, and X-rays, as well as protons, neutrons, and other particles such as pions (see Chapter 43). Because charged particles can ionize the atoms or molecules of any material they pass through, they are referred to as **ionizing radiation**. And because radiation produces ionization, it can cause considerable damage to materials, particularly to biological tissue.

Charged particles, such as α and β rays and protons, cause ionization because of electric forces. That is, when they pass through a material, they can attract or repel electrons strongly enough to remove them from the atoms of the material. Since the α and β rays emitted by radioactive substances have energies on the order of 1 MeV (10^4 to 10^7 eV), whereas ionization of atoms and molecules requires on the order of 10 eV, it is clear that a single α or β particle can cause thousands of ionizations.

Neutral particles also give rise to ionization when they pass through materials. For example, X-ray and γ-ray photons can ionize atoms by knocking out electrons by means of the photoelectric and Compton effects (Chapter 37). Furthermore, if a γ ray has sufficient energy (greater than 1.02 MeV), it can undergo pair production: an electron and a positron are produced (Section 37–5). The charged particles produced in all of these processes can themselves go on to produce further ionization. Neutrons, on the other hand, interact with matter mainly by collisions with nuclei, with which they interact strongly. Often the nucleus is broken apart by such a collision, altering the molecule of which it was a part. The fragments produced can in turn cause ionization.

Radiation passing through matter can do considerable damage. Metals and other structural materials become brittle and their strength can be weakened if the radiation is very intense, as in nuclear reactor power plants and for space vehicles that must pass through areas of intense cosmic radiation.

*Biological Damage

PHYSICS APPLIED
Biological radiation damage

The radiation damage produced in biological organisms is due primarily to ionization produced in cells. Several related processes can occur. Ions or radicals are produced that are highly reactive and take part in chemical reactions that interfere with the normal operation of the cell. All forms of radiation can ionize atoms by knocking out electrons. If these are bonding electrons, the molecule may break apart, or its structure may be altered so it does not perform its normal function or may perform a harmful function. In the case of proteins, the loss of one molecule is not serious if there are other copies of it in the cell and additional copies can be made from the gene that codes for it. However, large doses of radiation may damage so many molecules that new copies cannot be made quickly enough, and the cell dies. Damage to the DNA is more serious, since a cell may have only one copy. Each alteration in the DNA can affect a gene and alter the molecule it codes for (Section 40–3), so that needed proteins or other materials may not be made at all. Again the cell may die. The death of a single cell is not normally a problem, since the body can replace it with a new one. (There are exceptions, such as neurons, which are mostly not replaceable, so their loss is serious.) But if many cells die, the organism may not be able to recover. On the other hand, a cell may survive but be defective. It may go on dividing and produce many more defective cells, to the detriment of the whole organism. Thus radiation can cause cancer—the rapid uncontrolled production of cells.

The possible damage done by the medical use of X-rays and other radiation must be balanced against the medical benefits and prolongation of life as a result of their use.

42–6 Measurement of Radiation—Dosimetry

Although the passage of ionizing radiation through the human body can cause considerable damage, radiation can also be used to treat certain diseases, particularly cancer, often by using very narrow beams directed at a cancerous tumor in order to destroy it (Section 42–7). It is therefore important to be able to quantify the amount, or **dose**, of radiation. This is the subject of **dosimetry**.

PHYSICS APPLIED
Dosimetry

The strength of a source can be specified at a given time by stating the **source activity**: how many nuclear decays (or disintegrations) occur per second. The traditional unit is the **curie** (Ci), defined as

$$1\text{ Ci} = 3.70 \times 10^{10}\text{ decays per second.}$$

(This number comes from the original definition as the activity of exactly one gram of radium.) Although the curie is still in common use, the SI unit for source activity is the **becquerel** (Bq), defined as

$$1\text{ Bq} = 1\text{ decay/s.}$$

Commercial suppliers of **radionuclides** (radioactive nuclides) specify the activity at a given time. Since the activity decreases over time, more so for short-lived isotopes, it is important to take this into account.

The magnitude of the source activity $|dN/dt|$ is related to the number of radioactive nuclei present, N, and to the half-life, $T_{\frac{1}{2}}$, by (see Section 41–8):

$$\left|\frac{dN}{dt}\right| = \lambda N = \frac{0.693}{T_{\frac{1}{2}}} N.$$

EXAMPLE 42–11 Radioactivity taken up by cells. In a certain experiment, 0.016 μCi of ${}^{32}_{15}\text{P}$ is injected into a medium containing a culture of bacteria. After 1.0 h the cells are washed and a 70% efficient detector (counts 70% of emitted β rays) records 720 counts per minute from the cells. What percentage of the original ${}^{32}_{15}\text{P}$ was taken up by the cells?

APPROACH The half-life of ${}^{32}_{15}\text{P}$ is about 14 days (Appendix F), so we can ignore any loss of activity over 1 hour. From the given activity, we find how many β rays are emitted. We can compare 70% of this to the $(720/\text{min})/(60\text{ s/min}) = 12$ per second detected.

SOLUTION The total number of decays per second originally was $(0.016 \times 10^{-6})(3.7 \times 10^{10}) = 590$. The counter could be expected to count 70% of this, or 410 per second. Since it counted $720/60 = 12$ per second, then $12/410 = 0.029$ or 2.9% was incorporated into the cells.

Another type of measurement is the exposure or **absorbed dose**—that is, the effect the radiation has on the absorbing material. The earliest unit of dosage was the **roentgen** (R), defined in terms of the amount of ionization produced by the radiation ($1\text{ R} = 1.6 \times 10^{12}$ ion pairs per gram of dry air at standard conditions). Today, 1 R is defined as the amount of X or γ radiation that deposits 0.878×10^{-2} J of energy per kilogram of air. The roentgen was largely superseded by another unit of absorbed dose applicable to any type of radiation, the **rad**: *1 rad is that amount of radiation which deposits energy per unit mass of 1.00×10^{-2} J/kg in any absorbing material.* (This is quite close to the roentgen for X- and γ rays.) The proper SI unit for absorbed dose is the **gray** (Gy):

$$1\text{ Gy} = 1\text{ J/kg} = 100\text{ rad}. \tag{42–10}$$

The absorbed dose depends not only on the strength of a given radiation beam (number of particles per second) and the energy per particle, but also on the type of material that is absorbing the radiation. Bone, for example, absorbs more of the radiation normally used than does flesh, so the same beam passing through a human body deposits a greater dose (in rads or grays) in bone than in flesh.

The gray and the rad are physical units of dose—the energy deposited per unit mass of material. They are, however, not the most meaningful units for measuring the biological damage produced by radiation because equal doses of different types of radiation cause differing amounts of damage. For example, 1 rad of α radiation does 10 to 20 times the amount of damage as 1 rad of β or γ rays. This difference arises largely because α rays (and other heavy particles such as protons and neutrons) move much more slowly than β and γ rays of equal energy due to their greater mass. Hence, ionizing collisions occur closer together, so more irreparable damage can be done. The **relative biological effectiveness** (RBE) or **quality factor** (QF) of a given type of radiation is defined as the number of rads of X or γ radiation that produces the same biological damage as 1 rad of the given radiation. Table 42–1 gives the QF for several types of radiation. The numbers are approximate since they depend somewhat on the energy of the particles and on the type of damage that is used as the criterion.

TABLE 42–1 Quality Factor (QF) of Different Kinds of Radiation

Type	QF
X- and γ rays	1
β (electrons)	≈ 1
Fast protons	1
Slow neutrons	≈ 3
Fast neutrons	Up to 10
α particles and heavy ions	Up to 20

The **effective dose** can be given as the product of the dose in rads and the QF, and this unit is known as the **rem** (which stands for *rad equivalent man*):

$$\text{effective dose (in rem)} = \text{dose (in rad)} \times \text{QF}. \tag{42–11a}$$

This unit is being replaced by the SI unit for "effective dose," the **sievert** (Sv):

$$\text{effective dose (Sv)} = \text{dose (Gy)} \times \text{QF}. \tag{42–11b}$$

By these definitions, 1 rem (or 1 Sv) of any type of radiation does approximately the same amount of biological damage. For example, 50 rem of fast neutrons does the same damage as 50 rem of γ rays. But note that 50 rem of fast neutrons is only 5 rads, whereas 50 rem of γ rays is 50 rads.

EXERCISE E Return to the second Chapter-Opening Question, page 1131, and answer it again now. Try to explain why you may have answered it differently the first time.

Human Exposure to Radiation

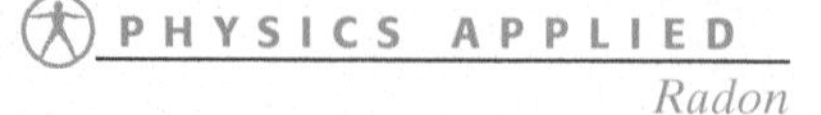

Radon

We are constantly exposed to low-level radiation from natural sources: cosmic rays, natural radioactivity in rocks and soil, and naturally occurring radioactive isotopes in our food, such as $^{40}_{19}$K. **Radon**, $^{222}_{86}$Rn, is of considerable concern today. It is the product of radium decay and is an intermediate in the decay series from uranium (see Fig. 41–12). Most intermediates remain in the rocks where formed, but radon is a gas that can escape from rock (and from building material like concrete) to enter the air we breathe, and attack the interior of the lung.

PHYSICS APPLIED
Human radiation exposure

The **natural radioactive background** averages about 0.30 rem (300 mrem) per year per person in the U.S., although there are large variations. From medical X-rays and scans, the average person receives about 50 to 60 mrem per year, giving an average total dose of about 360 mrem (3.6 mSv) per person. Government regulators suggest an upper limit of allowed radiation for an individual in the general populace at about 100 mrem (1 mSv) per year in addition to natural background. It is believed that even low doses of radiation increase the chances of cancer or genetic defects; there is no safe level or threshold of radiation exposure.

PHYSICS APPLIED
Radiation worker exposure
Film badge

The upper limit for people who work around radiation—in hospitals, in power plants, in research—has been set higher, a maximum of 5 rem (50 mSv) whole-body dose in any one year, and significantly less averaged over more years (below 2 rem/yr averaged over 5 years). To monitor exposure, those people who work around radiation generally carry some type of dosimeter, one common type being a **radiation film badge** which is a piece of film wrapped in light-tight material. The passage of ionizing radiation through the film changes it so that the film is darkened upon development, and thus indicates the received dose. Newer types include the *thermoluminescent dosimeter* (TLD). Dosimeters and badges do not protect the worker, but high levels detected suggest reassignment or modified work practices to reduce radiation exposure to acceptable levels.

PHYSICS APPLIED
Radiation sickness

Large doses of radiation can cause unpleasant symptoms such as nausea, fatigue, and loss of body hair. Such effects are sometimes referred to as **radiation sickness**. Large doses can be fatal, although the time span of the dose is important. A short dose of 1000 rem (10 Sv) is nearly always fatal. A 400-rem (4-Sv) dose in a short period of time is fatal in 50% of the cases. However, the body possesses remarkable repair processes, so that a 400-rem dose spread over several weeks is usually not fatal. It will, nonetheless, cause considerable damage to the body.

The effects of low doses over a long time are difficult to determine and are not well known as yet.

EXAMPLE 42–12 **Whole-body dose.** What whole-body dose is received by a 70-kg laboratory worker exposed to a 40-mCi ${}^{60}_{27}\mathrm{Co}$ source, assuming the person's body has cross-sectional area 1.5 m^2 and is normally about 4.0 m from the source for 4.0 h per day? ${}^{60}_{27}\mathrm{Co}$ emits γ rays of energy 1.33 MeV and 1.17 MeV in quick succession. Approximately 50% of the γ rays interact in the body and deposit all their energy. (The rest pass through.)

APPROACH Of the given energy emitted, only a fraction passes through the worker, equal to *her* area divided by the total area over a full sphere of radius 4.0 m (Fig. 42–16).

SOLUTION The total γ-ray energy per decay is $(1.33 + 1.17)\,\mathrm{MeV} = 2.50\,\mathrm{MeV}$, so the total energy emitted by the source per second is

$$(0.040\,\mathrm{Ci})(3.7\times10^{10}\,\mathrm{decays/Ci\cdot s})(2.50\,\mathrm{MeV}) = 3.7\times10^{9}\,\mathrm{MeV/s}.$$

The proportion of this energy intercepted by the body is its 1.5-m^2 area divided by the area of a sphere of radius 4.0 m (Fig. 42–16):

$$\frac{1.5\,\mathrm{m}^2}{4\pi r^2} = \frac{1.5\,\mathrm{m}^2}{4\pi(4.0\,\mathrm{m})^2} = 7.5\times10^{-3}.$$

So the rate energy is deposited in the body (remembering that only 50% of the γ rays interact in the body) is

$$E = \left(\tfrac{1}{2}\right)(7.5\times10^{-3})(3.7\times10^{9}\,\mathrm{MeV/s})(1.6\times10^{-13}\,\mathrm{J/MeV}) = 2.2\times10^{-6}\,\mathrm{J/s}.$$

Since $1\,\mathrm{Gy} = 1\,\mathrm{J/kg}$, the whole-body dose rate for this 70-kg person is $(2.2\times10^{-6}\,\mathrm{J/s})/(70\,\mathrm{kg}) = 3.1\times10^{-8}\,\mathrm{Gy/s}$. In 4.0 h, this amounts to a dose of

$$(4.0\,\mathrm{h})(3600\,\mathrm{s/h})(3.1\times10^{-8}\,\mathrm{Gy/s}) = 4.5\times10^{-4}\,\mathrm{Gy}.$$

Since QF ≈ 1 for gammas, the effective dose is 450 μSv (Eqs. 42–11b and 42–10) or:

$$(100\,\mathrm{rad/Gy})(4.5\times10^{-4}\,\mathrm{Gy})(1\,\mathrm{rem/rad}) = 45\,\mathrm{mrem} = 0.45\,\mathrm{mSv}.$$

NOTE This 45-mrem effective dose is almost 50% of the normal allowed dose for a whole year (100 mrem/yr), or 1% of the maximum one-year allowance for radiation workers. This worker should not receive such a large dose every day and should seek ways to reduce it (shield the source, vary the work, work farther from the source, work less time this close to source, etc.).

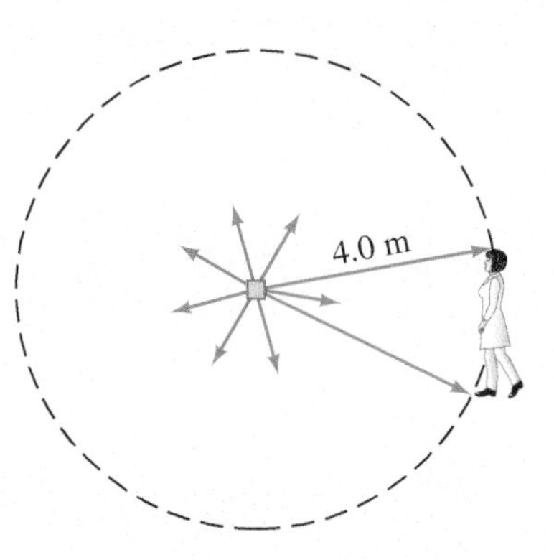

FIGURE 42–16 Radiation spreads out in all directions. A person 4.0 m away intercepts only a fraction: her cross-sectional area divided by the area of a sphere of radius 4.0 m. Example 42–12.

PHYSICS APPLIED
Radon exposure

EXAMPLE 42–13 **Radon exposure.** In the U.S., yearly deaths from radon exposure (the second leading cause of lung cancer) are estimated to be on the order of 20,000 and maybe much more. The Environmental Protection Agency recommends taking action to reduce the radon concentration in living areas if it exceeds 4 pCi/L of air. In some areas 50% of houses exceed this level from naturally occurring radon in the soil. Estimate (a) the number of decays/s in 1.0 m^3 of air and (b) the mass of radon that emits 4.0 pCi of $^{222}_{86}Rn$ radiation.

APPROACH We can use the definition of the curie to determine how many decays per second correspond to 4 pCi, then Eq. 41–7b to determine how many nuclei of radon it takes to have this activity $|dN/dt|$.

SOLUTION We saw at the start of this Section that $1\ \text{Ci} = 3.70 \times 10^{10}$ decays/s. Thus

$$\left|\frac{dN}{dt}\right| = 4.0\ \text{pCi}$$
$$= (4.0 \times 10^{-12}\ \text{Ci})(3.70 \times 10^{10}\ \text{decays/s/Ci})$$
$$= 0.148\ \text{s}^{-1}$$

per liter of air. In 1.0 m^3 of air ($1\ \text{m}^3 = 10^6\ \text{cm} = 10^3\ \text{L}$) there would be $(0.148\ \text{s}^{-1})(1000) \approx 150$ decays/s. (b) From Eqs. 41–7a and 41–8

$$\left|\frac{dN}{dt}\right| = \lambda N = \frac{0.693}{T_{\frac{1}{2}}} N.$$

Appendix F tells us $T_{\frac{1}{2}} = 3.8232$ days, so

$$N = \left|\frac{dN}{dt}\right| \frac{T_{\frac{1}{2}}}{0.693}$$
$$= (0.148\ \text{s}^{-1}) \frac{(3.8232\ \text{days})(8.64 \times 10^4\ \text{s/day})}{0.693}$$
$$= 7.05 \times 10^4\ \text{atoms of radon-222}.$$

The molar mass (222 u) and Avogadro's number are used to find the mass:

$$m = \frac{(7.05 \times 10^4\ \text{atoms})(222\ \text{g/mol})}{6.02 \times 10^{23}\ \text{atoms/mol}} = 2.6 \times 10^{-17}\ \text{g}$$

or 26 attograms in 1L of air at the limit of 4 pCi/L. This 2.6×10^{-17} g/L is 2.6×10^{-14} grams of radon per m^3 of air.

NOTE Each radon atom emits 4 α particles and 4 β particles, each one capable of causing many harmful ionizations, before the sequence of decays reaches a stable element.

*42–7 Radiation Therapy

The medical application of radioactivity and radiation to human beings involves two basic aspects: (1) **radiation therapy**—the treatment of disease (mainly cancer)—which we discuss in this Section; and (2) the *diagnosis* of disease, which we discuss in the following Sections of this Chapter.

PHYSICS APPLIED
Radiation therapy

Radiation can cause cancer. It can also be used to treat it. Rapidly growing cancer cells are especially susceptible to destruction by radiation. Nonetheless, large doses are needed to kill the cancer cells, and some of the surrounding normal cells are inevitably killed as well. It is for this reason that cancer patients receiving radiation therapy often suffer side effects characteristic of radiation sickness. To minimize the destruction of normal cells, a narrow beam of γ or X-rays is often used when a cancerous tumor is well localized. The beam is directed at the tumor, and the source (or body) is rotated so that the beam passes through various parts of the body to keep the dose at any one place as low as possible—except at the tumor and its immediate surroundings, where the beam passes at all times (Fig. 42–17). The radiation may be from a radioactive source such as $^{60}_{27}\text{Co}$, or it may be from an X-ray machine that produces photons in the range 200 keV to 5 MeV. Protons, neutrons, electrons, and pions, which are produced in particle accelerators (Section 43–1), are also being used in cancer therapy.

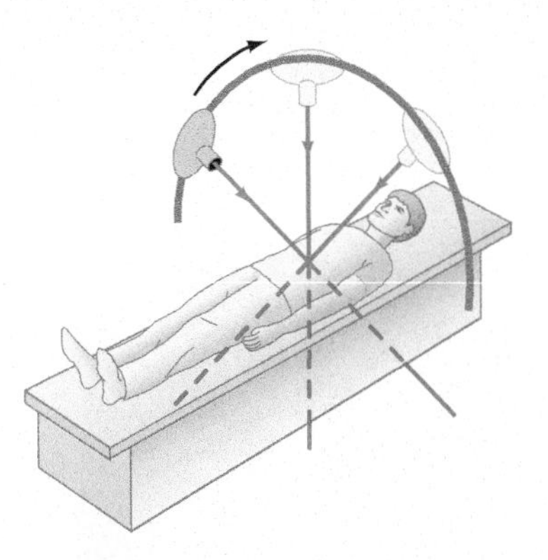

FIGURE 42–17 Radiation source rotates so that the beam always passes through the diseased tissue, but minimizes the dose in the rest of the body.

Protons used to kill tumors have a special property that makes them particularly useful. As shown in Fig. 42–18, when protons enter tissue, most of their energy is deposited at the end of their path. The protons' initial kinetic energy can be chosen so that most of the energy is deposited at the depth of the tumor itself, to destroy it. The incoming protons deposit only a small amount of energy in the tissue in front of the tumor, and none at all behind the tumor, thus having less negative effect on healthy tissue than X- or γ rays. Because tumors have physical size, even several centimeters in diameter, a range of proton energies is often used. Heavier ions, such as α particles or carbon ions, are similarly useful. This **proton therapy** technique is more than a half century old, but the necessity of having a large accelerator has meant that few hospitals have used the technique until now. Many such "proton centers" are now being built.

PHYSICS APPLIED
Proton therapy

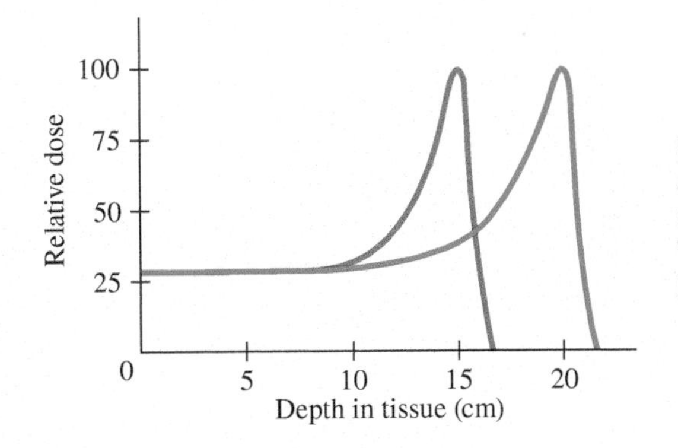

FIGURE 42–18 Energy deposited in tissue as a function of depth for 170-MeV protons (red curve) and 190-MeV protons (green). The peak of each curve is often called the Bragg peak.

Another form of treatment is to insert a tiny radioactive source directly inside a tumor, which will eventually kill the majority of the cells. A similar technique is used to treat cancer of the thyroid with the radioactive isotope $^{131}_{53}\text{I}$. The thyroid gland concentrates iodine present in the bloodstream, particularly in any area where abnormal growth is taking place. Its intense radioactivity can destroy the defective cells.

Another application of radiation is for sterilizing bandages, surgical equipment, and even packaged foods, since bacteria and viruses can be killed or deactivated by large doses of radiation.

*42–8 Tracers in Research and Medicine

Radioactive isotopes are commonly used in biological and medical research as **tracers**. A given compound is artificially synthesized using a radioactive isotope such as $^{14}_{6}\text{C}$ or $^{3}_{1}\text{H}$. Such "tagged" molecules can then be traced as they move through an organism or as they undergo chemical reactions. The presence of these tagged molecules (or parts of them, if they undergo chemical change) can be detected by a Geiger or scintillation counter, which detects emitted radiation (see Section 41–11). How food molecules are digested, and to what parts of the body they are diverted, can be traced in this way.

PHYSICS APPLIED
Tracers in medicine and biology

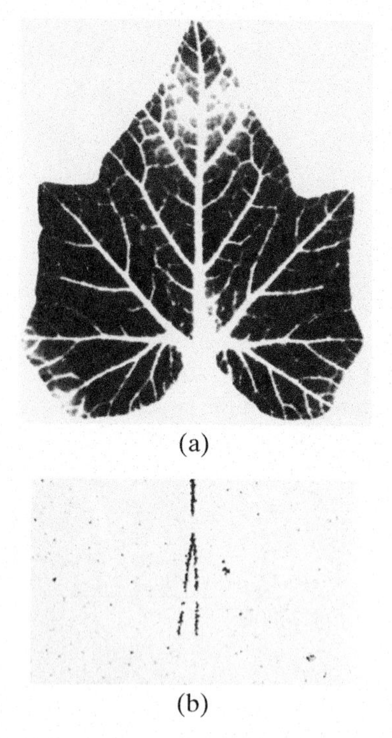

FIGURE 42–19 (a) Autoradiograph of a leaf exposed for 30 s to $^{14}CO_2$. The photosynthetic (green) tissue has become radioactive; the nonphotosynthetic tissue of the veins is free of $^{14}_{6}C$ and therefore does not blacken the X-ray sheet. This technique is useful in following patterns of nutrient transport in plants. (b) Autoradiograph of chromosomal DNA. The dashed arrays of film grains show the Y-shaped growing point of replicating DNA.

Radioactive tracers have been used to determine how amino acids and other essential compounds are synthesized by organisms. The permeability of cell walls to various molecules and ions can be determined using radioactive isotopes: the tagged molecule or ion is injected into the extracellular fluid, and the radioactivity present inside and outside the cells is measured as a function of time.

In a technique known as **autoradiography**, the position of the radioactive isotopes is detected on film. For example, the distribution of carbohydrates produced in the leaves of plants from absorbed CO_2 can be observed by keeping the plant in an atmosphere where the carbon atom in the CO_2 is $^{14}_{6}C$. After a time, a leaf is placed firmly on a photographic plate and the emitted radiation darkens the film most strongly where the isotope is most strongly concentrated (Fig. 42–19a). Autoradiography using labeled nucleotides (components of DNA) has revealed much about the details of DNA replication (Fig. 42–19b).

For medical diagnosis, the radionuclide commonly used today is $^{99m}_{43}Tc$, a long-lived excited state of technetium-99 (the "m" in the symbol stands for "metastable" state). It is formed when $^{99}_{42}Mo$ decays. The great usefulness of $^{99m}_{43}Tc$ derives from its convenient half-life of 6 h (short, but not too short) and the fact that it can combine with a large variety of compounds. The compound to be labeled with the radionuclide is so chosen because it concentrates in the organ or region of the anatomy to be studied. Detectors outside the body then record, or image, the distribution of the radioactively labeled compound. The detection could be done by a single detector (Fig. 42–20a) which is moved across the body, measuring the intensity of radioactivity at a large number of points. The image represents the relative intensity of radioactivity at each point. The relative radioactivity is a diagnostic tool. For example, high or low radioactivity may represent overactivity or underactivity of an organ or part of an organ, or in another case may represent a lesion or tumor. More complex **gamma cameras** make use of many detectors which simultaneously record the radioactivity at many points. The measured intensities can be displayed on a TV or computer monitor. The image is sometimes called a **scintigram** (after scintillator), Fig. 42–20b. Gamma cameras are relatively inexpensive, but their resolution is limited—by non-perfect collimation†; but they allow "dynamic" studies: images that change in time, like a movie.

†To "collimate" means to "make parallel," usually by blocking non-parallel rays with a narrow tube inside lead, as in Fig. 42–20a.

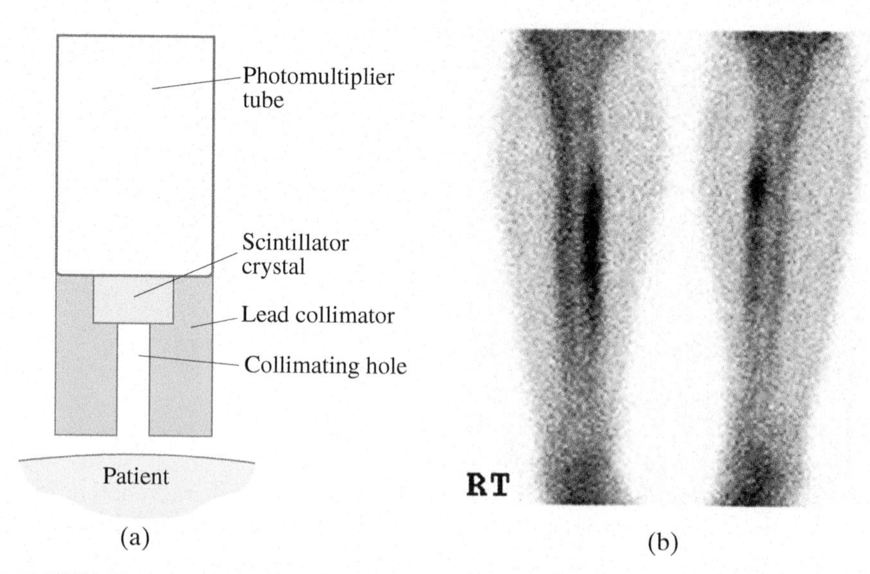

FIGURE 42–20 (a) Collimated gamma-ray detector for scanning (moving) over a patient. The collimator selects γ rays that come in a (nearly) straight line from the patient. Without the collimator, γ rays from all parts of the body could strike the scintillator, producing a poor image. Detectors today usually have many collimator tubes and are called *gamma cameras*. (b) Gamma camera image (scintigram), of both legs of a patient with shin splints, detecting γs from $^{99m}_{43}Tc$.

*42–9 Imaging by Tomography: CAT Scans and Emission Tomography

*Normal X-ray Image

For a conventional medical or dental X-ray photograph, the X-rays emerging from the tube (Section 35–10) pass through the body and are detected on photographic film or a fluorescent screen, Fig. 42–21. The rays travel in very nearly straight lines through the body with minimal deviation since at X-ray wavelengths there is little diffraction or refraction. There is absorption (and scattering), however; and the difference in absorption by different structures in the body is what gives rise to the image produced by the transmitted rays. The less the absorption, the greater the transmission and the darker the film. The image is, in a sense, a "shadow" of what the rays have passed through. The X-ray image is *not* produced by focusing rays with lenses as for the instruments discussed in Chapter 33.

PHYSICS APPLIED
Normal X-ray image is a sort of shadow (no lenses involved)

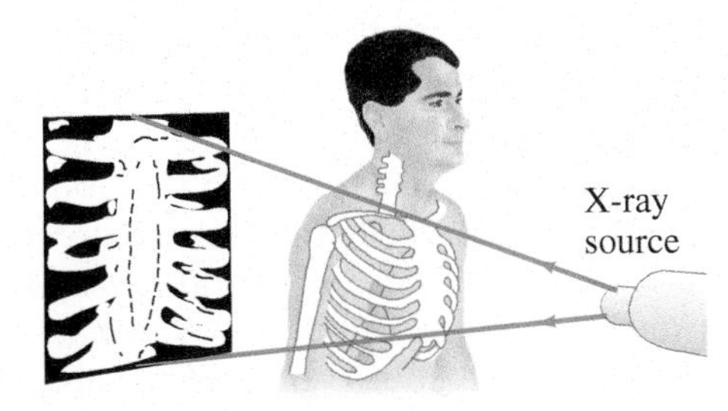

FIGURE 42–21 Conventional X-ray imaging, which is essentially shadowing.

*Tomography Images (CT)

In conventional X-ray images, the entire thickness of the body is projected onto the film; structures overlap and in many cases are difficult to distinguish. In the 1970s, a revolutionary new X-ray technique was developed called **computed tomography** (CT), which produces an image of a *slice* through the body. (The word **tomography** comes from the Greek: *tomos* = slice, *graph* = picture.) Structures and lesions previously impossible to visualize can now be seen with remarkable clarity. The principle behind CT is shown in Fig. 42–22: a thin collimated beam of X-rays (to "collimate" means to "make parallel") passes through the body to a detector that measures the transmitted intensity. Measurements are made at a large number of points as the source and detector are moved past the body together. The apparatus is then rotated slightly about the body axis and again scanned; this is repeated at (perhaps) 1° intervals for 180°. The intensity of the transmitted beam for the many points of each scan, and for each angle, are sent to a computer that reconstructs the image of the slice. Note that the imaged slice is perpendicular to the long axis of the body. For this reason, CT is sometimes called **computerized axial tomography** (CAT), although the abbreviation CAT, as in CAT scan, can also be read as **computer-assisted tomography**.

PHYSICS APPLIED
Computed tomography images

The use of a single detector as in Fig. 42–22 would require a few minutes for the many scans needed to form a complete image. Much faster scanners use

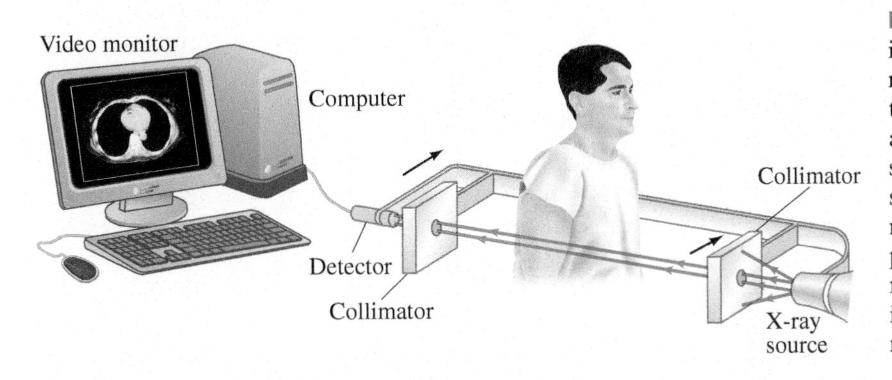

FIGURE 42–22 Tomographic imaging: the X-ray source and detector move together across the body, the transmitted intensity being measured at a large number of points. Then the source–detector assembly is rotated slightly (say, 1°) and another scan is made. This process is repeated for perhaps 180°. The computer reconstructs the image of the slice and it is presented on a TV or computer monitor.

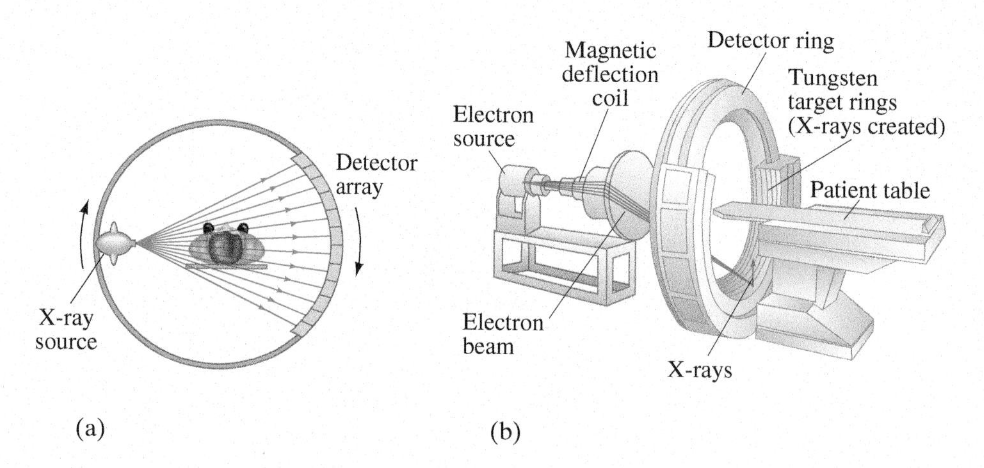

FIGURE 42–23 (a) Fan-beam scanner. Rays transmitted through the entire body are measured simultaneously at each angle. The source and detector rotate to take measurements at different angles. In another type of fan-beam scanner, there are detectors around the entire 360° of the circle which remain fixed as the source moves. (b) In still another type, a beam of electrons from a source is directed by magnetic fields at tungsten targets surrounding the patient.

a fan beam, Fig. 42–23a, in which beams passing through the entire cross section of the body are detected simultaneously by many detectors. The source and detectors are then rotated about the patient, and an image requires only a few seconds. Even faster, and therefore useful for heart scans, are fixed source machines wherein an electron beam is directed (by magnetic fields) to tungsten targets surrounding the patient, creating the X-rays. See Fig. 42–23b.

FIGURE 42–24 Two CT images, with different resolutions, each showing a cross section of a brain. Photo (a) is of low resolution. Photo (b), of higher resolution, shows a brain tumor, and uses false color to highlight it.

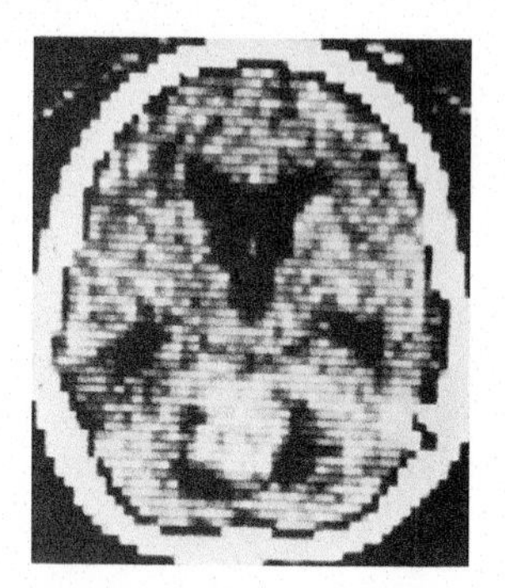

(a)

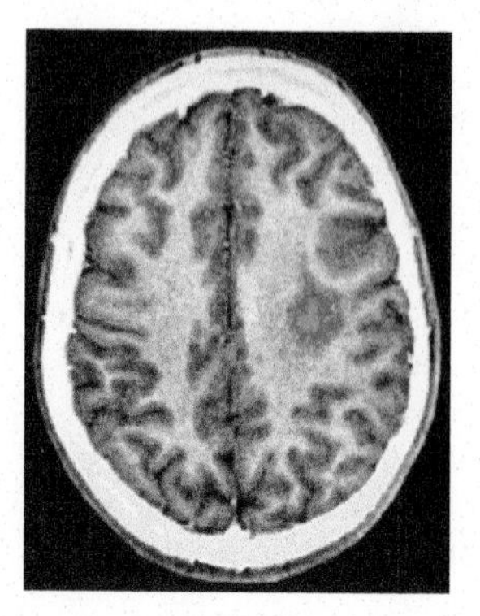

(b)

*Image Formation

But how is the image formed? We can think of the slice to be imaged as being divided into many tiny picture elements (or **pixels**), which could be squares. (See Fig. 35–42.) For CT, the width of each pixel is chosen according to the width of the detectors and/or the width of the X-ray beams, and this determines the resolution of the image, which might be 1 mm. An X-ray detector measures the intensity of the transmitted beam. When we subtract this value from the intensity of the beam at the source, we obtain the total absorption (called a "projection") along that beam line. Complicated mathematical techniques are used to analyze all the absorption projections for the huge number of beam scans measured (see the next Subsection), obtaining the absorption at each pixel and assigning each a "grayness value" according to how much radiation was absorbed. The image is made up of tiny spots (pixels) of varying shades of gray. Often the amount of absorption is color-coded. The colors in the resulting **false-color** image have nothing to do, however, with the actual color of the object. The real medical images are monochromatic (various shades of gray). Only *visible* light has color; X-rays and γ rays don't.

Figure 42–24 illustrates what actual CT images look like. It is generally agreed that CT scanning has revolutionized some areas of medicine by providing much less invasive, and/or more accurate, diagnosis.

Computed tomography can also be applied to ultrasound imaging (Section 16–9) and to emissions from radioisotopes and nuclear magnetic resonance, which we discuss in Section 42–10.

*Tomographic Image Reconstruction

How can the "grayness" of each pixel be determined even though all we can measure is the total absorption along each beam line in the slice? It can be done only by using the many beam scans made at a great many different angles. Suppose the image is to be an array of 100×100 elements for a total of 10^4 pixels. If we have 100 detectors and measure the absorption projections at 100 different angles, then we get 10^4 pieces of information. From this information, an image can be reconstructed, but not precisely. If more angles are measured, the reconstruction of the image can be done more accurately.

To suggest how mathematical reconstruction is done, we consider a very simple case using the "iterative" technique ("to iterate" is from the Latin "to repeat"). Suppose our sample slice is divided into the simple 2×2 pixels as shown in Fig. 42–25. The number inside each pixel represents the amount of absorption by the material in that area (say, in tenths of a percent): that is, 4 represents twice as much absorption as 2. But we cannot directly measure these values—they are the unknowns we want to solve for. All we can measure are the projections—the total absorption along each beam line—and these are shown in the diagram as the sum of the absorptions for the pixels along each line at four different angles. These projections (given at the tip of each arrow) are what we can measure, and we now want to work back from them to see how close we can get to the true absorption value for each pixel. We start our analysis with each pixel being assigned a zero value, Fig. 42–26a. In the iterative technique, we use the projections to estimate the absorption value in each square, and repeat for each angle. The angle 1 projections are 7 and 13. We divide each of these equally between their two squares: each square in the left column gets $3\frac{1}{2}$ (half of 7), and each square in the right column gets $6\frac{1}{2}$ (half of 13); see Fig. 42–26b. Next we use the projections at angle 2. We

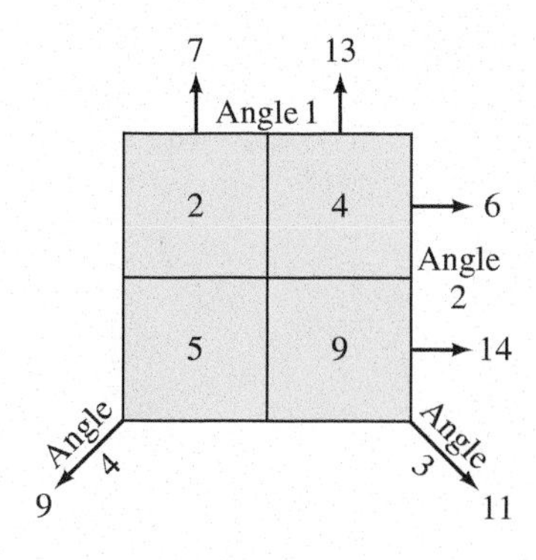

FIGURE 42–25 A simple 2×2 image showing true absorption values and measured projections.

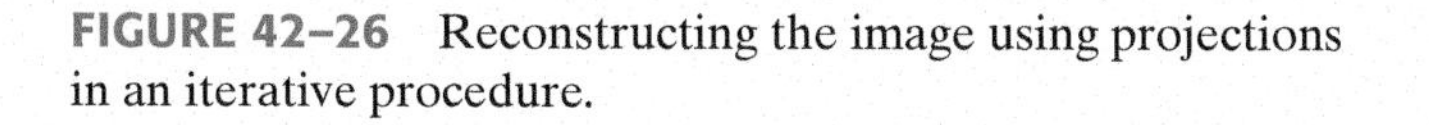
FIGURE 42–26 Reconstructing the image using projections in an iterative procedure.

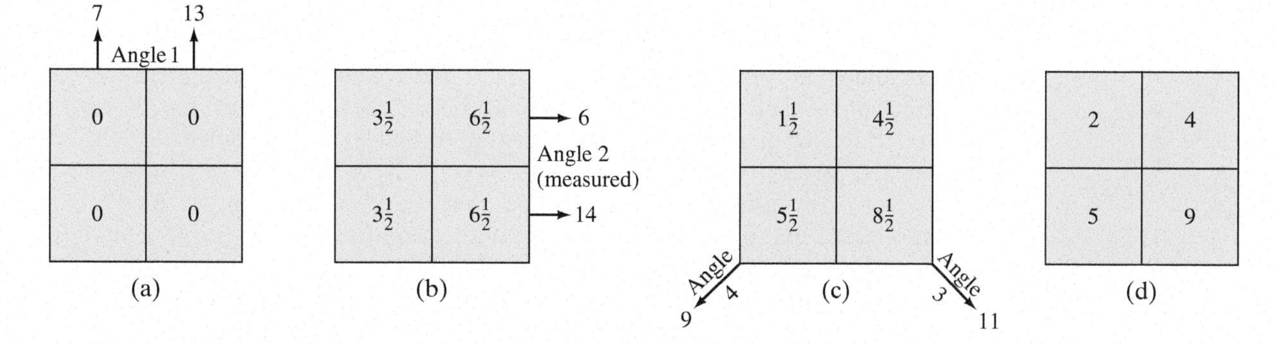

calculate the difference between the measured projections at angle 2 (6 and 14) and the projections based on the previous estimate (top row: $3\frac{1}{2} + 6\frac{1}{2} = 10$; same for bottom row). Then we distribute this difference equally to the squares in that row. For the top row, we have

$$3\tfrac{1}{2} + \frac{6-10}{2} = 1\tfrac{1}{2} \quad \text{and} \quad 6\tfrac{1}{2} + \frac{6-10}{2} = 4\tfrac{1}{2};$$

and for the bottom row,

$$3\tfrac{1}{2} + \frac{14-10}{2} = 5\tfrac{1}{2} \quad \text{and} \quad 6\tfrac{1}{2} + \frac{14-10}{2} = 8\tfrac{1}{2}.$$

These values are inserted as shown in Fig. 42–26c. Next, the projection at angle 3 gives

$$(\text{upper left})\ 1\tfrac{1}{2} + \frac{11-10}{2} = 2 \quad \text{and} \quad (\text{lower right})\ 8\tfrac{1}{2} + \frac{11-10}{2} = 9;$$

and that for angle 4 gives

$$(\text{lower left})\ 5\tfrac{1}{2} + \frac{9-10}{2} = 5 \quad \text{and} \quad (\text{upper right})\ 4\tfrac{1}{2} + \frac{9-10}{2} = 4.$$

The result, shown in Fig. 42–26d, corresponds exactly to the true values. (In real situations, the true values are not known, which is why these computer techniques are required.) To obtain these numbers exactly, we used six pieces of information (two each at angles 1 and 2, one each at angles 3 and 4). For the much larger number of pixels used for actual images, exact values are generally not attained. Many iterations may be needed, and the calculation is considered sufficiently precise when the difference between calculated and measured projections is sufficiently small. The above example illustrates the "convergence" of the process: the first iteration (b to c in Fig. 42–26) changed the values by 2, the last iteration (c to d) by only $\frac{1}{2}$.

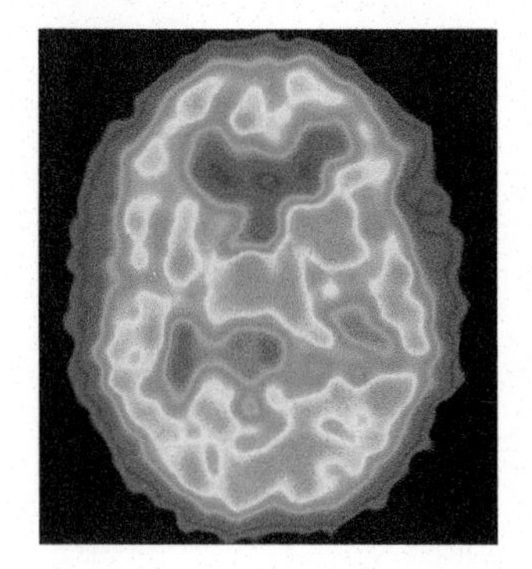

FIGURE 42–27 SPECT scan of brain (false color) with epilepsy, labeled with $^{99m}_{43}\text{Tc}$.

PHYSICS APPLIED
Emission tomography (SPECT, PET)

*Emission Tomography

It is possible to image the emissions of a radioactive tracer (see Section 42–8) in a single plane or slice through a body using computed tomography techniques. A basic gamma detector (Fig. 42–20a) can be moved around the patient to measure the radioactive intensity from the tracer at many points and angles; the data are processed in much the same way as for X-ray CT scans. This technique is referred to as **single photon emission tomography** (SPET), or SPECT (single photon emission computed tomography); see Fig. 42–27.

Another important technique is **positron emission tomography** (PET), which makes use of positron emitters such as $^{11}_{6}\text{C}$, $^{13}_{7}\text{N}$, $^{15}_{8}\text{O}$, and $^{18}_{9}\text{F}$. These isotopes are incorporated into molecules that, when inhaled or injected, accumulate in the organ or region of the body to be studied. When such a nuclide undergoes β^+ decay, the emitted positron travels at most a few millimeters before it collides with a normal electron. In this collision, the positron and electron are annihilated, producing two γ rays $(e^+ + e^- \rightarrow 2\gamma)$. The two γ rays fly off in opposite directions $(180° \pm 0.25°)$ since they must have almost exactly equal and opposite momenta to conserve momentum (the momenta of the initial e^+ and e^- are essentially zero compared to the momenta of the γ rays). Because the photons travel along the same line in opposite directions, their detection in coincidence by rings of detectors surrounding the patient (Fig. 42–28) readily establishes the line along which the emission took place. If the difference in time of arrival of the two photons could be determined accurately, the actual position of the emitting nuclide along that line could be calculated. Present-day electronics can measure times to at best $\pm 300\,\text{ps}$, so at the γ ray's speed $(c = 3 \times 10^8\,\text{m/s})$, the actual position could be determined to an accuracy on the order of about $d = vt \approx (3 \times 10^8\,\text{m/s})(300 \times 10^{-12}\,\text{s}) \approx 10\,\text{cm}$, which is not very useful. Although there may be future potential for time-of-flight measurements to determine position, today computed tomography techniques are used instead, similar to those for X-ray CT, which can reconstruct PET images with a resolution on the order of 3–5 mm. One big advantage of PET is that no collimators are needed (as for detection of a single photon—see Fig. 42–20a). Thus, fewer photons are "wasted" and lower doses can be administered to the patient with PET.

FIGURE 42–28 Positron emission tomography (PET) system showing a ring of detectors to detect the two annihilation γ rays $(e^+ + e^- \rightarrow 2\gamma)$ emitted at 180° to each other.

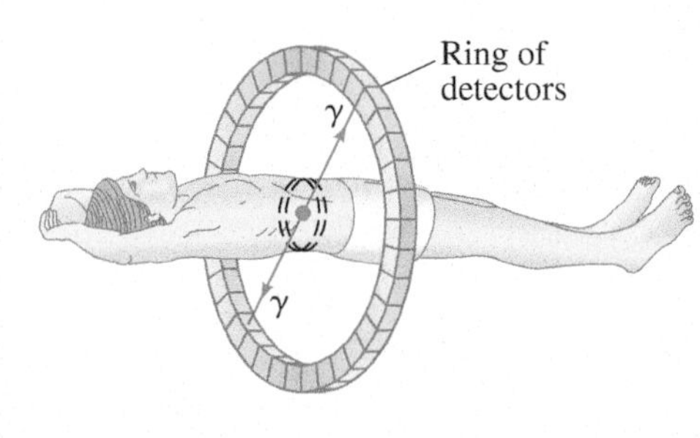

Both PET and SPET systems can give images related to biochemistry, metabolism, and function. This is to be compared to X-ray CT scans, whose images reflect shape and structure—that is, the anatomy of the imaged region.

*42–10 Nuclear Magnetic Resonance (NMR); Magnetic Resonance Imaging (MRI)

Nuclear magnetic resonance (NMR) is a phenomenon which soon after its discovery in 1946 became a powerful research tool in a variety of fields from physics to chemistry and biochemistry. It is also an important medical imaging technique. We first briefly discuss the phenomenon, and then look at its applications.

*Nuclear Magnetic Resonance (NMR)

FIGURE 42–29 Schematic picture of a proton in a magnetic field $\vec{\mathbf{B}}$ (pointing upward) with the two possible states of proton spin, up and down.

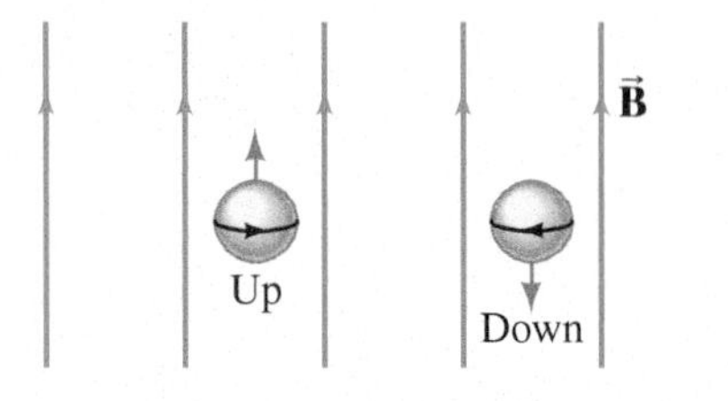

We saw in Chapter 39 (Section 39–2) that when atoms are placed in a magnetic field B, atomic energy levels are split into several closely spaced levels (the Zeeman effect) according to the angular momentum or spin of the state. The splitting is proportional to B and to the magnetic moment, μ. Nuclei too have magnetic moments (Section 41–1), and we examine mainly the simplest, the hydrogen (^1_1H) nucleus because it is the one most used, even for medical imagining. The ^1_1H nucleus consists of a single proton. Its spin angular momentum (and its magnetic moment), like that of the electron, can take on only two values when placed in a magnetic field: We call these "spin up" (parallel to the field) and "spin down" (antiparallel to the field) as suggested in Fig. 42–29.

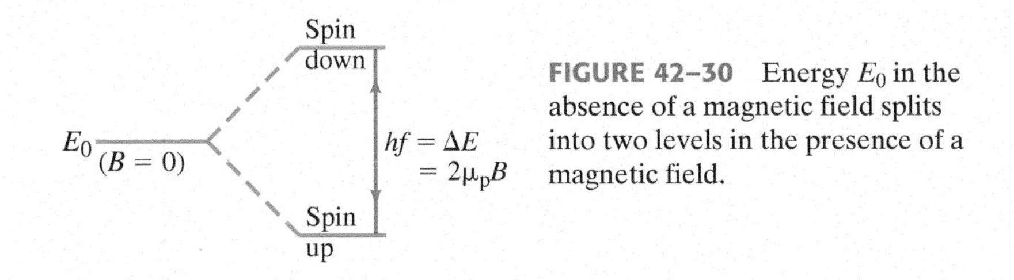

FIGURE 42–30 Energy E_0 in the absence of a magnetic field splits into two levels in the presence of a magnetic field.

When a magnetic field is present, an energy state of a nucleus splits into two levels as shown in Fig. 42–30 with the spin-up state (parallel to field) having the lower energy. The spin-down state acquires an additional energy $\mu_p B_T$ and the spin-up state has its energy changed by $-\mu_p B_T$ (see Eq. 27–12, Example 27–12, and Section 39–7), where B_T is the total magnetic field *at the nucleus*. The difference in energy between the two states (Fig. 42–30) is thus

$$\Delta E = 2\,\mu_p B_T,$$

where μ_p is the magnetic moment of the proton.

In a standard **nuclear magnetic resonance** (NMR) setup, the sample to be examined is placed in a static magnetic field. A radiofrequency (RF) pulse of electromagnetic radiation (that is, photons) is applied to the sample. If the frequency, f, of this pulse corresponds precisely to the energy difference between the two energy levels (Fig. 42–30), so that

$$hf = \Delta E = 2\mu_p B_T, \qquad \textbf{(42–12)}$$

then the photons of the RF beam will be absorbed, exciting many of the nuclei from the lower state to the upper state. This is a resonance phenomenon since there is significant absorption only if f is very near $f = 2\mu_p B_T/h$. Hence the name "nuclear magnetic resonance." For free ${}^1_1\mathrm{H}$ nuclei, the frequency is 42.58 MHz for a field $B_T = 1.0\,\mathrm{T}$ (Example 42–14). If the H atoms are bound in a molecule, the total magnetic field B_T at the H nuclei will be the sum of the external applied field (B_{ext}) plus the local magnetic field (B_{loc}) due to electrons and nuclei of neighboring atoms. Since f is proportional to B_T, the value of f for a given external field will be slightly different for bound H atoms than for free atoms:

$$hf = 2\mu_p(B_{ext} + B_{loc}).$$

This small change in frequency can be measured, and is called the "chemical shift." A great deal has been learned about the structure of molecules and bonds using this NMR technique.

EXAMPLE 42–14 **NMR for free protons.** Calculate the resonant frequency for free protons in a 1.000-T magnetic field.

APPROACH We use Eq. 42–12, where the magnetic moment of the proton (Section 41–1) is

$$\mu_p = 2.7928\mu_N = 2.7928\left(\frac{e\hbar}{2m_p}\right) = 2.7928\left(\frac{eh}{4\pi m_p}\right).$$

SOLUTION We solve for f in Eq. 42–12 and find

$$f = \frac{\Delta E}{h} = \frac{2\mu_p B}{h}$$

$$= (2.7928)\left(\frac{eB}{2\pi m_p}\right) = 2.7928\left[\frac{(1.6022\times10^{-19}\,\mathrm{C})(1.000\,\mathrm{T})}{2\pi(1.6726\times10^{-27}\,\mathrm{kg})}\right]$$

$$= 42.58\,\mathrm{MHz}.$$

* Magnetic Resonance Imaging (MRI)

PHYSICS APPLIED
NMR imaging (MRI)

For producing medically useful NMR images—now commonly called MRI, or **magnetic resonance imaging**—the element most used is hydrogen since it is the commonest element in the human body and gives the strongest NMR signals. The experimental apparatus is shown in Fig. 42–31. The large coils set up the static magnetic field, and the RF coils produce the RF pulse of electromagnetic waves (photons) that cause the nuclei to jump from the lower state to the upper one (Fig. 42–30). These same coils (or another coil) can detect the absorption of energy or the emitted radiation (also of frequency $f = \Delta E/h$) when the nuclei jump back down to the lower state.

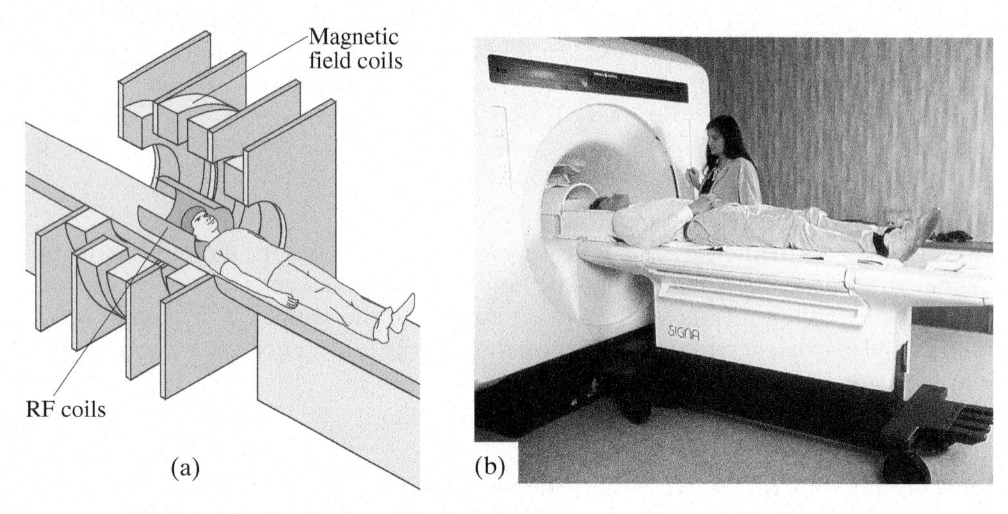

FIGURE 42–31 Typical MRI imaging setup: (a) diagram; (b) photograph.

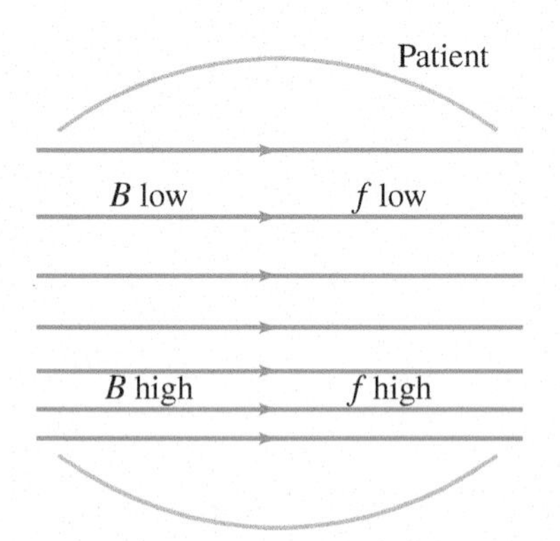

FIGURE 42–32 A static field that is stronger at the bottom than at the top. The frequency of absorbed or emitted radiation is proportional to B in NMR.

The formation of a two-dimensional or three-dimensional image can be done using techniques similar to those for computed tomography (Section 42–9). The simplest thing to measure for creating an image is the intensity of absorbed and/or reemitted radiation from many different points of the body, and this would be a measure of the density of H atoms at each point. But how do we determine from what part of the body a given photon comes? One technique is to give the static magnetic field a gradient; that is, instead of applying a uniform magnetic field, B_T, the field is made to vary with position across the width of the sample (or patient). Since the frequency absorbed by the H nuclei is proportional to B_T (Eq. 42–12), only one plane within the body will have the proper value of B_T to absorb photons of a particular frequency f. By varying f, absorption by different planes can be measured. Alternately, if the field gradient is applied *after* the RF pulse, the frequency of the emitted photons will be a measure of where they were emitted. See Fig. 42–32. If a magnetic field gradient in one direction is applied during excitation (absorption of photons) and photons of a single frequency are transmitted, only H nuclei in one thin slice will be excited. By applying a gradient during reemission in a direction perpendicular to the first, the frequency f of the reemitted radiation will represent depth in that slice. Other ways of varying the magnetic field throughout the volume of the body can be used in order to correlate NMR frequency with position.

A reconstructed image based on the density of H atoms (that is, the intensity of absorbed or emitted radiation) is not very interesting. More useful are images based on the rate at which the nuclei decay back to the ground state, and such images can produce resolution of 1 mm or better. This NMR technique (sometimes called **spin-echo**) produces images of great diagnostic value, both in the delineation of structure (anatomy) and in the study of metabolic processes. An NMR image is shown in Fig. 42–33, color enhanced—no medical imaging uses visible light, so the colors shown here are added. The original images, those looked at by your doctor, are various shades of gray, representing intensity (or counts).

FIGURE 42–33 False-color NMR image (MRI) of a vertical section through the head showing structures in the normal brain.

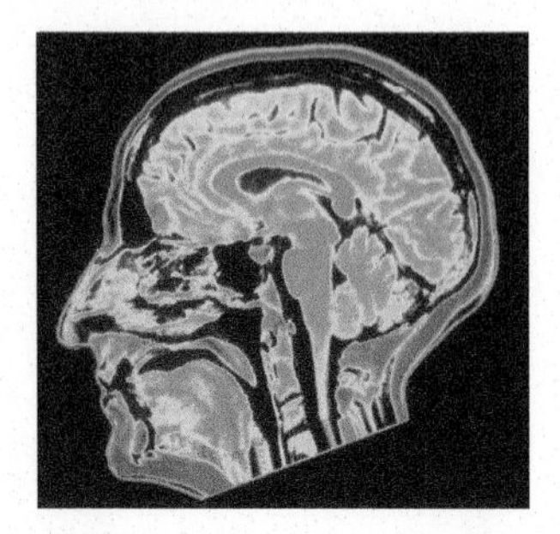

NMR imaging is considered to be noninvasive. We can calculate the energy of the photons involved: as determined in Example 42–14, in a 1.0-T magnetic field, $f = 42.58\,\text{MHz}$ for ${}^1_1\text{H}$. This corresponds to an energy of $hf = (6.6 \times 10^{-34}\,\text{J}\cdot\text{s})(43 \times 10^6\,\text{Hz}) \approx 3 \times 10^{-26}\,\text{J}$ or about $10^{-7}\,\text{eV}$. Since molecular bonds are on the order of 1 eV, it is clear that the RF photons can cause little cellular disruption. This should be compared to X- or γ rays whose energies are 10^4 to $10^6\,\text{eV}$ and thus can cause significant damage. The static magnetic fields, though often large (typically around 2.0 T), are believed to be harmless (except for people wearing heart pacemakers).

Table 42–2 lists the major techniques we have discussed for imaging the interior of the human body, along with the optimum resolution attainable today. Resolution is just one factor that must be considered, as the different imaging techniques provide different types of information, useful for different types of diagnosis.

TABLE 42–2 Medical Imaging Techniques

Technique	Resolution
Conventional X-ray	$\frac{1}{2}$ mm
CT scan, X-ray	0.2 mm
Nuclear medicine (tracers)	1 cm
SPET (single photon emission)	1 cm
PET (positron emission)	2–5 mm
NMR (MRI)	$\frac{1}{2}$–1 mm
Ultrasound (Section 16–9)	0.3–2 mm

Summary

A **nuclear reaction** occurs when two nuclei collide and two or more other nuclei (or particles) are produced. In this process, as in radioactivity, **transmutation** (change) of elements occurs.

The **reaction energy** or ***Q*-value** of a reaction $a + X \rightarrow Y + b$ is

$$Q = (M_a + M_X - M_b - M_Y)c^2 \qquad \text{(42–2a)}$$

$$= K_b + K_Y - K_a - K_X. \qquad \text{(42–2b)}$$

The effective **cross section** σ for a reaction is a measure of the reaction probability per target nucleus.

In **fission**, a heavy nucleus such as uranium splits into two intermediate-sized nuclei after being struck by a neutron. ${}^{235}_{92}\text{U}$ is fissionable by slow neutrons, whereas some fissionable nuclei require fast neutrons. Much energy is released in fission ($\approx 200\,\text{MeV}$ per fission) because the binding energy per nucleon is lower for heavy nuclei than it is for intermediate-sized nuclei, so the mass of a heavy nucleus is greater than the total mass of its fission products. The fission process releases neutrons, so that a **chain reaction** is possible. The **critical mass** is the minimum mass of fuel needed to sustain a chain reaction. In a **nuclear reactor** or nuclear bomb, a **moderator** is used to slow down the released neutrons.

The **fusion** process, in which small nuclei combine to form larger ones, also releases energy. The energy from our Sun originates in the fusion reactions known as the **proton–proton chain** in which four protons fuse to form a ${}^4_2\text{He}$ nucleus producing 25 MeV of energy. A useful fusion reactor for power generation has not yet proved possible because of the difficulty in containing the fuel (e.g., deuterium) long enough at the extremely high temperature required ($\approx 10^8\,\text{K}$). Nonetheless, great progress has been made in confining the collection of charged ions known as a **plasma**. The two main methods are **magnetic confinement**, using a magnetic field in a device such as the toroidal-shaped **tokamak**, and **inertial confinement** in which intense laser beams compress a fuel pellet of deuterium and tritium.

Radiation can cause damage to materials, including biological tissue. Quantifying amounts of radiation is the subject of **dosimetry**. The **curie** (Ci) and the **becquerel** (Bq) are units that measure the **source activity** or rate of decay of a sample: $1\,\text{Ci} = 3.70 \times 10^{10}$ decays per second, whereas $1\,\text{Bq} = 1\,\text{decay/s}$. The **absorbed dose**, often specified in **rads**, measures the amount of energy deposited per unit mass of absorbing material: 1 rad is the amount of radiation that deposits energy at the rate of $10^{-2}\,\text{J/kg}$ of material. The SI unit of absorbed dose is the **gray**: $1\,\text{Gy} = 1\,\text{J/kg} = 100\,\text{rad}$. The **effective dose** is often specified by the **rem** = rad × QF, where QF is the "quality factor" of a given type of radiation; 1 rem of any type of radiation does approximately the same amount of biological damage. The average dose received per person per year in the United States is about 360 mrem. The SI unit for effective dose is the **sievert**: $1\,\text{Sv} = 10^2\,\text{rem}$.

[*Nuclear radiation is used in medicine for cancer therapy, and for imaging of biological structure and processes. **Tomographic imaging** of the human body, which can provide 3-dimensional detail, includes several types: CT scans, PET, SPET (= SPECT), and MRI; the latter makes use of **nuclear magnetic resonance** (NMR).]

Questions

(NOTE: Masses are found in Appendix F.)

1. Fill in the missing particles or nuclei:
   (*a*) $n + {}^{137}_{56}\text{Ba} \rightarrow\ ? + \gamma$;
   (*b*) $n + {}^{137}_{56}\text{Ba} \rightarrow {}^{137}_{55}\text{Cs} + ?$;
   (*c*) $d + {}^2_1\text{H} \rightarrow {}^4_2\text{He} + ?$;
   (*d*) $\alpha + {}^{197}_{79}\text{Au} \rightarrow\ ? + d$
   where d stands for deuterium.
2. The isotope ${}^{32}_{15}\text{P}$ is produced by the reaction: $n + ? \rightarrow {}^{32}_{15}\text{P} + p$. What must be the target nucleus?
3. When ${}^{22}_{11}\text{Na}$ is bombarded by deuterons (${}^2_1\text{H}$), an α particle is emitted. What is the resulting nuclide?
4. Why are neutrons such good projectiles for producing nuclear reactions?
5. A proton strikes a ${}^{20}_{10}\text{Ne}$ nucleus, and an α particle is observed to emerge. What is the residual nucleus? Write down the reaction equation.
6. Are fission fragments β^+ or β^- emitters? Explain.
7. The energy from nuclear fission appears in the form of thermal energy—but the thermal energy of what?

8. $^{238}_{92}$U releases an average of 2.5 neutrons per fission compared to 2.9 for $^{239}_{94}$Pu. Which of these two nuclei do you think would have the smaller critical mass? Explain.

9. If $^{235}_{92}$U released only 1.5 neutrons per fission on the average, would a chain reaction be possible? If so, how would the chain reaction be different than if 3 neutrons were released per fission?

10. Why can't uranium be enriched by chemical means?

11. How can a neutron, with practically no kinetic energy, excite a nucleus to the extent shown in Fig. 42–4?

12. Why would a porous block of uranium be more likely to explode if kept under water rather than in air?

13. A reactor that uses highly enriched uranium can use ordinary water (instead of heavy water) as a moderator and still have a self-sustaining chain reaction. Explain.

14. Why must the fission process release neutrons if it is to be useful?

15. Why are neutrons released in a fission reaction?

16. What is the reason for the "secondary system" in a nuclear reactor, Fig. 42–9? That is, why is the water heated by the fuel in a nuclear reactor not used directly to drive the turbines?

17. What is the basic difference between fission and fusion?

18. Discuss the relative merits and disadvantages, including pollution and safety, of power generation by fossil fuels, nuclear fission, and nuclear fusion.

19. A higher temperature is required for deuterium–deuterium ignition than for deuterium–tritium. Explain.

20. Light energy emitted by the Sun and stars comes from the fusion process. What conditions in the interior of stars make this possible?

21. How do stars, and our Sun, maintain confinement of the plasma for fusion?

22. Why is the recommended maximum radiation dose higher for women beyond the child-bearing age than for younger women?

23. People who work around metals that emit alpha particles are trained that there is little danger from proximity or touching the material, but they must take extreme precautions against ingesting it. Why? (Eating and drinking while working are forbidden.)

24. What is the difference between absorbed dose and effective dose? What are the SI units for each?

25. Radiation is sometimes used to sterilize medical supplies and even food. Explain how it works.

*26. How might radioactive tracers be used to find a leak in a pipe?

Problems

(NOTE: Masses are found in Appendix F.)

42–1 Nuclear Reactions, Transmutation

1. (I) Natural aluminum is all $^{27}_{13}$Al. If it absorbs a neutron, what does it become? Does it decay by β^+ or β^-? What will be the product nucleus?

2. (I) Determine whether the reaction $^{2}_{1}\text{H} + ^{2}_{1}\text{H} \rightarrow ^{3}_{2}\text{He} + \text{n}$ requires a threshold energy.

3. (I) Is the reaction $\text{n} + ^{238}_{92}\text{U} \rightarrow ^{239}_{92}\text{U} + \gamma$ possible with slow neutrons? Explain.

4. (II) Does the reaction $\text{p} + ^{7}_{3}\text{Li} \rightarrow ^{4}_{2}\text{He} + \alpha$ require energy, or does it release energy? How much energy?

5. (II) Calculate the energy released (or energy input required) for the reaction $\alpha + ^{9}_{4}\text{Be} \rightarrow ^{12}_{6}\text{C} + \text{n}$.

6. (II) (*a*) Can the reaction $\text{n} + ^{24}_{12}\text{Mg} \rightarrow ^{23}_{11}\text{Na} + \text{d}$ occur if the bombarding particles have 16.00 MeV of kinetic energy? (d stands for deuterium, $^{2}_{1}$H.) (*b*) If so, how much energy is released? If not, what kinetic energy is needed?

7. (II) (*a*) Can the reaction $\text{p} + ^{7}_{3}\text{Li} \rightarrow ^{4}_{2}\text{He} + \alpha$ occur if the incident proton has kinetic energy = 3500 keV? (*b*) If so, what is the total kinetic energy of the products? If not, what kinetic energy is needed?

8. (II) In the reaction $\alpha + ^{14}_{7}\text{N} \rightarrow ^{17}_{8}\text{O} + \text{p}$, the incident α particles have 9.68 MeV of kinetic energy. The mass of $^{17}_{8}$O is 16.999132 u. (*a*) Can this reaction occur? (*b*) If so, what is the total kinetic energy of the products? If not, what kinetic energy is needed?

9. (II) Calculate the Q-value for the "capture" reaction $\alpha + ^{16}_{8}\text{O} \rightarrow ^{20}_{10}\text{Ne} + \gamma$.

10. (II) Calculate the total kinetic energy of the products of the reaction $\text{d} + ^{13}_{6}\text{C} \rightarrow ^{14}_{7}\text{N} + \text{n}$ if the incoming deuteron has kinetic energy $K = 44.4$ MeV.

11. (II) Radioactive $^{14}_{6}$C is produced in the atmosphere when a neutron is absorbed by $^{14}_{7}$N. Write the reaction and find its Q-value.

12. (II) An example of a **stripping** nuclear reaction is $\text{d} + ^{6}_{3}\text{Li} \rightarrow \text{X} + \text{p}$. (*a*) What is X, the resulting nucleus? (*b*) Why is it called a "stripping" reaction? (*c*) What is the Q-value of this reaction? Is the reaction endothermic or exothermic?

13. (II) An example of a **pick-up** nuclear reaction is $^{3}_{2}\text{He} + ^{12}_{6}\text{C} \rightarrow \text{X} + \alpha$. (*a*) Why is it called a "pick-up" reaction? (*b*) What is the resulting nucleus? (*c*) What is the Q-value of this reaction? Is the reaction endothermic or exothermic?

14. (II) (*a*) Complete the following nuclear reaction, $\text{p} + ? \rightarrow ^{32}_{16}\text{S} + \gamma$. (*b*) What is the Q-value?

15. (II) The reaction $\text{p} + ^{18}_{8}\text{O} \rightarrow ^{18}_{9}\text{F} + \text{n}$ requires an input of energy equal to 2.438 MeV. What is the mass of $^{18}_{9}$F?

16. (III) Use conservation of energy and momentum to show that a bombarding proton must have an energy of 3.23 MeV in order to make the reaction $^{13}_{6}\text{C}(\text{p}, \text{n})^{13}_{7}\text{N}$ occur. (See Example 42–3.)

17. (III) How much kinetic energy (if any) would the proton require for the reaction $^{14}_{6}\text{C}(\text{p}, \text{n})^{14}_{7}\text{N}$ to proceed?

42–2 Cross Section

18. (I) The cross section for the reaction $\text{n} + ^{10}_{5}\text{B} \rightarrow ^{7}_{3}\text{Li} + ^{4}_{2}\text{He}$ is about 40 bn for an incident neutron of low energy (kinetic energy ≈ 0). The boron is contained in a gas with $n = 1.7 \times 10^{21}$ nuclei/m^3 and the target has thickness $\ell = 12.0$ cm. What fraction of incident neutrons will be scattered?

19. (I) What is the effective cross section for the collision of two hard spheres of radius R_1 and R_2?

20. (II) When the target is thick, the rate at which projectile particles collide with nuclei in the rear of the target is less than in the front of the target, since some scattering (i.e., collisions) takes place in the front layers. Let R_0 be the rate at which incident particles strike the front of the target, and R_x be the rate at a distance x into the target ($R_x = R_0$ at $x = 0$). Then show that the rate at which particles are scattered (and therefore lost from the incident beam) in a thickness dx is $-dR_x = R_x n\sigma\, dx$, where the minus sign means that R_x is decreasing and n is the number of nuclei per unit volume. Then show that $R_x = R_0 e^{-n\sigma x}$, where σ is the total cross section. If the thickness of the target is ℓ, what does $R_x = R_0 e^{-n\sigma\ell}$ represent?

21. (II) A 1.0-cm-thick lead target reduces a beam of gamma rays to 25% of its original intensity. What thickness of lead will allow only one γ in 10^6 to penetrate (see Problem 20)?

22. (II) Use Fig. 42–3 to estimate what thickness of $^{114}_{48}\text{Cd}$ ($\rho = 8650\ \text{kg/m}^3$) will cause a 2.0% reaction rate ($R/R_0 = 0.020$) for (*a*) 0.10-eV neutrons (*b*) 5.0-eV neutrons.

42–3 Nuclear Fission

23. (I) What is the energy released in the fission reaction of Eq. 42–5? (The masses of $^{141}_{56}\text{Ba}$ and $^{92}_{36}\text{Kr}$ are 140.914411 u and 91.926156 u, respectively.)

24. (I) Calculate the energy released in the fission reaction $\text{n} + {}^{235}_{92}\text{U} \rightarrow {}^{88}_{38}\text{Sr} + {}^{136}_{54}\text{Xe} + 12\text{n}$. Use Appendix F, and assume the initial kinetic energy of the neutron is very small.

25. (I) How many fissions take place per second in a 200-MW reactor? Assume 200 MeV is released per fission.

26. (I) The energy produced by a fission reactor is about 200 MeV per fission. What fraction of the mass of a $^{235}_{92}\text{U}$ nucleus is this?

27. (II) Suppose that the average electric power consumption, day and night, in a typical house is 880 W. What initial mass of $^{235}_{92}\text{U}$ would have to undergo fission to supply the electrical needs of such a house for a year? (Assume 200 MeV is released per fission, as well as 100% efficiency.)

28. (II) Consider the fission reaction

$$^{235}_{92}\text{U} + \text{n} \rightarrow {}^{133}_{51}\text{Sb} + {}^{98}_{41}\text{Nb} + ?\text{n}.$$

(*a*) How many neutrons are produced in this reaction? (*b*) Calculate the energy release. The atomic masses for Sb and Nb isotopes are 132.915250 u and 97.910328 u, respectively.

29. (II) How much mass of $^{238}_{92}\text{U}$ is required to produce the same amount of energy as burning 1.0 kg of coal (about 3×10^7 J)?

30. (II) What initial mass of $^{235}_{92}\text{U}$ is required to operate a 950-MW reactor for 1 yr? Assume 38% efficiency.

31. (II) If a 1.0-MeV neutron emitted in a fission reaction loses one-half of its kinetic energy in each collision with moderator nuclei, how many collisions must it make to reach thermal energy ($\frac{3}{2}kT = 0.040\ \text{eV}$)?

32. (II) Assuming a fission of $^{236}_{92}\text{U}$ into two roughly equal fragments, estimate the electric potential energy just as the fragments separate from each other. Assume that the fragments are spherical (see Eq. 41–1) and compare your calculation to the nuclear fission energy released, about 200 MeV.

33. (II) Estimate the ratio of the height of the Coulomb barrier for α decay to that for fission of $^{236}_{92}\text{U}$. (Both are described by a potential energy diagram of the shape shown in Fig. 41–7.)

34. (II) Suppose that the neutron multiplication factor is 1.0004. If the average time between successive fissions in a chain of reactions is 1.0 ms, by what factor will the reaction rate increase in 1.0 s?

42–4 Nuclear Fusion

35. (I) What is the average kinetic energy of protons at the center of a star where the temperature is 2×10^7 K? [*Hint*: See Eq. 18–4.]

36. (II) Show that the energy released in the fusion reaction $^2_1\text{H} + {}^3_1\text{H} \rightarrow {}^4_2\text{He} + \text{n}$ is 17.57 MeV.

37. (II) Show that the energy released when two deuterium nuclei fuse to form ^3_2He with the release of a neutron is 3.23 MeV.

38. (II) Verify the Q-value stated for each of the reactions of Eqs. 42–7. [*Hint*: Be careful with electrons.]

39. (II) (*a*) Calculate the energy release per gram of fuel for the reactions of Eqs. 42–9a, b, and c. (*b*) Calculate the energy release per gram of uranium $^{235}_{92}\text{U}$ in fission, and give its ratio to each reaction in (*a*).

40. (II) How much energy is released when $^{238}_{92}\text{U}$ absorbs a slow neutron (kinetic energy ≈ 0) and becomes $^{239}_{92}\text{U}$?

41. (II) If a typical house requires 850 W of electric power on average, what minimum amount of deuterium fuel would have to be used in a year to supply these electrical needs? Assume the reaction of Eq. 42–9b.

42. (II) If ^6_3Li is struck by a slow neutron, it can form ^4_2He and another isotope. (*a*) What is the second isotope? (This is a method of generating this isotope.) (*b*) How much energy is released in the process?

43. (II) Suppose a fusion reactor ran on "d–d" reactions, Eqs. 42–9a and b in equal amounts. Estimate how much natural water, for fuel, would be needed per hour to run a 1250-MW reactor, assuming 33% efficiency.

44. (II) Show that the energies carried off by the ^4_2He nucleus and the neutron for the reaction of Eq. 42–9c are about 3.5 MeV and 14 MeV, respectively. Are these fixed values, independent of the plasma temperature?

45. (II) How much energy (J) is contained in 1.00 kg of water if its natural deuterium is used in the fusion reaction of Eq. 42–9a? Compare to the energy obtained from the burning of 1.0 kg of gasoline, about 5×10^7 J.

46. (III) (*a*) Give the ratio of the energy needed for the first reaction of the *carbon cycle* to the energy needed for a deuterium–tritium reaction (Example 42–10). (*b*) If a deuterium–tritium reaction requires $T \approx 3 \times 10^8$ K, estimate the temperature needed for the first carbon-cycle reaction.

47. (III) The energy output of massive stars is believed to be due to the *carbon cycle* (see text). (*a*) Show that no carbon is consumed in this cycle and that the net effect is the same as for the proton–proton cycle. (*b*) What is the total energy release? (*c*) Determine the energy output for each reaction and decay. (*d*) Why might the carbon cycle require a higher temperature ($\approx 2 \times 10^7$ K) than the proton–proton cycle ($\approx 1.5 \times 10^7$ K)?

42–6 Dosimetry

48. (I) 250 rads of α-particle radiation is equivalent to how many rads of X-rays in terms of biological damage?

49. (I) A dose of 4.0 Sv of γ rays in a short period would be lethal to about half the people subjected to it. How many grays is this?

50. (I) How much energy is deposited in the body of a 65-kg adult exposed to a 3.0-Gy dose?

51. (I) How many rads of slow neutrons will do as much biological damage as 65 rads of fast neutrons?

52. (II) A cancer patient is undergoing radiation therapy in which protons with an energy of 1.2 MeV are incident on a 0.25-kg tumor. (*a*) If the patient receives an effective dose of 1.0 rem, what is the absorbed dose? (*b*) How many protons are absorbed by the tumor? Assume QF $\approx$ 1.

53. (II) A 0.035-μCi sample of $^{32}_{15}\mathrm{P}$ is injected into an animal for tracer studies. If a Geiger counter intercepts 25% of the emitted β particles, what will be the counting rate, assumed 85% efficient?

54. (II) About 35 eV is required to produce one ion pair in air. Show that this is consistent with the two definitions of the roentgen given in the text.

55. (II) A 1.6-mCi source of $^{32}_{15}\mathrm{P}$ (in $NaHPO_4$), a β emitter, is implanted in a tumor where it is to administer 36 Gy. The half-life of $^{32}_{15}\mathrm{P}$ is 14.3 days, and 1.0 mCi delivers about 10 mGy/min. Approximately how long should the source remain implanted?

56. (II) What is the mass of a 2.00-μCi $^{14}_{6}\mathrm{C}$ source?

57. (II) Huge amounts of radioactive $^{131}_{53}\mathrm{I}$ were released in the accident at Chernobyl in 1986. Chemically, iodine goes to the human thyroid. (Doctors can use it for diagnosis and treatment of thyroid problems.) In a normal thyroid, $^{131}_{53}\mathrm{I}$ absorption can cause damage to the thyroid. (*a*) Write down the reaction for the decay of $^{131}_{53}\mathrm{I}$. (*b*) Its half-life is 8.0 d; how long would it take for ingested $^{131}_{53}\mathrm{I}$ to become 7.0% of the initial value? (*c*) Absorbing 1 mCi of $^{131}_{53}\mathrm{I}$ can be harmful; what mass of iodine is this?

58. (II) Assume a liter of milk typically has an activity of 2000 pCi due to $^{40}_{19}\mathrm{K}$. If a person drinks two glasses (0.5 L) per day, estimate the total effective dose (in Sv and in rem) received in a year. As a crude model, assume the milk stays in the stomach 12 hr and is then released. Assume also that very roughly 10% of the 1.5 MeV released per decay is absorbed by the body. Compare your result to the normal allowed dose of 100 mrem per year. Make your estimate for (*a*) a 60-kg adult, and (*b*) a 6-kg baby.

59. (II) $^{57}_{27}\mathrm{Co}$ emits 122-keV γ rays. If a 58-kg person swallowed 1.55 μCi of $^{57}_{27}\mathrm{Co}$, what would be the dose rate (Gy/day) averaged over the whole body? Assume that 50% of the γ-ray energy is deposited in the body. [*Hint*: Determine the rate of energy deposited in the body and use the definition of the gray.]

60. (II) Ionizing radiation can be used on meat products to reduce the levels of microbial pathogens. Refrigerated meat is limited to 4.5 kGy. If 1.2-MeV electrons irradiate 5 kg of beef, how many electrons would it take to reach the allowable limit?

61. (II) Radon gas, $^{222}_{86}\mathrm{Rn}$, is considered a serious health hazard (see discussion in text). It decays by α-emission. (*a*) What is the daughter nucleus? (*b*) Is the daughter nucleus stable or radioactive? If the latter, how does it decay, and what is its half-life? (See Fig. 41–12.) (*c*) Is the daughter nucleus also a noble gas, or is it chemically reactive? (*d*) Suppose 1.6 ng of $^{222}_{86}\mathrm{Rn}$ seeps into a basement. What will be its activity? If the basement is then sealed, what will be the activity 1 month later?

* 42–9 Imaging by Tomography

*62. (II) (*a*) Suppose for a conventional X-ray image that the X-ray beam consists of parallel rays. What would be the magnification of the image? (*b*) Suppose, instead, that the X-rays come from a point source (as in Fig. 42–21) that is 15 cm in front of a human body which is 25 cm thick, and the film is pressed against the person's back. Determine and discuss the range of magnifications that result.

* 42–10 NMR

*63. (I) Calculate the wavelength of photons needed to produce NMR transitions in free protons in a 1.000-T field. In what region of the spectrum is this wavelength?

*64. (II) Carbon-13 has a magnetic moment $\mu = 0.7023\mu_N$ (see Eq. 41–2). What magnetic field would be necessary if $^{13}_{6}\mathrm{C}$ were to be detected in a proton NMR spectrometer operating at 42.58 MHz? (This large field necessitates that a $^{13}_{6}\mathrm{C}$ spectrometer operate at a lower frequency.)

General Problems

65. J. Chadwick discovered the neutron by bombarding $^{9}_{4}\mathrm{Be}$ with the popular projectile of the day, alpha particles. (*a*) If one of the reaction products was the then unknown neutron, what was the other product? (*b*) What is the *Q*-value of this reaction?

66. Fusion temperatures are often given in keV. Determine the conversion factor from kelvins to keV using, as is common in this field, $\overline{K} = kT$ without the factor $\frac{3}{2}$.

67. One means of enriching uranium is by diffusion of the gas UF_6. Calculate the ratio of the speeds of molecules of this gas containing $^{235}_{92}\mathrm{U}$ and $^{238}_{92}\mathrm{U}$, on which this process depends.

68. (*a*) What mass of $^{235}_{92}\mathrm{U}$ was actually fissioned in the first atomic bomb, whose energy was the equivalent of about 20 kilotons of TNT (1 kiloton of TNT releases 5×10^{12} J)? (*b*) What was the actual mass transformed to energy?

69. The average yearly background radiation in a certain town consists of 29 mrad of X-rays and γ rays plus 3.6 mrad of particles having a QF of 10. How many rem will a person receive per year on the average?

70. Deuterium makes up 0.0115% of natural hydrogen on average. Make a rough estimate of the total deuterium in the Earth's oceans and estimate the total energy released if all of it were used in fusion reactors.

71. A shielded γ-ray source yields a dose rate of 0.052 rad/h at a distance of 1.0 m for an average-sized person. If workers are allowed a maximum dose of 5.0 rem in 1 year, how close to the source may they operate, assuming a 35-h work week? Assume that the intensity of radiation falls off as the square of the distance. (It actually falls off more rapidly than $1/r^2$ because of absorption in the air, so your answer will give a better-than-permissible value.)

72. Radon gas, $^{222}_{86}\mathrm{Rn}$, is formed by α decay. (*a*) Write the decay equation. (*b*) Ignoring the kinetic energy of the daughter nucleus (it's so massive), estimate the kinetic energy of the α particle produced. (*c*) Estimate the momentum of the alpha and of the daughter nucleus. (*d*) Estimate the kinetic energy of the daughter, and show that your approximation in (*b*) was valid.

73. Consider a system of nuclear power plants that produce 2400 MW. (*a*) What total mass of $^{235}_{92}U$ fuel would be required to operate these plants for 1 yr, assuming that 200 MeV is released per fission? (*b*) Typically 6% of the $^{235}_{92}U$ nuclei that fission produce $^{90}_{38}Sr$, a β^- emitter with a half-life of 29 yr. What is the total radioactivity of the $^{90}_{38}Sr$, in curies, produced in 1 yr? (Neglect the fact that some of it decays during the 1-yr period.)

74. In the net reaction, Eq. 42–8, for the proton–proton cycle in the Sun, the neutrinos escape from the Sun with energy of about 0.5 MeV. The remaining energy, 26.2 MeV, is available within the Sun. Use this value to calculate the "heat of combustion" per kilogram of hydrogen fuel and compare it to the heat of combustion of coal, about 3×10^7 J/kg.

75. Energy reaches Earth from the Sun at a rate of about 1300 W/m^2. Calculate (*a*) the total power output of the Sun, and (*b*) the number of protons consumed per second in the reaction of Eq. 42–8, assuming that this is the source of all the Sun's energy. (*c*) Assuming that the Sun's mass of 2.0×10^{30} kg was originally all protons and that all could be involved in nuclear reactions in the Sun's core, how long would you expect the Sun to "glow" at its present rate? See Problem 74.

76. Estimate how many solar neutrinos pass through a 180-m^2 ceiling of a room, at latitude 38°, for an hour around midnight on midsummer night. [*Hint*: See Problems 74 and 75.]

77. Estimate how much total energy would be released via fission if 2.0 kg of uranium were enriched to 5% of the isotope $^{235}_{92}U$.

78. Some stars, in a later stage of evolution, may begin to fuse two $^{12}_{6}C$ nuclei into one $^{24}_{12}Mg$ nucleus. (*a*) How much energy would be released in such a reaction? (*b*) What kinetic energy must two carbon nuclei each have when far apart, if they can then approach each other to within 6.0 fm, center-to-center? (*c*) Approximately what temperature would this require?

79. An average adult body contains about 0.10 μCi of $^{40}_{19}K$, which comes from food. (*a*) How many decays occur per second? (*b*) The potassium decay produces beta particles with energies of around 1.4 MeV. Estimate the dose per year in sieverts for a 55-kg adult. Is this a significant fraction of the 3.6-mSv/yr background rate?

80. When the nuclear reactor accident occurred at Chernobyl in 1986, 2.0×10^7 Ci were released into the atmosphere. Assuming that this radiation was distributed uniformly over the surface of the Earth, what was the activity per square meter? (The actual activity was not uniform; even within Europe wet areas received more radioactivity from rainfall.)

81. A star with a large helium abundance can burn helium in the reaction $^{4}_{2}He + ^{4}_{2}He + ^{4}_{2}He \rightarrow ^{12}_{6}C$. What is the Q-value for this reaction?

82. A 1.2-μCi $^{137}_{55}Cs$ source is used for 1.6 hours by a 65-kg worker. Radioactive $^{137}_{55}Cs$ decays by β^- decay with a half-life of 30 yr. The average energy of the emitted betas is about 190 keV per decay. The β decay is quickly followed by a γ with an energy of 660 keV. Assuming the person absorbs *all* emitted energy, what effective dose (in rem) is received?

83. A large amount of $^{90}_{38}Sr$ was released during the Chernobyl nuclear reactor accident in 1986. The $^{90}_{38}Sr$ enters the body through the food chain. How long will it take for 85% of the $^{90}_{38}Sr$ released during the accident to decay? See Appendix F.

84. Three radioactive sources have the same activity, 35 mCi. Source A emits 1.0-MeV γ rays, source B emits 2.0-MeV γ rays, and source C emits 2.0-MeV alphas. What is the relative danger of these sources?

85. A 60-kg patient is to be given a medical test involving the ingestion of $^{99m}_{43}Tc$ (Section 42–8) which decays by emitting a 140-keV gamma. The half-life for this decay is 6 hours. Assuming that about half the gamma photons exit the body without interacting with anything, what must be the initial activity of the Tc sample if the whole-body dose cannot exceed 50 mrem? Make the rough approximation that biological elimination of Tc can be ignored.

86. Centuries ago, paint generally contained a different amount of cobalt ($^{59}_{27}Co$) than paint today. A certain "old" painting is suspected of being a new forgery, and an examiner has decided to use **neutron activation analysis** to test this hypothesis. After placing the painting in a neutron flux of 5.0×10^{12} neutrons/cm^2/s for 5.0 minutes, an activity of 55 decays/s of $^{60}_{27}Co$ ($T_{\frac{1}{2}} = 5.27$ yr) is observed. Assuming $^{59}_{27}Co$ has a cross section of 19 bn, how much cobalt (in grams) does the paint contain?

87. Show, using the laws of conservation of energy and momentum, that for a nuclear reaction requiring energy, the minimum kinetic energy of the bombarding particle (the **threshold energy**) is equal to $[-Qm_{pr}/(m_{pr} - m_b)]$, where $-Q$ is the energy required (difference in total mass between products and reactants), m_b is the mass of the bombarding particle, and m_{pr} is the total mass of the products. Assume the target nucleus is at rest before an interaction takes place, and that all speeds are nonrelativistic.

88. The early scattering experiments performed around 1910 in Ernest Rutherford's laboratory in England produced the first evidence that an atom consists of a heavy nucleus surrounded by electrons. In one of these experiments, α particles struck a gold-foil target 4.0×10^{-5} cm thick in which there were 5.9×10^{28} gold atoms per cubic meter. Although most α particles either passed straight through the foil or were scattered at small angles, approximately 1.6×10^{-3} percent were scattered at angles greater than 90°—that is, in the backward direction. (*a*) Calculate the cross section, in barns, for backward scattering. (*b*) Rutherford concluded that such backward scattering could occur only if an atom consisted of a very tiny, massive, and positively charged nucleus with electrons orbiting some distance away. Assuming that backward scattering occurs for nearly direct collisions (i.e., $\sigma \approx$ area of nucleus), estimate the diameter of a gold nucleus.

Answers to Exercises

A: $^{138}_{56}Ba$.

B: 2×10^{17}.

C: (*e*).

D: (*b*).

E: (*a*).

This computer-generated reconstruction of a proton–antiproton collision at Fermilab (Fig. 43–3) occurred at a combined energy of nearly 2 TeV. It is one of the events that provided evidence for the top quark (1995). The wire drift chamber (Section 41–11) is in a magnetic field, and the radius of curvature of the charged particle tracks is a measure of each particle's momentum (Section 27–4).

The white dots represent the signals seen on the electric wires of the drift chamber. The colored lines are the particle paths.

The top quark (t) has too brief a lifetime ($\approx 10^{-23}$ s) to be detected itself, so we look for its possible decay products. Analysis indicates the following interaction and subsequent decays:

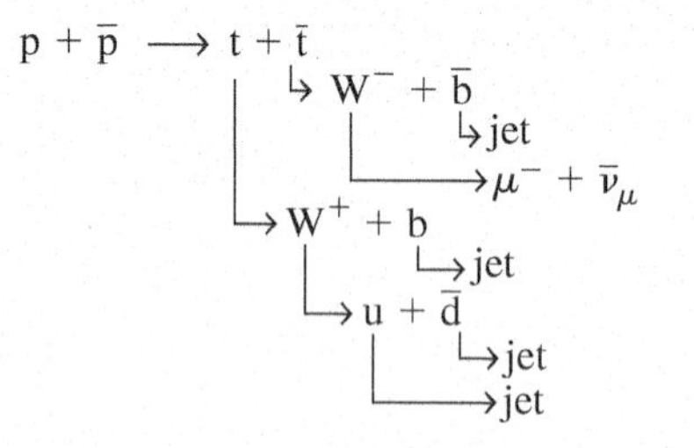

The tracks in the photo include jets (groups of particles moving in roughly the same direction), and a muon (μ^-) whose track is the pink one enclosed by a yellow rectangle to make it stand out. After reading this Chapter, try to name each symbol above and comment on whether all conservation laws hold.

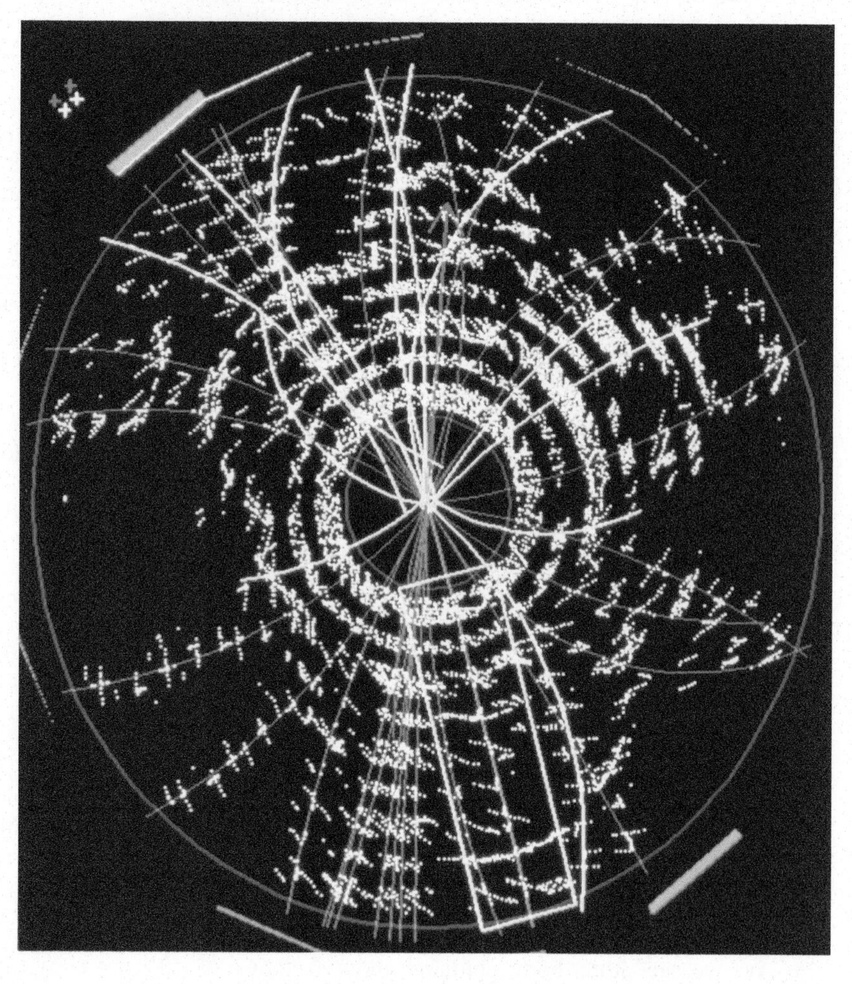

Elementary Particles

CONTENTS

CHAPTER-OPENING QUESTIONS—**Guess now!**

1. Electrons are still considered fundamental particles (in the group called leptons). But protons and neutrons are no longer considered fundamental; they have substructure and are made up of
   (a) pions.
   (b) leptons.
   (c) quarks.
   (d) bosons.
   (e) photons.
2. Thus the elementary particles as we see them today are
   (a) atoms and electrons.
   (b) protons, neutrons, and electrons.
   (c) protons, neutrons, electrons, and photons.
   (d) quarks, leptons, and gauge bosons.
   (e) hadrons, leptons, and gauge bosons.

In the final two Chapters of this book we discuss two of the most exciting areas of contemporary physics: elementary particles in this Chapter, and cosmology and astrophysics in Chapter 44. These are subjects at the forefront of knowledge—elementary particles treats the smallest objects in the universe; cosmology treats the largest (and oldest) aspects of the universe.

In this penultimate Chapter we discuss *elementary particle* physics, which represents the human endeavor to understand the basic building blocks of all matter, and the fundamental forces that govern their interactions. By the mid-1930s, it was recognized that all atoms can be considered to be made up of neutrons, protons, and electrons. The basic constituents of the universe were no longer considered to be atoms but rather the proton, neutron, and electron. Besides these three "elementary particles," several others were also known: the positron (a positive electron), the neutrino, and the γ particle (or photon), for a total of six elementary particles.

By the 1950s and 1960s many new types of particles similar to the neutron and proton were discovered, as well as many "midsized" particles called *mesons* whose masses were mostly less than nucleon masses but more than the electron mass. (Other mesons, found later, have masses greater than nucleons.) Physicists felt that all of these particles could not be fundamental, and must be made up of even smaller constituents (later confirmed by experiment), which were given the name *quarks*.

Today, the basic constituents of matter are considered to be **quarks** (they make up protons and neutrons as well as mesons) and **leptons** (a class that includes electrons, positrons, and neutrinos). In addition, there are the "carriers of force" including **gluons**, the photon, and other "gauge bosons." The theory that describes our present view is called the **Standard Model**. How we came to our present understanding of elementary particles is the subject of this Chapter.

One of the exciting recent developments of the last few years is an emerging synthesis between the study of elementary particles and astrophysics (Chapter 44). In fact, recent observations in astrophysics have led to the conclusion that the greater part of the mass–energy content of the universe is not ordinary matter but two mysterious and invisible forms known as "dark matter" and "dark energy" which cannot be explained by the Standard Model in its present form.

Indeed, we are now aware that the Standard Model is not sufficient. There are problems and important questions still unanswered, and we will mention some of them in this Chapter and how we hope to answer them.

43–1 High-Energy Particles and Accelerators

In the years after World War II, it was found that if the incoming particle in a nuclear reaction has sufficient energy, new types of particles can be produced. The earliest experiments used **cosmic rays**—particles that impinge on the Earth from space. In the laboratory, various types of particle accelerators have been constructed to accelerate protons or electrons to high energies, although heavy ions can also be accelerated. These **high-energy accelerators** have been used to probe more deeply into matter, to produce and study new particles, and to give us information about the basic forces and constituents of nature. Because the projectile particles are at high energy, this field is sometimes called **high-energy physics**.

Wavelength and Resolution

Particles accelerated to high energy can probe the interior of nuclei and nucleons or other particles they strike. An important factor is that faster-moving projectiles can reveal more detail. The wavelength of projectile particles is given by de Broglie's wavelength formula (Eq. 37–7),

$$\lambda = \frac{h}{p}, \qquad \textbf{(43–1)}$$

showing that the greater the momentum p of the bombarding particle, the shorter its wavelength. As discussed in Chapter 35 on diffraction, resolution of details in images is limited by the wavelength: the shorter the wavelength, the finer the detail that can be obtained. This is one reason why particle accelerators of higher and higher energy have been built in recent years: to probe ever deeper into the structure of matter, to smaller and smaller size.

EXAMPLE 43–1 **High resolution with electrons.** What is the wavelength, and hence the expected resolution, for 1.3-GeV electrons?

APPROACH Because 1.3 GeV is much larger than the electron mass, we must be dealing with relativistic speeds. The momentum of the electrons is found from Eq. 36–13, and the wavelength is $\lambda = h/p$.

SOLUTION Each electron has $K = 1.3\,\text{GeV} = 1300\,\text{MeV}$, which is about 2500 times the rest energy of the electron $(mc^2 = 0.51\,\text{MeV})$. Thus we can ignore the term $(mc^2)^2$ in Eq. 36–13, $E^2 = p^2c^2 + m^2c^4$, and we solve for p:

$$p = \sqrt{\frac{E^2 - m^2c^4}{c^2}} \approx \sqrt{\frac{E^2}{c^2}} = \frac{E}{c}.$$

Therefore the de Broglie wavelength is

$$\lambda = \frac{h}{p} = \frac{hc}{E},$$

where $E = 1.3\,\text{GeV}$. Hence

$$\lambda = \frac{(6.63 \times 10^{-34}\,\text{J}\cdot\text{s})(3.0 \times 10^8\,\text{m/s})}{(1.3 \times 10^9\,\text{eV})(1.6 \times 10^{-19}\,\text{J/eV})} = 0.96 \times 10^{-15}\,\text{m},$$

or 0.96 fm. This resolution of about 1 fm is on the order of the size of nuclei (see Eq. 41–1).

NOTE The maximum possible resolution of this beam of electrons is far greater than for a light beam in a light microscope ($\lambda \approx 500\,\text{nm}$).

FIGURE 43–1 Ernest O. Lawrence, around 1930, holding the first cyclotron (we see the vacuum chamber enclosing it).

EXERCISE A What is the wavelength of a proton with $K = 1.00\,\text{TeV}$?

Another major reason for building high-energy accelerators is that new particles of greater mass can be produced at higher energies, transforming the kinetic energy of the colliding particles into massive particles by $E = mc^2$, as we will discuss shortly. Now we look at particle accelerators.

Cyclotron

The cyclotron was developed in 1930 by E. O. Lawrence (1901–1958; Fig. 43–1) at the University of California, Berkeley. It uses a magnetic field to maintain charged ions—usually protons—in nearly circular paths. Although particle physicists no longer use simple cyclotrons, they are used in medicine for treating cancer, and their operating principles are useful for understanding modern accelerators. The protons move in a vacuum inside two D-shaped cavities, as shown in Fig. 43–2. Each time they pass into the gap between the "dees," a voltage accelerates them (the electric force), increasing their speed and increasing the radius of curvature of their path in the magnetic field. After many revolutions, the protons acquire high kinetic energy and reach the outer edge of the cyclotron where they strike a target. The protons speed up only when they are in the gap *between* the dees, and the voltage must be alternating. When protons are moving to the right across the gap in Fig. 43–2, the right dee must be electrically negative and the left one positive. A half-cycle later, the protons are moving to the left, so the left dee must be negative in order to accelerate them.

FIGURE 43–2 Diagram of a cyclotron. The magnetic field, applied by a large electromagnet, points into the page. The protons start at A, the ion source. The red electric field lines shown are for the alternating electric field in the gap at a certain moment.

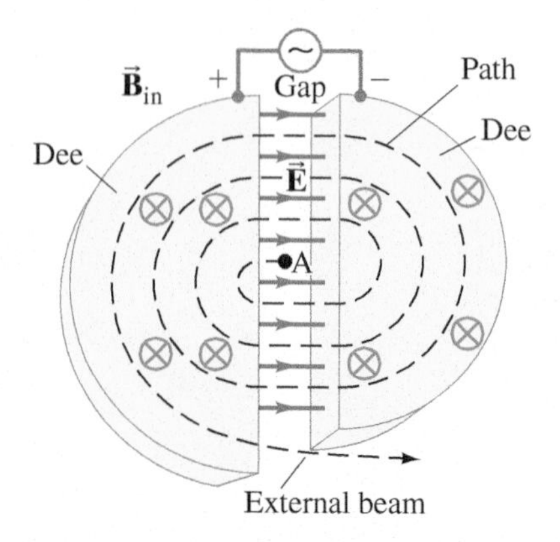

The frequency, f, of the applied voltage must be equal to that of the circulating protons. When ions of charge q are circulating *within* the hollow dees, the net force F on each is due to the magnetic field B, so $F = qvB$, where v is the speed of the ion at a given moment (Eq. 27–5). The magnetic force is perpendicular to both $\vec{v}$ and $\vec{B}$ and causes the ions to move in circles; the acceleration within the dees is thus centripetal and equals v^2/r, where r is the radius of the ion's path at a given moment.

We use Newton's second law, $F = ma$, and find that

$$F = ma$$
$$qvB = \frac{mv^2}{r}$$

when the protons are within the dees (not the gap), so

$$v = \frac{qBr}{m}.$$

The time required for a complete revolution is the period T and is equal to

$$T = \frac{\text{distance}}{\text{speed}} = \frac{2\pi r}{qBr/m} = \frac{2\pi m}{qB}.$$

Hence the frequency of revolution f is

$$f = \frac{1}{T} = \frac{qB}{2\pi m}. \quad \textbf{(43–2)}$$

This is known as the **cyclotron frequency**.

EXAMPLE 43–2 Cyclotron. A small cyclotron of maximum radius $R = 0.25\,\text{m}$ accelerates protons in a 1.7-T magnetic field. Calculate (*a*) the frequency needed for the applied alternating voltage, and (*b*) the kinetic energy of protons when they leave the cyclotron.

APPROACH The frequency of the protons revolving within the dees (Eq. 43–2) must equal the frequency of the voltage applied across the gap if the protons are going to increase in speed.

SOLUTION (*a*) From Eq. 43–2,

$$f = \frac{qB}{2\pi m}$$
$$= \frac{(1.6 \times 10^{-19}\,\text{C})(1.7\,\text{T})}{(6.28)(1.67 \times 10^{-27}\,\text{kg})} = 2.6 \times 10^7\,\text{Hz} = 26\,\text{MHz},$$

which is in the radio-wave region of the EM spectrum (Fig. 31–12).

(*b*) The protons leave the cyclotron at $r = R = 0.25\,\text{m}$. From $qvB = mv^2/r$ (see above), we have $v = qBr/m$, so their kinetic energy is

$$K = \frac{1}{2}mv^2 = \frac{1}{2}m\frac{q^2B^2R^2}{m^2} = \frac{q^2B^2R^2}{2m}$$
$$= \frac{(1.6 \times 10^{-19}\,\text{C})^2(1.7\,\text{T})^2(0.25\,\text{m})^2}{(2)(1.67 \times 10^{-27}\,\text{kg})} = 1.4 \times 10^{-12}\,\text{J} = 8.7\,\text{MeV}.$$

The kinetic energy is much less than the rest energy of the proton (938 MeV), so relativity is not needed.

NOTE The magnitude of the voltage applied to the dees does not appear in the formula for K, and so does not affect the final energy. But the higher this voltage, the fewer the revolutions required to bring the protons to full energy.

An important aspect of the cyclotron is that the frequency of the applied voltage, as given by Eq. 43–2, does not depend on the radius r of the particle's path. Thus the frequency does not have to be changed as the protons or ions start from the source and are accelerated to paths of larger and larger radii. But this is only true at nonrelativistic energies. At higher speeds, the momentum (Eq. 36–8) is $p = \gamma mv = mv/\sqrt{1 - v^2/c^2}$, so m in Eq. 43–2 has to be replaced by γm and the cyclotron frequency f (Eq. 43–2) depends on speed v because γ does. To keep the particles in sync, machines called **synchrocyclotrons** reduce the frequency in time to correspond to the increase of γm (in Eq. 43–2), as a packet of charged particles increases in speed more slowly at larger orbits.

(a) (b)

FIGURE 43–3 (a) Aerial view of Fermilab, near Chicago in Illinois; the main accelerator is a circular ring 1.0 km in radius. (b) The interior of the tunnel of the main accelerator at Fermilab, showing (red) the ring of superconducting magnets for the 1-TeV Tevatron.

Synchrotron

Another way to accelerate relativistic particles is to increase the magnetic field B in time so as to keep f (Eq. 43–2) constant as the particles speed up. Such devices are called **synchrotrons**; the particles move in a circle of fixed radius, which can be very large. At the European Center for Nuclear Research (CERN) in Geneva, Switzerland, the new Large Hadron Collider (LHC) is 4.3 km in radius and accelerates protons to 7 TeV. The *Tevatron* accelerator at Fermilab (the Fermi National Accelerator Laboratory) at Batavia, Illinois, has a radius of 1.0 km.† The Tevatron uses superconducting magnets to accelerate protons to about $1000\,\mathrm{GeV} = 1\,\mathrm{TeV}$ (hence its name); $1\,\mathrm{TeV} = 10^{12}\,\mathrm{eV}$. These large synchrotrons use a narrow ring of magnets (see Fig. 43–3) with each magnet placed at the same radius from the center of the circle. The magnets are interrupted by gaps where high voltage accelerates the particles. Another way to describe the acceleration is to say the particles "surf" on a traveling electromagnetic wave within radiofrequency (RF) cavities. (The particles are first given considerable energy in a smaller accelerator, "the injector," before being injected into the large ring of the large synchrotron.)

One problem of any accelerator is that accelerating electric charges radiate electromagnetic energy (see Chapter 31). Since ions or electrons are accelerated in an accelerator, we can expect considerable energy to be lost by radiation. The effect increases with energy and is especially important in circular machines where centripetal acceleration is present, such as synchrotrons, and hence is called **synchrotron radiation**. Synchrotron radiation can be useful, however. Intense beams of photons (γ rays) are sometimes needed, and they are usually obtained from an electron synchrotron.

EXERCISE B By what factor is the diameter of the Fermilab Tevatron (Fig. 43–3) greater than Lawrence's original cyclotron (estimate from Fig. 43–1)?

†Robert Wilson, who helped design the Tevatron, and founded the field of proton therapy (Section 42–7), expressed his views on accelerators and national security in this exchange with Senator John Pastore during testimony before a Congressional Committee in 1969:

Pastore: "Is there anything connected with the hopes of this accelerator [the Tevatron] that in any way involves the security of the country?"

Robert Wilson: "No sir, I don't believe so."

Pastore: "Nothing at all?"

Wilson: "Nothing at all. . . ."

Pastore: "It has no value in that respect?"

Wilson: "It has only to do with the respect with which we regard one another, the dignity of men, our love of culture. . . . It has to do with are we good painters, good sculptors, great poets? I mean all the things we really venerate in our country and are patriotic about . . . it has nothing to do directly with defending our country except to make it worth defending."

Linear Accelerators

In a **linear accelerator** (linac), electrons or ions are accelerated along a straight-line path, Fig. 43–4, passing through a series of tubular conductors. Voltage applied to the tubes is alternating so that when electrons (say) reach a gap, the tube in front of them is positive and the one they just left is negative. At low speeds, the particles cover less distance in the same amount of time, so the tubes are shorter at first. Electrons, with their small mass, get close to the speed of light quickly, $v \approx c$, and the tubes are nearly equal in length. Linear accelerators are particularly important for accelerating electrons to avoid loss of energy due to synchrotron radiation. The largest electron linear accelerator has been at Stanford University (Stanford Linear Accelerator Center, or SLAC), about 3 km (2 mi) long, accelerating electrons to 50 GeV. It is now being decommissioned. Linacs accelerating protons are used as injectors into circular machines to provide initial kinetic energy. Many hospitals have 10-MeV electron linacs that strike a metal foil to produce γ ray photons to irradiate tumors.

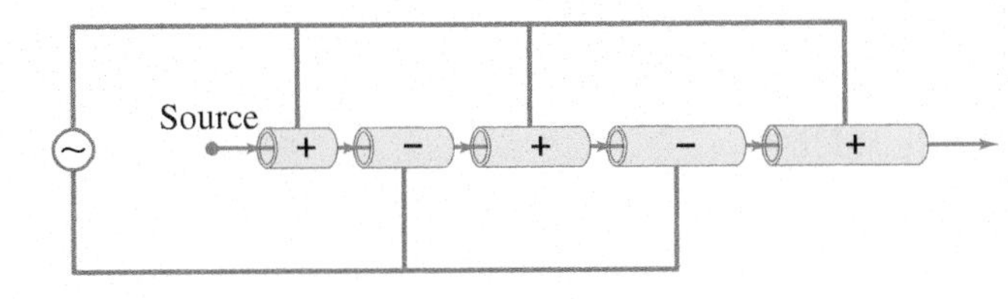

FIGURE 43–4 Diagram of a simple linear accelerator.

Colliding Beams

High-energy physics experiments were once done by aiming a beam of particles from an accelerator at a stationary target. But to obtain the maximum possible collision energy from a given accelerator, two beams of particles are now accelerated to very high energy and are steered so that they collide head-on. One way to accomplish such **colliding beams** with a single accelerator is through the use of **storage rings**, in which oppositely circulating beams can be repeatedly brought into collision with one another at particular points. For example, in the experiments that provided strong evidence for the top quark (Chapter-Opening Photo and Section 43–9), the Fermilab Tevatron accelerated protons and antiprotons each to 900 GeV, so that the combined energy of head-on collisions was 1.8 TeV.

The largest collider is the brand new Large Hadron Collider (LHC) at CERN, with a circumference of 26.7 km (Fig. 43–5). The two colliding beams each carry 7-TeV protons for a total interaction energy of 14 TeV.

FIGURE 43–5 The large circle represents the position of the tunnel, about 100 m below the ground at CERN (near Geneva) on the French-Swiss border, which houses the LHC. The smaller circle shows the position of the Super Proton Synchrotron that will be used for accelerating protons prior to injection into the LHC.

FIGURE 43–6 ATLAS, one of several large complex detectors at the LHC, is shown here as it was being built. It is hoped the LHC will provide evidence for the Higgs boson (to help understand the Standard Model), and perhaps to find supersymmetric particles which are candidates for the unknown dark matter that makes up a large part of the mass–energy of the universe. We will touch on these topics later in this Chapter.

Figure 43–6 shows part of one of the detectors (ATLAS) as it was being constructed at the LHC. The detectors within ATLAS include silicon semiconductor detectors with huge numbers of pixels used to track particle paths, to find their point of interaction, and to measure their radius of curvature in a magnetic field and thus determine their momentum (Section 27–4). Their energy is determined in "calorimeters" utilizing plastic, liquid, or dense metal compound crystal scintillators (Section 41–11).

In the planning stage is the International Linear Collider (ILC) which would have colliding beams of e^- and e^+ at around 0.3 to 1 TeV, with semiconductor detectors using CMOS (Section 33–5) with embedded transistors to allow fast readout.

EXAMPLE 43–3 **Protons at relativistic speeds.** Determine the energy required to accelerate a proton in a high-energy accelerator (*a*) from rest to $v = 0.900c$, and (*b*) from $v = 0.900c$ to $v = 0.999c$. (*c*) What is the kinetic energy achieved by the proton in each case?

APPROACH We use the work-energy principle, which is still valid relativistically as mentioned in Section 36–11: $W = \Delta K$.

SOLUTION The kinetic energy of a proton of mass m is given by Eq. 36–10,

$$K = (\gamma - 1)mc^2,$$

where the relativistic factor γ is

$$\gamma = \frac{1}{\sqrt{1 - v^2/c^2}}.$$

The work-energy theorem becomes

$$W = \Delta K = (\Delta\gamma)mc^2$$

since m and c are constant.

(*a*) For $v = 0, \gamma = 1$; and for $v = 0.900c$

$$\gamma = \frac{1}{\sqrt{1 - (0.900)^2}} = 2.29.$$

For a proton, $mc^2 = 938\text{ MeV}$, so the work (or energy) needed to accelerate it from rest to $v = 0.900c$ is

$$\begin{aligned} W = \Delta K &= (\Delta\gamma)mc^2 \\ &= (2.29 - 1.00)(938\text{ MeV}) = 1.21\text{ GeV}. \end{aligned}$$

(*b*) For $v = 0.999c$,

$$\gamma = \frac{1}{\sqrt{1 - (0.999)^2}} = 22.4.$$

So the work needed to accelerate a proton from $0.900c$ to $0.999c$ is

$$\begin{aligned} W = \Delta K &= (\Delta\gamma)mc^2 \\ &= (22.4 - 2.29)(938\text{ MeV}) = 18.9\text{ GeV}, \end{aligned}$$

which is 15 times as much.

(*c*) The kinetic energy reached by the proton in (*a*) is just equal to the work done on it, $K = 1.21\text{ GeV}$. The final kinetic energy of the proton in (*b*), moving at $v = 0.999c$, is

$$K = (\gamma - 1)mc^2 = (21.4)(938\text{ MeV}) = 20.1\text{ GeV},$$

which makes sense since, starting from rest, we did work $W = 1.21\text{ GeV} + 18.9\text{ GeV} = 20.1\text{ GeV}$ on it.

EXAMPLE 43–4 **Speed of a 1.0-TeV proton.** What is the speed of a 1.0-TeV proton produced at Fermilab?

APPROACH The kinetic energy $K = 1.0\,\text{TeV} = 1.0 \times 10^{12}\,\text{eV}$ is much greater than the mass of the proton, $0.938 \times 10^{9}\,\text{eV}$, so relativistic calculations must be used. In particular, we use Eq. 36–10:

$$K = (\gamma - 1)mc^2 = \frac{mc^2}{\sqrt{1 - v^2/c^2}} - mc^2.$$

SOLUTION Compared to $K = 1.0 \times 10^{12}\,\text{eV}$, the rest energy $(\approx 10^{-3}\,\text{TeV})$ can be neglected, so we write

$$K = \frac{mc^2}{\sqrt{1 - v^2/c^2}}.$$

Then

$$1 - \frac{v^2}{c^2} = \left(\frac{mc^2}{K}\right)^2$$

or

$$\frac{v}{c} = \sqrt{1 - \left(\frac{mc^2}{K}\right)^2} = \sqrt{1 - \left(\frac{938 \times 10^6\,\text{eV}}{1.0 \times 10^{12}\,\text{eV}}\right)^2}$$

$$v = 0.9999996c.$$

The proton is traveling at a speed extremely close to c, the speed of light.

43–2 Beginnings of Elementary Particle Physics—Particle Exchange

The accepted model for elementary particles today views *quarks* and *leptons* as the basic constituents of ordinary matter. To understand our present-day view of elementary particles, it is necessary to understand the ideas leading up to its formulation.

Elementary particle physics might be said to have begun in 1935 when the Japanese physicist Hideki Yukawa (1907–1981) predicted the existence of a new particle that would in some way mediate the strong nuclear force. To understand Yukawa's idea, we first consider the electromagnetic force. When we first discussed electricity, we saw that the electric force acts over a distance, without contact. To better perceive how a force can act over a distance, we used the idea of a **field**. The force that one charged particle exerts on a second can be said to be due to the electric field set up by the first. Similarly, the magnetic field can be said to carry the magnetic force. Later (Chapter 31), we saw that electromagnetic (EM) fields can travel through space as waves. Finally, in Chapter 37, we saw that electromagnetic radiation (light) can be considered as either a wave or as a collection of particles called *photons*. Because of this wave–particle duality, it is possible to imagine that the electromagnetic force between charged particles is due to

(1) the EM field set up by one charged particle and felt by the other, or
(2) an exchange of photons (γ particles) between them.

It is (2) that we want to concentrate on here, and a crude analogy for how an exchange of particles could give rise to a force is suggested in Fig. 43–7. In part (a), two children start throwing heavy pillows at each other; each throw and each catch results in the child being pushed backward by the impulse. This is the equivalent of a repulsive force. On the other hand, if the two children exchange pillows by grabbing them out of the other person's hand, they will be pulled toward each other, as when an attractive force acts.

FIGURE 43–7 Forces equivalent to particle exchange. (a) Repulsive force (children on roller skates throwing pillows at each other). (b) Attractive force (children grabbing pillows from each other's hands).

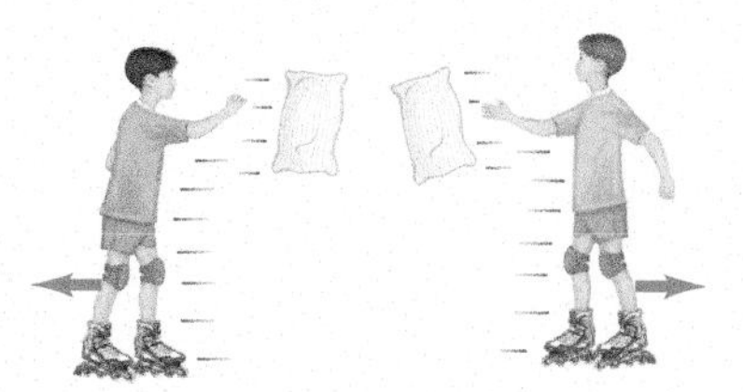

(a) Repulsive force (children throwing pillows)

(b) Attractive force (children grabbing pillows from each other's hands)

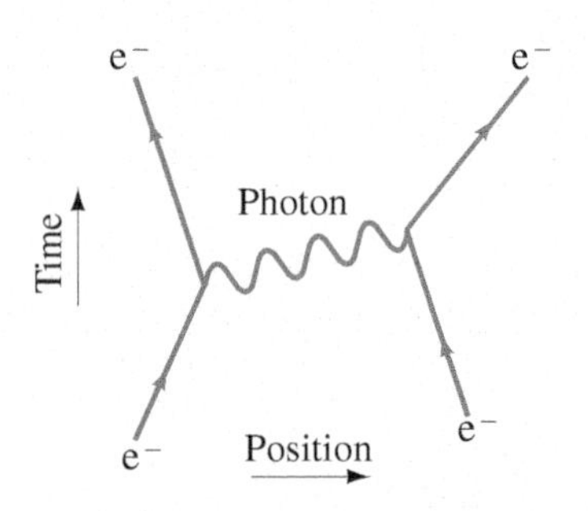

FIGURE 43–8 Feynman diagram showing a photon acting as the carrier of the electromagnetic force between two electrons. This is sort of an x vs. t graph, with t increasing upward. Starting at the bottom, two electrons approach each other (the distance between them decreases in time). As they get close, momentum and energy get transferred from one to the other, carried by a photon (or, perhaps, by more than one), and the two electrons bounce apart.

FIGURE 43–9 Early model showing meson exchange when a proton and neutron interact via the strong nuclear force. (Today, as we shall see shortly, we view the strong force as carried by gluons between quarks.)

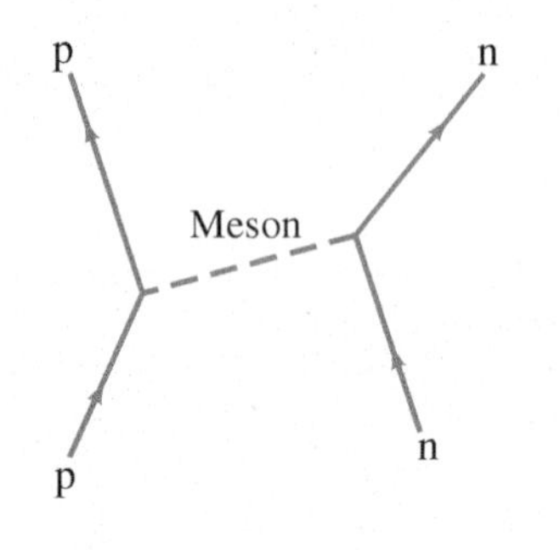

For the electromagnetic force, it is photons that are exchanged between two charged particles that give rise to the force between them. A simple diagram describing this photon exchange is shown in Fig. 43–8. Such a diagram, called a **Feynman diagram** after its inventor, the American physicist Richard Feynman (1918–1988), is based on the theory of **quantum electrodynamics** (QED).

Figure 43–8 represents the simplest case in QED, in which a single photon is exchanged. One of the charged particles emits the photon and recoils somewhat as a result; and the second particle absorbs the photon. In any collision or *interaction*, energy and momentum are transferred from one charged particle to the other, carried by the photon. The photon is absorbed by the second particle after it is emitted by the first and is not observable. Hence it is referred to as a **virtual** photon, in contrast to one that is free and can be detected by instruments. The photon is said to **mediate**, or **carry**, the electromagnetic force.

By analogy with photon exchange that mediates the electromagnetic force, Yukawa argued that there ought to be a particle that mediates the strong nuclear force—the force that holds nucleons together in the nucleus. Yukawa called this predicted particle a **meson** (meaning "medium mass"). Figure 43–9 is a Feynman diagram showing the original model of meson exchange: a meson carrying the strong force between a neutron and a proton.

A rough estimate of the mass of the meson can be made as follows. Suppose the proton on the left in Fig. 43–9 is at rest. For it to emit a meson would require energy (to make the meson's mass) which, coming from nowhere, would violate conservation of energy. But the uncertainty principle allows nonconservation of energy by an amount ΔE if it occurs only for a time Δt given by $(\Delta E)(\Delta t) \approx h/2\pi$. We set ΔE equal to the energy needed to create the mass m of the meson: $\Delta E = mc^2$. Conservation of energy is violated only as long as the meson exists, which is the time Δt required for the meson to pass from one nucleon to the other, where it is absorbed and disappears. If we assume the meson travels at relativistic speed, close to the speed of light c, then Δt need be at most about $\Delta t = d/c$, where d is the maximum distance that can separate the interacting nucleons. Thus we can write

$$\Delta E\,\Delta t \approx \frac{h}{2\pi}$$

$$mc^2\left(\frac{d}{c}\right) \approx \frac{h}{2\pi}$$

or

$$mc^2 \approx \frac{hc}{2\pi d}. \tag{43–3}$$

The range of the strong nuclear force (the maximum distance away it can be felt) is small—not much more than the size of a nucleon or small nucleus (see Eq. 41–1)—so let us take $d \approx 1.5 \times 10^{-15}$ m. Then from Eq. 43–3,

$$mc^2 \approx \frac{hc}{2\pi d} = \frac{(6.6 \times 10^{-34}\,\mathrm{J\cdot s})(3.0 \times 10^{8}\,\mathrm{m/s})}{(6.28)(1.5 \times 10^{-15}\,\mathrm{m})} \approx 2.1 \times 10^{-11}\,\mathrm{J} = 130\,\mathrm{MeV}.$$

The mass of the predicted meson, roughly $130\,\mathrm{MeV}/c^2$, is about 250 times the electron mass of $0.51\,\mathrm{MeV}/c^2$.

EXERCISE C What effect does an increase in the mass of the virtual exchange particle have on the range of the force it mediates? (*a*) Decreases it; (*b*) increases it; (*c*) has no appreciable effect; (*d*) decreases the range for charged particles and increases the range for neutral particles.

Note that since the electromagnetic force has infinite range, Eq. 43–3 with $d = \infty$ tells us that the exchanged particle for the electromagnetic force, the photon, will have zero mass, which it does.

The particle predicted by Yukawa was discovered in cosmic rays by C. F. Powell and G. Occhialini in 1947, and is called the "π" or pi meson, or simply the **pion**. It comes in three charge states: $+e$, $-e$, or 0, where $e = 1.6 \times 10^{-19}\,\mathrm{C}$. The π^+ and π^- have mass of $139.6\,\mathrm{MeV}/c^2$ and the π^0 a mass of $135.0\,\mathrm{MeV}/c^2$, all close to Yukawa's prediction. All three interact strongly with matter. Reactions observed in the laboratory, using a particle accelerator, include

$$\begin{aligned} \mathrm{p} + \mathrm{p} &\rightarrow \mathrm{p} + \mathrm{p} + \pi^0, \\ \mathrm{p} + \mathrm{p} &\rightarrow \mathrm{p} + \mathrm{n} + \pi^+. \end{aligned} \tag{43–4}$$

The incident proton from the accelerator must have sufficient energy to produce the additional mass of the free pion.

Yukawa's theory of pion exchange as carrier of the strong force has been superseded by *quantum chromodynamics* in which protons, neutrons, and other strongly interacting particles are made up of basic entities called *quarks*, and the basic carriers of the strong force are *gluons*, as we shall discuss shortly. But the basic idea of the earlier theory, that forces can be understood as the exchange of particles, remains valid.

There are four known types of force—or interactions—in nature. The electromagnetic force is carried by the photon, the strong force by gluons. What about the other two: the weak force and gravity? These too are believed to be mediated by particles. The particles that transmit the weak force are referred to as the W^+, W^-, and Z^0, and were detected in 1983 (Fig. 43–10). The quantum (or carrier) of the gravitational force is called the **graviton**, and if it exists it has not yet been observed.

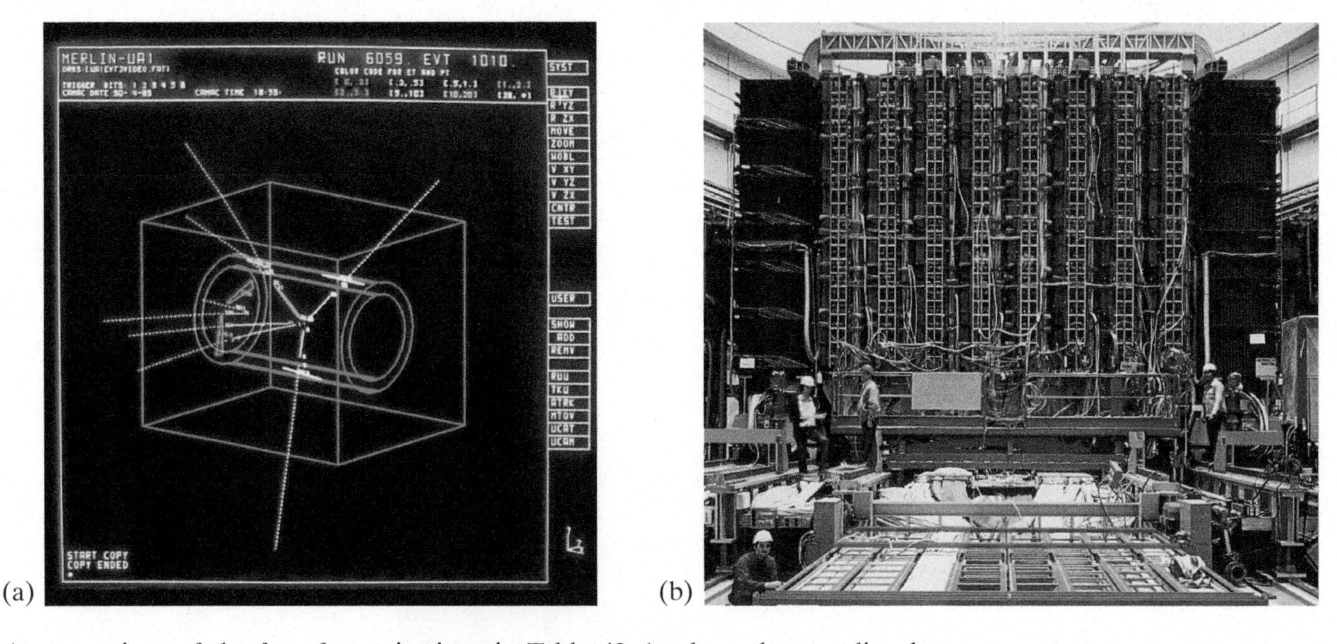

FIGURE 43–10 (a) Computer reconstruction of a Z-particle decay into an electron and a positron ($\mathrm{Z}^0 \rightarrow \mathrm{e}^+ + \mathrm{e}^-$) whose tracks are shown in white, which took place in the UA1 detector at CERN. (b) Photo of the UA1 detector at CERN as it was being built.

A comparison of the four forces is given in Table 43–1, where they are listed according to their (approximate) relative strengths. Notice that although gravity may be the most obvious force in daily life (because of the huge mass of the Earth), on a nuclear scale it is by far the weakest of the four forces, and its effect at the particle level can nearly always be ignored.

TABLE 43–1 The Four Forces in Nature

Type	Relative Strength (approx., for 2 protons in nucleus)	Field Particle
Strong	1	Gluons
Electromagnetic	10^{-2}	Photon
Weak	10^{-6}	$\mathrm{W}^\pm$ and Z^0
Gravitational	10^{-38}	Graviton (?)

43–3 Particles and Antiparticles

The positron, as we discussed in Sections 37–5 (pair production) and 41–5 (β^+ decay), is basically a positive electron. That is, many of its properties are the same as for the electron, such as mass, but it has the opposite electric charge $(+e)$. Other quantum numbers that we will discuss shortly are also reversed. The positron is said to be the **antiparticle** to the electron.

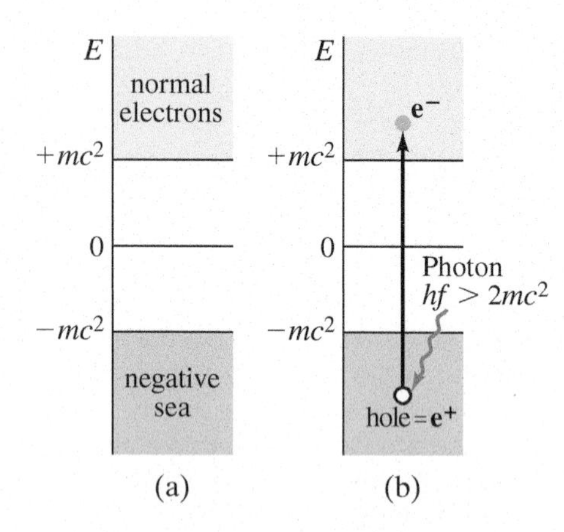

FIGURE 43–11 (a) Possible energy states for an electron. Note the vast sea of fully occupied electron states at $E < -mc^2$. (b) An electron in the negative sea is hit by a photon $(E > 2mc^2)$ and knocks it up to a normal positive energy state. The positive "hole" left behind acts like a positive electron—it is a positron.

The original idea for antiparticles came from a relativistic wave equation developed in 1928 by the Englishman P. A. M. Dirac (1902–1984). Recall that the non-relativistic Schrödinger equation took conservation of energy as a starting point. The Dirac equation too was based in part on conservation of energy. In Chapter 36, we saw that the total energy E of a particle with mass m and momentum p and zero potential energy is given by Eq. 36–13, $E^2 = p^2c^2 + m^2c^4$. Thus

$$E = \pm\sqrt{p^2c^2 + m^2c^4}.$$

Dirac applied his new equation and found that it included solutions with both + and − signs. He could not ignore the solution with the negative sign, which we might have thought unphysical. If those negative energy states are real, then we would expect normal free electrons to drop down into those states, emitting photons—never experimentally seen. To deal with this difficulty, Dirac postulated that all those negative energy states are *normally occupied*. That is, what we thought was the **vacuum** is instead a vast **sea of electrons** in negative energy states (Fig. 43–11a). These electrons are not normally observable. But if a photon strikes one of these negative energy electrons, that electron can be knocked up to a normal $(E > mc^2)$ energy state as shown in Fig. 43–11b. (Note in Fig. 43–11 that there are no energy states between $E = -mc^2$ and $E = +mc^2$ because p^2 cannot be negative in the equation $E = \pm\sqrt{p^2c^2 + m^2c^4}$.) The photon that knocks an e^- from the negative sea up to a normal state (Fig. 43–11b) must have an energy greater than $2mc^2$. What is left behind is a hole (as in semiconductors, Sections 40–7 and 40–8) with positive charge. We call that "hole" a **positron**, and it can move around as a free particle with positive energy. Thus Fig. 43–11b represents (Section 37–5) **pair production**: $\gamma \rightarrow e^-e^+$.

The positron was first detected as a curved path in a cloud chamber in a magnetic field by Carl Anderson in 1932. It was predicted that other particles also would have antiparticles. It was decades before another type was found. Finally, in 1955 the antiparticle to the proton, the **antiproton** $(\overline{p})$, which carries a negative charge (Fig. 43–12), was discovered at the University of California, Berkeley,

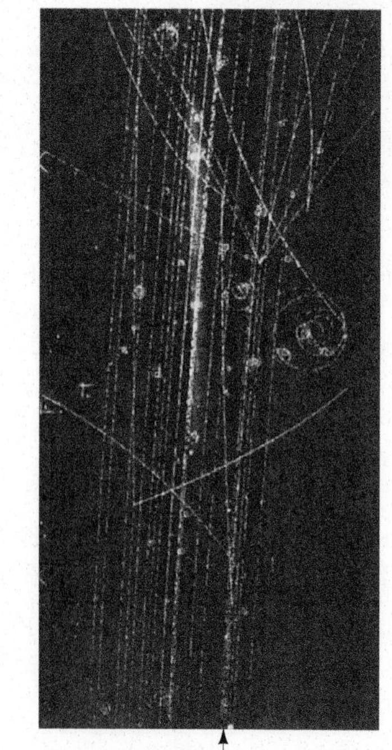

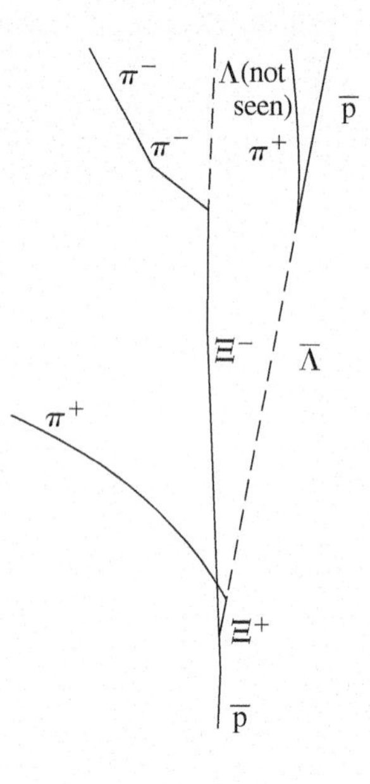

FIGURE 43–12 Liquid-hydrogen bubble-chamber photograph of an antiproton $(\overline{p})$ colliding with a proton at rest, producing a Xi–anti-Xi pair $(\overline{p} + p \rightarrow \Xi^- + \overline{\Xi}^+)$ that subsequently decay into other particles. The drawing indicates the assignment of particles to each track, which is based on how or if that particle decays, and on mass values estimated from measurement of momentum (curvature of track in magnetic field) and energy (thickness of track, for example). Neutral particle paths are shown by dashed lines since neutral particles produce no bubbles and hence no tracks.

by Emilio Segrè (1905–1989, Fig. 43–13) and Owen Chamberlain (1920–2006). A bar, such as over the p, is used in general to indicate the antiparticle ($\overline{\mathrm{p}}$). Soon after, the antineutron ($\overline{\mathrm{n}}$) was found. All particles have antiparticles. But a few, like the photon and the π^0, do not have distinct antiparticles—we say that they are their own antiparticles.

Antiparticles are produced in nuclear reactions when there is sufficient energy available to produce the required mass, and they do not live very long in the presence of matter. For example, a positron is stable when by itself; but if it encounters an electron, the two annihilate each other. The energy of their vanished mass, plus any kinetic energy they possessed, is converted into the energy of γ rays or of other particles. Annihilation also occurs for all other particle–antiparticle pairs.

The vast sea of electrons with negative energy in Fig. 43–11 is the vacuum (or **vacuum state**). According to quantum mechanics, the vacuum is not empty, but contains electrons and other particles as well. The uncertainty principle allows a particle to jump briefly up to a normal energy, thus creating a **particle–antiparticle** pair. It is possible that they could be the source of the recently discovered *dark energy* that fills the universe (Chapter 44). We still have a lot to learn.

Antimatter is a term referring to material that would be made up of "antiatoms" in which antiprotons and antineutrons would form the nucleus around which positrons (antielectrons) would move. The term is also used for antiparticles in general. If there were pockets of antimatter in the universe, a huge explosion would occur if it should encounter normal matter. It is believed that antimatter was prevalent in the very early universe (Section 44–7).

Emilio Segrè.
Feb. 20 1985
To enlist the help of the subjects of this book for the next edition of Giancoli's texts

FIGURE 43–13 Emilio Segrè: he worked with Fermi in the 1930s, later discovered the first "man-made" element, technetium, and other elements, and then the antiproton. The inscription below the photo is from a book by Segrè given to this book's author.

43–4 Particle Interactions and Conservation Laws

One of the important uses of high-energy accelerators is to study the interactions of elementary particles with each other. As a means of ordering this subnuclear world, the conservation laws are indispensable. The laws of conservation of energy, of momentum, of angular momentum, and of electric charge are found to hold precisely in all particle interactions.

A study of particle interactions has revealed a number of new conservation laws which (just like the old ones) are ordering principles: they help to explain why some reactions occur and others do not. For example, the following reaction has never been observed:

$$\mathrm{p} + \mathrm{n} \not\rightarrow \mathrm{p} + \mathrm{p} + \overline{\mathrm{p}}$$

even though charge, energy, and so on, are conserved ($\not\rightarrow$ means the reaction does not occur). To understand why such a reaction does not occur, physicists hypothesized a new conservation law, the conservation of **baryon number**. (Baryon number is a generalization of nucleon number, which we saw earlier is conserved in nuclear reactions and decays.) All nucleons are defined to have baryon number $B = +1$, and all antinucleons (antiprotons, antineutrons) have $B = -1$. All other types of particles, such as photons, mesons, and electrons and other leptons, have $B = 0$. The reaction shown at the start of this paragraph does not conserve baryon number since the left side has $B = (+1) + (+1) = +2$, and the right has $B = (+1) + (+1) + (-1) = +1$. On the other hand, the following reaction does conserve B and *does* occur if the incoming proton has sufficient energy:

$$\begin{aligned} \mathrm{p} + \mathrm{p} &\rightarrow \mathrm{p} + \mathrm{p} + \overline{\mathrm{p}} + \mathrm{p}, \\ B = +1 + 1 &= +1 + 1 - 1 + 1. \end{aligned}$$

As indicated, $B = +2$ on both sides of this equation. From these and other reactions, the **conservation of baryon number** has been established as a basic principle of physics.

Also useful are conservation laws for the three **lepton numbers**, associated with weak interactions including decays. In ordinary β decay, an electron or positron is emitted along with a neutrino or antineutrino. In another type of decay, a particle known as a "μ" or mu meson, or **muon**, can be emitted instead of an electron. The muon (discovered in 1937) seems to be much like an electron, except its mass is 207 times larger ($106\,\mathrm{MeV}/c^2$). The neutrino (ν_e) that accompanies an emitted electron is found to be different from the neutrino (ν_μ) that accompanies an emitted muon.

CAUTION
The different types of neutrinos are not identical

Each of these neutrinos has an antiparticle: $\bar{\nu}_e$ and $\bar{\nu}_\mu$. In ordinary β decay we have, for example,

$$\mathrm{n} \rightarrow \mathrm{p} + \mathrm{e}^- + \bar{\nu}_e$$

but not $\mathrm{n} \not\rightarrow \mathrm{p} + \mathrm{e}^- + \bar{\nu}_\mu$. To explain why these do not occur, the concept of **electron lepton number**, L_e, was invented. If the electron (e^-) and the electron neutrino (ν_e) are assigned $L_e = +1$, and e^+ and $\bar{\nu}_e$ are assigned $L_e = -1$, whereas all other particles have $L_e = 0$, then all observed decays conserve L_e. For example, in $\mathrm{n} \rightarrow \mathrm{p} + \mathrm{e}^- + \bar{\nu}_e$, initially $L_e = 0$, and afterward $L_e = 0 + (+1) + (-1) = 0$. Decays that do not conserve L_e, even though they would obey the other conservation laws, are not observed to occur.

In a decay involving muons, such as

$$\pi^+ \rightarrow \mu^+ + \nu_\mu,$$

a second quantum number, **muon lepton number** (L_μ), is conserved. The μ^- and ν_μ are assigned $L_\mu = +1$, and their antiparticles μ^+ and $\bar{\nu}_\mu$ have $L_\mu = -1$, whereas all other particles have $L_\mu = 0$. L_μ too is conserved in interactions and decays. Similar assignments can be made for the **tau lepton number**, L_τ, associated with the τ lepton (discovered in 1976 with mass more than 3000 times the electron mass) and its neutrino, ν_τ.

Keep in mind that antiparticles have not only opposite electric charge from their particles, but also opposite B, L_e, L_μ, and L_τ. For example, a neutron has $B = +1$, an antineutron has $B = -1$ (and all the L's are zero).

CONCEPTUAL EXAMPLE 43–5 **Lepton number in muon decay.** Which of the following decay schemes is possible for muon decay: (*a*) $\mu^- \rightarrow \mathrm{e}^- + \bar{\nu}_e$; (*b*) $\mu^- \rightarrow \mathrm{e}^- + \bar{\nu}_e + \nu_\mu$; (*c*) $\mu^- \rightarrow \mathrm{e}^- + \nu_e$? All of these particles have $L_\tau = 0$.

RESPONSE A μ^- has $L_\mu = +1$ and $L_e = 0$. This is the initial state for all decays given, and the final state must also have $L_\mu = +1$, $L_e = 0$. In (*a*), the final state has $L_\mu = 0 + 0 = 0$, and $L_e = +1 - 1 = 0$; L_μ would not be conserved and indeed this decay is not observed to occur. The final state of (*b*) has $L_\mu = 0 + 0 + 1 = +1$ and $L_e = +1 - 1 + 0 = 0$, so both L_μ and L_e are conserved. This is in fact the most common decay mode of the μ^-. Lastly, (*c*) does not occur because L_e ($= +2$ in the final state) is not conserved, nor is L_μ.

EXAMPLE 43–6 **Energy and momentum are conserved.** In addition to the "number" conservation laws which help explain the decay schemes of particles, we can also apply the laws of conservation of energy and momentum. The decay of a Σ^+ particle at rest with a mass of $1189\ \mathrm{MeV}/c^2$ (Table 43–2 in Section 43–6) commonly yields a proton (mass $= 938\ \mathrm{MeV}/c^2$) and a neutral pion, π^0 (mass $= 135\ \mathrm{MeV}/c^2$):

$$\Sigma^+ \rightarrow \mathrm{p} + \pi^0.$$

What are the kinetic energies of the decay products, assuming the Σ^+ parent particle was at rest?

APPROACH We find the energy release from the change in mass $(E = mc^2)$ as we did for nuclear processes (Eq. 41–3 or 42–2a), and apply conservation of energy and momentum.

SOLUTION The energy released, or Q-value, is the change in mass times c^2:

$$Q = \left[m_{\Sigma^+} - (m_\mathrm{p} + m_{\pi^0})\right]c^2 = \left[1189 - (938 + 135)\right]\mathrm{MeV} = 116\ \mathrm{MeV}.$$

This energy Q becomes the kinetic energy of the resulting decay particles, p and π^0:

$$Q = K_\mathrm{p} + K_{\pi^0}$$

with each particle's kinetic energy related to its momentum by (Eqs. 36–11 and 13):

$$K_\mathrm{p} = E_\mathrm{p} - m_\mathrm{p}c^2 = \sqrt{(p_\mathrm{p}c)^2 + (m_\mathrm{p}c^2)^2} - m_\mathrm{p}c^2,$$

and similarly for the pion. From momentum conservation, the proton and pion have the same magnitude of momentum since the original Σ^+ was at rest: $p_\mathrm{p} = p_{\pi^0} = p$. Then, $Q = K_\mathrm{p} + K_{\pi^0}$ gives $116\ \mathrm{MeV} = \left[\sqrt{(pc)^2 + (938\ \mathrm{MeV})^2} - 938\ \mathrm{MeV}\right] + \left[\sqrt{(pc)^2 + (135\ \mathrm{MeV})^2} - 135\ \mathrm{MeV}\right]$. We solve this for pc, which gives $pc = 189\ \mathrm{MeV}$. Substituting into the expression above for the kinetic energy, first for the proton, then for the pion, we obtain $K_\mathrm{p} = 19\ \mathrm{MeV}$ and $K_{\pi^0} = 97\ \mathrm{MeV}$.

43–5 Neutrinos—Recent Results

We first met neutrinos with regard to β^- decay in Section 41–5. The study of neutrinos is a "hot" subject today. Experiments are being carried out in deep underground laboratories, sometimes in deep mine shafts. The thick layer of earth above is meant to filter out all other "background" particles, leaving mainly the very weakly interacting neutrinos to arrive at the detectors.

Some very important results have come to the fore in recent years. First there was the **solar neutrino problem**. The energy output of the Sun is believed to be due to the nuclear fusion reactions discussed in Chapter 42, Eqs. 42–7 and 42–8. The neutrinos emitted in these reactions are all ν_e (accompanied by e^+). But the rate at which ν_e arrive at Earth is measured to be much less than expected based on the power output of the Sun. It was then proposed that perhaps, on the long trip between Sun and Earth, ν_e might turn into ν_μ or ν_τ. Subsequent experiments confirmed this hypothesis. Thus the three neutrinos, ν_e, ν_μ, ν_τ, can change into one another in certain circumstances, a phenomenon called **neutrino flavor oscillation** (each of the three neutrino types being called, whimsically, a different "flavor"). This result suggests that the lepton numbers L_e, L_μ, and L_τ are not perfectly conserved. But the sum, $L_e + L_\mu + L_\tau$, is believed to be always conserved.

The second exceptional result has long been speculated on: are neutrinos massless as originally thought, or do they have a nonzero mass? Rough upper limits on the masses have been made. Today astrophysical experiments show that the sum of all three neutrino masses is less than about $0.14\,\mathrm{eV}/c^2$. But can the masses be zero? Not if there are the flavor oscillations discussed above. It seems likely that at least one neutrino type has a mass of at least $0.04\,\mathrm{eV}/c^2$.

As a result of neutrino oscillations, the three types of neutrino may not be exactly what we thought they were (e, μ, τ). If not, the three basic neutrinos, called 1, 2, and 3, are linear combinations of ν_e, ν_μ, and ν_τ.

Another outstanding question is whether or not neutrinos are in the category called **Majorana particles**,† meaning they would be their own antiparticles. If so, a lot of other questions (and answers) would appear.

*Neutrino Mass Estimate from a Supernova

The supernova of 1987 offered an opportunity to estimate electron neutrino mass. If neutrinos do have mass, then $v < c$ and neutrinos of different energy would take different times to travel the 170,000 light-years from the supernova to Earth. To get an idea of how such a measurement could be done, suppose two neutrinos from "SN1987a" were emitted at the same time and detected on Earth (via the reaction $\bar{\nu}_e + \mathrm{p} \rightarrow \mathrm{n} + \mathrm{e}^+$) 10 seconds apart, with measured kinetic energies of about 20 MeV and 10 MeV. Since we expect the neutrino mass to be surely less than 100 eV (from other laboratory measurements), and since our neutrinos have kinetic energy of 20 MeV and 10 MeV, we can make the approximation $m_\nu c^2 \ll E$, so that E (the total energy) is essentially equal to the kinetic energy. We use Eq. 36–11, which tells us

$$E = \frac{m_\nu c^2}{\sqrt{1 - v^2/c^2}}.$$

We solve this for v, the velocity of a neutrino with energy E:

$$v = c\left(1 - \frac{m_\nu^2 c^4}{E^2}\right)^{\frac{1}{2}} = c\left(1 - \frac{m_\nu^2 c^4}{2E^2} + \cdots\right),$$

where we have used the binomial expansion $(1 - x)^{\frac{1}{2}} = 1 - \frac{1}{2}x + \cdots$, and we ignore higher-order terms since $m_\nu^2 c^4 \ll E^2$. The time t for a neutrino to travel a distance d (= 170,000 ly) is

$$t = \frac{d}{v} = \frac{d}{c\left(1 - \frac{m_\nu^2 c^4}{2E^2}\right)} \approx \frac{d}{c}\left(1 + \frac{m_\nu^2 c^4}{2E^2}\right),$$

where again we used the binomial expansion $\left[(1 - x)^{-1} = 1 + x + \cdots\right]$.

†The brilliant young physicist Ettore Majorana (1906–1938) disappeared from a ship under mysterious circumstances in 1938 at the age of 31.

The difference in arrival times for our two neutrinos of energies $E_1 = 20\,\text{MeV}$ and $E_2 = 10\,\text{MeV}$ is

$$t_2 - t_1 = \frac{d}{c}\frac{m_\nu^2 c^4}{2}\left(\frac{1}{E_2^2} - \frac{1}{E_1^2}\right).$$

We solve this for $m_\nu c^2$ and set $t_2 - t_1 = 10\,\text{s}$:

$$m_\nu c^2 = \left[\frac{2c(t_2 - t_1)}{d}\frac{E_1^2 E_2^2}{E_1^2 - E_2^2}\right]^{\frac{1}{2}}$$

$$= \left[\frac{2(3.0 \times 10^8\,\text{m/s})(10\,\text{s})}{(1.7 \times 10^5\,\text{ly})(1.0 \times 10^{16}\,\text{m/ly})}\frac{(400\,\text{MeV}^2)(100\,\text{MeV}^2)}{(400\,\text{MeV}^2 - 100\,\text{MeV}^2)}\right]^{\frac{1}{2}}$$

$$= 22 \times 10^{-6}\,\text{MeV} = 22\,\text{eV}.$$

We thus estimate the mass of the neutrino to be $22\,\text{eV}/c^2$, but there would of course be experimental uncertainties, not to mention the unwarranted assumption that the two neutrinos were emitted at the same time.

Theoretical models of supernova explosions suggest that the neutrinos are emitted in a burst that lasts from a second or two up to perhaps 10 s. If we assume the neutrinos are not emitted simultaneously but rather at any time over a 10-s interval, what then could we say about the neutrino mass based on the data given above? The 10-s difference in their arrival times could be due to a 10-s difference in their emission time. In this case our data would be consistent with zero mass, and it puts an approximate *upper limit* on the neutrino mass of $22\,\text{eV}/c^2$.

The actual detection of these neutrinos was brilliant—it was a rare event that allowed us to detect something other than EM radiation from beyond the solar system, and was an exceptional confirmation of theory. In the experiments, the most sensitive detector consisted of several thousand tons of water in an underground chamber. It detected 11 events in 12 seconds, probably via the reaction $\bar{\nu}_e + p \rightarrow n + e^+$. There was not a clear correlation between energy and time of arrival. Nonetheless, a careful analysis of that experiment set a rough upper limit on the electron anti-neutrino mass of about $4\,\text{eV}/c^2$. The more recent results mentioned above are much more definitive—they provide evidence that mass is much smaller, and that it is *not zero*.

43–6 Particle Classification

In the decades following the discovery of the π meson in the late 1940s, hundreds of other subnuclear particles were discovered. One way of arranging the particles in categories is according to their interactions, since not all particles interact by means of all four of the forces known in nature (though all interact via gravity). Table 43–2 (next page) lists some of the more common particles classified in this way along with many of their properties. At the top of Table 43–2 are the so-called "fundamental" particles which we believe have no internal structure. Below them are some of the "composite" particles which are made up of quarks, according to the Standard Model.

The fundamental particles include the **gauge bosons** (so-named after the theory that describes them, "gauge theory"), which include the gluons, the photon, and the W and Z particles; these are the particles that mediate the strong, electromagnetic, and weak interactions, respectively. Also fundamental are the **leptons**, which are particles that do not interact via the strong force but do interact via the weak nuclear force. Leptons that carry electric charge also interact via the electromagnetic force. The leptons include the electron, the muon, and the tau, and three types of neutrino: the electron neutrino (ν_e), the muon neutrino (ν_μ), and the tau neutrino (ν_τ). Each has an antiparticle.

TABLE 43–2 Particles (selected)[†]

Category	Forces involved	Particle name	Symbol	Anti-particle	Spin	Mass (MeV/c^2)	B	L_e	L_μ	L_τ	S	Mean life (s)	Principal Decay Modes
							[antiparticles have opposite sign]						
Fundamental													
Gauge bosons (force carriers)	str	Gluons	g	Self	1	0	0	0	0	0	0	Stable	
	em	Photon	γ	Self	1	0	0	0	0	0	0	Stable	
	w, em	W	W^+	W^-	1	80.40×10^3	0	0	0	0	0	$\approx 10^{-24}$	$e\nu_e$, $\mu\nu_\mu$, $\tau\nu_\tau$, hadrons
	w	Z	Z^0	Self	1	91.19×10^3	0	0	0	0	0	$\approx 10^{-24}$	e^+e^-, $\mu^+\mu^-$, $\tau^+\tau^-$, hadrons
Leptons	w, em[‡]	Electron	e^-	e^+	$\frac{1}{2}$	0.511	0	+1	0	0	0	Stable	
		Neutrino (e)	ν_e	$\bar{\nu}_e$	$\frac{1}{2}$	0 (<0.14 eV)[‡]	0	+1	0	0	0	Stable	
		Muon	μ^-	μ^+	$\frac{1}{2}$	105.7	0	0	+1	0	0	2.20×10^{-6}	$e^-\bar{\nu}_e\nu_\mu$
		Neutrino (μ)	ν_μ	$\bar{\nu}_\mu$	$\frac{1}{2}$	0 (<0.14 eV)[‡]	0	0	+1	0	0	Stable	
		Tau	τ^-	τ^+	$\frac{1}{2}$	1777	0	0	0	+1	0	2.91×10^{-13}	$\mu^-\bar{\nu}_\mu\nu_\tau$, $e^-\bar{\nu}_e\nu_\tau$, hadrons $+\nu_\tau$
		Neutrino (τ)	ν_τ	$\bar{\nu}_\tau$	$\frac{1}{2}$	0 (<0.14 eV)[‡]	0	0	0	+1	0	Stable	
Hadrons (composite), selected													
Mesons (quark–antiquark)	str, em, w	Pion	π^+	π^-	0	139.6	0	0	0	0	0	2.60×10^{-8}	$\mu^+\nu_\mu$
			π^0	Self	0	135.0	0	0	0	0	0	0.84×10^{-16}	2γ
		Kaon	K^+	K^-	0	493.7	0	0	0	0	+1	1.24×10^{-8}	$\mu^+\nu_\mu$, $\pi^+\pi^0$
			K^0_S	$\bar{K}^0_S$	0	497.7	0	0	0	0	+1	0.89×10^{-10}	$\pi^+\pi^-$, $2\pi^0$
			K^0_L	$\bar{K}^0_L$	0	497.7	0	0	0	0	+1	5.17×10^{-8}	$\pi^\pm e^\mp \overset{(-)}{\nu}_e$, $\pi^\pm\mu^\mp\overset{(-)}{\nu}_\mu$, 3π
		Eta	η^0	Self	0	547.5	0	0	0	0	0	$\approx 10^{-18}$	2γ, $3\pi^0$, $\pi^+\pi^-\pi^0$
		Rho	ρ^0	Self	1	775	0	0	0	0	0	$\approx 10^{-23}$	$\pi^+\pi^-$, $2\pi^0$
			ρ^+	ρ^-	1	775	0	0	0	0	0	$\approx 10^{-23}$	$\pi^+\pi^0$
		and others											
Baryons (3 quarks)	str, em, w	Proton	p	$\bar{p}$	$\frac{1}{2}$	938.3	+1	0	0	0	0	Stable	
		Neutron	n	$\bar{n}$	$\frac{1}{2}$	939.6	+1	0	0	0	0	886	$pe^-\bar{\nu}_e$
		Lambda	Λ^0	$\bar{\Lambda}^0$	$\frac{1}{2}$	1115.7	+1	0	0	0	−1	2.63×10^{-10}	$p\pi^-$, $n\pi^0$
		Sigma	Σ^+	$\bar{\Sigma}^-$	$\frac{1}{2}$	1189.4	+1	0	0	0	−1	0.80×10^{-10}	$p\pi^0$, $n\pi^+$
			Σ^0	$\bar{\Sigma}^0$	$\frac{1}{2}$	1192.6	+1	0	0	0	−1	7.4×10^{-20}	$\Lambda^0\gamma$
			Σ^-	$\bar{\Sigma}^+$	$\frac{1}{2}$	1197.4	+1	0	0	0	−1	1.48×10^{-10}	$n\pi^-$
		Xi	Ξ^0	$\bar{\Xi}^0$	$\frac{1}{2}$	1314.8	+1	0	0	0	−2	2.90×10^{-10}	$\Lambda^0\pi^0$
			Ξ^-	$\bar{\Xi}^+$	$\frac{1}{2}$	1321.3	+1	0	0	0	−2	1.64×10^{-10}	$\Lambda^0\pi^-$
		Omega	Ω^-	$\bar{\Omega}^+$	$\frac{3}{2}$	1672.5	+1	0	0	0	−3	0.82×10^{-10}	$\Xi^0\pi^-$, Λ^0K^-, $\Xi^-\pi^0$
		and others											

[†]See also Table 43–4 for particles with charm and bottomness. S in this Table stands for "strangeness" (see Section 43–8). More detail online at: pdg.lbl.gov.
[‡]Neutrinos partake only in the weak interaction. Experimental upper limits on neutrino masses are given in parentheses, as obtained mainly from the WMAP survey (Chapter 44). Detection of neutrino oscillations suggests that at least one type of neutrino has a nonzero mass greater than 0.04 eV.

The second category of particle in Table 43–2 is the **hadrons**, which are composite particles as we will discuss shortly. Hadrons are those particles that interact via the strong nuclear force. Hence they are said to be **strongly interacting particles**. They also interact via the other forces, but the strong force predominates at short distances. The hadrons include the proton, neutron, pion, and a large number of other particles. They are divided into two subgroups: **baryons**, which are those particles that have baryon number +1 (or −1 in the case of their antiparticles) and, as we shall see, are each made up of three quarks; and **mesons**, which have baryon number $= 0$, and are made up of a quark and an antiquark.

Only a few of the hundreds of hadrons (a veritable "zoo") are included in Table 43–2. Notice that the baryons Λ, Σ, Ξ, and Ω all decay to lighter-mass baryons, and eventually to a proton or neutron. All these processes conserve baryon number. Since there is no particle lighter than the proton with $B = +1$, if baryon number is strictly conserved, the proton itself cannot decay and is stable. (But see Section 43–11.) Note that Table 43–2 gives the *mean life* (τ) of each particle (as is done in particle physics), not the half-life ($T_{\frac{1}{2}}$). Recall that they differ by a factor 0.693: $\tau = T_{\frac{1}{2}}/\ln 2 = T_{\frac{1}{2}}/0.693$, Eq. 41–9. The term **lifetime** in particle physics means the mean life τ (= mean lifetime).

The baryon and lepton numbers (B, L_e, L_μ, L_τ), as well as strangeness S (Section 43–8), as given in Table 43–2 are for particles; their antiparticles have opposite sign for these numbers.

EXAMPLE 43–7 **Baryon decay.** Show that the decay modes of the Σ^+ baryon given in Table 43–2 do not violate the conservation laws we have studied up to now: energy, charge, baryon number, lepton numbers.

APPROACH Table 43–2 shows two possible decay modes, (*a*) $\Sigma^+ \rightarrow p + \pi^0$, (*b*) $\Sigma^+ \rightarrow n + \pi^+$. We check each for energy conservation, charge conservation, and conservation of baryon number. All the particles have lepton numbers equal to zero.

SOLUTION (*a*) Energy: for $\Sigma^+ \rightarrow p + \pi^0$ the change in mass-energy is

$$\begin{aligned} \Delta M &= m_\Sigma c^2 - m_p c^2 - m_{\pi^0} c^2 \\ &= 1189.4\,\text{MeV}/c^2 - 938.3\,\text{MeV}/c^2 - 135.0\,\text{MeV}/c^2 = +116.1\,\text{MeV}/c^2, \end{aligned}$$

so energy can be conserved with the resulting particles having kinetic energy.
Charge: $+e = +e + 0$, so charge is conserved.
Baryon number: $+1 = +1 + 0$, so baryon number is conserved.
(*b*) Energy: for $\Sigma^+ \rightarrow n + \pi^+$, the mass-energy change is

$$\begin{aligned} \Delta M &= m_\Sigma c^2 - m_n c^2 - m_{\pi^+} c^2 \\ &= 1189.4\,\text{MeV}/c^2 - 939.6\,\text{MeV}/c^2 - 139.6\,\text{MeV}/c^2 = 110.2\,\text{MeV}/c^2. \end{aligned}$$

This reaction releases 110.2 MeV of energy as kinetic energy of the products.
Charge: $+e = 0 + e$, so charge is conserved.
Baryon number: $+1 = +1 + 0$, so baryon number is conserved.

43–7 Particle Stability and Resonances

Many particles listed in Table 43–2 are unstable. The lifetime of an unstable particle depends on which force is most active in causing the decay. When a stronger force influences a decay, that decay occurs more quickly. Decays caused by the weak force typically have lifetimes of 10^{-13} s or longer (W and Z are exceptions). Decays via the electromagnetic force have much shorter lifetimes, typically about 10^{-16} to 10^{-19} s, and normally involve a γ (photon). The unstable particles listed in Table 43–2 decay either via the weak or the electromagnetic interaction.

Many particles have been found that decay via the strong interaction, with very short lifetimes, typically about 10^{-23} s. Their lifetimes are so short they do not travel far enough to be detected before decaying. The existence of such short-lived particles is inferred from their decay products. Consider the first such particle discovered (by Fermi), using a beam of π^+ particles with varying amounts of energy directed through a hydrogen target (protons). The number of interactions (π^+ scattered) plotted versus the pion's kinetic energy is shown in Fig. 43–14.

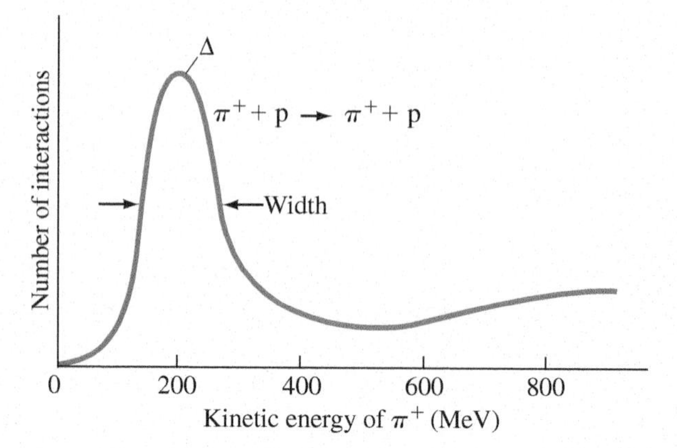

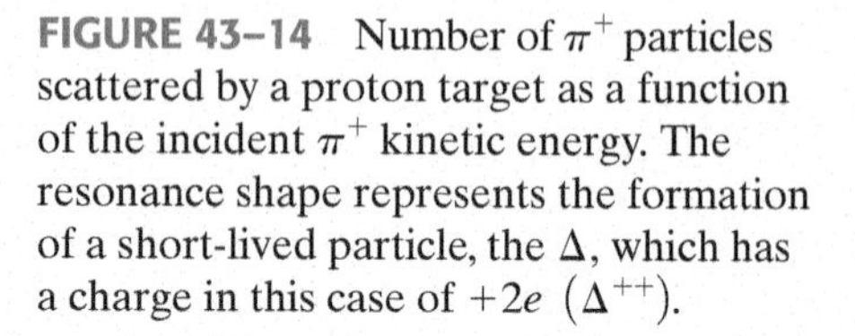

FIGURE 43–14 Number of π^+ particles scattered by a proton target as a function of the incident π^+ kinetic energy. The resonance shape represents the formation of a short-lived particle, the Δ, which has a charge in this case of $+2e$ (Δ^{++}).

The large number of interactions around 200 MeV led Fermi to conclude that the π^+ and proton combined momentarily to form a short-lived particle before coming apart again, or at least that they resonated together for a short time. Indeed, the large peak in Fig. 43–14 resembles a resonance curve (see Figs. 14–23, 14–26, and 30–22), and this new "particle"—now called the Δ—is referred to as a **resonance**. Hundreds of other resonances have been found, and are regarded as excited states of lighter mass particles such as the nucleon.

The **width** of a resonance—in Fig. 43–14 the full width of the Δ peak at half the peak height is on the order of 100 MeV—is an interesting application of the uncertainty principle. If a particle lives only 10^{-23} s, then its mass (i.e., its rest energy) will be uncertain by an amount

$$\Delta E \approx h/(2\pi\,\Delta t)$$

$$\approx (6.6\times10^{-34}\,\mathrm{J\cdot s})/(6)(10^{-23}\,\mathrm{s}) \approx 10^{-11}\,\mathrm{J} \approx 100\,\mathrm{MeV},$$

which is what is observed. Actually, the lifetimes of $\approx 10^{-23}\,\mathrm{s}$ for such resonances are inferred by the reverse process: from the measured width being ≈ 100 MeV.

43–8 Strange Particles? Charm? Towards a New Model

In the early 1950s, the newly found particles K, Λ, and Σ were found to behave rather strangely in two ways. First, they were always produced in pairs. For example, the reaction

$$\pi^- + \mathrm{p} \rightarrow \mathrm{K}^0 + \Lambda^0$$

occurred with high probability, but the similar reaction $\pi^- + \mathrm{p} \not\rightarrow \mathrm{K}^0 + \mathrm{n}$, was never observed to occur even though it did not violate any known conservation law. The second feature of these **strange particles**, as they came to be called, was that they were produced via the strong interaction (that is, at a high interaction rate), but did not decay at a fast rate characteristic of the strong interaction (even though they decayed into strongly interacting particles).

To explain these observations, a new quantum number, **strangeness**, and a new conservation law, **conservation of strangeness**, were introduced. By assigning the strangeness numbers (S) indicated in Table 43–2, the production of strange particles in pairs was explained. Antiparticles were assigned opposite strangeness from their particles. For example, in the reaction $\pi^- + \mathrm{p} \rightarrow \mathrm{K}^0 + \Lambda^0$, the initial state has strangeness $S = 0 + 0 = 0$, and the final state has $S = +1 - 1 = 0$, so strangeness is conserved. But for $\pi^- + \mathrm{p} \not\rightarrow \mathrm{K}^0 + \mathrm{n}$, the initial state has $S = 0$ and the final state has $S = +1 + 0 = +1$, so strangeness would not be conserved; and this reaction is not observed.

To explain the decay of strange particles, it is assumed that strangeness is conserved in the strong interaction but is *not conserved in the weak interaction.* Thus, strange particles were forbidden by strangeness conservation to decay to nonstrange particles of lower mass via the strong interaction, but could decay by means of the weak interaction at the observed longer lifetimes of 10^{-10} to 10^{-8} s.

CAUTION
Partially conserved quantities

The conservation of strangeness was the first example of a *partially conserved* quantity. In this case, the quantity strangeness is conserved by strong interactions but not by weak.

CONCEPTUAL EXAMPLE 43–8 **Guess the missing particle.** Using the conservation laws for particle interactions, determine the possibilities for the missing particle in the reaction

$$\pi^- + p \rightarrow K^0 + ?$$

in addition to $K^0 + \Lambda^0$ mentioned above.

RESPONSE We write equations for the conserved numbers in this reaction, with B, L_e, S, and Q as unknowns whose determination will reveal what the possible particle might be:

Baryon number:	$0 + 1 = 0 + B$
Lepton number:	$0 + 0 = 0 + L_e$
Charge:	$-1 + 1 = 0 + Q$
Strangeness:	$0 + 0 = 1 + S.$

The unknown product particle would have to have these characteristics:

$$B = +1 \qquad L_e = 0 \qquad Q = 0 \qquad S = -1.$$

In addition to Λ^0, a neutral sigma particle, Σ^0, is also consistent with these numbers.

In the next Section we will discuss another partially conserved quantity which was given the name **charm**. The discovery in 1974 of a particle with charm helped solidify a new theory involving quarks, which we now discuss.

43–9 Quarks

All particles, except the gauge bosons (Section 43–6), are either leptons or hadrons. One difference between these two groups is that the hadrons interact via the strong interaction, whereas the leptons do not.

There is another major difference. The six leptons $(e^-, \mu^-, \tau^-, \nu_e, \nu_\mu, \nu_\tau)$ are considered to be truly fundamental particles because they do not show any internal structure, and have no measurable size. (Attempts to determine the size of leptons have put an upper limit of about 10^{-18} m.) On the other hand, there are hundreds of hadrons, and experiments indicate they do have an internal structure.

In 1963, M. Gell-Mann and G. Zweig proposed that none of the hadrons, not even the proton and neutron, are truly fundamental, but instead are made up of combinations of three more fundamental pointlike entities called (somewhat whimsically) **quarks**.† Today, the quark theory is well-accepted, and quarks are considered truly fundamental particles, like leptons. The three quarks originally proposed were labeled u, d, s, and have the names **up**, **down**, and **strange**. The theory today has six quarks, just as there are six leptons—based on a presumed symmetry in nature. The other three quarks are called **charmed**, **bottom**, and **top**. The names apply also to new properties of each (quantum numbers c, b, t) that distinguish the new quarks from the old quarks (see Table 43–3), and which (like strangeness) are conserved in strong, but not weak, interactions.

†Gell-Mann chose the word from a phrase in James Joyce's *Finnegans Wake*.

TABLE 43–3 Properties of Quarks (Antiquarks have opposite sign Q, B, S, c, t, b)

Quarks								
Name	Symbol	Mass (MeV/c^2)	Charge Q	Baryon Number B	Strangeness S	Charm c	Bottomness b	Topness t
Up	u	2	$+\frac{2}{3}e$	$\frac{1}{3}$	0	0	0	0
Down	d	5	$-\frac{1}{3}e$	$\frac{1}{3}$	0	0	0	0
Strange	s	95	$-\frac{1}{3}e$	$\frac{1}{3}$	−1	0	0	0
Charmed	c	1250	$+\frac{2}{3}e$	$\frac{1}{3}$	0	+1	0	0
Bottom	b	4200	$-\frac{1}{3}e$	$\frac{1}{3}$	0	0	−1	0
Top	t	173,000	$+\frac{2}{3}e$	$\frac{1}{3}$	0	0	0	+1

TABLE 43–4 Partial List of Heavy Hadrons, with Charm and Bottomness ($L_e = L_\mu = L_\tau = 0$)

Category	Particle	Anti-particle	Spin	Mass (MeV/c^2)	Baryon Number B	Strangeness S	Charm c	Bottomness b	Mean life (s)	Principal Decay Modes
Mesons	D^+	D^-	0	1869.4	0	0	+1	0	10.6×10^{-13}	K + others, e + others
	D^0	$\overline{D}^0$	0	1864.5	0	0	+1	0	4.1×10^{-13}	K + others, μ or e + others
	D_S^+	D_S^-	0	1968	0	+1	+1	0	5.0×10^{-13}	K + others
	J/ψ (3097)	Self	1	3096.9	0	0	0	0	$\approx 10^{-20}$	Hadrons, e^+e^-, $\mu^+\mu^-$
	Υ (9460)	Self	1	9460	0	0	0	0	$\approx 10^{-20}$	Hadrons, $\mu^+\mu^-$, e^+e^-, $\tau^+\tau^-$
	B^-	B^+	0	5279	0	0	0	−1	1.6×10^{-12}	D^0 + others
	B^0	$\overline{B}^0$	0	5279	0	0	0	−1	1.5×10^{-12}	D^0 + others
Baryons	Λ_c^+	Λ_c^-	$\frac{1}{2}$	2286	+1	0	+1	0	2.0×10^{-13}	Hadrons (e.g., Λ + others)
	Σ_c^{++}	Σ_c^{--}	$\frac{1}{2}$	2454	+1	0	+1	0	$\approx 10^{-21}$	$\Lambda_c^+\pi^+$
	Σ_c^+	Σ_c^-	$\frac{1}{2}$	2453	+1	0	+1	0	$\approx 10^{-21}$	$\Lambda_c^+\pi^0$
	Σ_c^0	$\overline{\Sigma}_c^0$	$\frac{1}{2}$	2454	+1	0	+1	0	$\approx 10^{-21}$	$\Lambda_c^+\pi^-$
	Λ_b^0	$\overline{\Lambda}_b^0$	$\frac{1}{2}$	5620	+1	0	0	−1	1.2×10^{-12}	$J/\psi\Lambda^0$, $pD^0\pi^-$, $\Lambda_c^+\pi^+\pi^-\pi^-$

All quarks have spin $\frac{1}{2}$ and an electric charge of either $+\frac{2}{3}e$ or $-\frac{1}{3}e$ (that is, a fraction of the previously thought smallest charge e). Antiquarks have opposite sign of electric charge Q, baryon number B, strangeness S, charm c, bottomness b, and topness t. Other properties of quarks are shown in Table 43–3.

All hadrons are considered to be made up of combinations of quarks (plus the gluons that hold them together), and their properties are described by looking at their quark content. Mesons consist of a quark–antiquark pair. For example, a π^+ meson is a $u\overline{d}$ combination: note that for the $u\overline{d}$ pair (Table 43–3), $Q = \frac{2}{3}e + \frac{1}{3}e = +1e$, $B = \frac{1}{3} - \frac{1}{3} = 0$, $S = 0 + 0 = 0$, as they must for a π^+; and a $K^+ = u\overline{s}$, with $Q = +1$, $B = 0$, $S = +1$.

Baryons, on the other hand, consist of three quarks. For example, a neutron is $n = ddu$, whereas an antiproton is $\overline{p} = \overline{u}\,\overline{u}\,\overline{d}$. See Fig. 43–15. Strange particles all contain an s or $\overline{s}$ quark, whereas charmed particles contain a c or $\overline{c}$ quark. A few of these hadrons are listed in Table 43–4.

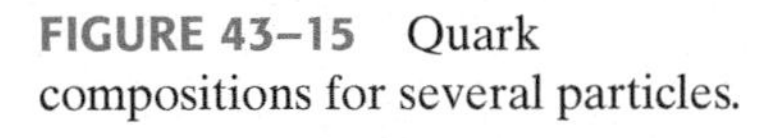

FIGURE 43–15 Quark compositions for several particles.

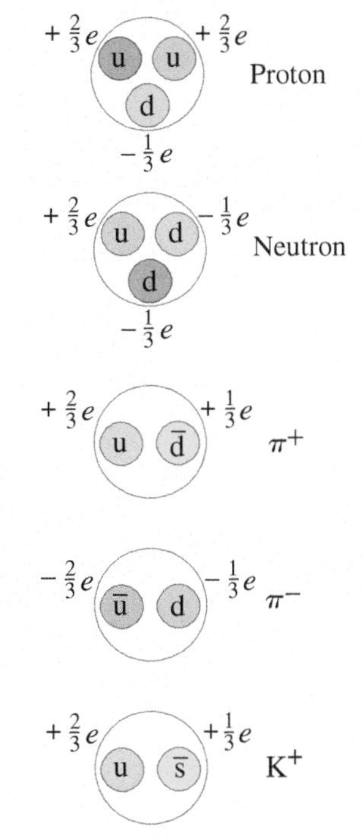

After the quark theory was proposed, physicists began looking for these fractionally charged particles, but direct detection has not been successful. Current models suggest that quarks may be so tightly bound together that they may not ever exist singly in the free state. But observations of very high energy electrons scattered off protons suggest that protons are indeed made up of constituents.

Today, the truly **fundamental particles** are considered to be the six quarks, the six leptons, and the gauge bosons that carry the fundamental forces. See Table 43–5, where the quarks and leptons are arranged in three "families" or "generations." Ordinary matter—atoms made of protons, neutrons, and electrons—is contained in the "first generation." The others are thought to have existed in the very early universe, but are seen by us today only at powerful accelerators or in cosmic rays. All of the hundreds of hadrons can be accounted for by combinations of the six quarks and six antiquarks.

EXERCISE D Return to the Chapter-Opening Questions, page 1164, and answer them again now. Try to explain why you may have answered differently the first time.

TABLE 43–5 The Fundamental Particles† as Seen Today

Gauge bosons	Force		First generation	Second generation	Third generation
Gluons	Strong	Quarks	u, d	s, c	b, t
$W^\pm$, Z_0	Weak	Leptons	e, ν_e	μ, ν_μ	τ, ν_τ
γ (photon)	EM				

†The quarks and leptons are arranged into three generations each.

CONCEPTUAL EXAMPLE 43–9 **Quark combinations.** Find the baryon number, charge, and strangeness for the following quark combinations, and identify the hadron particle that is made up of these quark combinations: (*a*) udd, (*b*) $u\bar{u}$, (*c*) uss, (*d*) sdd, and (*e*) $b\bar{u}$.

RESPONSE We use Table 43–3 to get the properties of the quarks, then Table 43–2 or 43–4 to find the particle that has these properties.

(*a*) udd has

$$Q = +\tfrac{2}{3}e - \tfrac{1}{3}e - \tfrac{1}{3}e = 0,$$
$$B = \tfrac{1}{3} + \tfrac{1}{3} + \tfrac{1}{3} = 1,$$
$$S = 0 + 0 + 0 = 0,$$

as well as $c = 0$, bottomness $= 0$, topness $= 0$. The only baryon $(B = +1)$ that has $Q = 0$, $S = 0$, etc., is the neutron (Table 43–2).

(*b*) $u\bar{u}$ has $Q = \tfrac{2}{3}e - \tfrac{2}{3}e = 0$, $B = \tfrac{1}{3} - \tfrac{1}{3} = 0$, and all other quantum numbers $= 0$. Sounds like a π^0 ($d\bar{d}$ also gives a π^0).

(*c*) uss has $Q = 0$, $B = +1$, $S = -2$, others $= 0$. This is a Ξ^0.

(*d*) sdd has $Q = -1$, $B = +1$, $S = -1$, so must be a Σ^-.

(*e*) $b\bar{u}$ has $Q = -1$, $B = 0$, $S = 0$, $c = 0$, bottomness $= -1$, topness $= 0$. This must be a B^- meson (Table 43–4).

EXERCISE E What is the quark composition of a K^- meson?

43–10 The "Standard Model": Quantum Chromodynamics (QCD) and Electroweak Theory

Not long after the quark theory was proposed, it was suggested that quarks have another property (or quality) called **color**, or "color charge" (analogous to electric charge). The distinction between the six types of quark (u, d, s, c, b, t) was referred to as **flavor**. According to theory, each of the flavors of quark can have three colors, usually designated red, green, and blue. (These are the three primary colors which, when added together in appropriate amounts, as on a TV screen, produce white.) Note that the names "color" and "flavor" have nothing to do with our senses, but are purely whimsical—as are other names, such as charm, in this new field. (We did, however, "color" the quarks in Fig. 43–15.) The antiquarks are colored antired, antigreen, and antiblue. Baryons are made up of three quarks, one of each color. Mesons consist of a quark–antiquark pair of a particular color and its anticolor. Both baryons and mesons are thus colorless or white.

Originally, the idea of quark color was proposed to preserve the Pauli exclusion principle (Section 39–4). Not all particles obey the exclusion principle. Those that do, such as electrons, protons, and neutrons, are called **fermions**. Those that don't are called **bosons**. These two categories are distinguished also in their spin (Section 39–2): bosons have integer spin (0, 1, 2, etc.) whereas fermions have half-integer spin, usually $\tfrac{1}{2}$ as for electrons and nucleons, but other fermions have spin $\tfrac{3}{2}, \tfrac{5}{2}$, etc. Matter is made up mainly of fermions, but the carriers of the forces (γ, W, Z, and gluons) are all bosons. Quarks are fermions (they have spin $\tfrac{1}{2}$) and therefore should obey the exclusion principle. Yet for three particular baryons (uuu, ddd, and sss), all three quarks would have the same quantum numbers, and at least two quarks have their spin in the same direction (since there are only two choices, spin up $\left[m_s = +\tfrac{1}{2}\right]$ or spin down $\left[m_s = -\tfrac{1}{2}\right]$). This would seem to violate the exclusion principle; but if quarks have an additional quantum number (color), which is different for each quark, it would serve to distinguish them and allow the exclusion principle to hold. Although quark color, and the resulting threefold increase in the number of quarks, was originally an *ad hoc* idea, it also served to bring the theory into better agreement with experiment, such as predicting the correct lifetime of the π^0 meson, and the measured rate of hadron production in observed e^+e^- collisions at accelerators. The idea of color soon became a central feature of the theory as determining the force binding quarks together in a hadron.

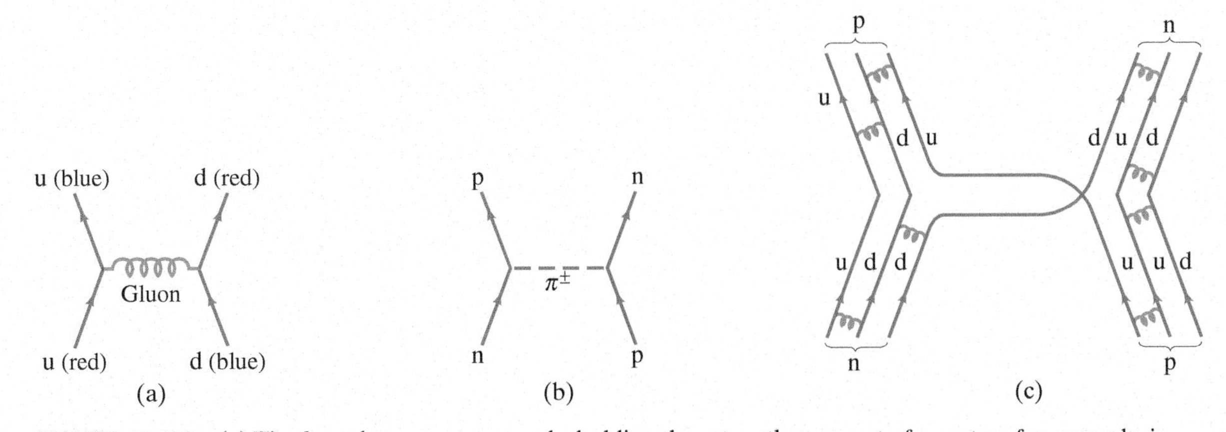

FIGURE 43–16 (a) The force between two quarks holding them together as part of a proton, for example, is carried by a gluon, which in this case involves a change in color. (b) Strong interaction $\mathrm{n} + \mathrm{p} \rightarrow \mathrm{n} + \mathrm{p}$ with the exchange of a charged π meson ($+$ or $-$, depending on whether it is considered moving to the left or to the right). (c) Quark representation of the same interaction $\mathrm{n} + \mathrm{p} \rightarrow \mathrm{n} + \mathrm{p}$. The blue coiled lines between quarks represent gluon exchanges holding the hadrons together. (The exchanged meson may be regarded as $\bar{\mathrm{u}}\mathrm{d}$ emitted by the n and absorbed by the p, or as $\mathrm{u}\bar{\mathrm{d}}$ emitted by p and absorbed by n, because a u (or d) quark going to the left in the diagram is equivalent to a $\bar{\mathrm{u}}$ (or $\bar{\mathrm{d}}$) going to the right.)

Each quark is assumed to carry a *color charge*, analogous to electric charge, and the strong force between quarks is referred to as the **color force**. This theory of the strong force is called **quantum chromodynamics** (*chroma* = color in Greek), or **QCD**, to indicate that the force acts between color charges (and not between, say, electric charges). The strong force between two hadrons is considered to be a force between the quarks that make them up, as suggested in Fig. 43–16. The particles that transmit the color force (analogous to photons for the EM force) are called **gluons** (a play on "glue"). They are included in Tables 43–2 and 43–5. There are eight gluons, according to the theory, all massless and all have color charge.†

You might ask what would happen if we try to see a single quark with color by reaching deep inside a hadron and extracting a single quark. Quarks are so tightly bound to other quarks that extracting one would require a tremendous amount of energy, so much that it would be sufficient to create more quarks $(E = mc^2)$. Indeed, such experiments are done at modern particle colliders and all we get is more hadrons (quark–antiquark pairs, or triplets, which we observe as mesons or baryons), never an isolated quark. This property of quarks, that they are always bound in groups that are colorless, is called **confinement**.

The color force has the interesting property that, as two quarks approach each other very closely (equivalently, have high energy), the force between them becomes small. This aspect is referred to as **asymptotic freedom**.

The weak force, as we have seen, is thought to be mediated by the W^+, W^-, and Z^0 particles. It acts between the "weak charges" that each particle has. Each elementary particle can thus have electric charge, weak charge, color charge, and gravitational mass, although one or more of these could be zero. For example, all leptons have color charge of zero, so they do not interact via the strong force.

CONCEPTUAL EXAMPLE 43–10 **Beta decay.** Draw a Feynman diagram, showing what happens in beta decay using quarks.

RESPONSE Beta decay is a result of the weak interaction, and the mediator is either a $W^{\pm}$ or Z^0 particle. What happens, in part, is that a neutron (udd quarks) decays into a proton (uud). Apparently a d quark (charge $-\frac{1}{3}e$) has turned into a u quark (charge $+\frac{2}{3}e$). Charge conservation means that a negatively charged particle, namely a W^-, was emitted by the d quark. Since an electron and an antineutrino appear in the final state, they must have come from the decay of the virtual W^-, as shown in Fig. 43–17.

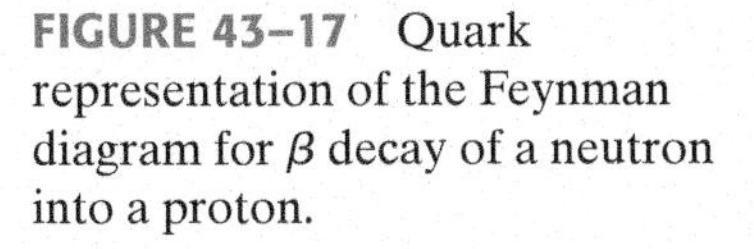
FIGURE 43–17 Quark representation of the Feynman diagram for β decay of a neutron into a proton.

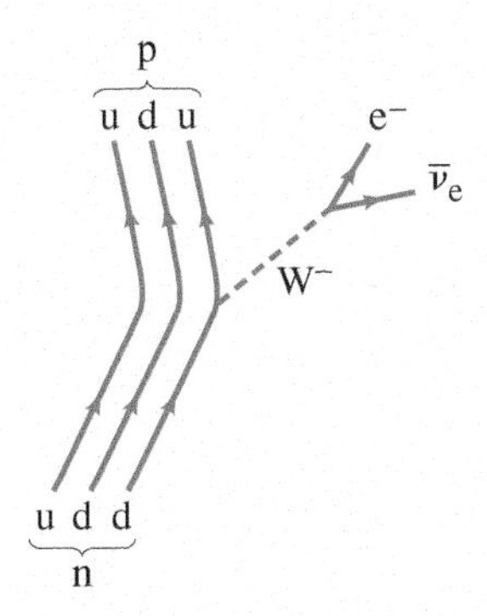

†Compare to the EM interaction, where the photon has no electric charge. Because gluons have color charge, they could attract each other and form composite particles (photons cannot). Such "glueballs" are being searched for.

To summarize, the Standard Model says that the truly fundamental particles (Table 43–5) are the leptons, the quarks, and the gauge bosons (photon, W and Z, and the gluons). The photon, leptons, W^+, W^-, and Z^0 have all been observed in experiments. But so far only combinations of quarks (baryons and mesons) have been observed, and it seems likely that free quarks and gluons are unobservable.

One important aspect of theoretical work is the attempt to find a **unified** basis for the different forces in nature. This was a long-held hope of Einstein, which he was never able to fulfill. A so-called **gauge theory** that unifies the weak and electromagnetic interactions was put forward in the 1960s by S. Weinberg, S. Glashow, and A. Salam. In this **electroweak theory**, the weak and electromagnetic forces are seen as two different manifestations of a single, more fundamental, *electroweak* interaction. The electroweak theory has had many successes, including the prediction of the $W^\pm$ particles as carriers of the weak force, with masses of $80.38 \pm 0.02\,\mathrm{GeV}/c^2$ in excellent agreement with the measured values of $80.403 \pm 0.029\,\mathrm{GeV}/c^2$ (and similar accuracy for the Z^0).

The combination of electroweak theory plus QCD for the strong interaction is often referred to today as the **Standard Model**.

EXAMPLE 43–11 ESTIMATE **Range of weak force.** The weak nuclear force is of very short range, meaning it acts over only a very short distance. Estimate its range using the masses (Table 43–2) of the $W^\pm$ and Z: $m \approx 80$ or $90\,\mathrm{GeV}/c^2 \approx 10^2\,\mathrm{GeV}/c^2$.

APPROACH We assume the $W^\pm$ or Z^0 exchange particles can exist for a time Δt given by the uncertainty principle, $\Delta t \approx \hbar/\Delta E$, where $\Delta E \approx mc^2$ is the energy needed to create the virtual particle ($W^\pm$, Z) that carries the weak force.

SOLUTION Let Δx be the distance the virtual W or Z can move before it must be reabsorbed within the time $\Delta t \approx \hbar/\Delta E$. To find an upper limit on Δx, and hence the maximum range of the weak force, we let the W or Z travel close to the speed of light, so $\Delta x \lesssim c\,\Delta t$. Recalling that $1\,\mathrm{GeV} = 1.6 \times 10^{-10}\,\mathrm{J}$, then

$$\Delta x \lesssim c\,\Delta t \approx \frac{c\hbar}{\Delta E} \approx \frac{(3 \times 10^8\,\mathrm{m/s})(10^{-34}\,\mathrm{J\cdot s})}{(10^2\,\mathrm{GeV})(1.6 \times 10^{-10}\,\mathrm{J/GeV})} \approx 10^{-18}\,\mathrm{m}.$$

This is indeed a very small range.

NOTE Compare this to the range of the electromagnetic force whose range is infinite ($1/r^2$ never becomes zero for any finite r), which makes sense because the mass of its virtual exchange particle, the photon, is zero (in the denominator of the above equation).

We did a similar calculation for the strong force in Section 43–2, and estimated the mass of the π meson as exchange particle between nucleons, based on the apparent range of $10^{-15}\,\mathrm{m}$ (size of nuclei). This is only one aspect of the strong force. In our deeper view, namely the color force between quarks within a nucleon, the gluons have zero mass, which implies (see the formula above in Example 43–11) infinite range. We might have expected a range of about $10^{-15}\,\mathrm{m}$; but according to the Standard Model, the color force is weak at very close distances and increases greatly with distance (causing quark confinement). Thus its range could be infinite.

Theoreticians have wondered why the W and Z have large masses rather than being massless like the photon. Electroweak theory suggests an explanation by means of an hypothesized **Higgs field** and its particle, the **Higgs boson**, which interact with the W and Z to "slow them down." In being forced to go slower than the speed of light, they would have to have mass ($m = 0$ only if $v = c$). Indeed, the Higgs is thought to permeate the vacuum ("empty space") and to perhaps confer mass on all particles with mass by slowing them down. The search for the Higgs boson is a priority for experimental particle physicists at CERN's Large Hadron Collider (Section 43–1). So far, searches suggest the Higgs mass is greater than $115\,\mathrm{GeV}/c^2$. Yet it is expected to have a mass no larger than $1\,\mathrm{TeV}/c^2$. We are narrowing in on it.

43–11 Grand Unified Theories

The Standard Model, for all its success, cannot explain some important issues—such as why the charge on the electron has *exactly* the same magnitude as the charge on the proton. This is crucial, because if the charge magnitudes were even a little different, atoms would not be neutral and the resulting large electric forces would surely have made life impossible. Indeed, the Standard Model is now considered to be a low-energy approximation to a more complete theory.

With the success of unified electroweak theory, theorists are trying to incorporate it and QCD for the strong (color) force into a so-called **grand unified theory** (GUT).

One type of such a grand unified theory of the electromagnetic, weak, and strong forces has been worked out in which there is only one class of particle—leptons and quarks belong to the same family and are able to change freely from one type to the other—and the three forces are different aspects of a single underlying force. The unity is predicted to occur, however, only on a scale of less than about 10^{-31} m, corresponding to a typical particle energy of about 10^{16} GeV. If two elementary particles (leptons or quarks) approach each other to within this **unification scale**, the apparently fundamental distinction between them would not exist at this level, and a quark could readily change to a lepton, or vice versa. Baryon and lepton numbers would not be conserved. The weak, electromagnetic, and strong (color) force would blend to a force of a single strength.

What happens between the unification distance of 10^{-31} m and more normal (larger) distances is referred to as **symmetry breaking**. As an analogy, consider an atom in a crystal. Deep within the atom, there is much symmetry—in the innermost regions the electron cloud is spherically symmetric (Chapter 39). Farther out, this symmetry breaks down—the electron clouds are distributed preferentially along the lines (bonds) joining the atoms in the crystal. In a similar way, at 10^{-31} m the force between elementary particles is theorized to be a single force—it is symmetrical and does not single out one type of "charge" over another. But at larger distances, that symmetry is broken and we see three distinct forces. (In the "Standard Model" of electroweak interactions, Section 43–10, the symmetry breaking between the electromagnetic and the weak interactions occurs at about 10^{-18} m.)

CONCEPTUAL EXAMPLE 43–12 **Symmetry.** The table in Fig. 43–18 has four identical place settings. Four people sit down to eat. Describe the symmetry of this table and what happens to it when someone starts the meal.

RESPONSE The table has several kinds of symmetry. It is symmetric to rotations of 90°: that is, the table will look the same if everyone moved one chair to the left or to the right. It is also north–south symmetric and east–west symmetric, so that swaps across the table don't affect the way the table looks. It also doesn't matter whether any person picks up the fork to the left of the plate or the fork to the right. But once that first person picks up either fork, the choice is set for all the rest at the table as well. The symmetry has been *broken*. The underlying symmetry is still there—the blue glasses could still be chosen either way—but some choice must get made and at that moment the symmetry of the diners is broken.

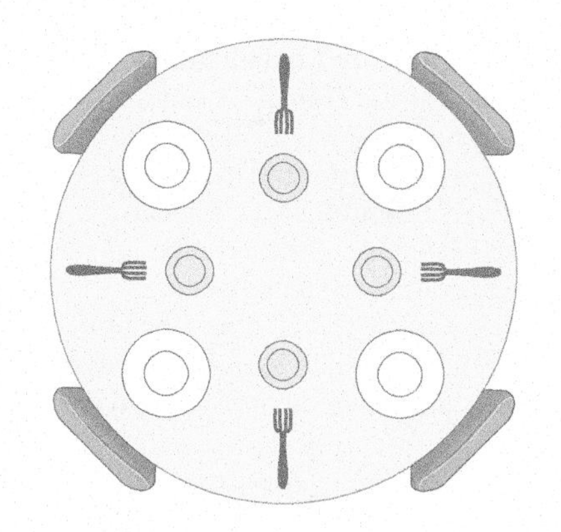

FIGURE 43–18 Symmetry around a table. Example 43–12.

Another example of symmetry breaking is a pencil standing on its point before falling. Standing, it looks the same from any horizontal direction. From above, it is a tiny circle. But when it falls to the table, it points in one particular direction—the symmetry is broken.

Proton Decay

Since unification is thought to occur at such tiny distances and huge energies, the theory is difficult to test experimentally. But it is not completely impossible. One testable prediction is the idea that the proton might decay (via, for example, $p \rightarrow \pi^0 + e^+$) and violate conservation of baryon number. This could happen if two quarks approached to within 10^{-31} m of each other. But it is very unlikely at normal temperature and energy, so the decay of a proton can only be an unlikely process.

In the simplest form of GUT, the theoretical estimate of the proton mean life for the decay mode $p \rightarrow \pi^0 + e^+$ is about 10^{31} yr, and this is now within the realm of testability.† Proton decays have still not been seen, and experiments put the lower limit on the proton mean life for the above mode to be about 10^{33} yr, somewhat greater than this prediction. This may seem a disappointment, but on the other hand, it presents a challenge. Indeed more complex GUTs are not affected by this result.

EXAMPLE 43–13 ESTIMATE **Proton decay.** An experiment uses 3300 tons of water waiting to see a proton decay of the type $p \rightarrow \pi^0 + e^+$. If the experiment is run for 4 years without detecting a decay, estimate the lower limit on the proton mean life.

APPROACH As with radioactive decay, the number of decays is proportional to the number of parent species (N), the time interval (Δt), and the decay constant (λ) which is related to the mean life τ by (see Eqs. 41–4 and 41–9a):

$$\Delta N = -\lambda N \Delta t = -\frac{N \Delta t}{\tau}.$$

SOLUTION Dealing only with magnitudes, we solve for τ:

$$\tau = \frac{N \Delta t}{\Delta N}.$$

Thus for $\Delta N < 1$ over the four-year trial,

$$\tau > N(4\,\text{yr}),$$

where N is the number of protons in 3300 tons of water. To determine N, we note that each molecule of H_2O contains $2 + 8 = 10$ protons. So one mole of water ($18\,\text{g}$, 6×10^{23} molecules) contains $10 \times 6 \times 10^{23}$ protons in 18 g of water, or about 3×10^{26} protons per kilogram. One ton is 10^3 kg, so the chamber contains $(3.3 \times 10^6\,\text{kg})(3 \times 10^{26}\,\text{protons/kg}) \approx 1 \times 10^{33}$ protons. Then our very rough estimate for a lower limit on the proton mean life is $\tau > (10^{33})(4\,\text{yr}) \approx 4 \times 10^{33}$ yr.

GUT and Cosmology

An interesting prediction of unified theories relates to cosmology (Chapter 44). It is thought that during the first 10^{-35} s after the theorized Big Bang that created the universe, the temperature was so extremely high that particles had energies corresponding to the unification scale. Baryon number would not have been conserved then, perhaps allowing an imbalance that might account for the observed predominance of matter ($B > 0$) over antimatter ($B < 0$) in the universe. The fact that we are surrounded by matter, with no significant antimatter in sight, is considered a problem in search of an explanation (not given by the Standard Model). See also Chapter 44. We call this the **matter–antimatter problem.** To understand it may require still undiscovered phenomena—perhaps related to quarks or neutrinos, or the Higgs boson or supersymmetry (next Section).

This last example is interesting, for it illustrates a deep connection between investigations at either end of the size scale: theories about the tiniest objects (elementary particles) have a strong bearing on the understanding of the universe on a large scale. We will look at this more in the next Chapter.

Figure 43–19 is a rough diagram indicating how the four fundamental forces in nature "condensed out" (a symmetry was broken) as time went on after the Big Bang (Chapter 44), and as the mean temperature of the universe and the typical particle energy decreased.

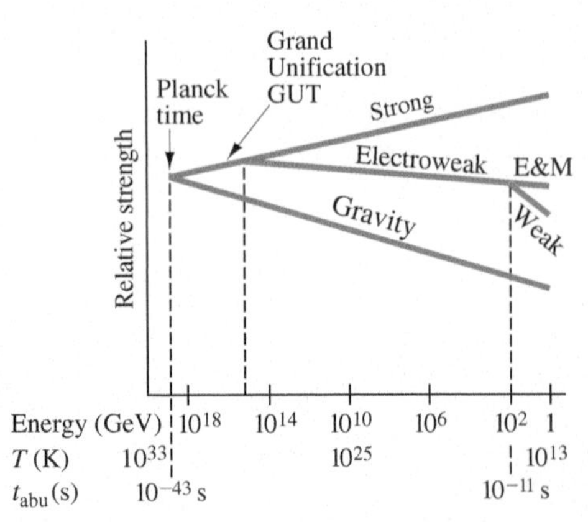

FIGURE 43–19 Time and energy plot of the four fundamental forces, unified at the Planck time, and how each condensed out. The symbol t_{abu} = time after the birth of the universe. Note that the typical particle energy (and average temperature of the universe) decreases to the right, as time after the Big Bang increases. We discuss the Big Bang in the next Chapter.

†This is much larger than the age of the universe ($\approx 14 \times 10^9$ yr). But we don't have to wait 10^{31} yr to see. Instead we can wait for one decay among 10^{31} protons over a year (see Eqs. 41–4 and 41–9a, $\Delta N = \lambda N \Delta t = N \Delta t/\tau$).

43–12 Strings and Supersymmetry

We have seen that the Standard Model is unable to address important experimental issues, and that theoreticians are attacking the problem as experimenters search for new data, new particles, new concepts.

Even more ambitious than grand unified theories are attempts to also incorporate gravity, and thus unify all four forces in nature into a single theory. (Such theories are sometimes referred to misleadingly as **theories of everything.**) There are consistent theories that attempt to unify all four forces called **string theories**, in which each fundamental particle (Table 43–5) is imagined not as a point but as a one-dimensional string, perhaps 10^{-35} m long, which vibrates in a particular standing wave pattern. (You might say each particle is a different note on a tiny stretched string.) More sophisticated theories propose the fundamental entities as being multidimensional **branes** (after 2-D membranes).

A related idea that also goes way beyond the Standard Model is **supersymmetry**, which applied to strings is known as **superstring theory**. Supersymmetry, developed by Bruno Zumino (1923–) and Julius Wess (1934–2007), predicts that interactions exist that would change fermions into bosons and vice versa, and that each known fermion would have a supersymmetric boson partner of the same mass. Thus, for each quark (a fermion), there would be a **squark** (a *boson*) or "supersymmetric" quark. For every lepton there would be a **slepton**. Likewise, for every known boson (photons and gluons, for example), there would be a supersymmetric fermion (**photinos** and **gluinos**). Supersymmetry predicts also that a *graviton*, which transmits the gravity force, has a partner, the **gravitino**. Supersymmetric particles are sometimes called "SUSYs" for short, and are a candidate for the "dark matter" of the universe (discussed in Chapter 44). But why hasn't this "missing part" of the universe ever been detected? The best guess is that supersymmetric particles might be heavier than their conventional counterparts, perhaps too heavy to have been produced in today's accelerators. A search for supersymmetric particles is already in the works for CERN's new Large Hadron Collider.

Versions of supersymmetry predict other interesting properties, such as that space has 11 dimensions, but 7 of them are "coiled up" so we normally only notice the 4-D of space–time. We would like to know if and how many extra dimensions there are, and how and why they are hidden. We hope to have some answers from the new LHC and the future ILC (Section 43–1).

The world of elementary particles is opening new vistas. What happens in the future is bound to be exciting.

Summary

Particle accelerators are used to accelerate charged particles, such as electrons and protons, to very high energy. High-energy particles have short wavelength and so can be used to probe the structure of matter in great detail (very small distances). High kinetic energy also allows the creation of new particles through collisions (via $E = mc^2$).

Cyclotrons and **synchrotrons** use a magnetic field to keep the particles in a circular path and accelerate them at intervals by high voltage. **Linear accelerators** accelerate particles along a line. **Colliding beams** allow higher interaction energy.

An **antiparticle** has the same mass as a particle but opposite charge. Certain other properties may also be opposite: for example, the antiproton has **baryon number** (nucleon number) opposite ($B = -1$) to that for the proton ($B = +1$).

In all nuclear and particle reactions, the following conservation laws hold: momentum, angular momentum, mass–energy, electric charge, baryon number, and **lepton numbers**.

Certain particles have a property called **strangeness**, which is conserved by the strong force but not by the weak force. The properties **charm, bottomness**, and **topness** also are conserved by the strong force but not by the weak force.

Just as the electromagnetic force can be said to be due to an exchange of photons, the strong nuclear force is carried by massless **gluons**. The W and Z particles carry the weak force. These fundamental force carriers (photon, W and Z, gluons) are called **gauge bosons**.

Other particles can be classified as either *leptons* or *hadrons*. **Leptons** participate only in gravity, the weak, and the electromagnetic interactions. **Hadrons**, which today are considered to be made up of **quarks**, participate in all four interactions, including the strong interaction. The hadrons can be classified as **mesons**, with baryon number zero, and **baryons**, with nonzero baryon number.

All particles, except for the photon, electron, neutrinos, and proton, decay with measurable mean lives varying from 10^{-25} s to 10^{3} s. The mean life depends on which force is predominant. Weak decays usually have mean lives greater than about 10^{-13} s. Electromagnetic decays typically have mean lives on the order of 10^{-16} to 10^{-19} s. The shortest lived particles, called **resonances**, decay via the strong interaction and live typically for only about 10^{-23} s.

Today's Standard Model of elementary particles considers **quarks** as the basic building blocks of the hadrons. The six quark "flavors" are called **up, down, strange, charmed, bottom**, and **top**. It is expected that there are the same number of quarks as leptons (six of each), and that quarks and leptons are the truly fundamental particles along with the gauge bosons (γ, W, Z, gluons). Quarks are said to have **color**, and, according to **quantum chromodynamics** (QCD), the strong color force acts between their color charges and is transmitted by **gluons**. **Electroweak theory** views the weak and electromagnetic forces as two aspects of a single underlying interaction. QCD plus the electroweak theory are referred to as the **Standard Model**.

Grand unified theories of forces suggest that at very short distance (10^{-31} m) and very high energy, the weak, electromagnetic, and strong forces appear as a single force, and the fundamental difference between quarks and leptons disappears.

According to **string theory**, the fundamental particles may be tiny strings, 10^{-35} m long, distinguished by their standing wave pattern. **Supersymmetry** hypothesizes that each fermion (or boson) has a corresponding boson (or fermion) partner.

Questions

1. Give a reaction between two nucleons, similar to Eq. 43–4, that could produce a π^-.
2. If a proton is moving at very high speed, so that its kinetic energy is much greater than its rest energy (mc^2), can it then decay via $p \rightarrow n + \pi^+$?
3. What would an "antiatom," made up of the antiparticles to the constituents of normal atoms, consist of? What might happen if *antimatter*, made of such antiatoms, came in contact with our normal world of matter?
4. What particle in a decay signals the electromagnetic interaction?
5. (*a*) Does the presence of a neutrino among the decay products of a particle necessarily mean that the decay occurs via the weak interaction? (*b*) Do all decays via the weak interaction produce a neutrino? Explain.
6. Why is it that a neutron decays via the weak interaction even though the neutron and one of its decay products (proton) are strongly interacting?
7. Which of the four interactions (strong, electromagnetic, weak, gravitational) does an electron take part in? A neutrino? A proton?
8. Check that charge and baryon number are conserved in each of the decays in Table 43–2.
9. Which of the particle decays listed in Table 43–2 occur via the electromagnetic interaction?
10. Which of the particle decays listed in Table 43–2 occur by the weak interaction?
11. The Δ baryon has spin $\frac{3}{2}$, baryon number 1, and charge $Q = +2, +1, 0,$ or -1. Why is there no charge state $Q = -2$?
12. Which of the particle decays in Table 43–4 occur via the electromagnetic interaction?
13. Which of the particle decays in Table 43–4 occur by the weak interaction?
14. Quarks have spin $\frac{1}{2}$. How do you account for the fact that baryons have spin $\frac{1}{2}$ or $\frac{3}{2}$, and mesons have spin 0 or 1?
15. Suppose there were a kind of "neutrinolet" that was massless, had no color charge or electrical charge, and did not feel the weak force. Could you say that this particle even exists?
16. Is it possible for a particle to be both (*a*) a lepton and a baryon? (*b*) a baryon and a hadron? (*c*) a meson and a quark? (*d*) a hadron and a lepton? Explain.
17. Using the ideas of quantum chromodynamics, would it be possible to find particles made up of two quarks and no antiquarks? What about two quarks and two antiquarks?
18. Why can neutrons decay when they are free, but not when they are inside a stable nucleus?
19. Is the reaction $e^- + p \rightarrow n + \bar{\nu}_e$ possible? Explain.
20. Occasionally, the Λ will decay by the following reaction: $\Lambda^0 \rightarrow p^+ + e^- + \bar{\nu}_e$. Which of the four forces in nature is responsible for this decay? How do you know?

Problems

43–1 Particles and Accelerators

1. (I) What is the total energy of a proton whose kinetic energy is 4.65 GeV?
2. (I) Calculate the wavelength of 28-GeV electrons.
3. (I) What strength of magnetic field is used in a cyclotron in which protons make 3.1×10^7 revolutions per second?
4. (I) What is the time for one complete revolution for a very high-energy proton in the 1.0-km-radius Fermilab accelerator?
5. (I) If α particles are accelerated by the cyclotron of Example 43–2, what must be the frequency of the voltage applied to the dees?
6. (II) (*a*) If the cyclotron of Example 43–2 accelerated α particles, what maximum energy could they attain? What would their speed be? (*b*) Repeat for deuterons ($^2_1$H). (*c*) In each case, what frequency of voltage is required?
7. (II) Which is better for resolving details of the nucleus: 25-MeV alpha particles or 25-MeV protons? Compare each of their wavelengths with the size of a nucleon in a nucleus.
8. (II) What magnetic field intensity is needed at the 1.0-km-radius Fermilab synchrotron for 1.0-TeV protons?
9. (II) What magnetic field is required for the 7.0-TeV protons in the 4.25-km-radius Large Hadron Collider (LHC)?
10. (II) A cyclotron with a radius of 1.0 m is to accelerate deuterons ($^2_1$H) to an energy of 12 MeV. (*a*) What is the required magnetic field? (*b*) What frequency is needed for the voltage between the dees? (*c*) If the potential difference between the dees averages 22 kV, how many revolutions will the particles make before exiting? (*d*) How much time does it take for one deuteron to go from start to exit? (*e*) Estimate how far it travels during this time.
11. (II) What is the wavelength (= minimum resolvable size) of 7.0-TeV protons?
12. (II) The 1.0-km radius Fermilab Tevatron takes about 20 seconds to bring the energies of the stored protons from 150 GeV to 1.0 TeV. The acceleration is done once per turn. Estimate the energy given to the protons on each turn. (You can assume that the speed of the protons is essentially *c* the whole time.)
13. (II) Show that the energy of a particle (charge *e*) in a synchrotron, in the relativistic limit ($v \approx c$), is given by E (in eV) $= Brc$, where B is the magnetic field and r is the radius of the orbit (SI units).

43–2 to 43–6 Particle Interactions, Particle Exchange

14. (I) About how much energy is released when a Λ^0 decays to $n + \pi^0$? (See Table 43–2.)
15. (I) How much energy is released in the decay
$$\pi^+ \rightarrow \mu^+ + \nu_\mu?$$
See Table 43–2.
16. (I) Estimate the range of the strong force if the mediating particle were the kaon in place of a pion.
17. (I) How much energy is required to produce a neutron–antineutron pair?
18. (II) Determine the energy released when Σ^0 decays to Λ^0 and then to a proton.
19. (II) Two protons are heading toward each other with equal speeds. What minimum kinetic energy must each have if a π^0 meson is to be created in the process? (See Table 43–2.)
20. (II) What minimum kinetic energy must two neutrons each have if they are traveling at the same speed toward each other, collide, and produce a K^+K^- pair in addition to themselves? (See Table 43–2.)
21. (II) For the decay $K^0 \rightarrow \pi^- + e^+ + \nu_e$, determine the maximum kinetic energy of (*a*) the positron, and (*b*) the π^-. Assume the K^0 is at rest.
22. (II) What are the wavelengths of the two photons produced when a proton and antiproton at rest annihilate?
23. (II) The Λ^0 cannot decay by the following reactions. What conservation laws are violated in each of the reactions?
(*a*) $\Lambda^0 \nrightarrow n + \pi^-$
(*b*) $\Lambda^0 \nrightarrow p + K^-$
(*c*) $\Lambda^0 \nrightarrow \pi^+ + \pi^-$
24. (II) For the decay $\Lambda^0 \rightarrow p + \pi^-$, calculate (*a*) the *Q*-value (energy released), and (*b*) the kinetic energy of the p and π^-, assuming the Λ^0 decays from rest. (Use relativistic formulas.)
25. (II) (*a*) Show, by conserving momentum and energy, that it is impossible for an isolated electron to radiate only a single photon. (*b*) With this result in mind, how can you defend the photon exchange diagram in Fig. 43–8?
26. (II) What would be the wavelengths of the two photons produced when an electron and a positron, each with 420 keV of kinetic energy, annihilate in a head-on collision?
27. (II) In the rare decay $\pi^+ \rightarrow e^+ + \nu_e$, what is the kinetic energy of the positron? Assume the π^+ decays from rest.
28. (II) Which of the following reactions and decays are possible? For those forbidden, explain what laws are violated.
(*a*) $\pi^- + p \rightarrow n + \eta^0$
(*b*) $\pi^+ + p \rightarrow n + \pi^0$
(*c*) $\pi^+ + p \rightarrow p + e^+$
(*d*) $p \rightarrow e^+ + \nu_e$
(*e*) $\mu^+ \rightarrow e^+ + \bar{\nu}_\mu$
(*f*) $p \rightarrow n + e^+ + \nu_e$
29. (II) Calculate the kinetic energy of each of the two products in the decay $\Xi^- \rightarrow \Lambda^0 + \pi^-$. Assume the Ξ^- decays from rest.
30. (II) Antiprotons can be produced when a proton with sufficient energy hits a stationary proton. Even if there is enough energy, which of the following reactions will not happen?
$$p + p \rightarrow p + \bar{p}$$
$$p + p \rightarrow p + p + \bar{p}$$
$$p + p \rightarrow p + p + p + \bar{p}$$
$$p + p \rightarrow p + e^+ + e^+ + \bar{p}$$
31. (III) Calculate the maximum kinetic energy of the electron when a muon decays from rest via $\mu^- \rightarrow e^- + \bar{\nu}_e + \nu_\mu$. [*Hint*: In what direction do the two neutrinos move relative to the electron in order to give the electron the maximum kinetic energy? Both energy and momentum are conserved; use relativistic formulas.]
32. (III) Could a π^+ meson be produced if a 110-MeV proton struck a proton at rest? What minimum kinetic energy must the incoming proton have?

43–7 to 43–11 Resonances, Standard Model, Quarks, QCD, GUT

33. (I) The mean life of the Σ^0 particle is 7×10^{-20} s. What is the uncertainty in its rest energy? Express your answer in MeV.
34. (I) The measured width of the ψ (3686) meson is about 300 keV. Estimate its mean life.
35. (I) The measured width of the J/ψ meson is 88 keV. Estimate its mean life.
36. (I) The B^- meson is a $b\bar{u}$ quark combination. (*a*) Show that this is consistent for all quantum numbers. (*b*) What are the quark combinations for B^+, B^0, $\bar{B}^0$?
37. (I) What is the energy width (or uncertainty) of (*a*) η^0, and (*b*) ρ^+? See Table 43–2.
38. (II) Which of the following decays are possible? For those that are forbidden, explain which laws are violated.
(*a*) $\Xi^0 \rightarrow \Sigma^+ + \pi^-$
(*b*) $\Omega^- \rightarrow \Sigma^0 + \pi^- + \nu$
(*c*) $\Sigma^0 \rightarrow \Lambda^0 + \gamma + \gamma$
39. (II) What quark combinations produce (*a*) a Ξ^0 baryon and (*b*) a Ξ^- baryon?
40. (II) What are the quark combinations that can form (*a*) a neutron, (*b*) an antineutron, (*c*) a Λ^0, (*d*) a $\bar{\Sigma}^0$?
41. (II) What particles do the following quark combinations produce: (*a*) uud, (*b*) $\bar{u}\bar{u}\bar{s}$, (*c*) $\bar{u}s$, (*d*) $d\bar{u}$, (*e*) $\bar{c}s$?
42. (II) What is the quark combination needed to produce a D^0 meson ($Q = B = S = 0$, $c = +1$)?
43. (II) The D_S^+ meson has $S = c = +1$, $B = 0$. What quark combination would produce it?
44. (II) Draw a possible Feynman diagram using quarks (as in Fig. 43–16c) for the reaction $\pi^- + p \rightarrow \pi^0 + n$.
45. (II) Draw a Feynman diagram for the reaction $n + \nu_\mu \rightarrow p + \mu^-$.

General Problems

46. The mean lifetimes listed in Table 43–2 are in terms of *proper time*, measured in a reference frame where the particle is at rest. If a tau lepton is created with a kinetic energy of 950 MeV, how long would its track be as measured in the lab, on average, ignoring any collisions?
47. Assume there are 5.0×10^{13} protons at 1.0 TeV stored in the 1.0-km-radius ring of the Tevatron. (*a*) How much current (amperes) is carried by this beam? (*b*) How fast would a 1500-kg car have to move to carry the same kinetic energy as this beam?

48. (*a*) How much energy is released when an electron and a positron annihilate each other? (*b*) How much energy is released when a proton and an antiproton annihilate each other? (All particles have $K \approx 0$.)

49. Protons are injected into the 1.0-km-radius Fermilab Tevatron with an energy of 150 GeV. If they are accelerated by 2.5 MV each revolution, how far do they travel and approximately how long does it take for them to reach 1.0 TeV?

50. Which of the following reactions are possible, and by what interaction could they occur? For those forbidden, explain why.
(*a*) $\pi^- + p \rightarrow K^0 + p + \pi^0$
(*b*) $K^- + p \rightarrow \Lambda^0 + \pi^0$
(*c*) $K^+ + n \rightarrow \Sigma^+ + \pi^0 + \gamma$
(*d*) $K^+ \rightarrow \pi^0 + \pi^0 + \pi^+$
(*e*) $\pi^+ \rightarrow e^+ + \nu_e$

51. Which of the following reactions are possible, and by what interaction could they occur? For those forbidden, explain why.
(*a*) $\pi^- + p \rightarrow K^+ + \Sigma^-$
(*b*) $\pi^+ + p \rightarrow K^+ + \Sigma^+$
(*c*) $\pi^- + p \rightarrow \Lambda^0 + K^0 + \pi^0$
(*d*) $\pi^+ + p \rightarrow \Sigma^0 + \pi^0$
(*e*) $\pi^- + p \rightarrow p + e^- + \bar{\nu}_e$

52. One decay mode for a π^+ is $\pi^+ \rightarrow \mu^+ + \nu_\mu$. What would be the equivalent decay for a π^-? Check conservation laws.

53. Symmetry breaking occurs in the electroweak theory at about 10^{-18} m. Show that this corresponds to an energy that is on the order of the mass of the $W^\pm$.

54. Calculate the *Q*-value for each of the reactions, Eq. 43–4, for producing a pion.

55. How many fundamental fermions are there in a water molecule?

56. The mass of a π^0 can be measured by observing the reaction $\pi^- + p \rightarrow \pi^0 + n$ at very low incident π^- kinetic energy (assume it is zero). The neutron is observed to be emitted with a kinetic energy of 0.60 MeV. Use conservation of energy and momentum to determine the π^0 mass.

57. (*a*) Show that the so-called unification distance of 10^{-31} m in grand unified theory is equivalent to an energy of about 10^{16} GeV. Use the uncertainty principle, and also de Broglie's wavelength formula, and explain how they apply. (*b*) Calculate the temperature corresponding to 10^{16} GeV.

58. Calculate the *Q*-value for the reaction $\pi^- + p \rightarrow \Lambda^0 + K^0$, when negative pions strike stationary protons. Estimate the minimum pion kinetic energy needed to produce this reaction. [*Hint*: Assume Λ^0 and K^0 move off with the same velocity.]

59. A proton and an antiproton annihilate each other at rest and produce two pions, π^- and π^+. What is the kinetic energy of each pion?

60. For the reaction $p + p \rightarrow 3p + \bar{p}$, where one of the initial protons is at rest, use relativistic formulas to show that the threshold energy is $6m_p c^2$, equal to three times the magnitude of the *Q*-value of the reaction, where m_p is the proton mass. [*Hint*: Assume all final particles have the same velocity.]

61. What is the total energy of a proton whose kinetic energy is 15 GeV? What is its wavelength?

62. At about what kinetic energy (in eV) can the rest energy of a proton be ignored when calculating its wavelength, if the wavelength is to be within 1.0% of its true value? What are the corresponding wavelength and speed of the proton?

63. Use the quark model to describe the reaction

$$\bar{p} + n \rightarrow \pi^- + \pi^0.$$

64. Identify the missing particle in the following reactions.
(*a*) $p + p \rightarrow p + n + \pi^+ + ?$
(*b*) $p + ? \rightarrow n + \mu^+$

65. What fraction of the speed of light *c* is the speed of a 7.0-TeV proton?

66. A particle at rest, with a rest energy of mc^2, decays into two fragments with rest energies of $m_1 c^2$ and $m_2 c^2$. Show that the kinetic energy of fragment 1 is

$$K_1 = \frac{1}{2mc^2}\left[(mc^2 - m_1 c^2)^2 - (m_2 c^2)^2\right].$$

* Numerical/Computer

*67. (II) In a particle physics experiment to determine the mean lifetime of muons, the muons enter a scintillator and decay. Students have sampled the individual lifetimes of muons decaying within a time interval between 1 μs and 10 μs after being stopped in the scintillator. It is assumed that the muons obey the radioactive decay law $R = R_0 e^{-t/\tau}$ where R_0 is the unknown activity at $t = 0$ and R is the activity (counts/μs) at time t. Here is their data:

Time (μs)	1.5	2.5	3.5	4.5	5.5	6.5
$R(t)$	55	35	23	18	12	5

Make a graph of $\ln(R/R_0)$ versus time t (μs), and from the best fit of the graph to a straight line find the mean life τ. The accepted value of the mean life of the muon is $\tau = 2.19703\ \mu\text{s} \pm 0.00004\ \mu\text{s}$. What is the percentage error of their result from the accepted value?

Answers to Exercises

A: 1.24×10^{-18} m = 1.24 am.

B: $\approx 2 \times 10^3$ m/0.1 m $\approx 10^4$.

C: (*a*).

D: (*c*); (*d*).

E: $s\bar{u}$.

This map of the entire sky (WMAP) is color-coded to represent slight temperature variations in the almost perfectly uniform 2.7-kelvin microwave background radiation that reaches us from all directions in the sky. This latest version (2006) is providing detailed information on the origins of our universe and its structures. The tiny temperature variations, red slightly hotter, blue slightly cooler (on the order of 1 part in 10^4) are "quantum fluctuations" that are the seeds on which galaxies and clusters of galaxies eventually grew.

To discuss the nature of the universe as we understand it today, we examine the latest theories on how stars and galaxies form and evolve, including the role of nucleosynthesis. We briefly discuss Einstein's general theory of relativity, which deals with gravity and curvature of space. We take a thorough look at the evidence for the expansion of the universe, and the Standard Model of the universe evolving from an initial Big Bang. Finally we point out some unsolved problems, including the nature of dark matter and dark energy that make up most of our universe.

CHAPTER 44

Astrophysics and Cosmology

CHAPTER-OPENING QUESTION—**Guess now!**

Until recently, astronomers expected the expansion rate of the universe would be decreasing. Why?

(a) Friction.
(b) The second law of thermodynamics.
(c) Gravity.
(d) The electromagnetic force.

In the previous Chapter, we studied the tiniest objects in the universe—the elementary particles. Now we leap to the grandest objects in the universe—stars, galaxies, and clusters of galaxies. These two extreme realms, elementary particles and the cosmos, are among the most intriguing and exciting subjects in science. And, surprisingly, these two extreme realms are related in a fundamental way, as already hinted in Chapter 43.

Use of the techniques and ideas of physics to study the heavens is often referred to as **astrophysics**. Central to our present theoretical understanding of the universe (or cosmos) is Einstein's *general theory of relativity* which represents our most complete understanding of gravitation. Many other aspects of physics are involved, from electromagnetism and thermodynamics to atomic and nuclear physics as well as elementary particles. General Relativity serves also as the foundation for modern **cosmology**, which is the study of the universe as a whole. Cosmology deals especially with the search for a theoretical framework to understand the observed universe, its origin, and its future.

CONTENTS

The questions posed by cosmology are profound and difficult; the possible answers stretch the imagination. They are questions like "Has the universe always existed, or did it have a beginning in time?" Either alternative is difficult to imagine: time going back indefinitely into the past, or an actual moment when the universe began (but, then, what was there before?). And what about the size of the universe? Is it infinite in size? It is hard to imagine infinity. Or is it finite in size? This is also hard to imagine, for if the universe is finite, it does not make sense to ask what is beyond it, because the universe is all there is.

In the last few years, so much progress has occurred in astrophysics and cosmology that many scientists are calling recent work a "Golden Age" for cosmology. Our survey will be qualitative, but we will nonetheless touch on the major ideas. We begin with a look at what can be seen beyond the Earth.

44–1 Stars and Galaxies

According to the ancients, the stars, except for the few that seemed to move relative to the others (the planets), were fixed on a sphere beyond the last planet. The universe was neatly self-contained, and we on Earth were at or near its center. But in the centuries following Galileo's first telescopic observations of the night sky in 1610, our view of the universe has changed dramatically. We no longer place ourselves at the center, and we view the universe as vastly larger. The distances involved are so great that we specify them in terms of the time it takes light to travel the given distance: for example, 1 light-second $= (3.0 \times 10^8\,\mathrm{m/s})(1.0\,\mathrm{s}) = 3.0 \times 10^8\,\mathrm{m} = 300{,}000\,\mathrm{km}$; 1 light-minute $= 18 \times 10^6\,\mathrm{km}$; and 1 **light-year** (ly) is

$$\begin{aligned} 1\,\mathrm{ly} &= (2.998 \times 10^8\,\mathrm{m/s})(3.156 \times 10^7\,\mathrm{s/yr}) \\ &= 9.46 \times 10^{15}\,\mathrm{m} \approx 10^{13}\,\mathrm{km}. \end{aligned}$$

For specifying distances to the Sun and Moon, we usually use meters or kilometers, but we could specify them in terms of light. The Earth–Moon distance is 384,000 km, which is 1.28 light-seconds. The Earth–Sun distance is $1.50 \times 10^{11}\,\mathrm{m}$, or 150,000,000 km; this is equal to 8.3 light-minutes. Far out in our solar system, Pluto is about $6 \times 10^9\,\mathrm{km}$ from the Sun, or $6 \times 10^{-4}\,\mathrm{ly}$. The nearest star to us, other than the Sun, is Proxima Centauri, about 4.3 ly away.

On a clear moonless night, thousands of stars of varying degrees of brightness can be seen, as well as the long cloudy stripe known as the Milky Way (Fig. 44–1). Galileo first observed, with his telescope, that the Milky Way is comprised of countless individual stars. A century and a half later (about 1750), Thomas Wright suggested that the Milky Way was a flat disk of stars extending to great distances in a plane, which we call the **Galaxy** (Greek for "milky way").

FIGURE 44–1 Sections of the Milky Way. In (a), the thin line is the trail of an artificial Earth satellite in this long time exposure. The dark diagonal area is due to dust absorption of visible light, blocking the view. In (b) the view is toward the center of the Galaxy (taken in summer from Arizona).

(a) (b)

Our Galaxy has a diameter of almost 100,000 light-years and a thickness of roughly 2000 ly. It has a bulging central "nucleus" and spiral arms (Fig. 44–2). Our Sun, which is a star like many others, is located about halfway from the galactic center to the edge, some 26,000 ly from the center. Our Galaxy contains roughly 100 billion (10^{11}) stars. The Sun orbits the galactic center approximately once every 250 million years, so its speed is about 200 km/s relative to the center of the Galaxy. The total mass of all the stars in our Galaxy is estimated to be about 3×10^{41} kg, which is ordinary matter. In addition, there is strong evidence that our Galaxy is surrounded by an invisible "halo" of "dark matter," which we discuss in Section 44–9.

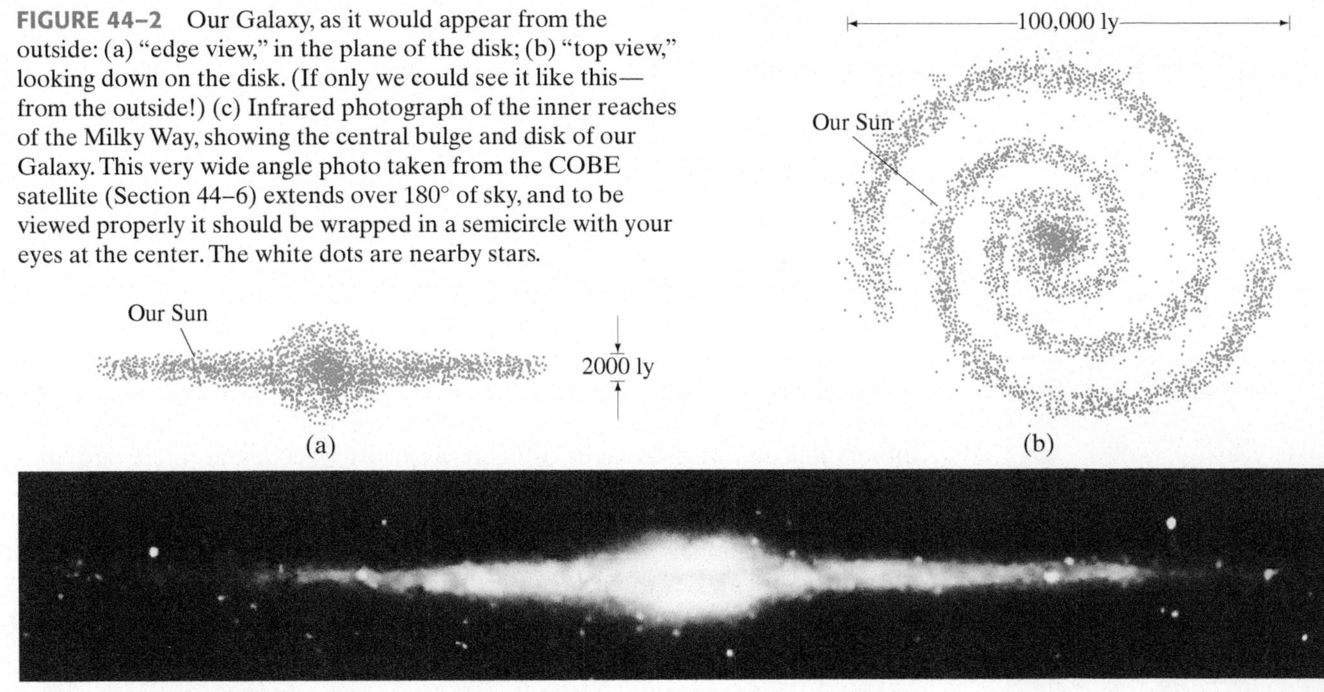

FIGURE 44–2 Our Galaxy, as it would appear from the outside: (a) "edge view," in the plane of the disk; (b) "top view," looking down on the disk. (If only we could see it like this—from the outside!) (c) Infrared photograph of the inner reaches of the Milky Way, showing the central bulge and disk of our Galaxy. This very wide angle photo taken from the COBE satellite (Section 44–6) extends over 180° of sky, and to be viewed properly it should be wrapped in a semicircle with your eyes at the center. The white dots are nearby stars.

EXAMPLE 44–1 ESTIMATE **Our Galaxy's mass.** Estimate the total mass of our Galaxy using the orbital data above for the Sun about the center of the Galaxy. Assume that most of the mass of the Galaxy is concentrated near the center of the Galaxy.

APPROACH We assume that the Sun (including our solar system) has total mass m and moves in a circular orbit about the center of the Galaxy (total mass M), and that the mass M can be considered as being located at the center of the Galaxy. We then apply Newton's second law, $F = ma$, with a being the centripetal acceleration, $a = v^2/r$, and F being the universal law of gravitation (Chapter 6).

SOLUTION Our Sun and solar system orbit the center of the Galaxy, according to the best measurements as mentioned above, with a speed of about $v = 200\text{ km/s}$ at a distance from the Galaxy center of about $r = 26{,}000$ ly. We use Newton's second law:

$$\begin{aligned} F &= ma \\ G\frac{Mm}{r^2} &= m\frac{v^2}{r} \end{aligned}$$

where M is the mass of the Galaxy and m is the mass of our Sun and solar system. Solving this, we find

$$M = \frac{rv^2}{G} \approx \frac{(26{,}000\text{ ly})(10^{16}\text{ m/ly})(2 \times 10^5\text{ m/s})^2}{6.67 \times 10^{-11}\text{ N}\cdot\text{m}^2/\text{kg}^2} \approx 2 \times 10^{41}\text{ kg}.$$

NOTE In terms of *numbers* of stars, if they are like our Sun $(m = 2.0 \times 10^{30}\text{ kg})$, there would be about $(2 \times 10^{41}\text{ kg})/(2 \times 10^{30}\text{ kg}) \approx 10^{11}$ or on the order of 100 billion stars.

FIGURE 44–3 This globular star cluster is located in the constellation Hercules.

FIGURE 44–4 This gaseous nebula, found in the constellation Carina, is about 9000 light-years from us.

In addition to stars both within and outside the Milky Way, we can see by telescope many faint cloudy patches in the sky which were all referred to once as "nebulae" (Latin for "clouds"). A few of these, such as those in the constellations Andromeda and Orion, can actually be discerned with the naked eye on a clear night. Some are **star clusters** (Fig. 44–3), groups of stars that are so numerous they appear to be a cloud. Others are glowing clouds of gas or dust (Fig. 44–4), and it is for these that we now mainly reserve the word **nebula**. Most fascinating are those that belong to a third category: they often have fairly regular elliptical shapes and seem to be a great distance beyond our Galaxy. Immanuel Kant (about 1755) seems to have been the first to suggest that these latter might be circular disks, but appear elliptical because we see them at an angle, and are faint because they are so distant. At first it was not universally accepted that these objects were **extragalactic**—that is, outside our Galaxy. The very large telescopes constructed in the twentieth century revealed that individual stars could be resolved within these extragalactic objects and that many contain spiral arms. Edwin Hubble (1889–1953) did much of this observational work in the 1920s using the 2.5-m (100-inch) telescope† on Mt. Wilson near Los Angeles, California, then the world's largest. Hubble demonstrated that these objects were indeed extragalactic because of their great distances. The distance to our nearest large galaxy,‡ Andromeda, is over 2 million light-years, a distance 20 times greater than the diameter of our Galaxy. It seemed logical that these nebulae must be **galaxies** similar to ours. (Note that it is usual to capitalize the word "galaxy" only when it refers to our own.) Today it is thought there are roughly 10^{11} galaxies in the observable universe—that is, roughly as many galaxies as there are stars in a galaxy. See Fig. 44–5.

Many galaxies tend to be grouped in **galaxy clusters** held together by their mutual gravitational attraction. There may be anywhere from a few to many thousands of galaxies in each cluster. Furthermore, clusters themselves seem to be

†2.5 m (= 100 inches) refers to the diameter of the curved objective mirror. The bigger the mirror, the more light it collects (greater intensity) and the less diffraction there is (better resolution), so more and fainter stars can be seen. See Chapters 33 and 35. Until recently, photographic films or plates were used to take long time exposures. Now large solid-state CCD or CMOS sensors (Section 33–5) are available containing hundreds of millions of pixels (compared to 10 million pixels in a good-quality digital camera).

‡The *Magellanic clouds* are much closer than Andromeda, but are small and are usually considered small satellite galaxies of our own Galaxy.

FIGURE 44–5 Photographs of galaxies. (a) Spiral galaxy in the constellation Hydra. (b) Two galaxies: the larger and more dramatic one is known as the Whirlpool galaxy. (c) An infrared image (given "false" colors) of the same galaxies as in (b), here showing the arms of the spiral as having more substance than in the visible light photo (b); the different colors correspond to different light intensities. Visible light is scattered and absorbed by interstellar dust much more than infrared is, so infrared gives us a clearer image.

(a) (b) (c)

organized into even larger aggregates: clusters of clusters of galaxies, or **superclusters**. The farthest detectable galaxies are more than 10^{10} ly distant. See Table 44–1.

TABLE 44–1 Astronomical Distances

Object	Approx. Distance from Earth (ly)
Moon	4×10^{-8}
Sun	1.6×10^{-5}
Size of solar system (distance to Pluto)	6×10^{-4}
Nearest star (Proxima Centauri)	4.3
Center of our Galaxy	2.6×10^4
Nearest large galaxy	2.4×10^6
Farthest galaxies	10^{10}

CONCEPTUAL EXAMPLE 44–2 **Looking back in time.** Astronomers often think of their telescopes as time machines, looking back toward the origin of the universe. How far back do they look?

RESPONSE The distance in light-years measures how long in years the light has been traveling to reach us, so Table 44–1 tells us also how far back in time we are looking. For example, if we saw Proxima Centauri explode into a supernova today, then the event would have really occurred 4.3 years ago. The most distant galaxies emitted the light we see now roughly 10^{10} years ago. What we see was how they were then, 10^{10} yr ago, or about 10^9 years after the universe was born in the Big Bang.

EXERCISE A Suppose we could place a huge mirror 1 light-year away from us. What would we see in this mirror if it is facing us on Earth? When did it take place? (This might be called a "time machine.")

Besides the usual stars, clusters of stars, galaxies, and clusters and superclusters of galaxies, the universe contains many other interesting objects. Among these are stars known as *red giants*, *white dwarfs*, *neutron stars*, exploding stars called *novae* and *supernovae*, and *black holes* whose gravity is so strong even light can not escape them. In addition, there is electromagnetic radiation that reaches the Earth but does not emanate from the bright pointlike objects we call stars: particularly important is the microwave background radiation that arrives nearly uniformly from all directions in the universe. We will discuss all these phenomena.

Finally, there are *active galactic nuclei* (AGN), which are very luminous pointlike sources of light in the centers of distant galaxies. The most dramatic examples of AGN are *quasars* ("quasistellar objects" or QSOs), which are so luminous that the surrounding starlight of the galaxy is drowned out. Their luminosity is thought to come from matter falling into a giant black hole at a galaxy's center.

44–2 Stellar Evolution: Nucleosynthesis, and the Birth and Death of Stars

The stars appear unchanging. Night after night the night sky reveals no significant variations. Indeed, on a human time scale, the vast majority of stars change very little (except for novae, supernovae, and certain variable stars). Although stars *seem* fixed in relation to each other, many move sufficiently for the motion to be detected. Speeds of stars relative to neighboring stars can be hundreds of km/s, but at their great distance from us, this motion is detectable only by careful measurement. Furthermore, there is a great range of brightness among stars. The differences in brightness are due both to differences in the rate at which stars emit energy and to their different distances from us.

Luminosity and Brightness of Stars

A useful parameter for a star or galaxy is its **intrinsic luminosity**, L (or simply **luminosity**), by which we mean the total power radiated in watts. Also important is the **apparent brightness**, b, defined as the power crossing unit area at the Earth perpendicular to the path of the light. Given that energy is conserved, and ignoring any absorption in space, the total emitted power L when it reaches a distance d from the star will be spread over a sphere of surface area $4\pi d^2$. If d is the distance from the star to the Earth, then L must be equal to $4\pi d^2$ times b (power per unit area at Earth). That is,

$$b = \frac{L}{4\pi d^2}. \tag{44–1}$$

EXAMPLE 44–3 Apparent brightness. Suppose a particular star has intrinsic luminosity equal to that of our Sun, but is 10 ly away from Earth. By what factor will it appear dimmer than the Sun?

APPROACH The luminosity L is the same for both stars, so the apparent brightness depends only on their relative distances. We use the inverse square law as stated in Eq. 44–1 to determine the relative brightness.

SOLUTION Using the inverse square law, the star appears dimmer by a factor

$$\frac{b_{\text{star}}}{b_{\text{Sun}}} = \frac{d_{\text{Sun}}^2}{d_{\text{star}}^2} = \frac{(1.5 \times 10^8\,\text{km})^2}{(10\,\text{ly})^2(10^{13}\,\text{km/ly})^2} \approx 2 \times 10^{-12}.$$

Careful study of nearby stars has shown that the luminosity for most stars depends on the mass: *the more massive the star, the greater its luminosity*[†]. Indeed, we might expect that more massive stars would have higher core temperature and pressure to counterbalance the greater gravitational attraction, and thus be more luminous. Another important parameter of a star is its surface temperature, which can be determined from the spectrum of electromagnetic frequencies it emits (stars are "good" blackbodies—see Section 37–1). As we saw in Chapter 37, as the temperature of a body increases, the spectrum shifts from predominantly lower frequencies (and longer wavelengths, such as red) to higher frequencies (and shorter wavelengths such as blue). Quantitatively, the relation is given by Wien's law (Eq. 37–1): the peak wavelength λ_P in the spectrum of light emitted by a blackbody (we often approximate stars as blackbodies) is inversely proportional to its Kelvin temperature T; that is, $\lambda_P T = 2.90 \times 10^{-3}\,\text{m}\cdot\text{K}$. The surface temperatures of stars typically range from about 3000 K (reddish) to about 50,000 K (UV).

EXAMPLE 44–4 Determining star temperature and star size. Suppose that the distances from Earth to two nearby stars can be reasonably estimated, and that their measured apparent brightnesses suggest the two stars have about the same luminosity, L. The spectrum of one of the stars peaks at about 700 nm (so it is reddish). The spectrum of the other peaks at about 350 nm (bluish). Use Wien's law (Eq. 37–1) and the Stefan-Boltzmann equation (Section 19–10) to determine (*a*) the surface temperature of each star, and (*b*) how much larger one star is than the other.

APPROACH We determine the surface temperature T for each star using Wien's law and each star's peak wavelength. Then, using the Stefan-Boltzmann equation (power output or luminosity $\propto AT^4$ where A = surface area of emitter), we can find the surface area ratio and relative sizes of the two stars.

SOLUTION (*a*) Wien's law (Eq. 37–1) states that $\lambda_P T = 2.90 \times 10^{-3}\,\text{m}\cdot\text{K}$. So the temperature of the reddish star is

$$T_r = \frac{2.90 \times 10^{-3}\,\text{m}\cdot\text{K}}{700 \times 10^{-9}\,\text{m}} = 4140\,\text{K}.$$

The temperature of the bluish star will be double this since its peak wavelength is half (350 nm vs. 700 nm):

$$T_b = 8280\,\text{K}.$$

(*b*) The Stefan-Boltzmann equation, Eq. 19–17, states that the power radiated *per unit area* of surface from a blackbody is proportional to the fourth power of the Kelvin temperature, T^4. The temperature of the bluish star is double that of the reddish star, so the bluish one must radiate $(2^4) = 16$ times as much energy per unit area. But we are given that they have the same luminosity (the same total power output); so the surface area of the blue star must be $\frac{1}{16}$ that of the red one. The surface area of a sphere is $4\pi r^2$, so the radius of the reddish star is $\sqrt{16} = 4$ times larger than the radius of the bluish star (or $4^3 = 64$ times the volume).

[†]Applies to "main-sequence" stars (see next page). The mass of a star can be determined by observing its gravitational effects. Many stars are part of a cluster, the simplest being a binary star in which two stars orbit around each other, allowing their masses to be determined using rotational mechanics.

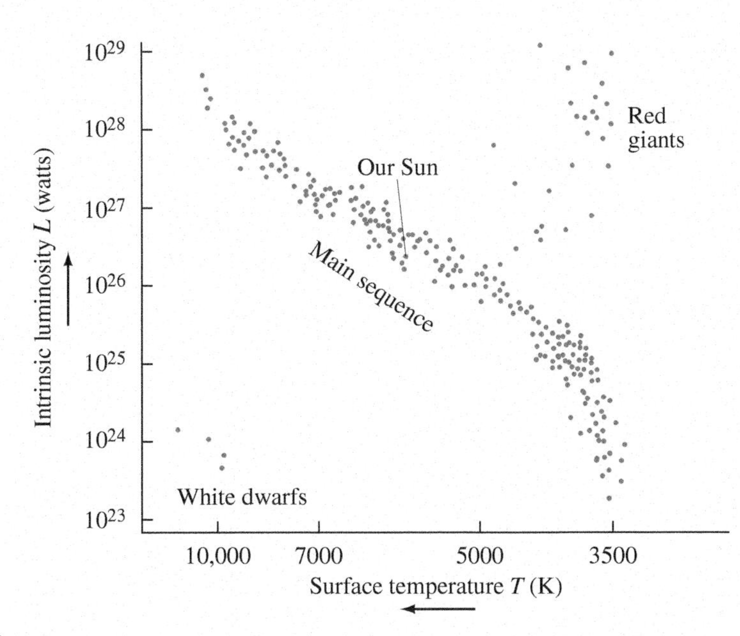

FIGURE 44–6 Hertzsprung–Russell (H–R) diagram is a logarithmic graph of luminosity vs. surface temperature T of stars (note that T increases to the left).

H–R Diagram

An important astronomical discovery, made around 1900, was that for most stars, the color is related to the intrinsic luminosity and therefore to the mass. A useful way to present this relationship is by the so-called Hertzsprung–Russell (H–R) diagram. On the H–R diagram, the horizontal axis shows the surface temperature T whereas the vertical axis is the luminosity L; each star is represented by a point on the diagram, Fig. 44–6. Most stars fall along the diagonal band termed the **main sequence**. Starting at the lower right we find the coolest stars, reddish in color; they are the least luminous and therefore of low mass. Farther up toward the left we find hotter and more luminous stars that are whitish, like our Sun. Still farther up we find even more massive and more luminous stars, bluish in color. Stars that fall on this diagonal band are called *main-sequence stars*. There are also stars that fall outside the main sequence. Above and to the right we find extremely large stars, with high luminosities but with low (reddish) color temperature: these are called **red giants**. At the lower left, there are a few stars of low luminosity but with high temperature: these are the **white dwarfs**.

EXAMPLE 44–5 ESTIMATE **Distance to a star using the H–R diagram and color.** Suppose that detailed study of a certain star suggests that it most likely fits on the main sequence of an H–R diagram. Its measured apparent brightness is $b = 1.0 \times 10^{-12}\,\mathrm{W/m^2}$, and the peak wavelength of its spectrum is $\lambda_P \approx 600\,\mathrm{nm}$. Estimate its distance from us.

APPROACH We find the temperature using Wien's law, Eq. 37–1. The luminosity is estimated for a main sequence star on the H–R diagram of Fig. 44–6, and then the distance is found using the relation between brightness and luminosity, Eq. 44–1.

SOLUTION The star's temperature, from Wien's law (Eq. 37–1), is

$$T \approx \frac{2.90 \times 10^{-3}\,\mathrm{m \cdot K}}{600 \times 10^{-9}\,\mathrm{m}} \approx 4800\,\mathrm{K}.$$

A star on the main sequence of an H–R diagram at this temperature has intrinsic luminosity of about $L \approx 1 \times 10^{26}\,\mathrm{W}$, read off of Fig. 44–6. Then, from Eq. 44–1,

$$d = \sqrt{\frac{L}{4\pi b}} \approx \sqrt{\frac{1 \times 10^{26}\,\mathrm{W}}{4(3.14)(1.0 \times 10^{-12}\,\mathrm{W/m^2})}} \approx 3 \times 10^{18}\,\mathrm{m}.$$

Its distance from us in light-years is

$$d = \frac{3 \times 10^{18}\,\mathrm{m}}{10^{16}\,\mathrm{m/ly}} \approx 300\,\mathrm{ly}.$$

EXERCISE B Estimate the distance to a 6000-K main-sequence star with an apparent brightness of $2.0 \times 10^{-12}\,\mathrm{W/m^2}$.

Stellar Evolution; Nucleosynthesis

Why are there different types of stars, such as red giants and white dwarfs, as well as main-sequence stars? Were they all born this way, in the beginning? Or might each different type represent a different age in the life cycle of a star? Astronomers and astrophysicists today believe the latter is the case. Note, however, that we cannot actually follow any but the tiniest part of the life cycle of any given star since they live for ages vastly greater than ours, on the order of millions or billions of years. Nonetheless, let us follow the process of **stellar evolution** from the birth to the death of a star, as astrophysicists have theoretically reconstructed it today.

Stars are born, it is believed, when gaseous clouds (mostly hydrogen) contract due to the pull of gravity. A huge gas cloud might fragment into numerous contracting masses, each mass centered in an area where the density was only slightly greater than that at nearby points. Once such "globules" formed, gravity would cause each to contract in toward its center of mass. As the particles of such a *protostar* accelerate inward, their kinetic energy increases. When the kinetic energy is sufficiently high, the Coulomb repulsion between the positive charges is not strong enough to keep the hydrogen nuclei apart, and nuclear fusion can take place.

In a star like our Sun, the fusion of hydrogen (sometimes referred to as "burning")† occurs via the *proton–proton cycle* (Section 42–4, Eqs. 42–7), in which four protons fuse to form a ${}^4_2\mathrm{He}$ nucleus with the release of γ rays, positrons, and neutrinos: $4\,{}^1_1\mathrm{H} \rightarrow {}^4_2\mathrm{He} + 2\,\mathrm{e}^+ + 2\nu_\mathrm{e} + 2\gamma$. These reactions require a temperature of about $10^7\,\mathrm{K}$, corresponding to an average kinetic energy ($\approx kT$) of about 1 keV (Eq. 18–4). In more massive stars, the carbon cycle produces the same net effect: four ${}^1_1\mathrm{H}$ produce a ${}^4_2\mathrm{He}$—see Section 42–4. The fusion reactions take place primarily in the core of a star, where T may be on the order of 10^7 to $10^8\,\mathrm{K}$. (The surface temperature is much lower—on the order of a few thousand kelvins.) The tremendous release of energy in these fusion reactions produces an outward pressure sufficient to halt the inward gravitational contraction. Our protostar, now really a young *star*, stabilizes on the main sequence. Exactly where the star falls along the main sequence depends on its mass. The more massive the star, the farther up (and to the left) it falls on the H–R diagram of Fig. 44–6. Our Sun required perhaps 30 million years to reach the main sequence, and is expected to remain there about 10 billion years ($10^{10}\,\mathrm{yr}$). Although most stars are billions of years old, evidence is strong that stars are actually being born at this moment. More massive stars have shorter lives, because they are hotter and the Coulomb repulsion is more easily overcome, so they use up their fuel faster. If our Sun remains on the main sequence for 10^{10} years, a star ten times more massive may reside there for only 10^7 years.

As hydrogen fuses to form helium, the helium that is formed is denser and tends to accumulate in the central core where it was formed. As the core of helium grows, hydrogen continues to fuse in a shell around it: see Fig. 44–7. When much of the hydrogen within the core has been consumed, the production of energy decreases at the center and is no longer sufficient to prevent the huge gravitational forces from once again causing the core to contract and heat up. The hydrogen in the shell around the core then fuses even more fiercely because of this rise in temperature, allowing the outer envelope of the star to expand and to cool. The surface temperature, thus reduced, produces a spectrum of light that peaks at longer wavelength (reddish).

FIGURE 44–7 A shell of "burning" hydrogen (fusing to become helium) surrounds the core where the newly formed helium gravitates.

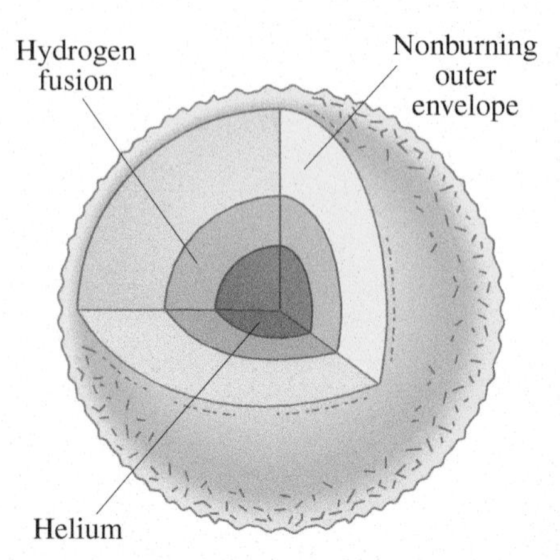

†The word "burn" is put in quotation marks because these high-temperature fusion reactions occur via a *nuclear* process, and must not be confused with ordinary burning (of, say, paper, wood, or coal) in air, which is a *chemical* reaction, occurring at the *atomic* level (and at a much lower temperature).

By this time the star has left the main sequence. It has become redder, and as it has grown in size, it has become more luminous. So it will have moved to the right and upward on the H–R diagram, as shown in Fig. 44–8. As it moves upward, it enters the **red giant** stage. Thus, theory explains the origin of red giants as a natural step in a star's evolution. Our Sun, for example, has been on the main sequence for about $4\frac{1}{2}$ billion years. It will probably remain there another 4 or 5 billion years. When our Sun leaves the main sequence, it is expected to grow in diameter (as it becomes a red giant) by a factor of 100 or more, possibly swallowing up inner planets such as Mercury.

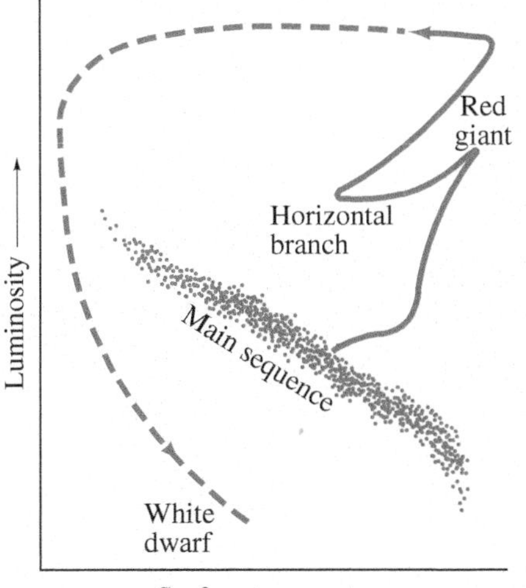

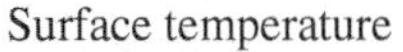

FIGURE 44–8 Evolutionary "track" of a star like our Sun represented on an H–R diagram.

If the star is like our Sun, or larger, further fusion can occur. As the star's outer envelope expands, its core continues to shrink and heat up. When the temperature reaches about 10^8 K, even helium nuclei, in spite of their greater charge and hence greater electrical repulsion, can come close enough to each other to undergo fusion. The reactions are

$$\begin{aligned} {}^4_2\text{He} + {}^4_2\text{He} &\rightarrow {}^8_4\text{Be} \\ {}^4_2\text{He} + {}^8_4\text{Be} &\rightarrow {}^{12}_{6}\text{C} \end{aligned} \qquad \textbf{(44–2)}$$

with the emission of two γ rays. These two reactions must occur in quick succession (because ${}^8_4$Be is very unstable), and the net effect is

$$3\,{}^4_2\text{He} \rightarrow {}^{12}_{6}\text{C}. \qquad (Q = 7.3\,\text{MeV})$$

This fusion of helium causes a change in the star which moves rapidly to the "horizontal branch" on the H–R diagram (Fig. 44–8). Further fusion reactions are possible, with ${}^4_2$He fusing with ${}^{12}_{6}$C to form ${}^{16}_{8}$O. In more massive stars, higher Z elements like ${}^{20}_{10}$Ne or ${}^{24}_{12}$Mg can be made. This process of creating heavier nuclei from lighter ones (or by absorption of neutrons which tends to occur at higher Z) is called **nucleosynthesis**.

The final fate of a star depends on its mass. Stars can lose mass as parts of their outer envelope move off into space. Stars born with a mass less than about 8 (or perhaps 10) solar masses eventually end up with a residual mass less than about 1.4 solar masses, which is known as the **Chandrasekhar limit**. For them, no further fusion energy can be obtained. The core of such a "low mass" star (original mass $\lesssim$ 8–10 solar masses) contracts under gravity; the outer envelope expands again and the star becomes an even larger red giant. Eventually the outer layers escape into space, the core shrinks, the star cools, and typically follows the dashed route shown in Fig. 44–8, descending downward, becoming a **white dwarf**. A white dwarf with a residual mass equal to that of the Sun would be about the size of the Earth. A white dwarf contracts to the point at which the electron clouds start to overlap, but no further because, by the Pauli exclusion principle, no two electrons can be in the same quantum state. At this point the star is supported against further collapse by this **electron degeneracy** pressure. A white dwarf continues to lose internal energy by radiation, decreasing in temperature and becoming dimmer until it glows no more. It has then become a cold dark chunk of extremely dense material.

Stars whose residual mass is greater than the Chandrasekhar limit of 1.4 solar masses (original mass greater than about 8 or 10 solar masses) are thought to follow a quite different scenario. A star with this great a mass can contract under gravity and heat up even further. In the range $T = (2.5\text{–}5) \times 10^9$ K, nuclei as heavy as ${}^{56}_{26}$Fe and ${}^{56}_{28}$Ni can be made. But here the formation of heavy nuclei from lighter ones, by fusion, ends. As we saw in Fig. 41–1, the average binding energy per nucleon begins to decrease for A greater than about 60. Further fusions would *require* energy, rather than release it.

Elements heavier than Ni are thought to form mainly by neutron capture, particularly in exploding stars called *supernovae* (singular is **supernova**). Large numbers of free neutrons, resulting from nuclear reactions, are present inside these highly evolved stars and they can readily combine with, say, a ${}^{56}_{26}$Fe nucleus to form (if three are captured) ${}^{59}_{26}$Fe, which decays to ${}^{59}_{27}$Co. The ${}^{59}_{27}$Co can capture neutrons, also becoming neutron rich and decaying by β^- to the next higher Z element, and so on to the highest Z elements.

Yet at these extremely high temperatures, well above 10^9 K, the kinetic energy of the nuclei is so high that fusion of elements heavier than iron is still possible even though the reactions require energy input. But the high-energy collisions can also cause the breaking apart of iron and nickel nuclei into He nuclei, and eventually into protons and neutrons:

$$^{56}_{26}\text{Fe} \rightarrow 13\,^{4}_{2}\text{He} + 4\text{n}$$

$$^{4}_{2}\text{He} \rightarrow 2\text{p} + 2\text{n}.$$

These are energy-requiring (endothermic) reactions, but at such extremely high temperature and pressure there is plenty of energy available, enough even to force electrons and protons together to form neutrons in **inverse β decay**:

$$\text{e}^- + \text{p} \rightarrow \text{n} + \nu.$$

As a result of these reactions, the pressure in the core drops precipitously. As the core collapses under the huge gravitational forces, the tremendous mass becomes essentially an enormous nucleus made up almost exclusively of neutrons. The size of the star is no longer limited by the exclusion principle applied to electrons, but rather by **neutron degeneracy** pressure, and the star contracts rapidly to form an enormously dense **neutron star**. The core of a neutron star contracts to the point at which all neutrons are as close together as they are in an atomic nucleus. That is, the density of a neutron star is on the order of 10^{14} times greater than normal solids and liquids on Earth. A cupful of such dense matter would weigh billions of tons. A neutron star that has a mass 1.5 times that of our Sun would have a diameter of only about 20 km. (Compare this to a white dwarf with 1 solar mass whose diameter would be $\approx 10^4$ km, as already mentioned.)

The contraction of the core of a massive star would mean a great reduction in gravitational potential energy. Somehow this energy would have to be released. Indeed, it was suggested in the 1930s that the final core collapse to a neutron star may be accompanied by a catastrophic explosion (a *supernova*—see previous page) whose tremendous energy could form virtually all elements of the Periodic Table and blow away the entire outer envelope of the star (Fig. 44–9), spreading its contents into interstellar space. The presence of heavy elements on Earth and in our solar system suggests that our solar system formed from the debris of such a supernova explosion.

(a) (b) (c)

FIGURE 44–9 The star indicated by the arrow in (a) exploded as a supernova, as shown in (b). It was detected in 1987, and named SN1987A. The large bright spot in (b) indicates a huge release of energy but does not represent the physical size. Part (c) is a photo taken a few years later, showing shock waves moving outward from where SN1987A was (blow-up in corner). Part (c) is magnified relative to (a) and (b).

If the final mass of a neutron star is less than about two or three solar masses, its subsequent evolution is thought to resemble that of a white dwarf. If the mass is greater than this, the star collapses under gravity, overcoming even the neutron exclusion principle. Gravity would then be so strong that even light emitted from the star could not escape—it would be pulled back in by the force of gravity. Since no radiation could escape from such a star, we could not see it—it would be black. An object may pass by it and be deflected by its gravitational field, but if it came too close it would be swallowed up, never to escape. This is a **black hole**.

Novae and Supernovae

Novae (singular is *nova*, meaning "new" in Latin) are faint stars that have suddenly increased in brightness by as much as a factor of 10^4 and last for a month or two before fading. Novae are thought to be faint white dwarfs that have pulled mass from a nearby companion (they make up a *binary* system), as illustrated in Fig. 44–10. The captured mass of hydrogen suddenly fuses into helium at a high rate for a few weeks. Many novae (maybe all) are *recurrent*—they repeat their bright glow years later.

Supernovae are also brief explosive events, but release millions of times more energy than novae, up to 10^{10} times more luminous than our Sun. The peak of brightness may exceed that of the entire galaxy in which they are located, but lasts only a few days or weeks. They slowly fade over a few months. Many supernovae form by core collapse to a neutron star as described above. See Fig. 44–9.

Type Ia supernovae are different. They all seem to have very nearly the same luminosity. They are believed to be binary stars, one of which is a white dwarf that pulls mass from its companion, much like for a nova, Fig. 44–10. The mass is higher, and as mass is captured and the total mass reaches the Chandrasekhar limit of 1.4 solar masses, it explodes as a supernova by undergoing a "thermonuclear runaway"—an uncontrolled chain of nuclear reactions. What is left is a neutron star or (if the mass is great enough) a black hole.

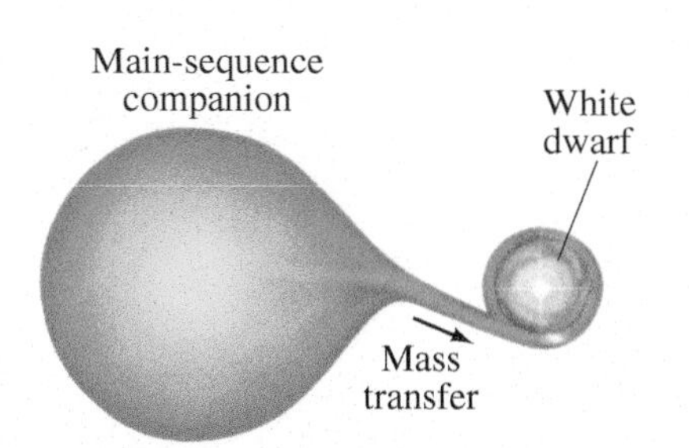

FIGURE 44–10 Hypothetical model for novae and type Ia supernovae, showing how a white dwarf could pull mass from its normal companion.

44–3 Distance Measurements

We have talked about the vast distances of objects in the universe. But how do we measure these distances? One basic technique employs simple geometry to measure the **parallax** of a star. By parallax we mean the apparent motion of a star, against the background of much more distant stars, due to the Earth's motion about the Sun. As shown in Fig. 44–11, the sighting angle of a star relative to the plane of Earth's orbit (angle θ) can be determined at different times of the year. Since we know the distance d from Earth to Sun, we can reconstruct the right triangles shown in Fig. 44–11 and can determine† the distance D to the star.

†This is essentially the way the heights of mountains are determined, by "triangulation." See Example 1–7.

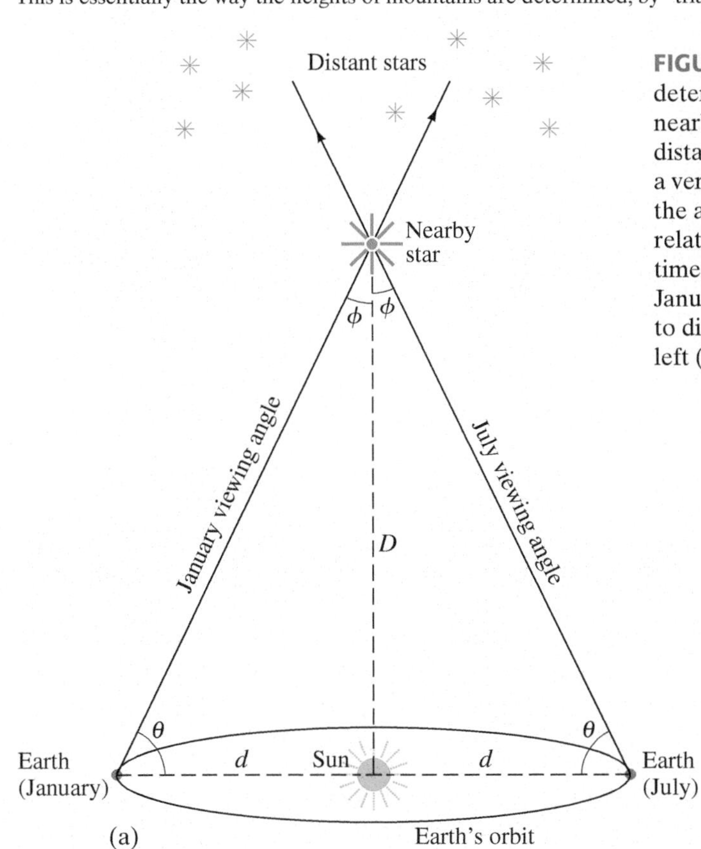

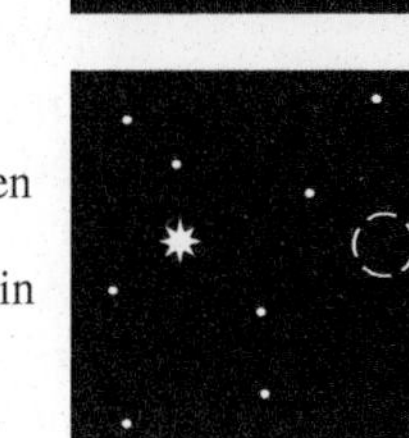

FIGURE 44–11 (a) Simple example of determining the distance D to a relatively nearby star using parallax. Horizontal distances are greatly exaggerated: in reality ϕ is a very small angle. (b) Diagram of the sky showing the apparent position of the "nearby" star relative to more distant stars, at two different times (January and July). The viewing angle in January puts the star more to the right relative to distant stars, whereas in July it is more to the left (dashed circle shows January location).

EXAMPLE 44–6 ESTIMATE **Distance to a star using parallax.** Estimate the distance D to a star if the angle θ in Fig. 44–11 is measured to be 89.99994°.

APPROACH From trigonometry, $\tan\phi = d/D$ in Fig. 44–11. The Sun–Earth distance is $d = 1.5 \times 10^8$ km.

SOLUTION The angle $\phi = 90° - 89.99994° = 0.00006°$, or about 1.0×10^{-6} radians. We can use $\tan\phi \approx \phi$ since ϕ is very small. We solve for D in $\tan\phi = d/D$. The distance D to the star is

$$D = \frac{d}{\tan\phi} \approx \frac{d}{\phi} = \frac{1.5 \times 10^8\,\text{km}}{1.0 \times 10^{-6}\,\text{rad}} = 1.5 \times 10^{14}\,\text{km},$$

or about 15 ly.

Distances to stars are often specified in terms of parallax angle (ϕ in Fig. 44–11a) given in seconds of arc: 1 second ($1''$) is $\frac{1}{60}$ of one minute ($1'$) of arc, which is $\frac{1}{60}$ of a degree, so $1'' = \frac{1}{3600}$ of a degree. The distance is then specified in **parsecs** (pc) (meaning *par*allax angle in *sec*onds of arc): $D = 1/\phi$ with ϕ in seconds of arc. In Example 44–6, $\phi = (6 \times 10^{-5})°(3600) = 0.22''$ of arc, so we would say the star is at a distance of $1/0.22'' = 4.5$ pc. One parsec is given by (recall $D = d/\phi$, and we set the Sun–Earth distance (Fig. 44–11a) as $d = 1.496 \times 10^{11}$ m):

$$1\,\text{pc} = \frac{d}{1''} = \frac{1.496 \times 10^{11}\,\text{m}}{(1'')\left(\frac{1'}{60''}\right)\left(\frac{1°}{60'}\right)\left(\frac{2\pi\,\text{rad}}{360°}\right)} = 3.086 \times 10^{16}\,\text{m}$$

$$1\,\text{pc} = (3.086 \times 10^{16}\,\text{m})\left(\frac{1\,\text{ly}}{9.46 \times 10^{15}\,\text{m}}\right) = 3.26\,\text{ly}.$$

Parallax can be used to determine the distance to stars as far away as about 100 light-years ($\approx$ 30 parsecs) from Earth, and from an orbiting spacecraft perhaps 5 to 10 times farther. Beyond that distance, parallax angles are too small to measure. For greater distances, more subtle techniques must be employed. We might compare the apparent brightnesses of two stars, or two galaxies, and use the inverse square law (apparent brightness drops off as the square of the distance) to roughly estimate their relative distances. We can't expect this technique to be very precise because we don't expect any two stars, or two galaxies, to have the same intrinsic luminosity. When comparing galaxies, a perhaps better estimate assumes the brightest stars in all galaxies (or the brightest galaxies in galaxy clusters) are similar and have about the same intrinsic luminosity. Consequently, their *apparent brightness* would be a measure of how far away they were.

Another technique makes use of the H–R diagram. Measurement of a star's surface temperature (from its spectrum) places it at a certain point (within 20%) on the H–R diagram, assuming it is a main-sequence star, and then its luminosity can be estimated off the vertical axis (Fig. 44–6). Its apparent brightness and Eq. 44–1 give its approximate distance; see Example 44–5.

A better estimate comes from comparing *variable stars*, especially *Cepheid variables* whose luminosity varies over time with a period that is found to be related to their average luminosity. Thus, from their period and apparent brightness we get their distance.

The largest distances are estimated by comparing the apparent brightnesses of type Ia supernovae (SNIa). Type Ia supernovae all have a similar origin (as described on the previous page, Fig. 44–10), and their brief explosive burst of light is expected to be of nearly the same luminosity. They are thus sometimes referred to as "standard candles."

Another important technique for estimating the distance of very distant stars is from the "redshift" in the line spectra of elements and compounds. The amount of redshift is related to the expansion of the universe, as we shall discuss in Section 44–5. It is useful for objects farther than 10^7 to 10^8 ly away.

As we look farther and farther away, the measurement techniques are less and less reliable, so there is more and more uncertainty in the measurements of large distances.

44–4 General Relativity: Gravity and the Curvature of Space

We have seen that the force of gravity plays an important role in the processes that occur in stars. Gravity too is important for the evolution of the universe as a whole. The reasons gravity plays a dominant role in the universe, and not one of the other of the four forces in nature, are (1) it is long-range and (2) it is always attractive. The strong and weak nuclear forces act over very short distances only, on the order of the size of a nucleus; hence they do not act over astronomical distances (they do act between nuclei and nucleons in stars to produce nuclear reactions). The electromagnetic force, like gravity, acts over great distances. But it can be either attractive or repulsive. And since the universe does not seem to contain large areas of net electric charge, a large net force does not occur. But gravity acts as an attractive force between *all* masses, and there are large accumulations in the universe of only the one "sign" of mass (not + and − as with electric charge). The force of gravity as Newton described it in his law of universal gravitation was modified by Einstein. In his general theory of relativity, Einstein developed a theory of gravity that now forms the basis of cosmological dynamics.

In the *special theory of relativity* (Chapter 36), Einstein concluded that there is no way for an observer to determine whether a given frame of reference is at rest or is moving at constant velocity in a straight line. Thus the laws of physics must be the same in different inertial reference frames. But what about the more general case of motion where reference frames can be *accelerating*?

Einstein tackled the problem of accelerating reference frames in his **general theory of relativity** and in it also developed a theory of gravity. The mathematics of General Relativity is complex, so our discussion will be mainly qualitative.

We begin with Einstein's **principle of equivalence**, which states that

no experiment can be performed that could distinguish between a uniform gravitational field and an equivalent uniform acceleration.

If observers sensed that they were accelerating (as in a vehicle speeding around a sharp curve), they could not prove by any experiment that in fact they weren't simply experiencing the pull of a gravitational field. Conversely, we might think we are being pulled by gravity when in fact we are undergoing an acceleration having nothing to do with gravity.

(a)

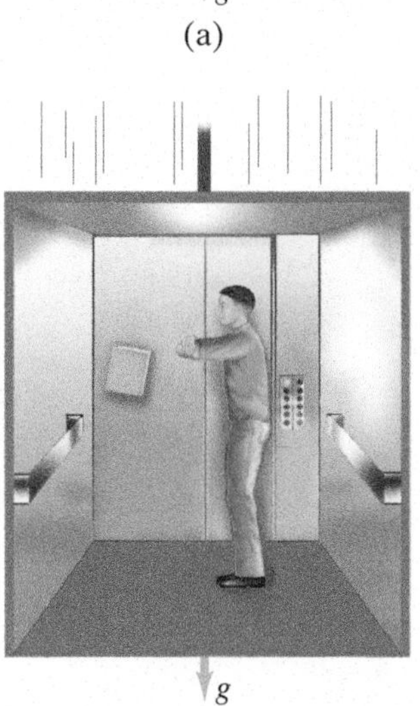

(b)

FIGURE 44–12 In an elevator falling freely under gravity, (a) a person releases a book; (b) the released book hovers next to the owner's hand; (b) is a few moments after (a).

As a thought experiment, consider a person in a freely falling elevator near the Earth's surface. If our observer held out a book and let go of it, what would happen? Gravity would pull it downward toward the Earth, but at the same rate ($g = 9.8\,\mathrm{m/s^2}$) at which the person and elevator were falling. So the book would hover right next to the person's hand (Fig. 44–12). The effect is exactly the same as if this reference frame was at rest and *no* forces were acting. On the other hand, if the elevator was out in space where the gravitational field is essentially zero, the released book would float, just as it does in Fig. 44–12. Next, if the elevator (out in space) is accelerating upward at an acceleration of $9.8\,\mathrm{m/s^2}$, the book as seen by our observer would fall to the floor with an acceleration of $9.8\,\mathrm{m/s^2}$, just as if it were falling due to gravity at the surface of the Earth. According to the principle of equivalence, the observer could not determine whether the book fell because the elevator was accelerating upward, or because a gravitational field was acting downward and the elevator was at rest. The two descriptions are equivalent.

The principle of equivalence is related to the concept that there are two types of mass. Newton's second law, $F = ma$, uses **inertial mass**. We might say that inertial mass represents "resistance" to any type of force. The second type of mass is **gravitational mass**. When one object attracts another by the gravitational force (Newton's law of universal gravitation, $F = Gm_1m_2/r^2$, Chapter 6), the strength of the force is proportional to the product of the *gravitational masses* of the two objects.

This is much like Coulomb's law for the electric force between two objects which is proportional to the product of their electric charges. The electric charge on an object is not related to its inertial mass; so why should we expect that an object's gravitational mass (call it gravitational charge if you like) be related to its inertial mass? All along we have assumed they were the same. Why? Because no experiment—not even of high precision—has been able to discern any measurable difference between inertial mass and gravitational mass. (For example, in the absence of air resistance, all objects fall at the same acceleration, g, on Earth.) This is another way to state the equivalence principle: *gravitational mass is equivalent to inertial mass.*

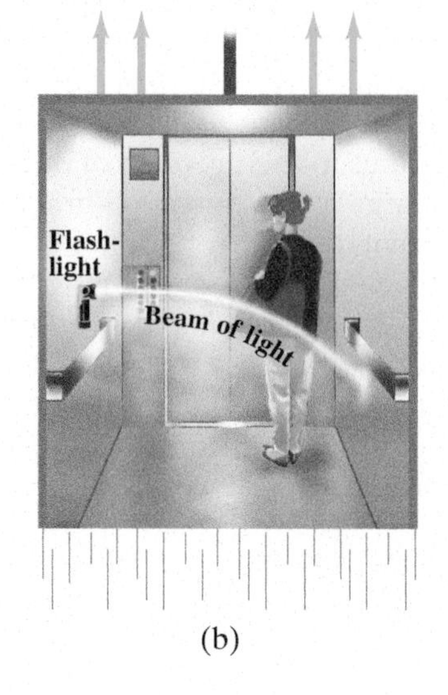

FIGURE 44–13 (a) Light beam goes straight across an elevator which is not accelerating. (b) The light beam bends (exaggerated) in an accelerating elevator whose speed increases in the upward direction. Both views are as seen by an outside observer in an inertial reference frame.

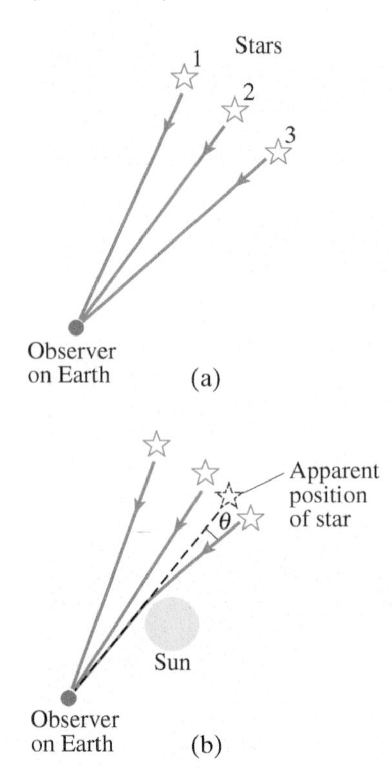

FIGURE 44–14 (a) Three stars in the sky observed from Earth. (b) If the light from one of these stars passes very near the Sun, whose gravity bends the rays, the star will appear higher than it actually is (follow the ray backwards).

The principle of equivalence can be used to show that light ought to be deflected due to the gravitational force of a massive object. Consider another thought experiment, in which an elevator is in free space where virtually no gravity acts. If a light beam is emitted by a flashlight attached to the side of the elevator, the beam travels straight across the elevator and makes a spot on the opposite side if the elevator is at rest or moving at constant velocity (Fig. 44–13a). If instead the elevator is accelerating upward, as in Fig. 44–13b, the light beam still travels straight across in a reference frame at rest. In the upwardly accelerating elevator, however, the beam is observed to curve downward. Why? Because during the time the light travels from one side of the elevator to the other, the elevator is moving upward at a vertical speed that is increasing relative to the light. Next we note that according to the equivalence principle, an upwardly accelerating reference frame is equivalent to a downward gravitational field. Hence, we can picture the curved light path in Fig. 44–13b as being due to the effect of a gravitational field. Thus, from the principle of equivalence, we expect gravity to exert a force on a beam of light and to bend it out of a straight-line path!

That light is affected by gravity is an important prediction of Einstein's general theory of relativity. And it can be tested. The amount a light beam would be deflected from a straight-line path must be small even when passing a massive object. (For example, light near the Earth's surface after traveling 1 km is predicted to drop only about 10^{-10} m, which is equal to the diameter of a small atom and not detectable.) The most massive object near us is the Sun, and it was calculated that light from a distant star would be deflected by 1.75″ of arc (tiny but detectable) as it passed by the edge of the Sun (Fig. 44–14). However, such a measurement could be made only during a total eclipse of the Sun, so that the Sun's tremendous brightness would not obscure the starlight passing near its edge. An opportune eclipse occurred in 1919, and scientists journeyed to the South Atlantic to observe it.

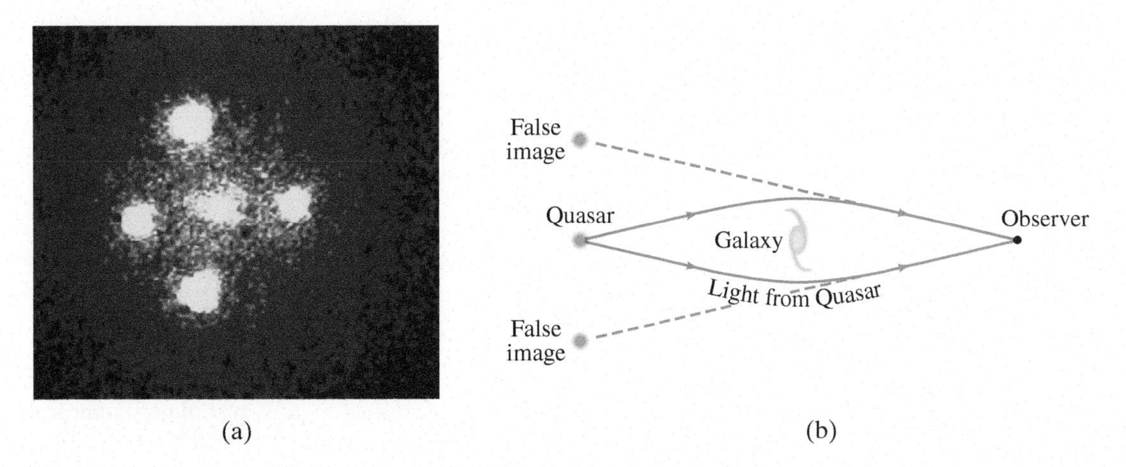

FIGURE 44–15 (a) Hubble Space Telescope photograph of the so-called "Einstein cross", thought to represent "gravitational lensing": the central spot is a relatively nearby galaxy, whereas the four other spots are thought to be images of a single quasar *behind* the galaxy. (b) Diagram showing how the galaxy could bend the light coming from the quasar behind it to produce the four images. See also Fig. 44–14. [If the shape of the nearby galaxy and distant quasar were perfect spheres, we would expect the "image" of the distant quasar to be a circular ring or halo instead of the four separate images seen here. Such a ring is called an "Einstein ring."]

Their photos of stars around the Sun revealed shifts in accordance with Einstein's prediction. Another example is gravitational lensing, as shown in Fig. 44–15.

Fermat showed that optical phenomena, including reflection, refraction, and effects of lenses, can be derived from a simple principle: that light traveling between two points follows the shortest path in space. Thus if gravity curves the path of light, then gravity must be able to curve space itself. That is, *space itself can be curved*, and it is gravitational mass that causes the curvature. Indeed, the curvature of space—or rather, of four-dimensional space-time—is a basic aspect of Einstein's General Relativity (GR).

What is meant by **curved space**? To understand, recall that our normal method of viewing the world is via Euclidean plane geometry. In Euclidean geometry, there are many axioms and theorems we take for granted, such as that the sum of the angles of any triangle is 180°. Non-Euclidean geometries, which involve curved space, have also been imagined by mathematicians. It is hard enough to imagine three-dimensional curved space, much less curved four-dimensional space-time. So let us try to understand the idea of curved space by using two-dimensional surfaces.

Consider, for example, the two-dimensional surface of a sphere. It is clearly curved, Fig. 44–16, at least to us who view it from the outside—from our three-dimensional world. But how would hypothetical two-dimensional creatures determine whether their two-dimensional space was flat (a plane) or curved? One way would be to measure the sum of the angles of a triangle. If the surface is a plane, the sum of the angles is 180°, as we learn in plane geometry. But if the space is curved, and a sufficiently large triangle is constructed, the sum of the angles will *not* be 180°. To construct a triangle on a curved surface, say the sphere of Fig. 44–16, we must use the equivalent of a straight line: that is, the shortest distance between two points, which is called a **geodesic**. On a sphere, a geodesic is an arc of a great circle (an arc in a plane passing through the center of the sphere) such as the Earth's equator and the Earth's longitude lines. Consider, for example, the large triangle of Fig. 44–16: its sides are two longitude lines passing from the north pole to the equator, and the third side is a section of the equator as shown. The two longitude lines make 90° angles with the equator (look at a world globe to see this more clearly). They make an angle with each other at the north pole, which could be, say, 90° as shown; the sum of these angles is $90° + 90° + 90° = 270°$. This is clearly *not* a Euclidean space. Note, however, that if the triangle is small in comparison to the radius of the sphere, the angles will add up to nearly 180°, and the triangle (and space) will seem flat.

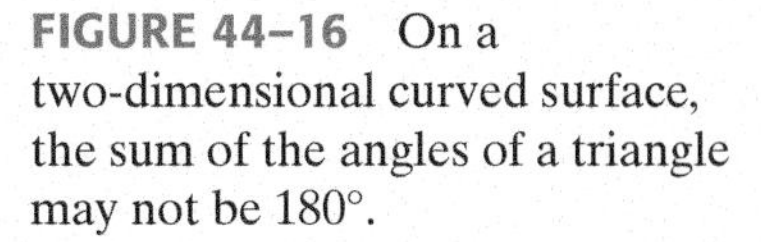

FIGURE 44–16 On a two-dimensional curved surface, the sum of the angles of a triangle may not be 180°.

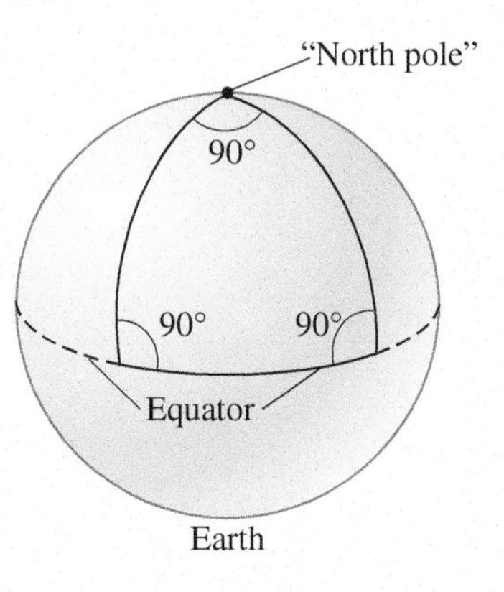

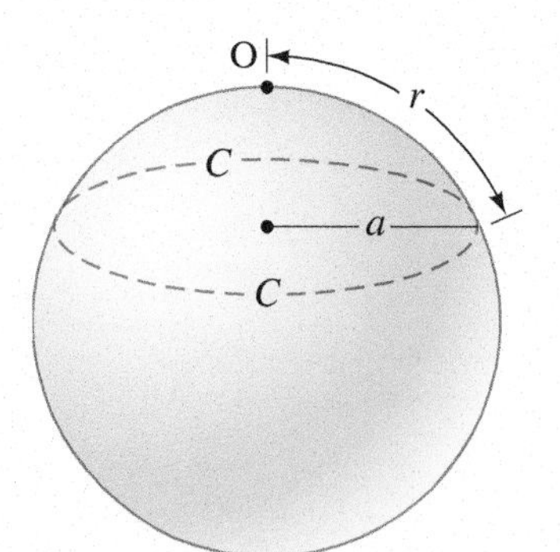

FIGURE 44–17 On a spherical surface (a two-dimensional world) a circle of circumference C is drawn (red) about point O as the center. The radius of the circle (not the sphere) is the distance r along the surface. (Note that in our three-dimensional view, we can tell that $C = 2\pi a$. Since $r > a$, then $C < 2\pi r$.)

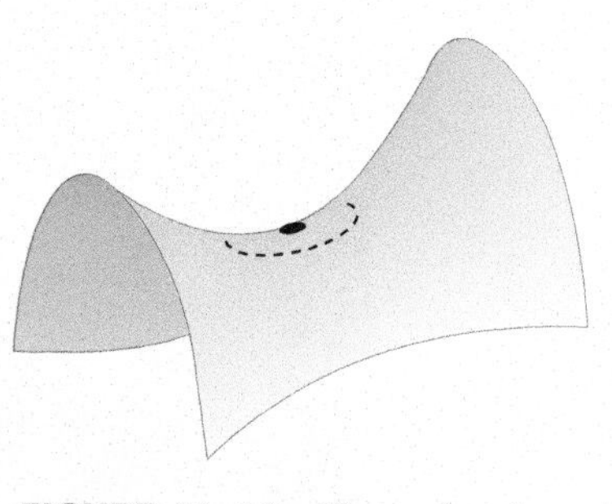

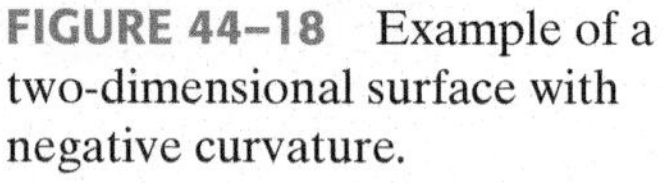

FIGURE 44–18 Example of a two-dimensional surface with negative curvature.

Another way to test the curvature of space is to measure the radius r and circumference C of a large circle. On a plane surface, $C = 2\pi r$. But on a two-dimensional spherical surface, C is *less* than $2\pi r$, as can be seen in Fig. 44–17. The proportionality between C and r is *less* than 2π. Such a surface is said to have *positive curvature*. On the saddlelike surface of Fig. 44–18, the circumference of a circle is greater than $2\pi r$, and the sum of the angles of a triangle is less than 180°. Such a surface is said to have a *negative curvature*.

Curvature of the Universe

What about our universe? On a large scale (not just near a large mass), what is the overall curvature of the universe? Does it have positive curvature, negative curvature, or is it flat (zero curvature)? We perceive our world as Euclidean (flat), but we can not exclude the possibility that space could have a curvature so slight that we don't normally notice it. This is a crucial question in cosmology, and it can be answered only by precise experimentation.

If the universe had a positive curvature, the universe would be *closed*, or *finite* in volume. This would *not* mean that the stars and galaxies extended out to a certain boundary, beyond which there is empty space. There is no boundary or edge in such a universe. The universe is all there is. If a particle were to move in a straight line in a particular direction, it would eventually return to the starting point—perhaps eons of time later.

On the other hand, if the curvature of space was zero or negative, the universe would be *open*. It could just go on forever. An open universe could be *infinite*; but according to recent research, even that may not necessarily be so.

Today the evidence is very strong that the universe on a large scale is very close to being flat. Indeed, it is so close to being flat that we can't tell if it might have very slightly positive or very slightly negative curvature.

Black Holes

FIGURE 44–19 Rubber-sheet analogy for space-time curved by matter.

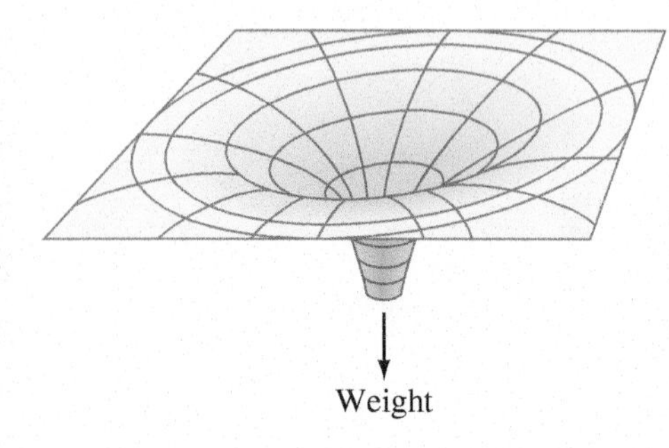

According to Einstein's theory, space-time is curved near massive objects. We might think of space as being like a thin rubber sheet: if a heavy weight is hung from it, it curves as shown in Fig. 44–19. The weight corresponds to a huge mass that causes space (space itself!) to curve. Thus, in Einstein's theory[†] we do not speak of the "force" of gravity acting on objects. Instead we say that objects and light rays move as they do because space-time is curved. An object starting at rest or moving slowly near the great mass of Fig. 44–19 would follow a geodesic (the equivalent of a straight line in plane geometry) toward that great mass.

[†]Alexander Pope (1688–1744) wrote an epitaph for Newton:
"Nature, and Nature's laws lay hid in night:
God said, *Let Newton be!* and all was light."
Sir John Squire (1884–1958), perhaps uncomfortable with Einstein's profound thoughts, added:
"It did not last: the Devil howling '*Ho!*
Let Einstein be!' restored the status quo."

The extreme curvature of space-time shown in Fig. 44–19 could be produced by a **black hole**. A black hole, as we mentioned in Section 44–2, is so dense that even light cannot escape from it. To become a black hole, an object of mass M must undergo **gravitational collapse**, contracting by gravitational self-attraction to within a radius called the **Schwarzschild radius**:

$$R = \frac{2GM}{c^2},$$

where G is the gravitational constant and c the speed of light. If an object collapses to within this radius, it is predicted by general relativity to rapidly $(\approx 10^{-5}\,\mathrm{s})$ collapse to a point at $r = 0$, forming an infinitely dense singularity. This prediction is uncertain, however, because in this realm we need to combine quantum mechanics with gravity, a unification of theories not yet achieved (Section 43–12).

EXERCISE C What is the Schwarzschild radius for an object with 2 solar masses?

The Schwarzschild radius also represents the event horizon of a black hole. By **event horizon** we mean the surface beyond which no emitted signals can ever reach us, and thus inform us of events that happen beyond that surface. As a star collapses toward a black hole, the light it emits is pulled harder and harder by gravity, but we can still see it. Once the matter passes within the event horizon, the emitted light cannot escape but is pulled back in by gravity.

All we can know about a black hole is its mass, its angular momentum (there could be rotating black holes), and its electric charge. No other information, no details of its structure or the kind of matter it was formed of, can be known because no information can escape.

How might we observe black holes? We cannot see them because no light can escape from them. They would be black objects against a black sky. But they do exert a gravitational force on nearby objects. The black hole believed to be at the center of our Galaxy ($M \approx 4 \times 10^6\, M_{\mathrm{Sun}}$) was discovered by examining the motion of matter in its vicinity. Another technique is to examine stars which appear to move as if they were one member of a *binary system* (two stars rotating about their common center of mass), but without a visible companion. If the unseen star is a black hole, it might be expected to pull off gaseous material from its visible companion (as in Fig. 44–10). As this matter approached the black hole, it would be highly accelerated and should emit X-rays of a characteristic type before plunging inside the event horizon. Such X-rays, plus a sufficiently high mass estimate from the rotational motion, can provide evidence for a black hole. One of the many candidates for a black hole is in the binary-star system Cygnus X-1. It is widely believed that the center of most galaxies is occupied by a black hole with a mass 10^6 to 10^9 times the mass of a typical star like our Sun.

EXERCISE D A black hole has radius R. Its mass is proportional to (*a*) R, (*b*) R^2, (*c*) R^3. Justify your answer.

44–5 The Expanding Universe: Redshift and Hubble's Law

We discussed in Section 44–2 how individual stars evolve from their birth to their death as white dwarfs, neutron stars, and black holes. But what about the universe as a whole: is it static, or does it change? One of the most important scientific discoveries of the twentieth century was that distant galaxies are racing away from us, and that the farther they are from us, the faster they are moving away. How astronomers arrived at this astonishing idea, and what it means for the past history of the universe as well as its future, will occupy us for the remainder of the book.

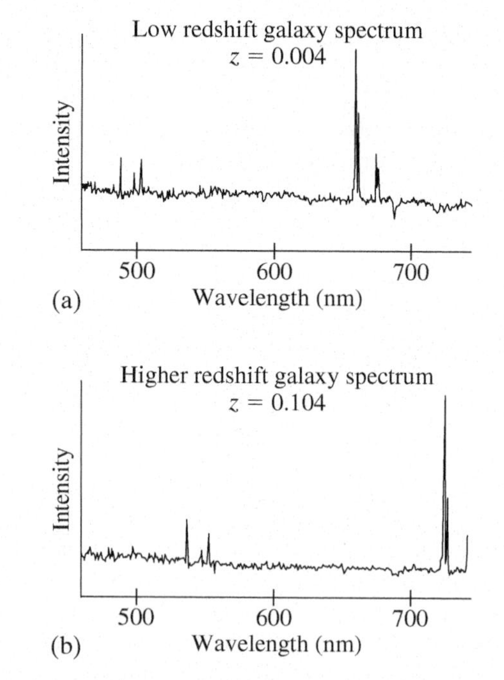

FIGURE 44–20 Atoms and molecules emit and absorb light of particular frequencies depending on the spacing of their energy levels, as we saw in Chapters 37 to 40. (a) The spectrum of light received from a relatively slow-moving galaxy. (b) Spectrum of a galaxy moving away from us at a much higher speed. Note how the peaks (or lines) in the spectrum have moved to longer wavelengths. The redshift is $z = (\lambda_{\text{obs}} - \lambda_{\text{rest}})/\lambda_{\text{rest}}$.

That the universe is expanding was first put forth by Edwin Hubble in 1929. This idea was based on distance measurements of galaxies (Section 44–3), and determination of their velocities by the Doppler shift of spectral lines in the light received from them (Fig. 44–20). In Chapter 16 we saw how the frequency of sound is higher and the wavelength shorter if the source and observer move toward each other. If the source moves away from the observer, the frequency is lower and the wavelength longer. The **Doppler effect** occurs also for light, and we saw in Section 36–12 (Eq. 36–15) that according to special relativity, the Doppler shift is given by

$$\lambda_{\text{obs}} = \lambda_{\text{rest}}\sqrt{\frac{1 + v/c}{1 - v/c}}, \quad \left[\begin{array}{l}\text{source and observer moving}\\ \text{away from each other}\end{array}\right] \qquad \textbf{(44–3)}$$

where λ_{rest} is the emitted wavelength as seen in a reference frame at rest with respect to the source, and λ_{obs} is the wavelength observed in a frame moving with velocity v away from the source along the line of sight. (For relative motion *toward* each other, $v < 0$ in this formula.) When a distant source emits light of a particular wavelength, and the source is moving away from us, the wavelength appears longer to us: the color of the light (if it is visible) is shifted toward the red end of the visible spectrum, an effect known as a **redshift**. (If the source moves toward us, the color shifts toward the blue or shorter wavelength.)

In the spectra of stars in other galaxies, lines are observed that correspond to lines in the known spectra of particular atoms (see Section 37–10 and Figs. 35–22 and 37–20). What Hubble found was that the lines seen in the spectra from distant galaxies were generally *redshifted*, and that the amount of shift seemed to be approximately proportional to the distance of the galaxy from us. That is, the velocity v of a galaxy moving away from us is proportional to its distance d from us:

HUBBLE'S LAW

$$v = Hd. \qquad \textbf{(44–4)}$$

This is **Hubble's law**, one of the most fundamental astronomical ideas. The constant H is called the **Hubble parameter**.

The value of H until recently was uncertain by over 20%, and thought to be between 50 and 80 km/s/Mpc. But recent measurements now put its value more precisely at

$$H = 71\ \text{km/s/Mpc}$$

(that is, 71 km/s per megaparsec of distance). The current uncertainty is about 5%, or ± 4 km/s/Mpc. If we use light-years for distance, then $H = 22$ km/s per million light-years of distance:

$$H = 22\ \text{km/s/Mly}$$

with an estimated uncertainty of ± 1 km/s/Mly.

Redshift Origins

Galaxies very near us seem to be moving randomly relative to us: some move towards us (blueshifted), others away from us (redshifted); their speeds are on the order of $0.001c$. But for more distant galaxies, the velocity of recession is much greater than the velocity of local random motion, and so is dominant and Hubble's law (Eq. 44–4) holds very well. More distant galaxies have higher recession velocity and a larger redshift, and we call their redshift a **cosmological redshift**. We interpret this redshift today as due to the *expansion of space* itself. We can think of the originally emitted wavelength λ_{rest} as being stretched out (becoming longer) along with the expanding space around it, as suggested in Fig. 44–21. Although Hubble thought of the redshift as a Doppler shift, now we understand it in this sense of expanding space.

Contrast the cosmological redshift, due to the expansion of space itself, with an ordinary *Doppler redshift* which is due to the relative motion of emitter and observer in a space that can be considered fixed over the time interval of observation.

There is a third way to produce a redshift, which we mention for completeness: a **gravitational redshift**. Light leaving a massive star is gaining in gravitational potential energy (just like a stone thrown upward from Earth). So the kinetic energy of each photon, hf, must be getting smaller (to conserve energy). A smaller frequency f means a larger (longer) wavelength λ $(= c/f)$, which is a redshift.

The amount of a redshift is specified by the **redshift parameter**, z, defined as

$$z = \frac{\lambda_{obs} - \lambda_{rest}}{\lambda_{rest}} = \frac{\Delta\lambda}{\lambda_{rest}}, \quad \textbf{(44–5a)}$$

where λ_{rest} is a wavelength as seen by an observer at rest relative to the source, and λ_{obs} is the wavelength measured by a moving observer. Equation 44–5a can also be written as

$$z = \frac{\lambda_{obs}}{\lambda_{rest}} - 1 \quad \textbf{(44–5b)}$$

and

$$z + 1 = \frac{\lambda_{obs}}{\lambda_{rest}}. \quad \textbf{(44–5c)}$$

For low speeds not close to the speed of light $(v \lesssim 0.1c)$, the Doppler formula (Eq. 44–3) can be used to show (Problem 29) that z is proportional to the speed of the source toward or away from us:

$$z = \frac{\lambda_{obs} - \lambda_{rest}}{\lambda_{rest}} \approx \frac{v}{c}. \qquad [v \ll c] \quad \textbf{(44–6)}$$

But redshifts are not always small, in which case the approximation of Eq. 44–6 is not valid. Modern telescopes regularly observe galaxies with $z \approx 5$ (Fig. 44–22); for large z galaxies, not even Eq. 44–3 applies because the redshift is due to the expansion of space (cosmological redshift), not the Doppler effect.

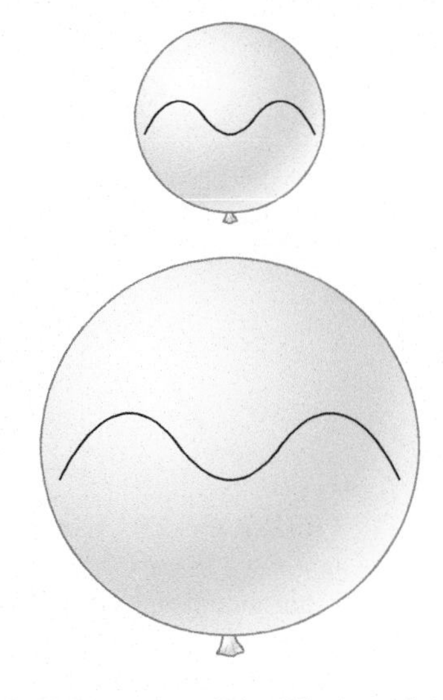

FIGURE 44–21 Simplified model of a 2-dimensional universe, imagined as a balloon. As you blow up the balloon (= expanding universe), the wavelength of a wave on its surface gets longer (redshifted).

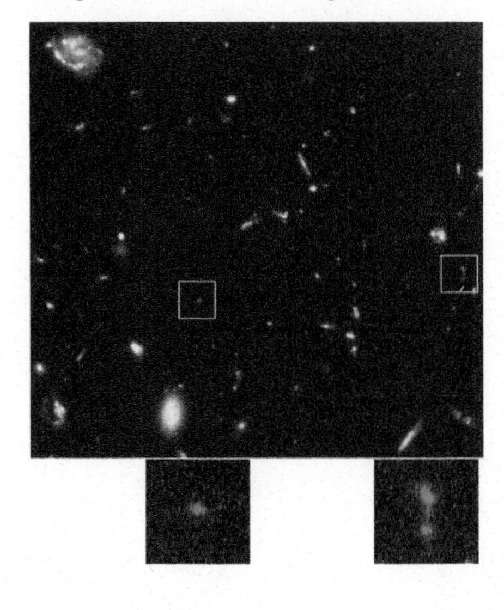

FIGURE 44–22 Hubble Ultra Deep Field photograph showing what may be among the most distant galaxies from us (small red dots, highlighted by squares), with $z \approx$ 5 or 6, existing when the universe was only about 800 million years old. The two distant galaxies in this photo are shown enlarged below.

Scale Factor

The expansion of space can be described as a simple scaling of the typical distance between two points or objects in the universe. If two distant galaxies are a distance d_0 apart at some initial time, then a time t later they will be separated by a greater distance $d(t)$. The **scale factor** is the same as for light, expressed in Eq. 44–5a. That is,

$$\frac{d(t) - d_0}{d_0} = \frac{\Delta\lambda}{\lambda} = z$$

or

$$\frac{d(t)}{d_0} = 1 + z.$$

Thus, for example, if a galaxy has $z = 3$, then the scale factor is now $(1 + 3) = 4$ times larger than when the light was emitted from that galaxy. That is, the average distance between galaxies has become 4 times larger. Thus the factor by which the wavelength has increased since it was emitted tells us by what factor the universe (or the typical distance between objects) has increased.

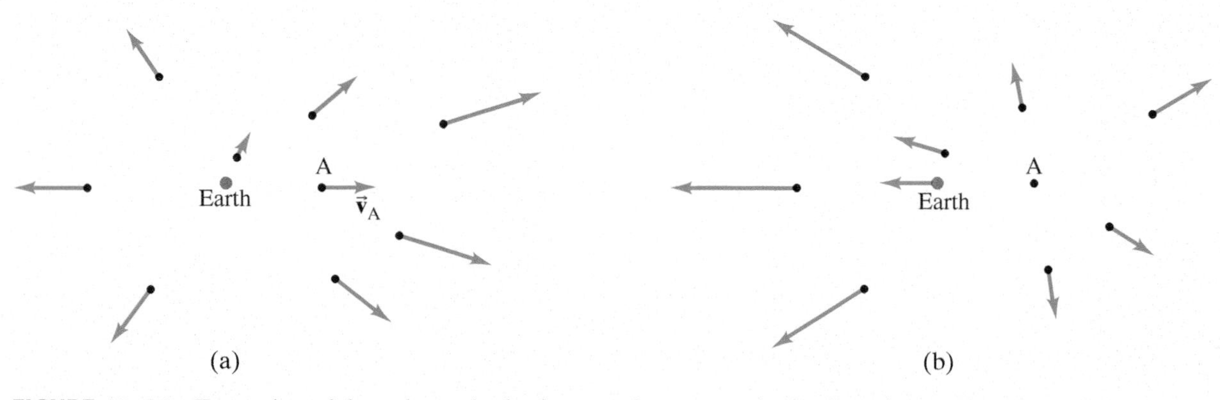

FIGURE 44–23 Expansion of the universe looks the same from any point in the universe. If you are on Earth as shown in part (a) or you are instead at point A (which is at rest in the reference frame shown in (b)), all other galaxies appear to be racing away from you.

Expansion, and the Cosmological Principle

What does it mean that distant galaxies are all moving away from us, and with ever greater speed the farther they are from us? It seems to suggest some kind of explosive expansion that started at some very distant time in the past. And at first sight we seem to be in the middle of it all. But we aren't. The expansion appears the same from any other point in the universe. To understand why, see Fig. 44–23. In Fig. 44–23a we have the view from Earth (or from our Galaxy). The velocities of surrounding galaxies are indicated by arrows, pointing away from us, and the arrows are longer for galaxies more distant from us. Now, what if we were on the galaxy labeled A in Fig. 44–23a? From Earth, galaxy A appears to be moving to the right at a velocity, call it $\vec{\mathbf{v}}_A$, represented by the arrow pointing to the right. If we were *on* galaxy A, Earth would appear to be moving to the left at velocity $-\vec{\mathbf{v}}_A$. To determine the velocities of other galaxies relative to A, we vectorially add the velocity vector, $-\vec{\mathbf{v}}_A$, to all the velocity arrows shown in Fig. 44–23a. This yields Fig. 44–23b, where we see clearly that the universe is expanding away from galaxy A as well; and the velocities of galaxies receding from A are proportional to their current distance from A. The universe looks pretty much the same from different points.

Thus the expansion of the universe can be stated as follows: all galaxies are racing away from *each other* at an average rate of about 22 km/s per million light-years of distance between them. The ramifications of this idea are profound, and we discuss them in a moment.

A basic assumption in cosmology has been that on a large scale, the universe would look the same to observers at different places at the same time. In other words, the universe is both *isotropic* (looks the same in all directions) and *homogeneous* (would look the same if we were located elsewhere, say in another galaxy). This assumption is called the **cosmological principle**. On a local scale, say in our solar system or within our Galaxy, it clearly does not apply (the sky looks different in different directions). But it has long been thought to be valid if we look on a large enough scale, so that the average population density of galaxies and clusters of galaxies ought to be the same in different areas of the sky. This seems to be valid on distances greater than about 200 Mpc (700 Mly). The expansion of the universe (Fig. 44–23) is consistent with the cosmological principle; and the near uniformity of the cosmic microwave background radiation (discussed in Section 44–6) supports it. Another way to state the cosmological principle is that *our place in the universe is not special.*

The expansion of the universe, as described by Hubble's law, strongly suggests that galaxies must have been closer together in the past than they are now. This is, in fact, the basis of the *Big Bang* theory of the origin of the universe, which pictures the universe as a relentless expansion starting from a very hot and compressed beginning. We discuss the Big Bang in detail shortly, but first let us see what can be said about the age of the universe.

One way to estimate the age of the universe uses the Hubble parameter. With $H \approx 22\text{ km/s}$ per 10^6 light-years, the time required for the galaxies to arrive at their present separations would be approximately (starting with $v = d/t$ and using Hubble's law, Eq. 44–4),

$$t = \frac{d}{v} = \frac{d}{Hd} = \frac{1}{H} \approx \frac{(10^6\text{ ly})(0.95 \times 10^{13}\text{ km/ly})}{(22\text{ km/s})(3.16 \times 10^7\text{ s/yr})} \approx 14 \times 10^9\text{ yr},$$

or 14 billion years. The age of the universe calculated in this way is called the *characteristic expansion time* or "Hubble age." It is a very rough estimate and assumes the rate of expansion of the universe was constant (which today we are quite sure is not true). Today's best measurements give the age of the universe as 13.7×10^9 yr, in remarkable agreement with the rough Hubble age estimate.

* Steady-State Model

Before discussing the Big Bang in detail, we mention one alternative to the Big Bang—the **steady-state model**—which assumed that the universe is infinitely old and on average looks the same now as it always has. (This assumed uniformity in time as well as space was called the *perfect cosmological principle.*) According to the steady-state model, no large-scale changes have taken place in the universe as a whole, particularly no Big Bang. To maintain this view in the face of the recession of galaxies away from each other, matter must be created continuously to maintain the assumption of uniformity. The rate of mass creation required is very small—about one nucleon per cubic meter every 10^9 years.

The steady-state model provided the Big Bang model with healthy competition in the mid-twentieth century. But the discovery of the cosmic microwave background radiation (next Section), as well as other observations of the universe, has made the Big Bang model universally accepted.

44–6 The Big Bang and the Cosmic Microwave Background

The expansion of the universe suggests that typical objects in the universe were once much closer together than they are now. This is the basis for the idea that the universe began about 14 billion years ago as an expansion from a state of very high density and temperature known affectionately as the **Big Bang**.

The birth of the universe was not an explosion, because an explosion blows pieces out into the surrounding space. Instead, the Big Bang was the start of an expansion of space itself. The observable universe was very small at the start and has been expanding ever since. The initial tiny universe of extremely dense matter is not to be thought of as a concentrated mass in the midst of a much larger space around it. The initial tiny but dense universe was the entire universe. There wouldn't have been anything else. When we say that the universe was once smaller than it is now, we mean that the average separation between objects (such as galaxies) was less. It is thought the universe was infinite in extent then, and it still is (only bigger). The observable universe, however, is finite.

A major piece of evidence supporting the Big Bang is the **cosmic microwave background** radiation (or CMB) whose discovery came about as follows.

In 1964, Arno Penzias and Robert Wilson pointed their radiotelescope (a large antenna device for detecting radio waves) into the night sky (Fig. 44–24). With it they detected widespread emission, and became convinced that it was coming from outside our Galaxy. They made precise measurements at a wavelength $\lambda = 7.35$ cm, in the microwave region of the electromagnetic spectrum (Fig. 31–12). The intensity of this radiation was found initially not to vary by day or night or time of year, nor to depend on direction. It came from all directions in the universe with equal intensity, to a precision of better than 1%. It could only be concluded that this radiation came from the universe as a whole.

FIGURE 44–24 Arno Penzias (right) and Robert Wilson, and behind them their "horn antenna."

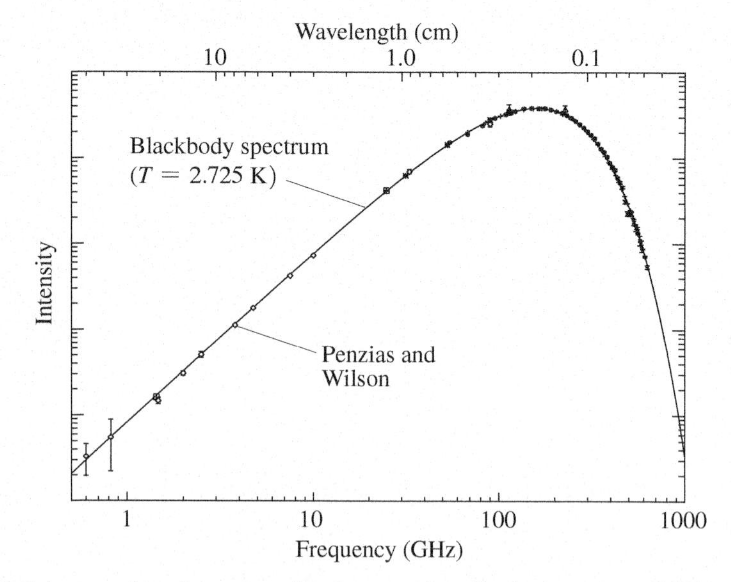

FIGURE 44–25 Spectrum of cosmic microwave background radiation, showing blackbody curve and experimental measurements including at the frequency detected by Penzias and Wilson. (Thanks to G. F. Smoot and D. Scott. The vertical bars represent the most recent experimental uncertainty in a measurement.)

They measured this cosmic microwave background radiation at $\lambda = 7.35$ cm, and its intensity corresponds to blackbody radiation (see Section 37–1) at a temperature of about 3 K. When radiation at other wavelengths was measured by the COBE satellite (COsmic Background Explorer), the intensities were found to fall on a nearly perfect blackbody curve as shown in Fig. 44–25, corresponding to a temperature of 2.725 K ($\pm$ 0.002 K).

The remarkable uniformity of the cosmic microwave background radiation was in accordance with the cosmological principle. But theorists felt that there needed to be some small inhomogeneities, or "anisotropies," in the CMB that would have provided "seeds" around which galaxy formation could have started. Small areas of slightly higher density, which could have contracted under gravity to form stars and galaxies, were indeed found. These tiny inhomogeneities in density and temperature were detected first by the COBE satellite experiment in 1992, led by John Mather and George Smoot (Fig. 44–26).

FIGURE 44–26 COBE scientists John Mather (left, chief scientist and responsible for measuring the blackbody form of the spectrum) and George Smoot (chief investigator for anisotropy experiment) shown here during celebrations for their Dec. 2006 Nobel Prize, given for their discovery of the spectrum and anisotropy of the CMB using the COBE instrument.

This discovery of the **anisotropy** of the CMB ranks with the discovery of the CMB itself in the history of cosmology. It was the culmination of decades of research by pioneers such as Paul Richards and David Wilkinson. Subsequent experiments with greater detail culminated in 2003 with the WMAP (Wilkinson Microwave Anisotropy Probe) results. See Fig. 44–27 which presents the latest (2006) results.

The CMB provides strong evidence in support of the Big Bang, and gives us information about conditions in the very early universe. In fact, in the late 1940s, George Gamow and his collaborators calculated that a Big Bang origin of the universe should have generated just such a microwave background radiation.

To understand why, let us look at what a Big Bang might have been like. (Today we usually use the term "Big Bang" to refer to the *process*, starting from the birth of the universe through the subsequent expansion.) The temperature must have been extremely high at the start, so high that there could not have been any atoms in the very early stages of the universe. Instead, the universe

FIGURE 44–27 Measurements of the cosmic microwave background radiation over the entire sky, color-coded to represent differences in temperature from the average 2.725 K: the color scale ranges from $+200\,\mu$K (red) to $-200\,\mu$K (dark blue), representing slightly hotter and colder spots (associated with variations in density). Results are from the WMAP satellite in 2006: the angular resolution is 0.2°. The white lines are added to show the measured polarization direction of the earliest light, which gives further clues to the early universe.

would have consisted solely of radiation (photons) and a plasma of charged electrons and other elementary particles. The universe would have been opaque—the photons in a sense "trapped," traveling very short distances before being scattered again, primarily by electrons. Indeed, the details of the microwave background radiation is strong evidence that matter and radiation were once in equilibrium at a very high temperature. As the universe expanded, the energy spread out over an increasingly larger volume and the temperature dropped. Only when the temperature had fallen to about 3000 K, some 380,000 years later, could nuclei and electrons combine together as atoms. With the disappearance of free electrons, as they combined with nuclei to form atoms, the radiation would have been freed—**decoupled** from matter, we say. The universe became *transparent* because photons were now free to travel nearly unimpeded straight through the universe.

It is this radiation, from 380,000 years after the birth of the universe, that we now see as the CMB. As the universe expanded, so too the wavelengths of the radiation lengthened, thus redshifting to longer wavelengths that correspond to lower temperature (recall Wien's law, $\lambda_P T = \text{constant}$, Section 37–1), until they would have reached the 2.7-K background radiation we observe today.

Looking Back toward the Big Bang—Lookback Time

Figure 44–28 shows our Earth point of view, looking out in all directions back toward the Big Bang and the brief (380,000-year-long) period when radiation was trapped in the early plasma (yellow band). The time it takes light to reach us from an event is called its **lookback time**. The "close-up" insert in Fig. 44–28 shows a photon scattering repeatedly inside that early plasma and then exiting the plasma in a straight line. No matter what direction we look, our view of the very early universe is blocked by this wall of plasma. It is like trying to look into a very thick fog or into the surface of the Sun—we can see only as far as its surface, called the **surface of last scattering**, but not into it. Wavelengths from there are redshifted by $z \approx 1100$. Time $\Delta t'$ in Fig. 44–28 is the lookback time (not real time that goes forward).

Recall that when we view an object far away, we are seeing it as it was then, when the light was emitted, not as it would appear today.

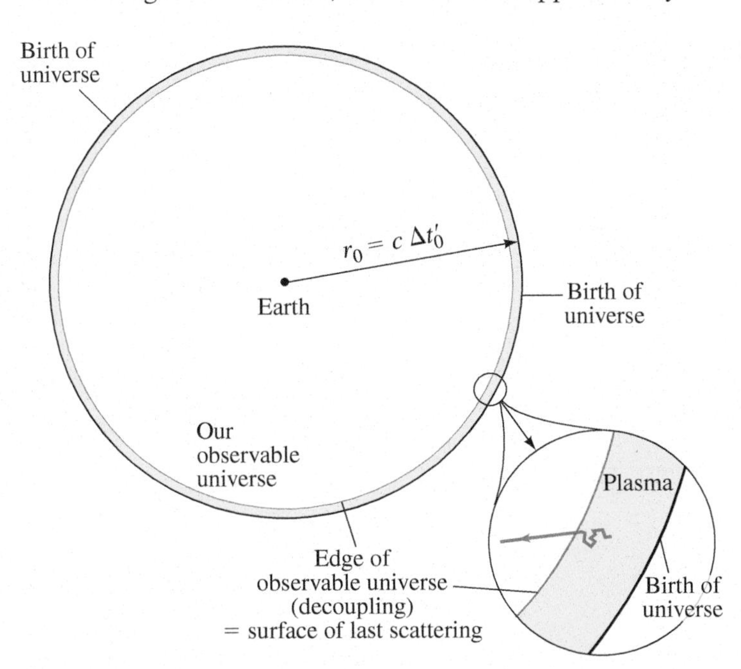

FIGURE 44–28 When we look out from the Earth, we look back in time. Any other observer in the universe would see more or less the same thing. The farther an object is from us, the longer ago the light we see had to have left it. We cannot see quite as far as the Big Bang; we can see only as far as the "surface of last scattering," which radiated the CMB. The insert on the lower right shows the earliest 380,000 years of the universe when it was opaque: a photon is shown scattering many times and then (at decoupling, 380,000 yr after the birth of the universe) becoming free to travel in a straight line. If this photon wasn't heading our way when "liberated," many others were. Galaxies are not shown, but would be concentrated close to Earth in this diagram because they were created relatively recently. *Note:* This diagram is not a normal map. Maps show a section of the world as might be seen all *at a given time*. This diagram shows space (like a map), but each point is *not* at the same time. The light coming from a point a distance r from Earth took a time $\Delta t' = r/c$ to reach Earth, and thus shows an event that took place long ago, a time $\Delta t' = r/c$ in the past, which we call its "lookback time." The universe began $\Delta t'_0 = 13.7$ Gyr ago.

The Observable Universe

Figure 44–28 is a bit dangerous: it is not a picture of the universe at a given instant, but is intended to suggest how we look out in all directions from our observation point (the Earth, or near it). Be careful not to think that the birth of the universe took place in a circle or a sphere surrounding us as if Fig. 44–28 were a photo taken at a given moment. What Fig. 44–28 does show is what we can see, the *observable universe*. Better yet, it shows the *most* we could see.

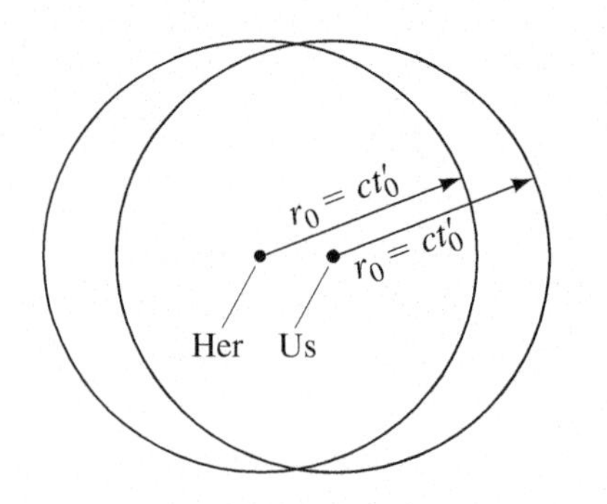

FIGURE 44–29 Two observers, on widely separated galaxies, have different horizons, different observable universes.

We would undoubtedly be arrogant to think that we could see the entire universe. Indeed, theories assume that we cannot see everything, that the **entire universe** is greater than the **observable universe**, which is a sphere of radius $r_0 = ct_0$ centered on the observer, with t_0 being the age of the universe. We can never see further back than the time it takes light to reach us.

Consider, for example, an observer in another galaxy, very far from us, located to the left of our observation point in Fig. 44–28. That observer would not yet have seen light coming from the far right of the large circle in Fig. 44–28 that we see—it will take some time for that light to reach her. But she will have already, some time ago, seen the light coming from the left that we are seeing now. In fact, her observable universe, superimposed on ours, is suggested by Fig. 44–29.

The edge of our observable universe is called the **horizon**. We could, in principle, see as far as the horizon, but not beyond it. An observer in another galaxy, far from us, will have a different horizon.

44–7 The Standard Cosmological Model: Early History of the Universe

In the last decade or two, a convincing theory of the origin and evolution of the universe has been developed, now called the **Standard Cosmological Model**, or (sometimes) the *concordance model*. Part of this theory is based on recent theoretical and experimental advances in elementary particle physics, and part from observations of the universe including COBE and WMAP. Indeed, cosmology and elementary particle physics have cross-fertilized to a surprising extent.

Let us go back to the earliest of times—as close as possible to the Big Bang—and follow a Standard Model scenario of events as the universe expanded and cooled after the Big Bang. Initially we talk of extremely small time intervals as well as extremely high temperatures, far beyond anything in the universe today. Figure 44–30 is a compressed graphical representation of the events, and it may be helpful to consult it as we go along.

FIGURE 44–30 Compressed graphical representation of the development of the universe after the Big Bang, according to modern cosmology. [The time scale is mostly logarithmic (each factor of 10 in time gets equal treatment), except at the start (there can be no $t = 0$ on a log scale), and just after $t = 10^{-35}$ s (to save space). The vertical height is a rough indication of the size of the universe, mainly to suggest expansion of the universe.]

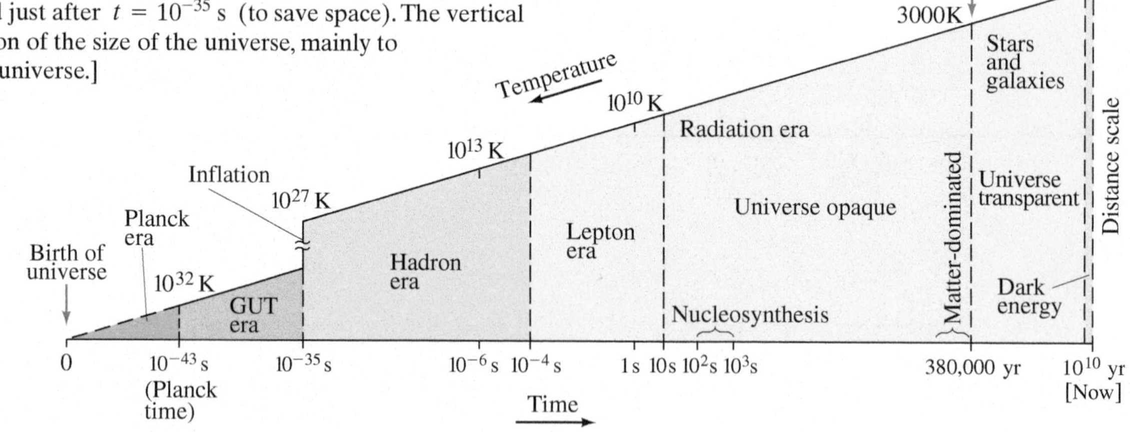

The History

We begin at a time only a minuscule fraction of a second after the birth of the universe, 10^{-43} s. This time is sometimes referred to as the **Planck time**, which is a value determined by the fundamental constants. It is related to the *Planck length* λ_P which we obtained in Chapter 1 (Example 1–10) by dimensional analysis: $\lambda_P = \sqrt{Gh/c^3} \approx 4 \times 10^{-35}$ m. The Planck time is the time it takes light to travel the Planck length: $t_P = \lambda_P/c \approx (4 \times 10^{-35}\,\text{m})/(3 \times 10^8\,\text{m/s}) \approx 10^{-43}$ s. This is an unimaginably short time, and predictions can be only speculative. Earlier, we can say nothing because we do not have a theory of quantum gravity which would be needed for the incredibly high densities and temperatures during this "Planck era." It is thought that, perhaps as early as 10^{-43} s, the four forces in nature were unified—there was only one force (Chapter 43, Fig. 43–19). The temperature would have been about 10^{32} K, corresponding to randomly moving particles with an average kinetic energy K of 10^{19} GeV (see Eq. 18–4):

$$K \approx kT \approx \frac{(1.4 \times 10^{-23}\,\mathrm{J/K})(10^{32}\,\mathrm{K})}{1.6 \times 10^{-19}\,\mathrm{J/eV}} \approx 10^{28}\,\mathrm{eV} = 10^{19}\,\mathrm{GeV}.$$

(Note that the factor $\frac{3}{2}$ in Eq. 18–4 is usually ignored in such order of magnitude calculations.) At $t = 10^{-43}$ s, a kind of "phase transition" is believed to have occurred during which the gravitational force, in effect, "condensed out" as a separate force. This, and subsequent phase transitions, are analogous to the phase transitions water undergoes as it cools from a gas, condenses into a liquid, and with further cooling freezes into ice.† The *symmetry* of the four forces was broken, but the strong, weak, and electromagnetic forces were still unified, and the universe entered the **grand unified era** (GUT—see Chapter 43). There was no distinction between quarks and leptons; baryon and lepton numbers were not conserved. Very shortly thereafter, as the universe expanded considerably and the temperature had dropped to about 10^{27} K, there was another phase transition and the strong force condensed out at about 10^{-35} s after the Big Bang. Now the universe was filled with a "soup" of leptons and quarks. The quarks were initially free, but soon began to "condense" into more normal particles: nucleons and the other hadrons and their antiparticles. With this **confinement of quarks**, the universe entered the **hadron era**.

About this time, when the universe was only 10^{-35} s old, a strange thing may have happened, according to theorists. A brilliant idea, proposed around 1980, suggests that the universe underwent an incredible exponential expansion, increasing in size by a factor of 10^{40} or maybe much more, in a tiny fraction of a second, perhaps 10^{-35} s. The usefulness of this **inflationary scenario** is that it solved major problems with earlier Big Bang models, such as explaining why the universe is flat, as well as the thermal equilibrium to provide the nearly uniform CMB, as discussed below.

After the very brief inflationary period, the universe would have settled back into its more regular expansion. The universe was now a "soup" of leptons and hadrons. We can think of this "soup" as a plasma of particles and antiparticles, as well as photons—all in roughly equal numbers—colliding with one another frequently and exchanging energy.

By the time the universe was only about a microsecond $(10^{-6}\,\mathrm{s})$ old, it had cooled to about 10^{13} K, corresponding to an average kinetic energy of 1 GeV, and the vast majority of hadrons disappeared. To see why, let us focus on the most familiar hadrons: nucleons and their antiparticles. When the average kinetic energy of particles was somewhat higher than 1 GeV, protons, neutrons, and their antiparticles were continually being created out of the energies of collisions involving photons and other particles, such as

$$\text{photons} \rightarrow \mathrm{p} + \bar{\mathrm{p}}$$
$$\rightarrow \mathrm{n} + \bar{\mathrm{n}}.$$

But just as quickly, particles and antiparticles would annihilate: for example

$$\mathrm{p} + \bar{\mathrm{p}} \rightarrow \text{photons or leptons}.$$

So the processes of creation and annihilation of nucleons were in equilibrium. The numbers of nucleons and antinucleons were high—roughly as many as there were electrons, positrons, or photons. But as the universe expanded and cooled, and the average kinetic energy of particles dropped below about 1 GeV, which is the minimum energy needed in a typical collision to create nucleons and antinucleons (about 940 MeV each), the process of nucleon creation could not continue. The process of annihilation could continue, however, with antinucleons annihilating nucleons, until there were almost no nucleons left. But not quite zero. Somehow we need to explain our present world of matter (nucleons and electrons) with very little antimatter in sight.

†It may be interesting to note that our story of origins here bears some resemblance to ancient accounts that mention the "void," "formless wasteland" (or "darkness over the deep"), "abyss," "divide the waters" (possibly a phase transition?), not to mention the sudden appearance of light.

To explain our world of matter, we might suppose that earlier in the universe, perhaps around 10^{-35} s after the Big Bang, a slight excess of quarks over antiquarks was formed.† This would have resulted in a slight excess of nucleons over antinucleons. And it is these "leftover" nucleons that we are made of today. The excess of nucleons over antinucleons was probably about one part in 10^9. During the hadron era, there should have been about as many nucleons as photons. After it ended, the "leftover" nucleons thus numbered only about one nucleon per 10^9 photons, and this ratio has persisted to this day. Protons, neutrons, and all other heavier particles were thus tremendously reduced in number by about 10^{-6} s after the Big Bang. The lightest hadrons, the pions, soon disappeared, about 10^{-4} s after the Big Bang; because they are the lightest mass hadrons (140 MeV), they were the last hadrons to be able to be created as the temperature (and average kinetic energy) dropped. Lighter particles, including electrons and neutrinos, were the dominant form of matter, and the universe entered the **lepton era**.

By the time the first full second had passed (clearly the most eventful second in history!), the universe had cooled to about 10 billion degrees, 10^{10} K. The average kinetic energy was about 1 MeV. This was still sufficient energy to create electrons and positrons and balance their annihilation reactions, since their masses correspond to about 0.5 MeV. So there were about as many e^+ and e^- as there were photons. But within a few more seconds, the temperature had dropped sufficiently so that e^+ and e^- could no longer be formed. Annihilation $(e^+ + e^- \rightarrow \text{photons})$ continued. And, like nucleons before them, electrons and positrons all but disappeared from the universe—except for a slight excess of electrons over positrons (later to join with nuclei to form atoms). Thus, about $t = 10$ s after the Big Bang, the universe entered the **radiation era** (Fig. 44–30). Its major constituents were photons and neutrinos. But the neutrinos, partaking only in the weak force, rarely interacted. So the universe, until then experiencing significant amounts of energy in matter and in radiation, now became **radiation-dominated**: much more energy was contained in radiation than in matter, a situation that would last more than 50,000 years.

FIGURE 44–30 (Repeated.) Compressed graphical representation of the development of the universe after the Big Bang, according to modern cosmology.

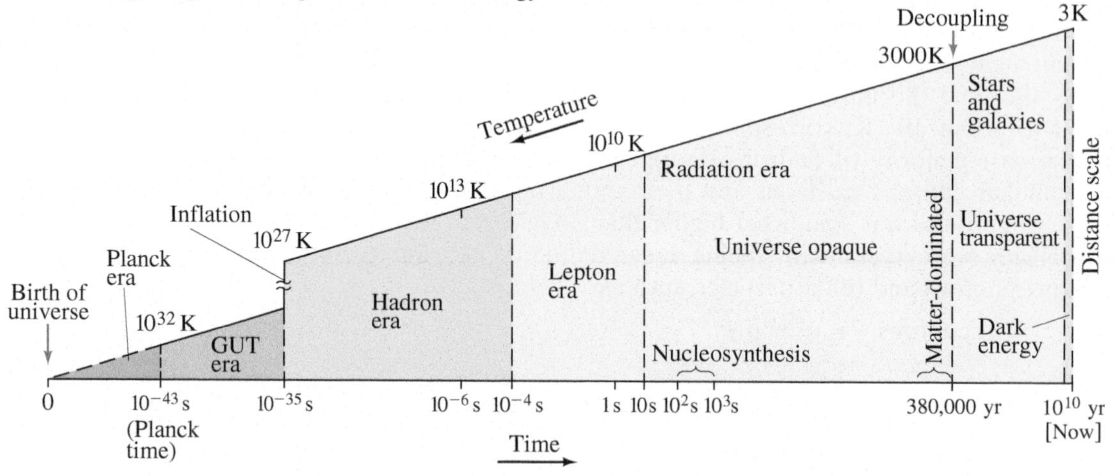

Meanwhile, during the next few minutes, crucial events were taking place. Beginning about 2 or 3 minutes after the Big Bang, nuclear fusion began to occur. The temperature had dropped to about 10^9 K, corresponding to an average kinetic energy $\bar{K} \approx 100$ keV, where nucleons could strike each other and be able to fuse (Section 42–4), but now cool enough so newly formed nuclei would not be immediately broken apart by subsequent collisions. Deuterium, helium, and very tiny amounts of lithium nuclei were made. But the universe was cooling too quickly, and larger nuclei were not made. After only a few minutes, probably not even a quarter of an hour after the Big Bang, the temperature dropped far enough that nucleosynthesis stopped, not to start again for millions of years (in stars).

†Why this could have happened is a question for which we are seeking an answer today.

Thus, after the first quarter hour or so of the universe, matter consisted mainly of bare nuclei of hydrogen (about 75%) and helium (about 25%)† and electrons. But radiation (photons) continued to dominate.

Our story is almost complete. The next important event is thought to have occurred 380,000 years later. The universe had expanded to about $\frac{1}{1000}$ of its present scale, and the temperature had cooled to about 3000 K. The average kinetic energy of nuclei, electrons, and photons was less than an electron volt. Since ionization energies of atoms are on the order of eV, then as the temperature dropped below this point, electrons could orbit the bare nuclei and remain there (without being ejected by collisions), thus forming atoms. This period is often called the **recombination** epoch (a misnomer since electrons had never before been combined with nuclei to form atoms). With the disappearance of free electrons and the birth of atoms, the photons—which had been continually scattering from the free electrons—now became free to spread throughout the universe. As mentioned in the previous Section, we say that the photons became **decoupled** from matter. Thus *decoupling* occurred at *recombination*. The total energy contained in radiation had been decreasing (lengthening in wavelength as the universe expanded); and at about $t = 56{,}000\,\text{yr}$ (even before decoupling) the total energy contained in matter became dominant over radiation. The universe was said to have become **matter-dominated** (marked on Fig. 44–30). As the universe continued to expand, the electromagnetic radiation cooled further, to 2.7 K today, forming the cosmic microwave background radiation we detect from everywhere in the universe.

After the birth of atoms, then stars and galaxies could begin to form: by self-gravitation around mass concentrations (inhomogeneities). Stars began to form about 200 million years after the Big Bang, galaxies after almost 10^9 years. The universe continued to evolve until today, some 14 billion years after it started.

* * *

This scenario, like other scientific models, cannot be said to be "proven." Yet this model is remarkably effective in explaining the evolution of the universe we live in, and makes predictions which can be tested against the next generation of observations.

A major event, and something only discovered very recently, is that when the universe was more than half as old as it is now (about 5 Gyr ago), its expansion began to accelerate. This was a big surprise because it was assumed the expansion of the universe would slow down due to gravitational attraction of all objects towards each other. Another major recent discovery is that ordinary matter makes up very little of the total mass–energy of the universe ($\approx 5\%$). Instead, as we discuss in Section 44–9, the major contributors to the energy density of the universe are *dark matter* and *dark energy*. On the right in Fig. 44–30 is a narrow vertical strip that represents the most recent 5 billion years of the universe, during which *dark energy* seems to have dominated.

44–8 Inflation: Explaining Flatness, Uniformity, and Structure

The idea that the universe underwent a period of exponential inflation early in its life, expanding by a factor of 10^{40} or more (previous Section), was first put forth by Alan Guth in 1981. Many more sophisticated models have since been proposed. The energy required for this wild expansion may have been released when the electroweak force separated from the strong force (end of GUT era, Fig. 43–19). So far, the evidence for inflation is indirect; yet it is a feature of most viable cosmological models because it is able to provide natural explanations for several remarkable features of our universe.

†This Standard Model prediction of a 25% primordial production of helium agrees with what we observe today—the universe *does* contain about 25% He—and it is strong evidence in support of the Standard Big Bang Model. Furthermore, the theory says that 25% He abundance is fully consistent with there being three neutrino types, which is the number we observe. And it sets an upper limit of four to the maximum number of possible neutrino types. This is a striking example of the exciting interface between particle physics and cosmology.

FIGURE 44–31 (a) Simple 2-D model of the entire universe; the observable universe is suggested by the small circle centered on us (blue dot). (b) Edge of universe is essentially flat after the 10^{40}-fold expansion during inflation.

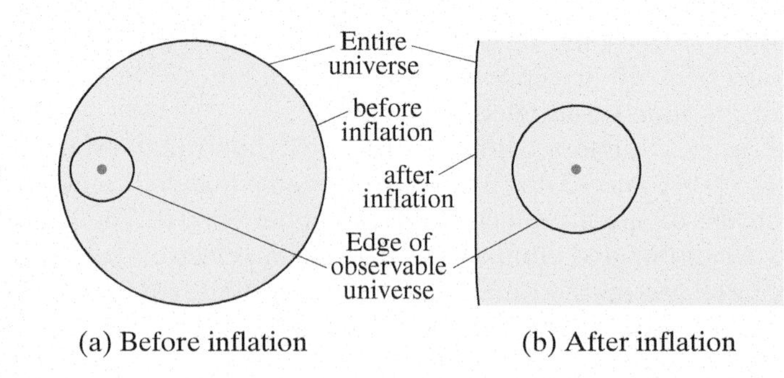

Flatness

First of all, our best measurements suggest that the universe is flat, that it has zero curvature. As scientists, we would like some reason for this remarkable result. To see how inflation explains flatness, let us consider a simple 2-dimensional model of the universe (as we did earlier in Figs. 44–16 and 44–21). A circle in this 2-dimensional universe (= surface of a sphere, Fig. 44–31a) represents the *observable* universe as seen by an observer at the circle's center. A possible hypothesis is that inflation occurred over a time interval that very roughly doubled the age of the universe, from let us say, $t = 1 \times 10^{-35}$ s to $t = 2 \times 10^{-35}$ s. The size of the observable universe $(r = ct)$ would have increased by a factor of two during inflation, while the radius of curvature of the entire universe increased by an enormous factor of 10^{40} or more. Thus the edge of our 2-D sphere representing the entire universe would have seemed flat to a high degree of precision, Fig. 44–31b. Even if the time of inflation was a factor of 10 or 100 (instead of 2), the expansion factor of 10^{40} or more would have blotted out any possibility of observing anything but a flat universe.

CMB Uniformity

Inflation also explains why the CMB is so uniform. Without inflation, the tiny universe at 10^{-35} s was too large for all parts of it to have been in contact so as to reach the same temperature (information cannot travel faster than c). Imagine a universe about 1 cm in diameter at $t = 10^{-36}$ s, as per original Big Bang theory. In that 10^{-36} s, light could have traveled $d = ct = (3 \times 10^8\,\mathrm{m/s})(10^{-36}\,\mathrm{s}) = 10^{-27}\,\mathrm{m}$, way too small for opposite sides of a 1-cm-wide universe to have been in communication. But if the universe had been 10^{40} times smaller $(= 10^{-42}\,\mathrm{m})$, as proposed by the inflation model, there could have been contact and thermal equilibrium to produce the observed nearly uniform CMB. Inflation, by making the early universe very small, assures that all parts could have been in thermal equilibrium; and after inflation the universe could be large enough to give us today's observable universe.

Galaxy Seeds, Fluctuations, Magnetic Monopoles

Inflation also gives us a clue as to how the present structure of the universe (galaxies and clusters of galaxies) came about. We saw earlier that, according to the uncertainty principle, energy may not be conserved by an amount ΔE for a time $\Delta t \approx \hbar/\Delta E$. Forces, whether electromagnetic or other types, can undergo such tiny **quantum fluctuations** according to quantum theory, but they are so tiny they are not detectable unless magnified in some way. That is what inflation might have done: it could have magnified those fluctuations perhaps 10^{40} times in size, which would give us the density irregularities seen in the cosmic microwave background (WMAP, Fig. 44–27). That would be very nice, because the density variations we see in the CMB are what we believe were the seeds that later coalesced under gravity into galaxies and galaxy clusters, including their substructures (stars and planets), and our models fit the data extremely well.

Sometimes it is said that the quantum fluctuations occurred in the **vacuum state** or vacuum energy. This could be possible because the vacuum is no longer considered to be empty, as we discussed in Section 43–3 relative to positrons and a negative energy sea of electrons. Indeed, the vacuum is thought to be filled with fields and particles occupying all the possible negative energy states. Also, the virtual exchange particles that carry the forces, as discussed in Chapter 43, could leave their brief virtual states and actually become real as a result of the 10^{40} magnification of space (according to inflation) and the very short time over which it occurred $(\Delta t = \hbar/\Delta E)$.

Inflation helps us too with the puzzle of why **magnetic monopoles** have never been observed, yet isolated magnetic poles may well have been copiously produced at the start. After inflation, they would have been so far apart that we have never stumbled on one.

Some theorists have proposed that inflation may not have occurred in the entire universe. Perhaps only some regions of that tiny early universe became unstable (maybe it was a quantum fluctuation) and inflated into cosmic "bubbles." If so, we would be living in one of the bubbles. The universe outside the bubble would be hopelessly **unobservable** to us.

Inflation may solve outstanding problems, but it needs to be confirmed and we may need new physics just to understand how inflation occurred.

44–9 Dark Matter and Dark Energy

According to the Standard Big Bang Model, the universe is evolving and changing. Individual stars are being created, evolving, and then dying to become white dwarfs, neutron stars, black holes. At the same time, the universe as a whole is expanding. One important question is whether the universe will continue to expand forever. Until the late 1990s, the universe was thought to be dominated by matter which interacts by gravity, and this question was connected to the curvature of space-time (Section 44–4). If the universe had *negative* curvature, the expansion of the universe would never stop, although the rate of expansion would decrease due to the gravitational attraction of its parts. Such a universe would be *open* and infinite. If the universe is *flat* (no curvature), it would still be open and infinite but its expansion would slowly approach a zero rate. Finally, if the universe had *positive* curvature, it would be *closed* and finite; the effect of gravity would be strong enough that the expansion would eventually stop and the universe would begin to contract, collapsing back onto itself in a **big crunch**.

EXERCISE E Return to the Chapter-Opening Question, page 1193, and answer it again. Try to explain why you may have answered differently the first time.

Critical Density

According to the above scenario (which does not include inflation or the recently discovered acceleration of the universe), the fate of the universe would depend on the average mass–energy density in the universe. For an average mass density greater than a critical value known as the **critical density**, estimated to be about

$$\rho_c \approx 10^{-26}\,\mathrm{kg/m^3}$$

(i.e., a few nucleons/$\mathrm{m^3}$ on average throughout the universe), gravity would prevent expansion from continuing forever. Eventually (if $\rho > \rho_c$) gravity would pull the universe back into a big crunch and space-time would have a positive curvature. If instead the actual density was equal to the critical density, $\rho = \rho_c$, the universe would be flat and open. If the actual density was less than the critical density, $\rho < \rho_c$, the universe would have negative curvature. See Fig. 44–32. Today we believe the universe is very close to flat. But recent evidence suggests the universe is expanding at an *accelerating* rate, as discussed below.

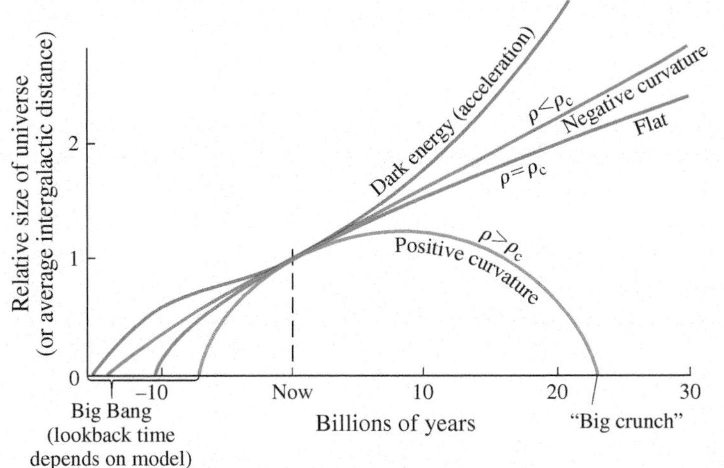

FIGURE 44–32 Three future possibilities for the universe, depending on the density ρ of ordinary matter, plus a fourth possibility that includes dark energy. Note that all curves have been chosen to have the same slope ($= H$, the Hubble parameter) right now. Looking back in time, the Big Bang occurs where each curve touches the horizontal (time) axis.

EXAMPLE 44–7 ESTIMATE **Critical density of the universe.** Use energy conservation and escape velocity (Section 8–7) to estimate the critical density of the universe.

APPROACH At the critical density, ρ_c, any given galaxy of mass m will just be able to "escape" away from our Galaxy. As we saw in Section 8–7, escape can just occur if the total energy E of the galaxy satisfies

$$E = K + U = \tfrac{1}{2}mv^2 - G\frac{mM}{R} = 0.$$

Here R is the distance of that galaxy m from us. We approximate the total mass M that pulls inward on m as the total mass within a sphere of radius R around us (Appendix D). If we assume the density of galaxies is roughly constant, then $M = \frac{4}{3}\pi\rho_c R^3$.

SOLUTION Substituting this M into the equation above, and setting $v = HR$ (Hubble's law, Eq. 44–4), we obtain

$$\frac{GM}{R} = \tfrac{1}{2}v^2$$

or

$$\frac{G\left(\frac{4}{3}\pi\rho_c R^3\right)}{R} = \tfrac{1}{2}(HR)^2.$$

We solve for ρ_c:

$$\rho_c = \frac{3H^2}{8\pi G} \approx \frac{3\left[(22\,\mathrm{km/s/Mly})(1\,\mathrm{Mly}/10^{19}\,\mathrm{km})\right]^2}{8(3.14)(6.67\times10^{-11}\,\mathrm{N\cdot m^2/kg^2})} \approx 10^{-26}\,\mathrm{kg/m^3}.$$

Dark Matter

WMAP and other experiments have convinced scientists that the universe is flat and $\rho = \rho_c$. But this ρ cannot be only normal baryonic matter (atoms are 99.9% baryons—protons and neutrons—by weight). These recent experiments put the amount of normal baryonic matter in the universe at only about 5% of the critical density. What is the other 95%? There is strong evidence for a significant amount of nonluminous matter in the universe referred to as **dark matter**, which acts normally under gravity, but does not absorb or radiate light. For example, observations of the rotation of galaxies suggest that they rotate as if they had considerably more mass than we can see. Recall from Chapter 6, Eq. 6–5, that for a satellite of mass m revolving around Earth (mass M)

$$m\frac{v^2}{r} = G\frac{mM}{r^2}$$

and hence $v = \sqrt{GM/r}$. If we apply this equation to stars in a galaxy, we see that their speed depends on galactic mass. Observations show that stars farther from the galactic center revolve much faster than expected if there is only the pull of visible matter, suggesting a great deal of invisible matter. Similarly, observations of the motion of galaxies within clusters also suggest that they have considerably more mass than can be seen. Without dark matter, galaxies and stars probably would not have formed and would not exist; it would seem to hold the universe together. But what might this nonluminous matter in the universe be? We don't know yet. But we hope to find out soon. It cannot be made of ordinary (baryonic) matter, so it must consist of some other sort of elementary particle, perhaps created at a very early time. Perhaps it is a supersymmetric particle (Section 43–12), maybe the lightest one. We are anxiously awaiting details both from particle accelerators and the cosmos.

Dark matter makes up about 20% of the mass–energy of the universe, according to the latest observations and models. Thus the total mass–energy is 20% dark matter plus 5% baryons for a total of about 25%, which does not bring ρ up to ρ_c. What is the other 75%? We are not sure about that either, but we have given it a name: "dark energy."

Dark Energy—Cosmic Acceleration

In 1998, just before the turn of the millennium, cosmologists received a huge surprise. Gravity was assumed to be the predominant force on a large scale in the universe, and it was thought that the expansion of the universe ought to be slowing down in time because gravity acts as an attractive force between objects. But measurements of type Ia supernovae (SNIa, our best standard candles—see Section 44–3) unexpectedly showed that very distant (high z) SNIa's were dimmer than expected. That is, given their great distance d as determined from their low brightness, their speed v as determined from the measured z was less than expected according to Hubble's law. This result suggests that nearer galaxies are moving away from us relatively faster than those very distant ones, meaning the expansion of the universe in more recent epochs has sped up. This **acceleration** in the expansion of the universe (in place of the expected deceleration due to gravitational attraction between masses) seems to have begun roughly 5 billion years ago (5 Gyr, which would be 8 to 9 Gyr after the Big Bang).

What could be causing the universe to accelerate in its expansion, against the attractive force of gravity? Does our understanding of gravity need to be revised? We don't yet know the answers to these questions. There are several speculations. Somehow there seems to be a long-range *repulsive* effect on space, like a negative gravity, causing objects to speed away from each other ever faster. Whatever it is, it has been given the name **dark energy**. Many scientists say dark energy is the biggest mystery facing physical science today.

One idea is a sort of quantum field given the name **quintessence**. Another possibility suggests an energy latent in space itself (**vacuum energy**) and relates to an aspect of General Relativity known as the **cosmological constant** (symbol Λ). When Einstein developed his equations, he found that they offered no solutions for a static universe. In those days (1917) it was thought the universe was static—unchanging and everlasting. Einstein added an arbitrary constant to his equations to provide solutions for a static universe. A decade later, when Hubble showed us an expanding universe, Einstein discarded his cosmological constant as no longer needed $(\Lambda = 0)$. But today, measurements are consistent with dark energy being due to a nonzero cosmological constant, although further measurements are needed to see subtle differences among theories.

There is increasing evidence that the effects of some form of dark energy are very real. Observations of the CMB, supernovae, and large-scale structure (Section 44–10) agree well with theories and computer models when they input dark energy as providing 75% of the mass–energy in the universe, and when the total mass–energy density equals the critical density ρ_c.

Today's best estimate of how the mass–energy in the universe is distributed is approximately (Fig. 44–33):

75% dark energy

25% matter, subject to the known gravitational force.

Of this 25%, about

20% is dark matter

5% is baryons (what atoms are made of); of this 5% only $\frac{1}{10}$ is readily visible matter—stars and galaxies (that is, 0.5% of the total); the other $\frac{9}{10}$ of ordinary matter, which is not visible, is mainly gaseous plasma.

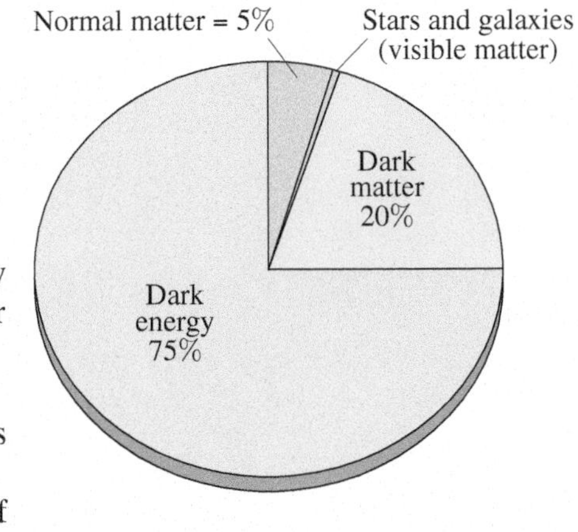

FIGURE 44–33 Portions of total mass–energy in the universe.

It is remarkable that only 0.5% of all the mass–energy in the universe is visible as stars and galaxies.

The idea that the universe is dominated by completely unknown forms of energy seems bizarre. Nonetheless, the ability of our present model to precisely explain observations of the CMB anisotropy, cosmic expansion, and large-scale structure (next Section) presents a compelling case.

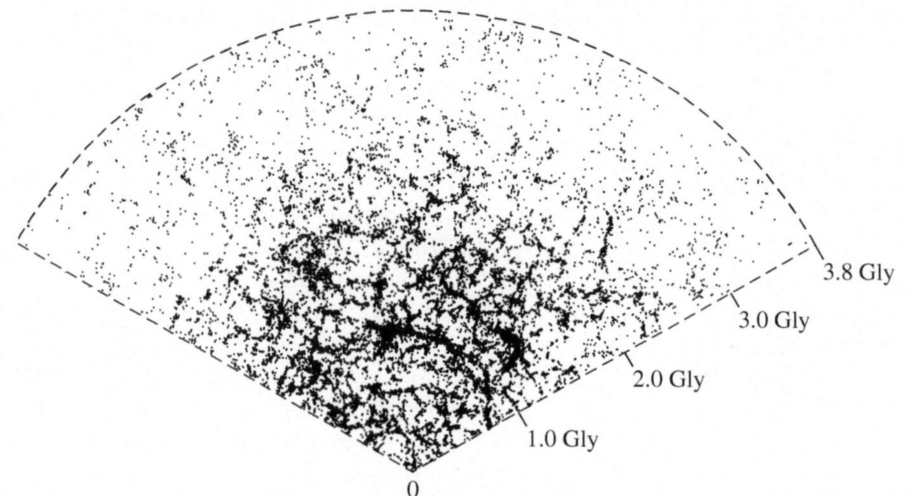

FIGURE 44–34 Distribution of some 50,000 galaxies in a 2.5° slice through almost half of the sky above the equator, as measured by the Sloan Digital Sky Survey (SDSS). Each dot represents a galaxy. The distance from us is obtained from the redshift and Hubble's law, and is given in units of 10^9 light-years (Gly). At greater distances, fewer galaxies are bright enough to be detected, thus resulting in an apparent thinning out of galaxies. The point 0 represents us, our observation point. Note the "walls" and "voids" of galaxies.

44–10 Large-Scale Structure of the Universe

The beautiful WMAP pictures of the sky (Fig. 44–27 and Chapter-Opening Photo) show small but significant inhomogeneities in the temperature of the CMB. These anisotropies reflect compressions and expansions in the primordial plasma just before decoupling, from which stars, galaxies, and clusters of galaxies formed. Analysis of the irregularities in the CMB using mammoth computer simulations predict a large-scale distribution of galaxies very similar to what is seen today (Fig. 44–34). These simulations are very successful if they contain dark energy and dark matter; and the dark matter needs to be *cold* (slow speed—think of Eq. 18–4, $\frac{1}{2}m\overline{v^2} = \frac{3}{2}kT$ where T is temperature), rather than "hot" dark matter such as neutrinos which move at or very near the speed of light. Indeed, the modern **cosmological model** is called the **ΛCDM** model, where lambda (Λ) stands for the cosmological constant, and CDM is **cold dark matter**.

Cosmologists have gained substantial confidence in this cosmological model from such a precise fit between observations and theory. They can also extract very precise values for cosmological parameters which previously were only known with low accuracy. The CMB is such an important cosmological observable that every effort is being made to extract all of the information it contains. A new generation of ground, balloon, and satellite experiments will observe the CMB with greater resolution and sensitivity. They may detect interaction of **gravity waves** (produced in the inflationary epoch) with the CMB and thereby provide direct evidence for cosmic inflation, and also provide information about elementary particle physics at energies far beyond the reach of man-made accelerators.

44–11 Finally . . .

When we look up into the night sky, we see stars; and with the best telescopes, we see galaxies and the exotic objects we discussed earlier, including rare supernovae. But even with our best instruments we do not see the processes going on inside stars and supernovae that we hypothesized (and believe). We are dependent on brilliant theorists who come up with viable theories and verifiable models. We depend on complicated computer models whose parameters are varied until the outputs compare favorably with our observations and analyses of WMAP and other experiments. And we now have a surprisingly precise idea about some aspects of our universe: it is flat, it is about 14 billion years old, it contains only 5% "normal" baryonic matter (for atoms), and so on.

The questions raised by cosmology are difficult and profound, and may seem removed from everyday "reality." We can always say, "the Sun is shining, it's going to burn on for an unimaginably long time, all is well." Nonetheless, the questions of cosmology are deep ones that fascinate the human intellect. One aspect that is especially intriguing is this: calculations on the formation and evolution of the universe have been performed that deliberately varied the values—just slightly—of certain fundamental physical constants. The result? A universe in which life as we know it could not exist. [For example, if the difference in mass between proton and neutron were zero, or small (less than the mass of the electron, $0.511\ \mathrm{MeV}/c^2$), there would be no atoms: electrons would be captured by protons never to be freed again.] Such results have contributed to a philosophical idea called the **Anthropic principle**, which says that if the universe were even a little different than it is, we could not be here. We physicists are trying to find out if there are some undiscovered fundamental laws that determined those conditions that allowed us to exist. A poet might say that the universe is exquisitely tuned, almost as if to accommodate us.

Summary

The night sky contains myriads of stars including those in the Milky Way, which is a "side view" of our **Galaxy** looking along the plane of the disk. Our Galaxy includes over 10^{11} stars. Beyond our Galaxy are billions of other galaxies.

Astronomical distances are measured in **light-years** ($1\ \mathrm{ly} \approx 10^{13}\ \mathrm{km}$). The nearest star is about 4 ly away and the nearest large galaxy is 2 million ly away. Our Galactic disk has a diameter of about 100,000 ly. Distances are often specified in **parsecs**, where 1 parsec = 3.26 ly.

Stars are believed to begin life as collapsing masses of gas (protostars), largely hydrogen. As they contract, they heat up (potential energy is transformed to kinetic energy). When the temperature reaches about 10 million degrees, nuclear fusion begins and forms heavier elements (**nucleosynthesis**), mainly helium at first. The energy released during these reactions heats the gas so its outward pressure balances the inward gravitational force, and the young star stabilizes as a **main-sequence** star. The tremendous luminosity of stars comes from the energy released during these thermonuclear reactions. After billions of years, as helium is collected in the core and hydrogen is used up, the core contracts and heats further. The envelope expands and cools, and the star becomes a **red giant** (larger diameter, redder color).

The next stage of stellar evolution depends on the mass of the star, which may have lost much of its original mass as its outer envelope escaped into space. Stars of residual mass less than about 1.4 solar masses cool further and become **white dwarfs**, eventually fading and going out altogether. Heavier stars contract further due to their greater gravity: the density approaches nuclear density, the huge pressure forces electrons to combine with protons to form neutrons, and the star becomes essentially a huge nucleus of neutrons. This is a **neutron star**, and the energy released from its final core collapse is believed to produce **supernova** explosions. If the star is very massive, it may contract even further and form a **black hole**, which is so dense that no matter or light can escape from it.

In the **general theory of relativity**, the **equivalence principle** states that an observer cannot distinguish acceleration from a gravitational field. Said another way, gravitational and inertial masses are the same. The theory predicts gravitational bending of light rays to a degree consistent with experiment. Gravity is treated as a curvature in space and time, the curvature being greater near massive objects. The universe as a whole may be curved. With sufficient mass, the curvature of the universe would be positive, and the universe is *closed* and *finite*; otherwise, it would be *open* and *infinite*. Today we believe the universe is **flat**.

Distant galaxies display a **redshift** in their spectral lines, originally interpreted as a Doppler shift. The universe seems to be **expanding**, its galaxies racing away from each other at speeds (v) proportional to the distance (d) between them:

$$v = Hd, \tag{44–4}$$

which is known as **Hubble's law** (H is the **Hubble parameter**). This expansion of the universe suggests an explosive origin, the **Big Bang**, which occurred about 13.7 billion years ago. It is not like an ordinary explosion, but rather an expansion of space itself.

The **cosmological principle** assumes that the universe, on a large scale, is homogeneous and isotropic.

Important evidence for the Big Bang model of the universe was the discovery of the **cosmic microwave background** radiation (CMB), which conforms to a blackbody radiation curve at a temperature of 2.725 K.

The **Standard Model** of the Big Bang provides a possible scenario as to how the universe developed as it expanded and cooled after the Big Bang. Starting at 10^{-43} seconds after the Big Bang, according to this model, there were a series of **phase transitions** during which previously unified forces of nature "condensed out" one by one. The **inflationary scenario** assumes that during one of these phase transitions, the universe underwent a brief but rapid exponential expansion. Until about 10^{-35} s, there was no distinction between quarks and leptons. Shortly thereafter, quarks were **confined** into hadrons (the **hadron era**). About 10^{-4} s after the Big Bang, the majority of hadrons disappeared, having combined with anti-hadrons, producing photons, leptons, and energy, leaving mainly photons and leptons to freely move, thus introducing the **lepton era**. By the time the universe was about 10 s old, the electrons too had mostly disappeared, having combined with their antiparticles; the universe was **radiation-dominated**. A couple of minutes later, nucleosynthesis began, but lasted only a few minutes. It then took almost four hundred thousand years before the universe was cool enough for electrons to combine with nuclei to form atoms (**recombination**). Photons, up to then continually being scattered off of free electrons, could now move freely—they were **decoupled** from matter and the universe became transparent. The background radiation had expanded and cooled so much that its total energy became less than the energy in matter, and **matter dominated** increasingly over radiation. Then stars and galaxies formed, producing a universe not much different than it is today—some 14 billion years later.

Recent observations indicate that the universe is flat, that it contains an as-yet unknown type of **dark matter**, and that it is dominated by a mysterious **dark energy** which exerts a sort of negative gravity causing the expansion of the universe to accelerate. The total contributions of baryonic (normal) matter, dark matter, and dark energy sum up to the **critical density**.

Questions

1. The Milky Way was once thought to be "murky" or "milky" but is now considered to be made up of point sources. Explain.
2. A star is in equilibrium when it radiates at its surface all the energy generated in its core. What happens when it begins to generate more energy than it radiates? Less energy? Explain.
3. Describe a red giant star. List some of its properties.
4. Select a point on the H–R diagram. Mark several directions away from this point. Now describe the changes that would take place in a star moving in each of these directions.
5. Does the H–R diagram reveal anything about the core of a star?
6. Why do some stars end up as white dwarfs, and others as neutron stars or black holes?
7. Can we tell, by looking at the population on the H–R diagram, that hotter main-sequence stars have shorter lives? Explain.
8. If you were measuring star parallaxes from the Moon instead of Earth, what corrections would you have to make? What changes would occur if you were measuring parallaxes from Mars?
9. *Cepheid variable* stars change in luminosity with a typical period of several days. The period has been found to have a definite relationship with the intrinsic luminosity of the star. How could these stars be used to measure the distance to galaxies?
10. What is a geodesic? What is its role in General Relativity?
11. If it were discovered that the redshift of spectral lines of galaxies was due to something other than expansion, how might our view of the universe change? Would there be conflicting evidence? Discuss.
12. All galaxies appear to be moving away from us. Are we therefore at the center of the universe? Explain.
13. If you were located in a galaxy near the boundary of our observable universe, would galaxies in the direction of the Milky Way appear to be approaching you or receding from you? Explain.
14. Compare an explosion on Earth to the Big Bang. Consider such questions as: Would the debris spread at a higher speed for more distant particles, as in the Big Bang? Would the debris come to rest? What type of universe would this correspond to, open or closed?
15. If nothing, not even light, escapes from a black hole, then how can we tell if one is there?
16. What mass will give a Schwarzschild radius equal to that of the hydrogen atom in its ground state?
17. The Earth's age is often given as about 4 billion years. Find that time on Fig. 44–30. People have lived on Earth on the order of a million years. Where is that on Fig. 44–30?
18. Explain what the 2.7-K cosmic microwave background radiation is. Where does it come from? Why is its temperature now so low?
19. Why were atoms, as opposed to bare nuclei, unable to exist until hundreds of thousands of years after the Big Bang?
20. (*a*) Why are type Ia supernovae so useful for determining the distances of galaxies? (*b*) How are their distances actually measured?
21. Explain why the CMB radiation should not be that of a perfect blackbody. (The deviations from a blackbody spectrum are slightly less than one part in 10^4.)
22. Under what circumstances would the universe eventually collapse in on itself?
23. When stable nuclei first formed, about 3 minutes after the Big Bang, there were about 7 times more protons than neutrons. Explain how this leads to a ratio of the mass of hydrogen to the mass of helium of 3:1. This is about the actual ratio observed in the universe.
24. (*a*) Why did astronomers expect that the expansion rate of the universe would be decreasing (decelerating) with time? (*b*) How, in principle, could astronomers hope to determine whether the universe used to expand faster than it does now?

Problems

44–1 to 44–3 Stars, Galaxies, Stellar Evolution, Distances

1. (I) The parallax angle of a star is 0.00029°. How far away is the star?
2. (I) A star exhibits a parallax of 0.27 seconds of arc. How far away is it?
3. (I) What is the parallax angle for a star that is 65 ly away? How many parsecs is this?
4. (I) A star is 56 pc away. What is its parallax angle? State (*a*) in seconds of arc, and (*b*) in degrees.
5. (I) If one star is twice as far away from us as a second star, will the parallax angle of the farther star be greater or less than that of the nearer star? By what factor?
6. (II) A star is 85 pc away. How long does it take for its light to reach us?
7. (II) What is the relative brightness of the Sun as seen from Jupiter, as compared to its brightness from Earth? (Jupiter is 5.2 times farther from the Sun than the Earth is.)
8. (II) We saw earlier (Chapter 19) that the rate energy reaches the Earth from the Sun (the "solar constant") is about $1.3 \times 10^3\ \mathrm{W/m^2}$. What is (*a*) the apparent brightness b of the Sun, and (*b*) the intrinsic luminosity L of the Sun?
9. (II) When our Sun becomes a red giant, what will be its average density if it expands out to the orbit of Mercury (6×10^{10} m, from the Sun)?
10. (II) Estimate the angular width that our Galaxy would subtend if observed from the nearest galaxy to us (Table 44–1). Compare to the angular width of the Moon from Earth.
11. (II) Calculate the Q-values for the He burning reactions of Eq. 44–2. (The mass of the very unstable ${}^{8}_{4}\mathrm{Be}$ is 8.005305 u.)
12. (II) When our Sun becomes a white dwarf, it is expected to be about the size of the Moon. What angular width would it subtend from the present distance to Earth?
13. (II) Calculate the density of a white dwarf whose mass is equal to the Sun's and whose radius is equal to the Earth's. How many times larger than Earth's density is this?

14. (II) A neutron star whose mass is 1.5 solar masses has a radius of about 11 km. Calculate its average density and compare to that for a white dwarf (Problem 13) and to that of nuclear matter.

15. (III) Stars located in a certain cluster are assumed to be about the same distance from us. Two such stars have spectra that peak at $\lambda_1 = 470\,\text{nm}$ and $\lambda_2 = 720\,\text{nm}$, and the ratio of their apparent brightness is $b_1/b_2 = 0.091$. Estimate their relative sizes (give ratio of their diameters) using Wien's law and the Stefan-Boltzmann equation, Eq. 19–17.

16. (III) Suppose two stars of the same apparent brightness b are also believed to be the same size. The spectrum of one star peaks at 750 nm whereas that of the other peaks at 450 nm. Use Wien's law and the Stefan-Boltzmann equation (Eq. 19–17) to estimate their relative distances from us.

44–4 General Relativity, Gravity and Curved Space

17. (I) Show that the Schwarzschild radius for a star with mass equal to that of Earth is 8.9 mm.

18. (II) What is the Schwarzschild radius for a typical galaxy (like ours)?

19. (II) What is the maximum sum-of-the-angles for a triangle on a sphere?

20. (II) Calculate the escape velocity, using Newtonian mechanics, from an object that has collapsed to its Schwarzschild radius.

21. (II) What is the apparent deflection of a light beam in an elevator (Fig. 44–13) which is 2.4 m wide if the elevator is accelerating downward at $9.8\,\text{m/s}^2$?

44–5 Redshift, Hubble's Law

22. (I) The redshift of a galaxy indicates a velocity of 1850 km/s. How far away is it?

23. (I) If a galaxy is traveling away from us at 1.5% of the speed of light, roughly how far away is it?

24. (II) A galaxy is moving away from Earth. The "blue" hydrogen line at 434 nm emitted from the galaxy is measured on Earth to be 455 nm. (*a*) How fast is the galaxy moving? (*b*) How far is it from Earth?

25. (II) Estimate the wavelength shift for the 656-nm line in the Balmer series of hydrogen emitted from a galaxy whose distance from us is (*a*) 7.0×10^6 ly, (*b*) 7.0×10^7 ly.

26. (II) If an absorption line of calcium is normally found at a wavelength of 393.4 nm in a laboratory gas, and you measure it to be at 423.4 nm in the spectrum of a galaxy, what is the approximate distance to the galaxy?

27. (II) What is the speed of a galaxy with $z = 0.060$?

28. (II) What would be the redshift parameter z for a galaxy traveling away from us at $v = 0.075\,c$?

29. (II) Starting from Eq. 44–3, show that the Doppler shift in wavelength is $\Delta\lambda/\lambda_{\text{rest}} \approx v/c$ (Eq. 44–6) for $v \ll c$. [*Hint*: Use the binomial expansion.]

30. (II) Radiotelescopes are designed to observe 21-cm waves emitted by atomic hydrogen gas. A signal from a distant radio-emitting galaxy is found to have a wavelength that is 0.10 cm longer than the normal 21-cm wavelength. Estimate the distance to this galaxy.

44–6 to 44–8 The Big Bang, CMB, Universe Expansion

31. (I) Calculate the wavelength at the peak of the blackbody radiation distribution at 2.7 K using Wien's law.

32. (II) Calculate the peak wavelength of the CMB at 1.0 s after the birth of the universe. In what part of the EM spectrum is this radiation?

33. (II) The critical density for closure of the universe is $\rho_c \approx 10^{-26}\,\text{kg/m}^3$. State ρ_c in terms of the average number of nucleons per cubic meter.

34. (II) The scale factor of the universe (average distance between galaxies) at any given time is believed to have been inversely proportional to the absolute temperature. Estimate the size of the universe, compared to today, at (*a*) $t = 10^6$ yr, (*b*) $t = 1$ s, (*c*) $t = 10^{-6}$ s, and (*d*) $t = 10^{-35}$ s.

35. (II) At approximately what time had the universe cooled below the threshold temperature for producing (*a*) kaons ($M \approx 500\,\text{MeV}/c^2$), (*b*) Υ ($M \approx 9500\,\text{MeV}/c^2$), and (*c*) muons ($M \approx 100\,\text{MeV}/c^2$)?

44–9 Dark Matter, Dark Energy

36. (II) Only about 5% of the energy in the universe is composed of baryonic matter. (*a*) Estimate the average density of baryonic matter in the observable universe with a radius of 14 billion light-years that contains 10^{11} galaxies, each with about 10^{11} stars like our Sun. (*b*) Estimate the density of dark matter in the universe.

General Problems

37. The evolution of stars, as discussed in Section 44–2, can lead to a white dwarf, a neutron star, or even a black hole, depending on the mass. (*a*) Referring to Sections 44–2 and 44–4, give the radius of (i) a white dwarf of 1 solar mass, (ii) a neutron star of 1.5 solar masses, and (iii) a black hole of 3 solar masses. (*b*) Express these three radii as ratios ($r_{\text{i}} : r_{\text{ii}} : r_{\text{iii}}$).

38. Use conservation of angular momentum to estimate the angular velocity of a neutron star which has collapsed to a diameter of 16 km, from a star whose radius was equal to that of our Sun (7×10^8 m). Assume its mass is 1.5 times that of the Sun, and that it rotated (like our Sun) about once a month.

39. By what factor does the rotational kinetic energy change when the star in Problem 38 collapses to a neutron star?

40. Assume that the nearest stars to us have an intrinsic luminosity about the same as the Sun's. Their apparent brightness, however, is about 10^{11} times fainter than the Sun. From this, estimate the distance to the nearest stars. (Newton did this calculation, although he made a numerical error of a factor of 100.)

41. Suppose that three main-sequence stars could undergo the three changes represented by the three arrows, A, B, and C, in the H–R diagram of Fig. 44–35. For each case, describe the changes in temperature, intrinsic luminosity, and size.

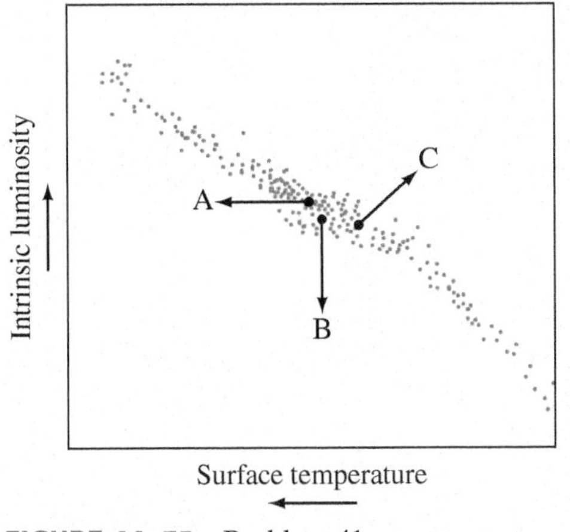

FIGURE 44–35 Problem 41.

42. A certain pulsar, believed to be a neutron star of mass 1.5 times that of the Sun, with diameter 16 km, is observed to have a rotation speed of 1.0 rev/s. If it loses rotational kinetic energy at the rate of 1 part in 10^9 per day, which is all transformed into radiation, what is the power output of the star?

43. The nearest large galaxy to our Galaxy is about 2×10^6 ly away. If both galaxies have a mass of 3×10^{41} kg, with what gravitational force does each galaxy attract the other?

44. Estimate what neutrino mass (in eV/c^2) would provide the critical density to close the universe. Assume the neutrino density is, like photons, about 10^9 times that of nucleons, and that nucleons make up only (*a*) 2% of the mass needed, or (*b*) 5% of the mass needed.

45. Two stars, whose spectra peak at 660 nm and 480 nm, respectively, both lie on the main sequence. Use Wien's law, the Stefan-Boltzmann equation, and the H–R diagram (Fig. 44–6) to estimate the ratio of their diameters.

46. (*a*) In order to measure distances with parallax at 100 parsecs, what minimum angular resolution (in degrees) is needed? (*b*) What diameter mirror or lens would be needed?

47. What is the temperature that corresponds to 1.96-TeV collisions at the Fermilab collider? To what era in cosmological history does this correspond? [*Hint*: See Fig. 44–30.]

48. Astronomers have measured the rotation of gas around a possible supermassive black hole of about 2 billion solar masses at the center of a galaxy. If the radius from the galactic center to the gas clouds is 68 light-years, estimate the value of z.

49. In the later stages of stellar evolution, a star (if massive enough) will begin fusing carbon nuclei to form, for example, magnesium:

$$ {}^{12}_{6}\mathrm{C} + {}^{12}_{6}\mathrm{C} \rightarrow {}^{24}_{12}\mathrm{Mg} + \gamma. $$

(*a*) How much energy is released in this reaction (see Appendix F)? (*b*) How much kinetic energy must each carbon nucleus have (assume equal) in a head-on collision if they are just to "touch" (use Eq. 41–1) so that the strong force can come into play? (*c*) What temperature does this kinetic energy correspond to?

50. Consider the reaction

$$ {}^{16}_{8}\mathrm{O} + {}^{16}_{8}\mathrm{O} \rightarrow {}^{28}_{14}\mathrm{Si} + {}^{4}_{2}\mathrm{He}, $$

and answer the same questions as in Problem 49.

51. Calculate the Schwarzschild radius using a semi-classical (Newtonian) gravitational theory, by calculating the minimum radius R for a sphere of mass M such that a photon can escape from the surface. (General Relativity gives $R = 2GM/c^2$.)

52. How large would the Sun be if its density equaled the critical density of the universe, $\rho_c \approx 10^{-26}\,\mathrm{kg/m^3}$? Express your answer in light-years and compare with the Earth–Sun distance and the diameter of our Galaxy.

53. The Large Hadron Collider in Geneva, Switzerland, can collide two beams of protons at an energy of 14 TeV. Estimate the time after the Big Bang probed by this energy.

54. (*a*) Use special relativity and Newton's law of gravitation to show that a photon of mass $m = E/c^2$ just grazing the Sun will be deflected by an angle $\Delta\theta$ given by

$$ \Delta\theta = \frac{2GM}{c^2 R} $$

where G is the gravitational constant, R and M are the radius and mass of the Sun, and c is the speed of light. (*b*) Put in values and show $\Delta\theta = 0.87''$. (General Relativity predicts an angle twice as large, 1.74''.)

55. Astronomers use an **apparent magnitude** (m) scale to describe apparent brightness. It uses a logarithmic scale, where a higher number corresponds to a less bright star. (For example, the Sun has magnitude −27, whereas most stars have positive magnitudes.) On this scale, a change in apparent magnitude by +5 corresponds to a decrease in apparent brightness by a factor of 100. If Venus has an apparent magnitude of −4.4, whereas Sirius has an apparent magnitude of −1.4, which is brighter? What is the ratio of the apparent brightness of these two objects?

56. Estimate the radius of a white dwarf whose mass is equal to that of the Sun by the following method, assuming there are N nucleons and $\frac{1}{2}N$ electrons (why $\frac{1}{2}$?): (*a*) Use Fermi-Dirac statistics (Section 40–6) to show that the total energy of all the electrons is

$$ E_e = \frac{3}{5}\left(\frac{1}{2}N\right)\frac{h^2}{8m_e}\left(\frac{3}{\pi}\frac{N}{2V}\right)^{\frac{2}{3}}. $$

[*Hint*: See Eqs. 40–12 and 40–13; we assume electrons fill energy levels from 0 up to the Fermi energy.] (*b*) The nucleons contribute to the total energy mainly via the gravitational force (note that the Fermi energy for nucleons is negligible compared to that for electrons—why?). Use a gravitational form of Gauss's law to show that the total gravitational potential energy of a uniform sphere of radius R is

$$ -\frac{3}{5}\frac{GM^2}{R}, $$

by considering the potential energy of a spherical shell of radius r due only to the mass inside it (why?) and integrate from $r = 0$ to $r = R$. (See also Appendix D.) (*c*) Write the total energy as a sum of these two terms, and set $dE/dR = 0$ to find the equilibrium radius, and evaluate it for a mass equal to the Sun's (2.0×10^{30} kg).

57. Determine the radius of a neutron star using the same argument as in Problem 56 but for N neutrons only. Show that the radius of a neutron star, of 1.5 solar masses, is about 11 km.

58. Use *dimensional analysis* with the fundamental constants c, G, and $\hbar$ to estimate the value of the so-called *Planck time*. It is thought that physics as we know it can say nothing about the universe before this time.

Answers to Exercises

A: Ourselves, 2 years ago.

B: 600 ly (estimating L from Fig. 44–6 as $L \approx 8 \times 10^{26}$ W; note that on a log scale, 6000 K is closer to 7000 K than it is to 5000 K).

C: 6 km.

D: (*a*); not the usual R^3, but R: see formula for the Schwarzschild radius.

E: (*c*).

APPENDIX A Mathematical Formulas

A–1 Quadratic Formula

If $$ax^2 + bx + c = 0$$

then $$x = \frac{-b \pm \sqrt{b^2 - 4ac}}{2a}$$

A–2 Binomial Expansion

$$(1 \pm x)^n = 1 \pm nx + \frac{n(n-1)}{2!}x^2 \pm \frac{n(n-1)(n-2)}{3!}x^3 + \cdots$$

$$(x + y)^n = x^n\left(1 + \frac{y}{x}\right)^n = x^n\left(1 + n\frac{y}{x} + \frac{n(n-1)}{2!}\frac{y^2}{x^2} + \cdots\right)$$

A–3 Other Expansions

$$e^x = 1 + x + \frac{x^2}{2!} + \frac{x^3}{3!} + \cdots$$

$$\ln(1 + x) = x - \frac{x^2}{2} + \frac{x^3}{3} - \frac{x^4}{4} + \cdots$$

$$\sin\theta = \theta - \frac{\theta^3}{3!} + \frac{\theta^5}{5!} - \cdots$$

$$\cos\theta = 1 - \frac{\theta^2}{2!} + \frac{\theta^4}{4!} - \cdots$$

$$\tan\theta = \theta + \frac{\theta^3}{3} + \frac{2}{15}\theta^5 + \cdots \qquad |\theta| < \frac{\pi}{2}$$

In general: $$f(x) = f(0) + \left(\frac{df}{dx}\right)_0 x + \left(\frac{d^2f}{dx^2}\right)_0 \frac{x^2}{2!} + \cdots$$

A–4 Exponents

$$(a^n)(a^m) = a^{n+m}$$

$$(a^n)(b^n) = (ab)^n$$

$$(a^n)^m = a^{nm}$$

$$\frac{1}{a^n} = a^{-n}$$

$$a^n a^{-n} = a^0 = 1$$

$$a^{\frac{1}{2}} = \sqrt{a}$$

A–5 Areas and Volumes

Object	Surface area	Volume
Circle, radius r	πr^2	—
Sphere, radius r	$4\pi r^2$	$\frac{4}{3}\pi r^3$
Right circular cylinder, radius r, height h	$2\pi r^2 + 2\pi rh$	$\pi r^2 h$
Right circular cone, radius r, height h	$\pi r^2 + \pi r\sqrt{r^2 + h^2}$	$\frac{1}{3}\pi r^2 h$

A–6 Plane Geometry

1. *Equal angles*:

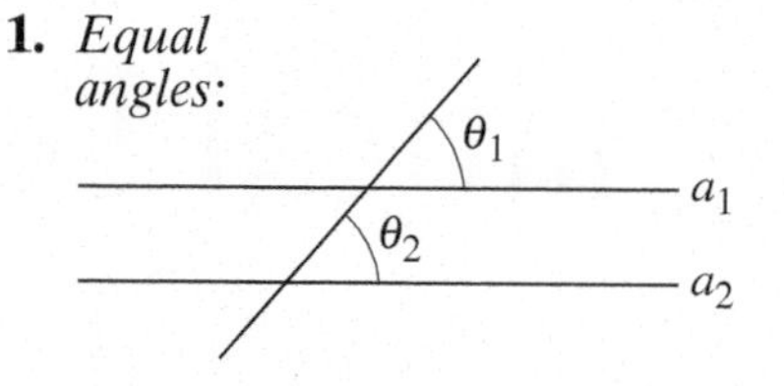

FIGURE A–1 If line a_1 is parallel to line a_2, then $\theta_1 = \theta_2$.

2. *Equal angles*:

FIGURE A–2 If $a_1 \perp a_2$ and $b_1 \perp b_2$, then $\theta_1 = \theta_2$.

3. The sum of the angles in any plane triangle is 180°.
4. *Pythagorean theorem*:

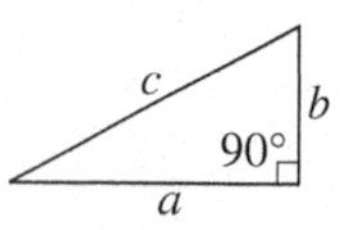

FIGURE A–3

In any right triangle (one angle = 90°) of sides a, b, and c:

$$a^2 + b^2 = c^2$$

where c is the length of the hypotenuse (opposite the 90° angle).

5. *Similar triangles*: Two triangles are said to be similar if all three of their angles are equal (in Fig. A–4, $\theta_1 = \phi_1, \theta_2 = \phi_2$, and $\theta_3 = \phi_3$). Similar triangles can have different sizes and different orientations.
   (*a*) Two triangles are similar if any two of their angles are equal. (This follows because the third angles must also be equal since the sum of the angles of a triangle is 180°.)
   (*b*) The ratios of corresponding sides of two similar triangles are equal (Fig. A–4):

$$\frac{a_1}{b_1} = \frac{a_2}{b_2} = \frac{a_3}{b_3}.$$

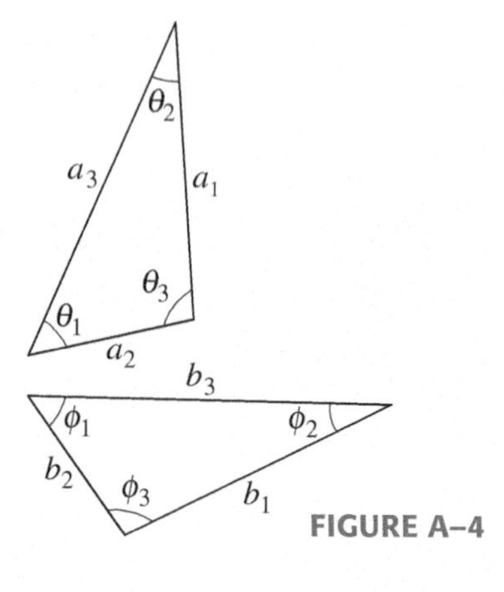

FIGURE A–4

6. *Congruent triangles*: Two triangles are congruent if one can be placed precisely on top of the other. That is, they are similar triangles and they have the same size. Two triangles are congruent if any of the following holds:
   (*a*) The three corresponding sides are equal.
   (*b*) Two sides and the enclosed angle are equal ("side-angle-side").
   (*c*) Two angles and the enclosed side are equal ("angle-side-angle").

A–7 Logarithms

Logarithms are defined in the following way:

$$\text{if } y = A^x, \text{ then } x = \log_A y.$$

That is, the logarithm of a number y to the base A is that number which, as the exponent of A, gives back the number y. For **common logarithms**, the base is 10, so

$$\text{if } y = 10^x, \text{ then } x = \log y.$$

The subscript 10 on $\log_{10}$ is usually omitted when dealing with common logs. Another important base is the exponential base $e = 2.718\ldots$, a natural number. Such logarithms are called **natural logarithms** and are written ln. Thus,

$$\text{if } y = e^x, \text{ then } x = \ln y.$$

For any number y, the two types of logarithm are related by

$$\ln y = 2.3026 \log y.$$

Some simple rules for logarithms are as follows:

$$\log(ab) = \log a + \log b, \tag{i}$$

which is true because if $a = 10^n$ and $b = 10^m$, then $ab = 10^{n+m}$. From the definition of logarithm, $\log a = n$, $\log b = m$, and $\log(ab) = n + m$; hence, $\log(ab) = n + m = \log a + \log b$. In a similar way, we can show that

$$\log\left(\frac{a}{b}\right) = \log a - \log b \qquad \textbf{(ii)}$$

and

$$\log a^n = n \log a. \qquad \textbf{(iii)}$$

These three rules apply to any kind of logarithm.

If you do not have a calculator that calculates logs, you can easily use a **log table**, such as the small one shown here (Table A–1): the number N whose log we want is given to two digits. The first digit is in the vertical column to the left, the second digit is in the horizontal row across the top. For example, Table A–1 tells us that $\log 1.0 = 0.000$, $\log 1.1 = 0.041$, and $\log 4.1 = 0.613$. Table A–1 does not include the decimal point. The Table gives logs for numbers between 1.0 and 9.9. For larger or smaller numbers, we use rule (i) above, $\log(ab) = \log a + \log b$. For example, $\log(380) = \log(3.8 \times 10^2) = \log(3.8) + \log(10^2)$. From the Table, $\log 3.8 = 0.580$; and from rule (iii) above $\log(10^2) = 2\log(10) = 2$, since $\log(10) = 1$. [This follows from the definition of the logarithm: if $10 = 10^1$, then $1 = \log(10)$.] Thus,

$$\begin{aligned}\log(380) &= \log(3.8) + \log(10^2)\\ &= 0.580 + 2\\ &= 2.580.\end{aligned}$$

Similarly,

$$\begin{aligned}\log(0.081) &= \log(8.1) + \log(10^{-2})\\ &= 0.908 - 2 = -1.092.\end{aligned}$$

The reverse process of finding the number N whose log is, say, 2.670, is called "taking the **antilogarithm**." To do so, we separate our number 2.670 into two parts, making the separation at the decimal point:

$$\begin{aligned}\log N = 2.670 &= 2 + 0.670\\ &= \log 10^2 + 0.670.\end{aligned}$$

We now look at Table A–1 to see what number has its log equal to 0.670; none does, so we must **interpolate**: we see that $\log 4.6 = 0.663$ and $\log 4.7 = 0.672$. So the number we want is between 4.6 and 4.7, and closer to the latter by $\frac{7}{9}$. Approximately we can say that $\log 4.68 = 0.670$. Thus

$$\begin{aligned}\log N &= 2 + 0.670\\ &= \log(10^2) + \log(4.68) = \log(4.68 \times 10^2),\end{aligned}$$

so $N = 4.68 \times 10^2 = 468$.

If the given logarithm is negative, say, -2.180, we proceed as follows:

$$\begin{aligned}\log N = -2.180 &= -3 + 0.820\\ &= \log 10^{-3} + \log 6.6 = \log 6.6 \times 10^{-3},\end{aligned}$$

so $N = 6.6 \times 10^{-3}$. Notice that we added to our given logarithm the next largest integer (3 in this case) so that we have an integer, plus a decimal number between 0 and 1.0 whose antilogarithm can be looked up in the Table.

TABLE A–1 Short Table of Common Logarithms

N	0.0	0.1	0.2	0.3	0.4	0.5	0.6	0.7	0.8	0.9
1	000	041	079	114	146	176	204	230	255	279
2	301	322	342	362	380	398	415	431	447	462
3	477	491	505	519	531	544	556	568	580	591
4	602	613	623	633	643	653	663	672	681	690
5	699	708	716	724	732	740	748	756	763	771
6	778	785	792	799	806	813	820	826	833	839
7	845	851	857	863	869	875	881	886	892	898
8	903	908	914	919	924	929	935	940	944	949
9	954	959	964	968	973	978	982	987	991	996

A–8 Vectors

Vector addition is covered in Sections 3–2 to 3–5.
Vector multiplication is covered in Sections 3–3, 7–2, and 11–2.

A–9 Trigonometric Functions and Identities

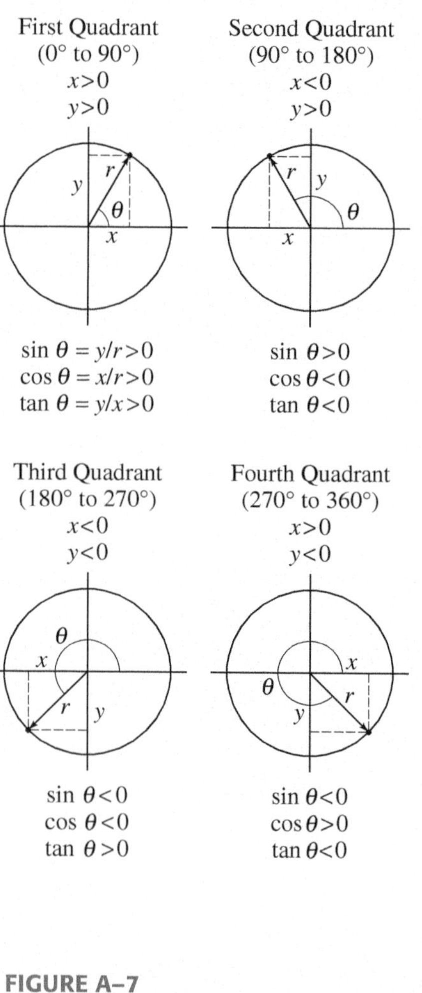

FIGURE A–5

FIGURE A–6

FIGURE A–7

The trigonometric functions are defined as follows (see Fig. A–5, o = side opposite, a = side adjacent, h = hypotenuse. Values are given in Table A–2):

$$\sin\theta = \frac{o}{h} \qquad \csc\theta = \frac{1}{\sin\theta} = \frac{h}{o}$$

$$\cos\theta = \frac{a}{h} \qquad \sec\theta = \frac{1}{\cos\theta} = \frac{h}{a}$$

$$\tan\theta = \frac{o}{a} = \frac{\sin\theta}{\cos\theta} \qquad \cot\theta = \frac{1}{\tan\theta} = \frac{a}{o}$$

and recall that

$$a^2 + o^2 = h^2 \qquad \text{[Pythagorean theorem].}$$

Figure A–6 shows the signs (+ or −) that cosine, sine, and tangent take on for angles θ in the four quadrants (0° to 360°). Note that angles are measured counterclockwise from the x axis as shown; negative angles are measured from *below* the x axis, clockwise: for example, $-30° = +330°$, and so on.

The following are some useful identities among the trigonometric functions:

$$\sin^2\theta + \cos^2\theta = 1$$

$$\sec^2\theta - \tan^2\theta = 1, \quad \csc^2\theta - \cot^2\theta = 1$$

$$\sin 2\theta = 2\sin\theta\cos\theta$$

$$\cos 2\theta = \cos^2\theta - \sin^2\theta = 2\cos^2\theta - 1 = 1 - 2\sin^2\theta$$

$$\tan 2\theta = \frac{2\tan\theta}{1 - \tan^2\theta}$$

$$\sin(A \pm B) = \sin A\cos B \pm \cos A\sin B$$

$$\cos(A \pm B) = \cos A\cos B \mp \sin A\sin B$$

$$\tan(A \pm B) = \frac{\tan A \pm \tan B}{1 \mp \tan A\tan B}$$

$$\sin(180° - \theta) = \sin\theta$$

$$\cos(180° - \theta) = -\cos\theta$$

$$\sin(90° - \theta) = \cos\theta$$

$$\cos(90° - \theta) = \sin\theta$$

$$\sin(-\theta) = -\sin\theta$$

$$\cos(-\theta) = \cos\theta$$

$$\tan(-\theta) = -\tan\theta$$

$$\sin\tfrac{1}{2}\theta = \sqrt{\frac{1-\cos\theta}{2}}, \quad \cos\tfrac{1}{2}\theta = \sqrt{\frac{1+\cos\theta}{2}}, \quad \tan\tfrac{1}{2}\theta = \sqrt{\frac{1-\cos\theta}{1+\cos\theta}}$$

$$\sin A \pm \sin B = 2\sin\left(\frac{A \pm B}{2}\right)\cos\left(\frac{A \mp B}{2}\right).$$

For any triangle (see Fig. A–7):

$$\frac{\sin\alpha}{a} = \frac{\sin\beta}{b} = \frac{\sin\gamma}{c} \qquad \text{[Law of sines]}$$

$$c^2 = a^2 + b^2 - 2ab\cos\gamma. \qquad \text{[Law of cosines]}$$

Values of sine, cosine, tangent are given in Table A–2.

TABLE A–2 Trigonometric Table: Numerical Values of Sin, Cos, Tan

Angle in Degrees	Angle in Radians	Sine	Cosine	Tangent	Angle in Degrees	Angle in Radians	Sine	Cosine	Tangent
0°	0.000	0.000	1.000	0.000					
1°	0.017	0.017	1.000	0.017	46°	0.803	0.719	0.695	1.036
2°	0.035	0.035	0.999	0.035	47°	0.820	0.731	0.682	1.072
3°	0.052	0.052	0.999	0.052	48°	0.838	0.743	0.669	1.111
4°	0.070	0.070	0.998	0.070	49°	0.855	0.755	0.656	1.150
5°	0.087	0.087	0.996	0.087	50°	0.873	0.766	0.643	1.192
6°	0.105	0.105	0.995	0.105	51°	0.890	0.777	0.629	1.235
7°	0.122	0.122	0.993	0.123	52°	0.908	0.788	0.616	1.280
8°	0.140	0.139	0.990	0.141	53°	0.925	0.799	0.602	1.327
9°	0.157	0.156	0.988	0.158	54°	0.942	0.809	0.588	1.376
10°	0.175	0.174	0.985	0.176	55°	0.960	0.819	0.574	1.428
11°	0.192	0.191	0.982	0.194	56°	0.977	0.829	0.559	1.483
12°	0.209	0.208	0.978	0.213	57°	0.995	0.839	0.545	1.540
13°	0.227	0.225	0.974	0.231	58°	1.012	0.848	0.530	1.600
14°	0.244	0.242	0.970	0.249	59°	1.030	0.857	0.515	1.664
15°	0.262	0.259	0.966	0.268	60°	1.047	0.866	0.500	1.732
16°	0.279	0.276	0.961	0.287	61°	1.065	0.875	0.485	1.804
17°	0.297	0.292	0.956	0.306	62°	1.082	0.883	0.469	1.881
18°	0.314	0.309	0.951	0.325	63°	1.100	0.891	0.454	1.963
19°	0.332	0.326	0.946	0.344	64°	1.117	0.899	0.438	2.050
20°	0.349	0.342	0.940	0.364	65°	1.134	0.906	0.423	2.145
21°	0.367	0.358	0.934	0.384	66°	1.152	0.914	0.407	2.246
22°	0.384	0.375	0.927	0.404	67°	1.169	0.921	0.391	2.356
23°	0.401	0.391	0.921	0.424	68°	1.187	0.927	0.375	2.475
24°	0.419	0.407	0.914	0.445	69°	1.204	0.934	0.358	2.605
25°	0.436	0.423	0.906	0.466	70°	1.222	0.940	0.342	2.747
26°	0.454	0.438	0.899	0.488	71°	1.239	0.946	0.326	2.904
27°	0.471	0.454	0.891	0.510	72°	1.257	0.951	0.309	3.078
28°	0.489	0.469	0.883	0.532	73°	1.274	0.956	0.292	3.271
29°	0.506	0.485	0.875	0.554	74°	1.292	0.961	0.276	3.487
30°	0.524	0.500	0.866	0.577	75°	1.309	0.966	0.259	3.732
31°	0.541	0.515	0.857	0.601	76°	1.326	0.970	0.242	4.011
32°	0.559	0.530	0.848	0.625	77°	1.344	0.974	0.225	4.331
33°	0.576	0.545	0.839	0.649	78°	1.361	0.978	0.208	4.705
34°	0.593	0.559	0.829	0.675	79°	1.379	0.982	0.191	5.145
35°	0.611	0.574	0.819	0.700	80°	1.396	0.985	0.174	5.671
36°	0.628	0.588	0.809	0.727	81°	1.414	0.988	0.156	6.314
37°	0.646	0.602	0.799	0.754	82°	1.431	0.990	0.139	7.115
38°	0.663	0.616	0.788	0.781	83°	1.449	0.993	0.122	8.144
39°	0.681	0.629	0.777	0.810	84°	1.466	0.995	0.105	9.514
40°	0.698	0.643	0.766	0.839	85°	1.484	0.996	0.087	11.43
41°	0.716	0.656	0.755	0.869	86°	1.501	0.998	0.070	14.301
42°	0.733	0.669	0.743	0.900	87°	1.518	0.999	0.052	19.081
43°	0.750	0.682	0.731	0.933	88°	1.536	0.999	0.035	28.636
44°	0.768	0.695	0.719	0.966	89°	1.553	1.000	0.017	57.290
45°	0.785	0.707	0.707	1.000	90°	1.571	1.000	0.000	∞

APPENDIX B

Derivatives and Integrals

B–1 Derivatives: General Rules

(See also Section 2–3.)

$$\frac{dx}{dx} = 1$$

$$\frac{d}{dx}[af(x)] = a\frac{df}{dx} \qquad [a = \text{constant}]$$

$$\frac{d}{dx}[f(x) + g(x)] = \frac{df}{dx} + \frac{dg}{dx}$$

$$\frac{d}{dx}[f(x)g(x)] = \frac{df}{dx}g + f\frac{dg}{dx}$$

$$\frac{d}{dx}[f(y)] = \frac{df}{dy}\frac{dy}{dx} \qquad [\text{chain rule}]$$

$$\frac{dx}{dy} = \frac{1}{\left(\frac{dy}{dx}\right)} \qquad \text{if } \frac{dy}{dx} \neq 0.$$

B–2 Derivatives: Particular Functions

$$\frac{da}{dx} = 0 \qquad [a = \text{constant}]$$

$$\frac{d}{dx}x^n = nx^{n-1}$$

$$\frac{d}{dx}\sin ax = a\cos ax$$

$$\frac{d}{dx}\cos ax = -a\sin ax$$

$$\frac{d}{dx}\tan ax = a\sec^2 ax$$

$$\frac{d}{dx}\ln ax = \frac{1}{x}$$

$$\frac{d}{dx}e^{ax} = ae^{ax}$$

B–3 Indefinite Integrals: General Rules

(See also Section 7–3.)

$$\int dx = x$$

$$\int af(x)\,dx = a\int f(x)\,dx \qquad [a = \text{constant}]$$

$$\int [f(x) + g(x)]\,dx = \int f(x)\,dx + \int g(x)\,dx$$

$$\int u\,dv = uv - \int v\,du \qquad [\text{integration by parts: see also B–6}]$$

B–4 Indefinite Integrals: Particular Functions

(An arbitrary constant can be added to the right side of each equation.)

$$\int a\,dx = ax \qquad [a = \text{constant}]$$

$$\int x^m\,dx = \frac{1}{m+1}x^{m+1} \qquad [m \neq -1]$$

$$\int \sin ax\,dx = -\frac{1}{a}\cos ax$$

$$\int \cos ax\,dx = \frac{1}{a}\sin ax$$

$$\int \tan ax\,dx = \frac{1}{a}\ln|\sec ax|$$

$$\int \frac{1}{x}\,dx = \ln x$$

$$\int e^{ax}\,dx = \frac{1}{a}e^{ax}$$

$$\int \frac{dx}{\sqrt{x^2 \pm a^2}} = \ln\left(x + \sqrt{x^2 \pm a^2}\right)$$

$$\int \frac{dx}{\sqrt{a^2 - x^2}} = \sin^{-1}\left(\frac{x}{a}\right) = -\cos^{-1}\left(\frac{x}{a}\right) \qquad [\text{if } x^2 \leq a^2]$$

$$\int \frac{dx}{(x^2 \pm a^2)^{\frac{3}{2}}} = \frac{\pm x}{a^2\sqrt{x^2 \pm a^2}}$$

$$\int \frac{x\,dx}{(x^2 \pm a^2)^{\frac{3}{2}}} = \frac{-1}{\sqrt{x^2 \pm a^2}}$$

$$\int \sin^2 ax\,dx = \frac{x}{2} - \frac{\sin 2ax}{4a}$$

$$\int xe^{-ax}\,dx = -\frac{e^{-ax}}{a^2}(ax + 1)$$

$$\int x^2e^{-ax}\,dx = -\frac{e^{-ax}}{a^3}(a^2x^2 + 2ax + 2)$$

$$\int \frac{dx}{x^2 + a^2} = \frac{1}{a}\tan^{-1}\frac{x}{a}$$

$$\int \frac{dx}{x^2 - a^2} = \frac{1}{2a}\ln\left(\frac{x-a}{x+a}\right) \qquad [x^2 > a^2]$$

$$= -\frac{1}{2a}\ln\left(\frac{a+x}{a-x}\right) \qquad [x^2 < a^2]$$

B–5 A Few Definite Integrals

$$\int_0^\infty x^n e^{-ax}\,dx = \frac{n!}{a^{n+1}}$$

$$\int_0^\infty e^{-ax^2}\,dx = \sqrt{\frac{\pi}{4a}}$$

$$\int_0^\infty xe^{-ax^2}\,dx = \frac{1}{2a}$$

$$\int_0^\infty x^2e^{-ax^2}\,dx = \sqrt{\frac{\pi}{16a^3}}$$

$$\int_0^\infty x^3e^{-ax^2}\,dx = \frac{1}{2a^2}$$

$$\int_0^\infty x^{2n}e^{-ax^2}\,dx = \frac{1\cdot3\cdot5\cdots(2n-1)}{2^{n+1}a^n}\sqrt{\frac{\pi}{a}}$$

B–6 Integration by Parts

Sometimes a difficult integral can be simplified by carefully choosing the functions u and v in the identity:

$$\int u\,dv = uv - \int v\,du. \qquad [\text{Integration by parts}]$$

This identity follows from the property of derivatives

$$\frac{d}{dx}(uv) = u\frac{dv}{dx} + v\frac{du}{dx}$$

or as differentials: $d(uv) = u\,dv + v\,du$.

For example $\int xe^{-x}\,dx$ can be integrated by choosing $u = x$ and $dv = e^{-x}\,dx$ in the "integration by parts" equation above:

$$\int xe^{-x}\,dx = (x)(-e^{-x}) + \int e^{-x}\,dx$$

$$= -xe^{-x} - e^{-x} = -(x+1)e^{-x}.$$

APPENDIX C More on Dimensional Analysis

An important use of dimensional analysis (Section 1–7) is to obtain the *form* of an equation: how one quantity depends on others. To take a concrete example, let us try to find an expression for the period T of a simple pendulum. First, we try to figure out what T could depend on, and make a list of these variables. It might depend on its length ℓ, on the mass m of the bob, on the angle of swing θ, and on the acceleration due to gravity, g. It might also depend on air resistance (we would use the viscosity of air), the gravitational pull of the Moon, and so on; but everyday experience suggests that the Earth's gravity is the major force involved, so we ignore the other possible forces. So let us assume that T is a function of ℓ, m, θ, and g, and that each of these factors is present to some power:

$$T = C\ell^w m^x \theta^y g^z.$$

C is a dimensionless constant, and w, x, y, and z are exponents we want to solve for. We now write down the dimensional equation (Section 1–7) for this relationship:

$$[T] = [L]^w[M]^x[L/T^2]^z.$$

Because θ has no dimensions (a radian is a length divided by a length—see Eq. 10–1a), it does not appear. We simplify and obtain

$$[T] = [L]^{w+z}[M]^x[T]^{-2z}$$

To have dimensional consistency, we must have

$$1 = -2z$$
$$0 = w + z$$
$$0 = x.$$

We solve these equations and find that $z = -\frac{1}{2}$, $w = \frac{1}{2}$, and $x = 0$. Thus our desired equation must be

$$T = C\sqrt{\ell/g}\, f(\theta), \qquad \textbf{(C–1)}$$

where $f(\theta)$ is some function of θ that we cannot determine using this technique. Nor can we determine in this way the dimensionless constant C. (To obtain C and f, we would have to do an analysis such as that in Chapter 14 using Newton's laws, which reveals that $C = 2\pi$ and $f \approx 1$ for small θ). But look what we *have* found, using only dimensional consistency. We obtained the form of the expression that relates the period of a simple pendulum to the major variables of the situation, ℓ and g (see Eq. 14–12c), and saw that it does not depend on the mass m.

How did we do it? And how useful is this technique? Basically, we had to use our intuition as to which variables were important and which were not. This is not always easy, and often requires a lot of insight. As to usefulness, the final result in our example could have been obtained from Newton's laws, as in Chapter 14. But in many physical situations, such a derivation from other laws cannot be done. In those situations, dimensional analysis can be a powerful tool.

In the end, any expression derived by the use of dimensional analysis (or by any other means, for that matter) must be checked against experiment. For example, in our derivation of Eq. C–1, we can compare the periods of two pendulums of different lengths, ℓ_1 and ℓ_2, whose amplitudes (θ) are the same. For, using Eq. C–1, we would have

$$\frac{T_1}{T_2} = \frac{C\sqrt{\ell_1/g}\, f(\theta)}{C\sqrt{\ell_2/g}\, f(\theta)} = \sqrt{\frac{\ell_1}{\ell_2}}.$$

Because C and $f(\theta)$ are the same for both pendula, they cancel out, so we can experimentally determine if the ratio of the periods varies as the ratio of the square roots of the lengths. This comparison to experiment checks our derivation, at least in part; C and $f(\theta)$ could be determined by further experiments.

APPENDIX

D Gravitational Force due to a Spherical Mass Distribution

In Chapter 6 we stated that the gravitational force exerted by or on a uniform sphere acts as if all the mass of the sphere were concentrated at its center, if the other object (exerting or feeling the force) is outside the sphere. In other words, the gravitational force that a uniform sphere exerts on a particle outside it is

$$F = G\frac{mM}{r^2}, \qquad [m \text{ outside sphere of mass } M]$$

where m is the mass of the particle, M the mass of the sphere, and r the distance of m from the center of the sphere. Now we will derive this result. We will use the concepts of infinitesimally small quantities and integration.

First we consider a very thin, uniform spherical shell (like a thin-walled basketball) of mass M whose thickness t is small compared to its radius R (Fig. D–1). The force on a particle of mass m at a distance r from the center of the shell can be calculated as the vector sum of the forces due to all the particles of the shell. We imagine the shell divided up into thin (infinitesimal) circular strips so that all points on a strip are equidistant from our particle m. One of these circular strips, labeled AB, is shown in Fig. D–1. It is $R\,d\theta$ wide, t thick, and has a radius $R\sin\theta$. The force on our particle m due to a tiny piece of the strip at point A is represented by the vector $\vec{\mathbf{F}}_A$ shown. The force due to a tiny piece of the strip at point B, which is diametrically opposite A, is the force $\vec{\mathbf{F}}_B$. We take the two pieces at A and B to be of equal mass, so $F_A = F_B$. The horizontal components of $\vec{\mathbf{F}}_A$ and $\vec{\mathbf{F}}_B$ are each equal to

$$F_A \cos\phi$$

and point toward the center of the shell. The vertical components of $\vec{\mathbf{F}}_A$ and $\vec{\mathbf{F}}_B$ are of equal magnitude and point in opposite directions, and so cancel. Since for every point on the strip there is a corresponding point diametrically opposite (as with A and B), we see that the net force due to the entire strip points toward the center of the shell. Its magnitude will be

$$dF = G\frac{m\,dM}{\ell^2}\cos\phi,$$

where dM is the mass of the entire circular strip and ℓ is the distance from all points on the strip to m, as shown. We write dM in terms of the density ρ; by density we mean the mass per unit volume (Section 13–2). Hence, $dM = \rho\,dV$, where dV is the volume of the strip and equals $(2\pi R\sin\theta)(t)(R\,d\theta)$. Then the force dF due to the circular strip shown is

$$dF = G\frac{m\rho 2\pi R^2 t\sin\theta\,d\theta}{\ell^2}\cos\phi. \qquad \textbf{(D–1)}$$

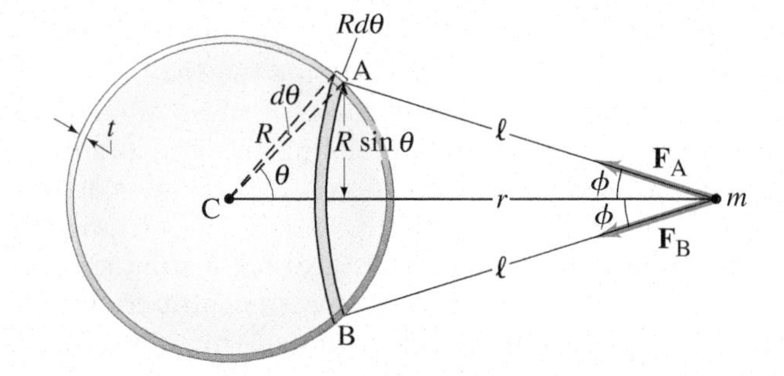

FIGURE D–1 Calculating the gravitational force on a particle of mass m due to a uniform spherical shell of radius R and mass M.

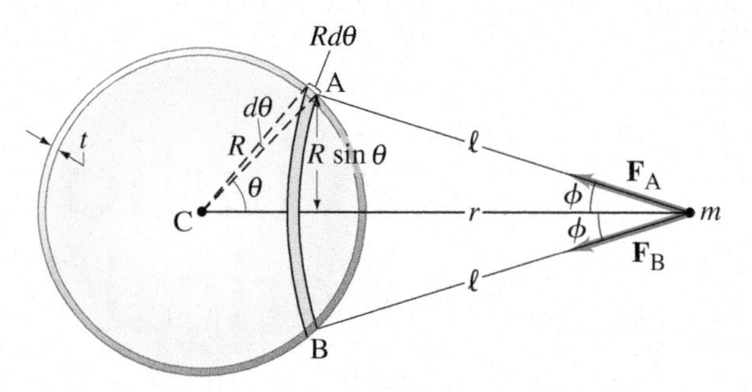

FIGURE D–1 (repeated) Calculating the gravitational force on a particle of mass m due to a uniform spherical shell of radius R and mass M.

To get the total force F that the entire shell exerts on the particle m, we must integrate over all the circular strips: that is, we integrate

$$dF = G\frac{m\rho 2\pi R^2 t \sin\theta\, d\theta}{\ell^2}\cos\phi \qquad \textbf{(D–1)}$$

from $\theta = 0°$ to $\theta = 180°$. But our expression for dF contains ℓ and ϕ, which are functions of θ. From Fig. D–1 we can see that

$$\ell\cos\phi = r - R\cos\theta.$$

Furthermore, we can write the law of cosines for triangle CmA:

$$\cos\theta = \frac{r^2 + R^2 - \ell^2}{2rR}. \qquad \textbf{(D–2)}$$

With these two expressions we can reduce our three variables (ℓ, θ, ϕ) to only one, which we take to be ℓ. We do two things with Eq. D–2: (1) We put it into the equation for $\ell\cos\phi$ above:

$$\cos\phi = \frac{1}{\ell}(r - R\cos\theta) = \frac{r^2 + \ell^2 - R^2}{2r\ell}.$$

and (2) we take the differential of both sides of Eq. D–2 (because $\sin\theta\, d\theta$ appears in the expression for dF, Eq. D–1), considering r and R to be constants when summing over the strips:

$$-\sin\theta\, d\theta = -\frac{2\ell\, d\ell}{2rR} \quad \text{or} \quad \sin\theta\, d\theta = \frac{\ell\, d\ell}{rR}.$$

We insert these into Eq. D–1 for dF and find

$$dF = Gm\rho\pi t\frac{R}{r^2}\left(1 + \frac{r^2 - R^2}{\ell^2}\right)d\ell.$$

Now we integrate to get the net force on our thin shell of radius R. To integrate over all the strips ($\theta = 0°$ to $180°$), we must go from $\ell = r - R$ to $\ell = r + R$ (see Fig. D–1). Thus,

$$F = Gm\rho\pi t\frac{R}{r^2}\left[\ell - \frac{r^2 - R^2}{\ell}\right]_{\ell=r-R}^{\ell=r+R}$$

$$= Gm\rho\pi t\frac{R}{r^2}(4R).$$

The volume V of the spherical shell is its area $(4\pi R^2)$ times the thickness t. Hence the mass $M = \rho V = \rho 4\pi R^2 t$, and finally

$$F = G\frac{mM}{r^2}. \qquad \begin{bmatrix}\text{particle of mass } m \text{ outside a} \\ \text{thin uniform spherical shell of mass } M\end{bmatrix}$$

This result gives us the force a thin shell exerts on a particle of mass m a distance r from the center of the shell, and *outside* the shell. We see that the force is the same as that between m and a particle of mass M at the center of the shell. In other words, for purposes of calculating the gravitational force exerted on or by a uniform spherical shell, we can consider all its mass concentrated at its center.

What we have derived for a shell holds also for a solid sphere, since a solid sphere can be considered as made up of many concentric shells, from $R = 0$ to $R = R_0$, where R_0 is the radius of the solid sphere. Why? Because if each shell has mass dM, we write for each shell, $dF = Gm\,dM/r^2$, where r is the distance from the center C to mass m and is the same for all shells. Then the total force equals the sum or integral over dM, which gives the total mass M. Thus the result

$$F = G\frac{mM}{r^2} \qquad \begin{bmatrix}\text{particle of mass } m \text{ outside} \\ \text{solid sphere of mass } M\end{bmatrix} \qquad \textbf{(D–3)}$$

is valid for a solid sphere of mass M even if the density varies with distance from the center. (It is not valid if the density varies within each shell—that is, depends not only on R.) Thus the gravitational force exerted on or by spherical objects, including nearly spherical objects like the Earth, Sun, and Moon, can be considered to act as if the objects were point particles.

This result, Eq. D–3, is true only if the mass m is outside the sphere. Let us next consider a point mass m that is located inside the spherical shell of Fig. D–1. Here, r would be less than R, and the integration over ℓ would be from $\ell = R - r$ to $\ell = R + r$, so

$$\left[\ell - \frac{r^2 - R^2}{\ell}\right]_{R-r}^{R+r} = 0.$$

Thus the force on any mass inside the shell would be zero. This result has particular importance for the electrostatic force, which is also an inverse square law. For the gravitational situation, we see that at points within a solid sphere, say 1000 km below the Earth's surface, only the mass up to that radius contributes to the net force. The outer shells beyond the point in question contribute zero net gravitational effect.

The results we have obtained here can also be reached using the gravitational analog of Gauss's law for electrostatics (Chapter 22).

APPENDIX

E Differential Form of Maxwell's Equations

Maxwell's equations can be written in another form that is often more convenient than Eqs. 31–5. This material is usually covered in more advanced courses, and is included here simply for completeness.

We quote here two theorems, without proof, that are derived in vector analysis textbooks. The first is called **Gauss's theorem** or the **divergence theorem**. It relates the integral over a surface of any vector function $\vec{\mathbf{F}}$ to a volume integral over the volume enclosed by the surface:

$$\oint_{\text{Area } A} \vec{\mathbf{F}} \cdot d\vec{\mathbf{A}} = \int_{\text{Volume } V} \vec{\nabla} \cdot \vec{\mathbf{F}}\, dV.$$

The operator $\vec{\nabla}$ is the **del operator**, defined in Cartesian coordinates as

$$\vec{\nabla} = \hat{\mathbf{i}}\frac{\partial}{\partial x} + \hat{\mathbf{j}}\frac{\partial}{\partial y} + \hat{\mathbf{k}}\frac{\partial}{\partial z}.$$

The quantity

$$\vec{\nabla} \cdot \vec{\mathbf{F}} = \frac{\partial F_x}{\partial x} + \frac{\partial F_y}{\partial y} + \frac{\partial F_z}{\partial z}$$

is called the **divergence** of $\vec{\mathbf{F}}$. The second theorem is **Stokes's theorem**, and relates a line integral around a closed path to a surface integral over any surface enclosed by that path:

$$\oint_{\text{Line}} \vec{\mathbf{F}} \cdot d\vec{\boldsymbol{\ell}} = \int_{\text{Area } A} \vec{\nabla} \times \vec{\mathbf{F}} \cdot d\vec{\mathbf{A}}.$$

The quantity $\vec{\nabla} \times \vec{\mathbf{F}}$ is called the **curl** of $\vec{\mathbf{F}}$. (See Section 11–2 on the vector product.)

We now use these two theorems to obtain the differential form of Maxwell's equations in free space. We apply Gauss's theorem to Eq. 31–5a (Gauss's law):

$$\oint_A \vec{\mathbf{E}} \cdot d\vec{\mathbf{A}} = \int \vec{\nabla} \cdot \vec{\mathbf{E}}\, dV = \frac{Q}{\epsilon_0}.$$

Now the charge Q can be written as a volume integral over the charge density ρ: $Q = \int \rho\, dV$. Then

$$\int \vec{\nabla} \cdot \vec{\mathbf{E}}\, dV = \frac{1}{\epsilon_0}\int \rho\, dV.$$

Both sides contain volume integrals over the same volume, and for this to be true over *any* volume, whatever its size or shape, the integrands must be equal:

$$\vec{\nabla} \cdot \vec{\mathbf{E}} = \frac{\rho}{\epsilon_0}. \qquad \textbf{(E–1)}$$

This is the differential form of Gauss's law. The second of Maxwell's equations, $\oint \vec{\mathbf{B}} \cdot d\vec{\mathbf{A}} = 0,$ is treated in the same way, and we obtain

$$\vec{\nabla} \cdot \vec{\mathbf{B}} = 0. \qquad \textbf{(E–2)}$$

Next, we apply Stokes's theorem to the third of Maxwell's equations,

$$\oint \vec{\mathbf{E}} \cdot d\vec{\boldsymbol{\ell}} = \int \vec{\nabla} \times \vec{\mathbf{E}} \cdot d\vec{\mathbf{A}} = -\frac{d\Phi_B}{dt}.$$

Since the magnetic flux $\Phi_B = \int \vec{\mathbf{B}} \cdot d\vec{\mathbf{A}}$, we have

$$\int \vec{\nabla} \times \vec{\mathbf{E}} \cdot d\vec{\mathbf{A}} = -\frac{\partial}{\partial t} \int \vec{\mathbf{B}} \cdot d\vec{\mathbf{A}}$$

where we use the partial derivative, $\partial \vec{\mathbf{B}}/\partial t$, since B may also depend on position. These are surface integrals over the same area, and to be true over any area, even a very small one, we must have

$$\vec{\nabla} \times \vec{\mathbf{E}} = -\frac{\partial \vec{\mathbf{B}}}{\partial t}. \qquad \textbf{(E–3)}$$

This is the third of Maxwell's equations in differential form. Finally, to the last of Maxwell's equations,

$$\oint \vec{\mathbf{B}} \cdot d\vec{\boldsymbol{\ell}} = \mu_0 I + \mu_0 \epsilon_0 \frac{d\Phi_E}{dt},$$

we apply Stokes's theorem and write $\Phi_E = \int \vec{\mathbf{E}} \cdot d\vec{\mathbf{A}}$:

$$\int \vec{\nabla} \times \vec{\mathbf{B}} \cdot d\vec{\mathbf{A}} = \mu_0 I + \mu_0 \epsilon_0 \frac{\partial}{\partial t} \int \vec{\mathbf{E}} \cdot d\vec{\mathbf{A}}.$$

The conduction current I can be written in terms of the current density $\vec{\mathbf{j}}$, using Eq. 25–12:

$$I = \int \vec{\mathbf{j}} \cdot d\vec{\mathbf{A}}.$$

Then Maxwell's fourth equation becomes:

$$\int \vec{\nabla} \times \vec{\mathbf{B}} \cdot d\vec{\mathbf{A}} = \mu_0 \int \vec{\mathbf{j}} \cdot d\vec{\mathbf{A}} + \mu_0 \epsilon_0 \frac{\partial}{\partial t} \int \vec{\mathbf{E}} \cdot d\vec{\mathbf{A}}.$$

For this to be true over any area A, whatever its size or shape, the integrands on each side of the equation must be equal:

$$\vec{\nabla} \times \vec{\mathbf{B}} = \mu_0 \vec{\mathbf{j}} + \mu_0 \epsilon_0 \frac{\partial \vec{\mathbf{E}}}{\partial t}. \qquad \textbf{(E–4)}$$

Equations E–1, 2, 3, and 4 are Maxwell's equations in differential form for free space. They are summarized in Table E–1.

TABLE E–1 Maxwell's Equations in Free Space†

Integral form	Differential form
$\oint \vec{\mathbf{E}} \cdot d\vec{\mathbf{A}} = \frac{Q}{\epsilon_0}$	$\vec{\nabla} \cdot \vec{\mathbf{E}} = \frac{\rho}{\epsilon_0}$
$\oint \vec{\mathbf{B}} \cdot d\vec{\mathbf{A}} = 0$	$\vec{\nabla} \cdot \vec{\mathbf{B}} = 0$
$\oint \vec{\mathbf{E}} \cdot d\vec{\boldsymbol{\ell}} = -\frac{d\Phi_B}{dt}$	$\vec{\nabla} \times \vec{\mathbf{E}} = -\frac{\partial \vec{\mathbf{B}}}{\partial t}$
$\oint \vec{\mathbf{B}} \cdot d\vec{\boldsymbol{\ell}} = \mu_0 I + \mu_0 \epsilon_0 \frac{d\Phi_E}{dt}$	$\vec{\nabla} \times \vec{\mathbf{B}} = \mu_0 \vec{\mathbf{j}} + \mu_0 \epsilon_0 \frac{\partial \vec{\mathbf{E}}}{\partial t}$

† $\vec{\nabla}$ stands for the *del operator* $\vec{\nabla} = \hat{\mathbf{i}}\frac{\partial}{\partial x} + \hat{\mathbf{j}}\frac{\partial}{\partial y} + \hat{\mathbf{k}}\frac{\partial}{\partial z}$ in Cartesian coordinates.

APPENDIX F Selected Isotopes

(1) Atomic Number Z	(2) Element	(3) Symbol	(4) Mass Number A	(5) Atomic Mass[†]	(6) % Abundance (or Radioactive Decay[‡] Mode)	(7) Half-life (if radioactive)
0	(Neutron)	n	1	1.008665	β^-	10.23 min
1	Hydrogen	H	1	1.007825	99.9885%	
	Deuterium	d or D	2	2.014082	0.0115%	
	Tritium	t or T	3	3.016049	β^-	12.312 yr
2	Helium	He	3	3.016029	0.000137%	
			4	4.002603	99.999863%	
3	Lithium	Li	6	6.015123	7.59%	
			7	7.016005	92.41%	
4	Beryllium	Be	7	7.016930	EC, γ	53.22 days
			9	9.012182	100%	
5	Boron	B	10	10.012937	19.9%	
			11	11.009305	80.1%	
6	Carbon	C	11	11.011434	β^+, EC	20.370 min
			12	12.000000	98.93%	
			13	13.003355	1.07%	
			14	14.003242	β^-	5730 yr
7	Nitrogen	N	13	13.005739	β^+, EC	9.9670 min
			14	14.003074	99.632%	
			15	15.000109	0.368%	
8	Oxygen	O	15	15.003066	β^+, EC	122.5 min
			16	15.994915	99.757%	
			18	17.999161	0.205%	
9	Fluorine	F	19	18.998403	100%	
10	Neon	Ne	20	19.992440	90.48%	
			22	21.991385	9.25%	
11	Sodium	Na	22	21.994436	β^+, EC, γ	2.6027 yr
			23	22.989769	100%	
			24	23.990963	β^-, γ	14.9574 h
12	Magnesium	Mg	24	23.985042	78.99%	
13	Aluminum	Al	27	26.981539	100%	
14	Silicon	Si	28	27.976927	92.2297%	
			31	30.975363	β^-, γ	157.3 min
15	Phosphorus	P	31	30.973762	100%	
			32	31.973907	β^-	14.284 days

[†] The masses given in column (5) are those for the neutral atom, including the Z electrons.

[‡] Chapter 41; EC = electron capture.

(1) Atomic Number Z	(2) Element	(3) Symbol	(4) Mass Number A	(5) Atomic Mass	(6) % Abundance (or Radioactive Decay Mode)	(7) Half-life (if radioactive)
16	Sulfur	S	32	31.972071	94.9%	
			35	34.969032	β^-	87.32 days
17	Chlorine	Cl	35	34.968853	75.78%	
			37	36.965903	24.22%	
18	Argon	Ar	40	39.962383	99.600%	
19	Potassium	K	39	38.963707	93.258%	
			40	39.963998	0.0117%	
					β^-, EC, γ, β^+	1.265×10^9 yr
20	Calcium	Ca	40	39.962591	96.94%	
21	Scandium	Sc	45	44.955912	100%	
22	Titanium	Ti	48	47.947946	73.72%	
23	Vanadium	V	51	50.943960	99.750%	
24	Chromium	Cr	52	51.940508	83.789%	
25	Manganese	Mn	55	54.938045	100%	
26	Iron	Fe	56	55.934938	91.75%	
27	Cobalt	Co	59	58.933195	100%	
			60	59.933817	β^-, γ	5.2710 yr
28	Nickel	Ni	58	57.935343	68.077%	
			60	59.930786	26.223%	
29	Copper	Cu	63	62.929598	69.17%	
			65	64.927790	30.83%	
30	Zinc	Zn	64	63.929142	48.6%	
			66	65.926033	27.9%	
31	Gallium	Ga	69	68.925574	60.108%	
32	Germanium	Ge	72	71.922076	27.5%	
			74	73.921178	36.3%	
33	Arsenic	As	75	74.921596	100%	
34	Selenium	Se	80	79.916521	49.6%	
35	Bromine	Br	79	78.918337	50.69%	
36	Krypton	Kr	84	83.911507	57.00%	
37	Rubidium	Rb	85	84.911790	72.17%	
38	Strontium	Sr	86	85.909260	9.86%	
			88	87.905612	82.58%	
			90	89.907738	β^-	28.80 yr
39	Yttrium	Y	89	88.905848	100%	
40	Zirconium	Zr	90	89.904704	51.4%	
41	Niobium	Nb	93	92.906378	100%	
42	Molybdenum	Mo	98	97.905408	24.1%	
43	Technetium	Tc	98	97.907216	β^-, γ	4.2×10^6 yr
44	Ruthenium	Ru	102	101.904349	31.55%	
45	Rhodium	Rh	103	102.905504	100%	
46	Palladium	Pd	106	105.903486	27.33%	
47	Silver	Ag	107	106.905097	51.839%	
			109	108.904752	48.161%	
48	Cadmium	Cd	114	113.903359	28.7%	
49	Indium	In	115	114.903878	95.71%; β^-	4.41×10^{14} yr
50	Tin	Sn	120	119.902195	32.58%	
51	Antimony	Sb	121	120.903816	57.21%	

(1) Atomic Number *Z*	(2) Element	(3) Symbol	(4) Mass Number *A*	(5) Atomic Mass	(6) % Abundance (or Radioactive Decay Mode)	(7) Half-life (if radioactive)
52	Tellurium	Te	130	129.906224	34.1%; $\beta^-\beta^-$	$>9.7 \times 10^{22}$ yr
53	Iodine	I	127	126.904473	100%	
			131	130.906125	β^-, γ	8.0233 days
54	Xenon	Xe	132	131.904154	26.89%	
			136	135.907219	8.87%; $\beta^-\beta^-$	$>8.5 \times 10^{21}$ yr
55	Cesium	Cs	133	132.905452	100%	
56	Barium	Ba	137	136.905827	11.232%	
			138	137.905247	71.70%	
57	Lanthanum	La	139	138.906353	99.910%	
58	Cerium	Ce	140	139.905439	88.45%	
59	Praseodymium	Pr	141	140.907653	100%	
60	Neodymium	Nd	142	141.907723	27.2%	
61	Promethium	Pm	145	144.912749	EC, α	17.7 yr
62	Samarium	Sm	152	151.919732	26.75%	
63	Europium	Eu	153	152.921230	52.19%	
64	Gadolinium	Gd	158	157.924104	24.84%	
65	Terbium	Tb	159	158.925347	100%	
66	Dysprosium	Dy	164	163.929175	28.2%	
67	Holmium	Ho	165	164.930322	100%	
68	Erbium	Er	166	165.930293	33.6%	
69	Thulium	Tm	169	168.934213	100%	
70	Ytterbium	Yb	174	173.938862	31.8%	
71	Lutetium	Lu	175	174.940772	97.41%	
72	Hafnium	Hf	180	179.946550	35.08%	
73	Tantalum	Ta	181	180.947996	99.988%	
74	Tungsten (wolfram)	W	184	183.950931	30.64%; α	$>8.9 \times 10^{21}$ yr
75	Rhenium	Re	187	186.955753	62.60%; β^-	4.35×10^{10} yr
76	Osmium	Os	191	190.960930	β^-, γ	15.4 days
			192	191.961481	40.78%	
77	Iridium	Ir	191	190.960594	37.3%	
			193	192.962926	62.7%	
78	Platinum	Pt	195	194.964791	33.832%	
79	Gold	Au	197	196.966569	100%	
80	Mercury	Hg	199	198.968280	16.87%	
			202	201.970643	29.9%	
81	Thallium	Tl	205	204.974428	70.476%	
82	Lead	Pb	206	205.974465	24.1%	
			207	206.975897	22.1%	
			208	207.976652	52.4%	
			210	209.984188	β^-, γ, α	22.23 yr
			211	210.988737	β^-, γ	36.1 min
			212	211.991898	β^-, γ	10.64 h
			214	213.999805	β^-, γ	26.8 min
83	Bismuth	Bi	209	208.980399	100%	
			211	210.987269	α, γ, β^-	2.14 min
84	Polonium	Po	210	209.982874	α, γ, EC	138.376 days
			214	213.995201	α, γ	162.3 μs
85	Astatine	At	218	218.008694	α, β^-	1.4 s

(1) Atomic Number Z	(2) Element	(3) Symbol	(4) Mass Number A	(5) Atomic Mass	(6) % Abundance (or Radioactive Decay Mode)	(7) Half-life (if radioactive)
86	Radon	Rn	222	222.017578	α, γ	3.8232 days
87	Francium	Fr	223	223.019736	β^-, γ, α	22.00 min
88	Radium	Ra	226	226.025410	α, γ	1600 yr
89	Actinium	Ac	227	227.027752	β^-, γ, α	21.772 yr
90	Thorium	Th	228	228.028741	α, γ	698.60 days
			232	232.038055	100%; α, γ	1.405×10^{10} yr
91	Protactinium	Pa	231	231.035884	α, γ	3.276×10^4 yr
92	Uranium	U	232	232.037156	α, γ	68.9 yr
			233	233.039635	α, γ	1.592×10^5 yr
			235	235.043930	0.720%; α, γ	7.04×10^8 yr
			236	236.045568	α, γ	2.342×10^7 yr
			238	238.050788	99.274%; α, γ	4.468×10^9 yr
			239	239.054293	β^-, γ	23.46 min
93	Neptunium	Np	237	237.048173	α, γ	2.144×10^6 yr
			239	239.052939	β^-, γ	2.356 days
94	Plutonium	Pu	239	239.052163	α, γ	24,100 yr
			244	244.064204	α	8.00×10^7 yr
95	Americium	Am	243	243.061381	α, γ	7370 yr
96	Curium	Cm	247	247.070354	α, γ	1.56×10^7 yr
97	Berkelium	Bk	247	247.070307	α, γ	1380 yr
98	Californium	Cf	251	251.079587	α, γ	898 yr
99	Einsteinium	Es	252	252.082980	α, EC, γ	471.7 days
100	Fermium	Fm	257	257.095105	α, γ	100.5 days
101	Mendelevium	Md	258	258.098431	α, γ	51.5 days
102	Nobelium	No	259	259.10103	α, EC	58 min
103	Lawrencium	Lr	262	262.10963	α, EC, fission	≈ 4 h
104	Rutherfordium	Rf	263	263.11255	fission	10 min
105	Dubnium	Db	262	262.11408	α, fission, EC	35 s
106	Seaborgium	Sg	266	266.12210	α, fission	≈ 21 s
107	Bohrium	Bh	264	264.12460	α	≈ 0.44 s
108	Hassium	Hs	269	269.13406	α	≈ 10 s
109	Meitnerium	Mt	268	268.13870	α	21 ms
110	Darmstadtium	Ds	271	271.14606	α	≈ 70 ms
111	Roentgenium	Rg	272	272.15360	α	3.8 ms
112		Uub	277	277.16394	α	≈ 0.7 ms

Preliminary evidence (unconfirmed) has been reported for elements 113, 114, 115, 116 and 118.

Answers to Odd-Numbered Problems

CHAPTER 36

1. 72.5 m.

3. 1.00, 1.00, 1.01, 1.02, 1.05, 1.09, 1.15, 1.25, 1.40, 1.67, 2.29, 7.09.

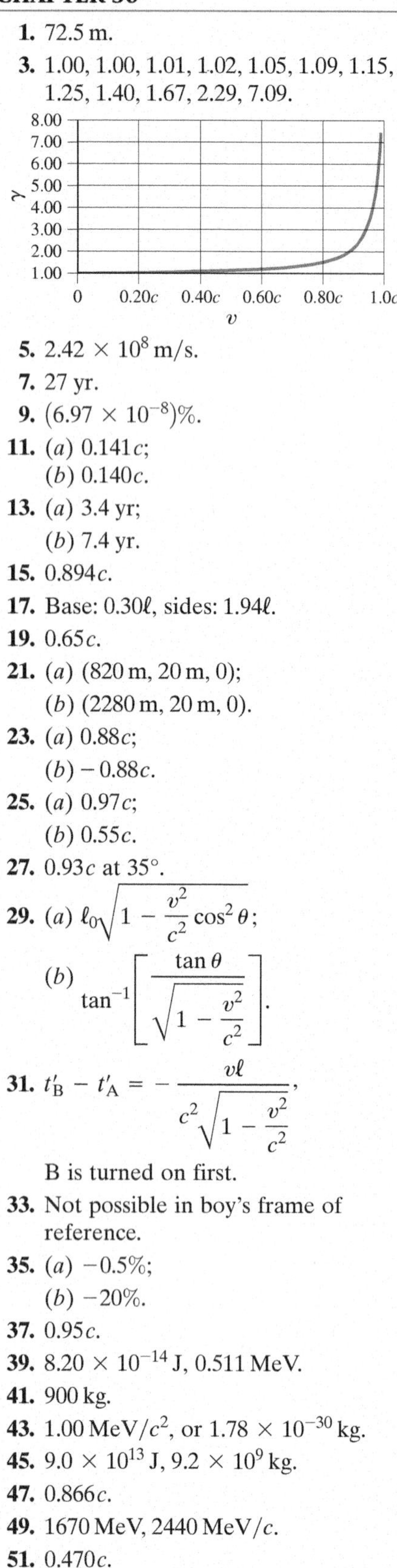

5. 2.42×10^8 m/s.

7. 27 yr.

9. $(6.97 \times 10^{-8})\%$.

11. (*a*) $0.141c$;
(*b*) $0.140c$.

13. (*a*) 3.4 yr;
(*b*) 7.4 yr.

15. $0.894c$.

17. Base: 0.30ℓ, sides: 1.94ℓ.

19. $0.65c$.

21. (*a*) (820 m, 20 m, 0);
(*b*) (2280 m, 20 m, 0).

23. (*a*) $0.88c$;
(*b*) $-0.88c$.

25. (*a*) $0.97c$;
(*b*) $0.55c$.

27. $0.93c$ at 35°.

29. (*a*) $\ell_0\sqrt{1 - \frac{v^2}{c^2}\cos^2\theta}$;
(*b*) $\tan^{-1}\left[\frac{\tan\theta}{\sqrt{1 - \frac{v^2}{c^2}}}\right]$.

31. $t'_B - t'_A = -\frac{v\ell}{c^2\sqrt{1 - \frac{v^2}{c^2}}}$,
B is turned on first.

33. Not possible in boy's frame of reference.

35. (*a*) -0.5%;
(*b*) -20%.

37. $0.95c$.

39. 8.20×10^{-14} J, 0.511 MeV.

41. 900 kg.

43. 1.00 MeV/c^2, or 1.78×10^{-30} kg.

45. 9.0×10^{13} J, 9.2×10^9 kg.

47. $0.866c$.

49. 1670 MeV, 2440 MeV/c.

51. $0.470c$.

53. $0.32c$.

55. $0.866c$, $0.745c$.

57. (*a*) 2.5×10^{19} J;
(*b*) -2.4%.

59. 237.04832 u.

61. 240 MeV.

65. 230 MHz.

67. (*a*) 1.00×10^2 km/h;
(*b*) 67 Hz.

69. 75 μs.

71. 8.0×10^{-8} s.

73. (*a*) $0.067c$;
(*b*) $0.070c$.

75. (*a*) $\tan^{-1}\sqrt{\frac{c^2}{v^2} - 1}$;
(*c*) $\tan^{-1}\frac{c}{v}$, $u = \sqrt{c^2 + v^2}$.

77. (*a*) 0.77 m/s;
(*b*) 0.21 m.

79. 1.022 MeV.

83. (*a*) 4×10^9 kg/s;
(*b*) 4×10^7 yr;
(*c*) 1×10^{13} yr.

85. 28.32 MeV.

87. (*a*) 2.86×10^{-18} kg·m/s;
(*b*) 0;
(*c*) 3.31×10^{-17} kg·m/s.

89. 3×10^7 kg.

91. $0.987c$.

93. 5.3×10^{21} J, 53 times as great.

95. (*a*) 6.5 yr;
(*b*) 2.3 ly.

99.

CHAPTER 37

1. (*a*) 10.6 μm, far infrared;
(*b*) 829 nm, infrared;
(*c*) 0.69 mm, microwave;
(*d*) 1.06 mm, microwave.

3. 5.4×10^{-20} J, 0.34 eV.

5. (*b*) 6.62×10^{-34} J·s.

7. $2.7 \times 10^{-19}\text{ J} < E < 4.9 \times 10^{-19}$ J,
$1.7\text{ eV} < E < 3.0$ eV.

9. 2×10^{13} Hz, 1×10^{-5} m.

11. 7.2×10^{14} Hz.

13. 3.05×10^{-27} m.

15. Copper and iron.

17. 0.55 eV.

19. 2.66 eV.

21. 3.56 eV.

23. (*a*) 1.66 eV;
(*b*) 3.03 eV.

25. (*a*) 1.66 eV;
(*b*) 3.03 eV.

27. 0.004, or 0.4%.

29. (*a*) 2.43 pm;
(*b*) 1.32 fm.

31. (*a*) 8.8×10^{-6};
(*b*) 0.049.

33. (*a*) 229 eV;
(*b*) 0.165 nm.

35. 1.65 MeV.

37. 212 MeV, 5.86 fm.

39. 1.772 MeV, 702 fm.

41. 4.7 pm.

43. 4.0 pm.

45. 1840.

47. (*a*) 1.1×10^{-24} kg·m/s;
(*b*) 1.2×10^6 m/s;
(*c*) 4.2 V.

51. 590 m/s.

53. 20.9 pm.

55. 1.51 eV

57. 122 eV.

59. 91.4 nm.

61. 37.0 nm.

63.

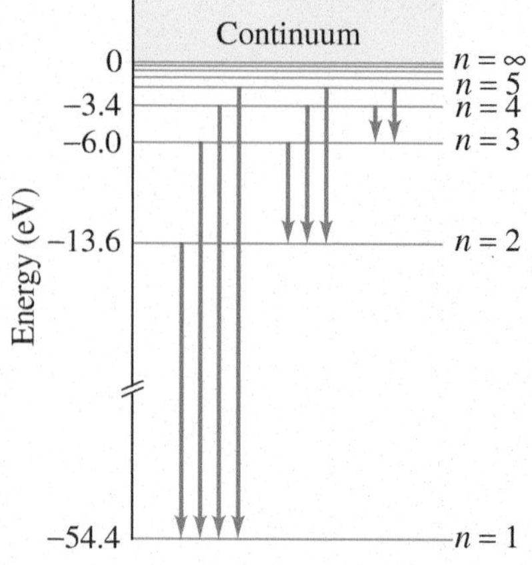

65. -27.2 eV, 13.6 eV.

67. Yes: $v = 7 \times 10^{-3}\,c$; $1/\gamma = 0.99997$.

69. 97.23 nm, 102.6 nm, 121.5 nm, 486.2 nm, 656.3 nm, 1875 nm.

71. Yes.

73. 3.28×10^{15} Hz.

75. 5.3×10^{26} photons/s.

77. 6.2×10^{18} photons/s.

79. 0.244 MeV for both.

81. 28 fm.

83. 4.4×10^{-40}, yes.

85. 2.25 V.

87. 9.0 N.

89. 1.2 nm.

91. (*a*)

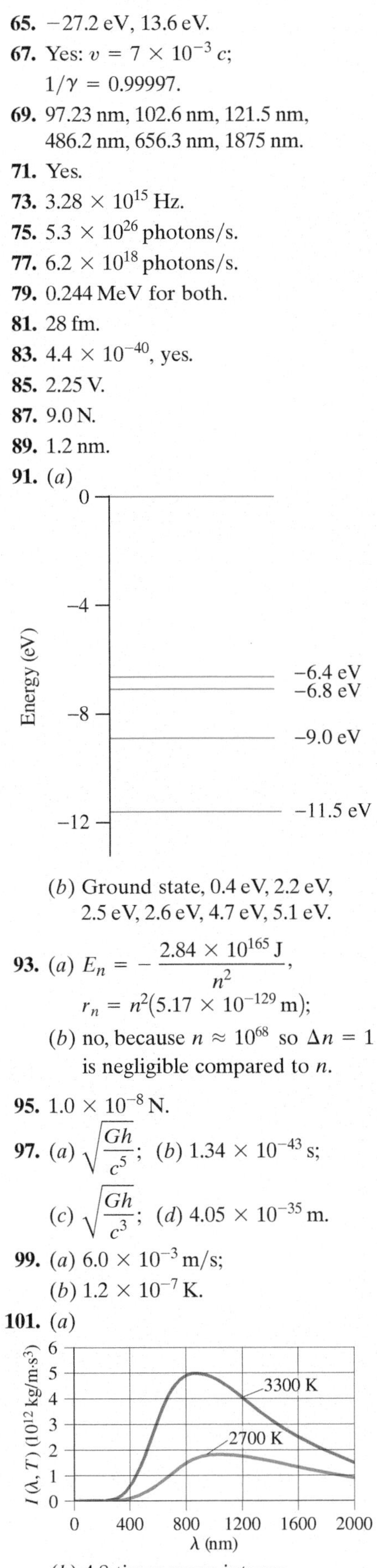

(*b*) Ground state, 0.4 eV, 2.2 eV, 2.5 eV, 2.6 eV, 4.7 eV, 5.1 eV.

93. (*a*) $E_n = -\dfrac{2.84 \times 10^{165}\,\mathrm{J}}{n^2}$, $r_n = n^2(5.17 \times 10^{-129}\,\mathrm{m})$;
(*b*) no, because $n \approx 10^{68}$ so $\Delta n = 1$ is negligible compared to n.

95. 1.0×10^{-8} N.

97. (*a*) $\sqrt{\dfrac{Gh}{c^5}}$; (*b*) 1.34×10^{-43} s;
(*c*) $\sqrt{\dfrac{Gh}{c^3}}$; (*d*) 4.05×10^{-35} m.

99. (*a*) 6.0×10^{-3} m/s;
(*b*) 1.2×10^{-7} K.

101. (*a*)

(*b*) 4.8 times more intense.

103. (*a*) $\dfrac{hc}{e}, -\dfrac{W_0}{e}$;
(*b*)

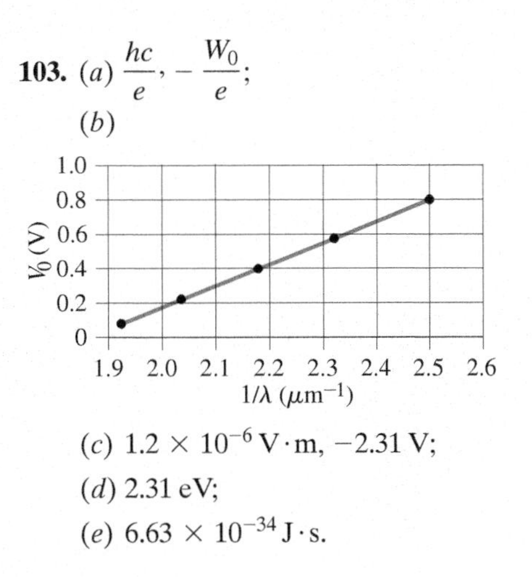

(*c*) 1.2×10^{-6} V·m, -2.31 V;
(*d*) 2.31 eV;
(*e*) 6.63×10^{-34} J·s.

CHAPTER 38

1. 2.8×10^{-7} m.

3. 5.3×10^{-11} m.

5. 4500 m/s.

7. 1.0×10^{-14}.

9. $\Delta x_{\mathrm{electron}} \geq 1.4 \times 10^{-3}$ m,
$\Delta x_{\mathrm{baseball}} \geq 9.3 \times 10^{-33}$ m,
$\dfrac{\Delta x_{\mathrm{electron}}}{\Delta x_{\mathrm{baseball}}} = 1.5 \times 10^{29}$.

11. 1.3×10^{-54} kg.

13. (*a*) 10^{-7} eV;
(*b*) $1/10^8$;
(*c*) 100 nm, 10^{-6} nm.

19. (*a*) $A\sin[(2.6 \times 10^9\,\mathrm{m}^{-1})x] + B\cos[(2.6 \times 10^9\,\mathrm{m}^{-1})x]$;
(*b*) $A\sin[(4.7 \times 10^{12}\,\mathrm{m}^{-1})x] + B\cos[(4.7 \times 10^{12}\,\mathrm{m}^{-1})x]$.

21. 1.8×10^6 m/s.

23. (*a*) 46 nm;
(*b*) 0.20 nm.

25. $\Delta p\,\Delta x \approx h$, which is consistent with the uncertainty principle.

27. $n = 1$: 0.094 eV, $(1.0\,\mathrm{nm}^{-1/2})\sin[(1.6\,\mathrm{nm}^{-1})x]$;
$n = 2$: 0.38 eV, $(1.0\,\mathrm{nm}^{-1/2})\sin[(3.1\,\mathrm{nm}^{-1})x]$;
$n = 3$: 0.85 eV, $(1.0\,\mathrm{nm}^{-1/2})\sin[(4.7\,\mathrm{nm}^{-1})x]$;
$n = 4$: 1.5 eV, $(1.0\,\mathrm{nm}^{-1/2})\sin[(6.3\,\mathrm{nm}^{-1})x]$.

29. (*a*) 940 MeV;
(*b*) 0.51 MeV;
(*c*) 0.51 MeV.

31. (*a*) 4.0×10^{-19} eV;
(*b*) 2×10^8;
(*c*) 1.4×10^{-10} eV.

33. n odd:

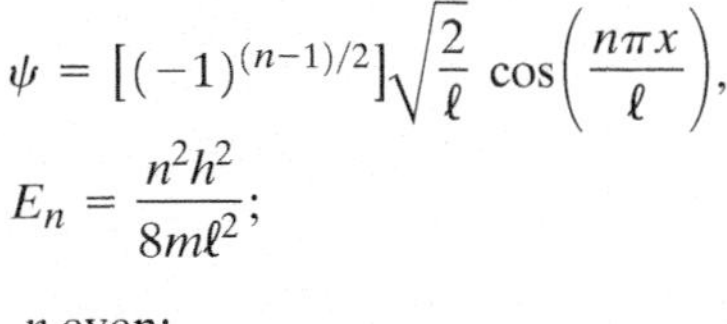

$$\psi = [(-1)^{(n-1)/2}]\sqrt{\frac{2}{\ell}}\cos\left(\frac{n\pi x}{\ell}\right),\quad E_n = \frac{n^2h^2}{8m\ell^2};$$

n even:

$$\psi = [(-1)^{n/2}]\sqrt{\frac{2}{\ell}}\sin\left(\frac{n\pi x}{\ell}\right),\quad E_n = \frac{n^2h^2}{8m\ell^2}.$$

35.

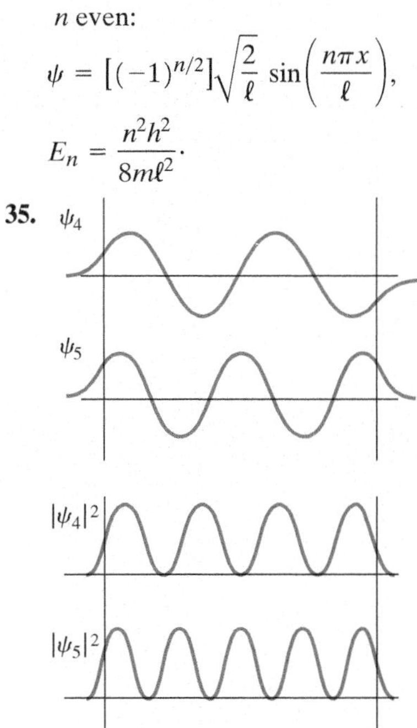

37. 0.020 nm.

39. 17 eV.

41. (*a*) 6.1%;
(*b*) 93.9%.

43. (*a*) 12% decrease;
(*b*) 6.2% decrease.

45. (*a*) 32 MeV;
(*b*) 57 fm;
(*c*) 1.4×10^7 m/s, 8.6×10^{20} Hz, 7×10^9 yr.

47. 14 MeV.

49. 25 nm.

51. $\Delta x = r_1$ (the Bohr radius).

53. 0.23 MeV, 3.3×10^6 m/s.

55. 27% decrease.

57.

59. (*a*) $\Delta\phi > 0$ so $\phi \neq 0$ exactly;
(*b*) 4 s.

61. (*a*)

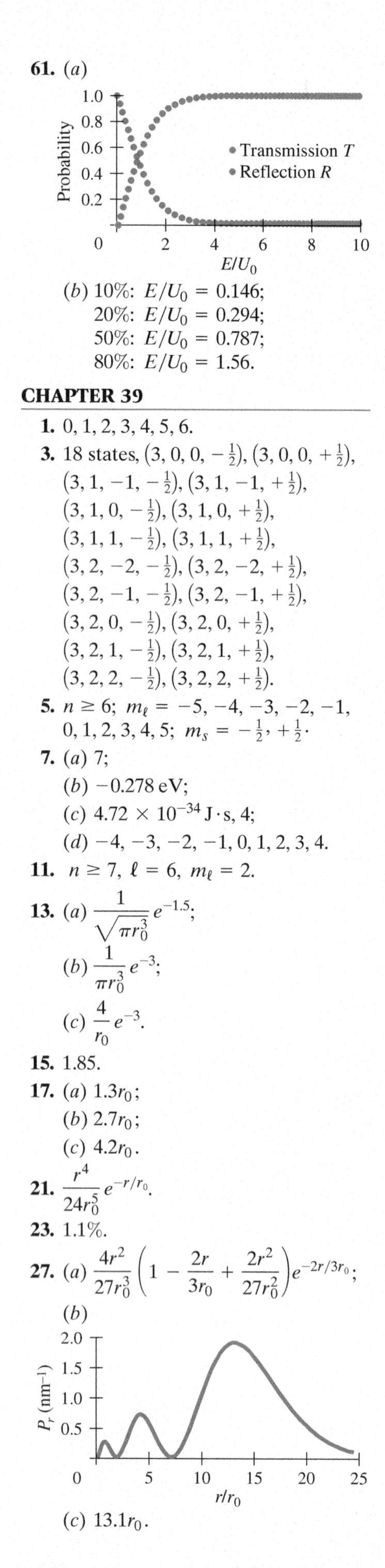

(*b*) 10%: $E/U_0 = 0.146$;
20%: $E/U_0 = 0.294$;
50%: $E/U_0 = 0.787$;
80%: $E/U_0 = 1.56$.

CHAPTER 39

1. 0, 1, 2, 3, 4, 5, 6.

3. 18 states, $(3, 0, 0, -\frac{1}{2}), (3, 0, 0, +\frac{1}{2}),$
$(3, 1, -1, -\frac{1}{2}), (3, 1, -1, +\frac{1}{2}),$
$(3, 1, 0, -\frac{1}{2}), (3, 1, 0, +\frac{1}{2}),$
$(3, 1, 1, -\frac{1}{2}), (3, 1, 1, +\frac{1}{2}),$
$(3, 2, -2, -\frac{1}{2}), (3, 2, -2, +\frac{1}{2}),$
$(3, 2, -1, -\frac{1}{2}), (3, 2, -1, +\frac{1}{2}),$
$(3, 2, 0, -\frac{1}{2}), (3, 2, 0, +\frac{1}{2}),$
$(3, 2, 1, -\frac{1}{2}), (3, 2, 1, +\frac{1}{2}),$
$(3, 2, 2, -\frac{1}{2}), (3, 2, 2, +\frac{1}{2}).$

5. $n \geq 6$; $m_\ell = -5, -4, -3, -2, -1, 0, 1, 2, 3, 4, 5$; $m_s = -\frac{1}{2}, +\frac{1}{2}$.

7. (*a*) 7;
(*b*) −0.278 eV;
(*c*) 4.72×10^{-34} J·s, 4;
(*d*) −4, −3, −2, −1, 0, 1, 2, 3, 4.

11. $n \geq 7$, $\ell = 6$, $m_\ell = 2$.

13. (*a*) $\dfrac{1}{\sqrt{\pi r_0^3}} e^{-1.5}$;
(*b*) $\dfrac{1}{\pi r_0^3} e^{-3}$;
(*c*) $\dfrac{4}{r_0} e^{-3}$.

15. 1.85.

17. (*a*) $1.3r_0$;
(*b*) $2.7r_0$;
(*c*) $4.2r_0$.

21. $\dfrac{r^4}{24r_0^5} e^{-r/r_0}$.

23. 1.1%.

27. (*a*) $\dfrac{4r^2}{27r_0^3}\left(1 - \dfrac{2r}{3r_0} + \dfrac{2r^2}{27r_0^2}\right) e^{-2r/3r_0}$;
(*b*)

(*c*) $13.1r_0$.

29. (*a*) $(1, 0, 0, -\frac{1}{2}), (1, 0, 0, +\frac{1}{2}),$
$(2, 0, 0, -\frac{1}{2}), (2, 0, 0, +\frac{1}{2}),$
$(2, 1, -1, -\frac{1}{2}), (2, 1, -1, +\frac{1}{2})$;
(*b*) $(1, 0, 0, -\frac{1}{2}), (1, 0, 0, +\frac{1}{2}),$
$(2, 0, 0, -\frac{1}{2}), (2, 0, 0, +\frac{1}{2}),$
$(2, 1, -1, -\frac{1}{2}), (2, 1, -1, +\frac{1}{2}),$
$(2, 1, 0, -\frac{1}{2}), (2, 1, 0, +\frac{1}{2}),$
$(2, 1, 1, -\frac{1}{2}), (2, 1, 1, +\frac{1}{2}),$
$(3, 0, 0, -\frac{1}{2}), (3, 0, 0, +\frac{1}{2}),$
$(3, 1, -1, -\frac{1}{2})$.

31. $n = 3$, $\ell = 2$.

33. (*a*) $1s^22s^22p^63s^23p^63d^84s^2$;
(*b*) $1s^22s^22p^63s^23p^63d^{10}4s^24p^64d^{10}5s^1$;
(*c*) $1s^22s^22p^63s^23p^63d^{10}4s^24p^64d^{10}$-$4f^{14}5s^25p^65d^{10}6s^26p^65f^36d^17s^2$.

35. 5.75×10^{-13} m, 115 keV.

39. 0.0383 nm, 1 nm.

41. 0.194 nm.

43. Chromium.

47. 2.9×10^{-4} eV.

49. (*a*) 0.38 mm; (*b*) 0.19 mm.

51. (*a*) $\frac{1}{2}, \frac{3}{2}$; (*b*) $\frac{5}{2}, \frac{7}{2}$; (*c*) $\frac{3}{2}, \frac{5}{2}$;
(*d*) 4*p*: $\frac{\sqrt{3}}{2}\hbar, \frac{\sqrt{15}}{2}\hbar$; 4*f*: $\frac{\sqrt{35}}{2}\hbar, \frac{\sqrt{63}}{2}\hbar$;
3*d*: $\frac{\sqrt{15}}{2}\hbar, \frac{\sqrt{35}}{2}\hbar$.

53. (*a*) 0.4 T;
(*b*) 0.5 T.

55. 4.7×10^{-4} rad; (*a*) 180 m;
(*b*) 1.8×10^5 m.

57. 634 nm.

59. 3.7×10^4 K.

61. (*a*) 1.56;
(*b*) 1.36×10^{-10} m.

63. (*a*) $1s^22s^22p^63s^23p^63d^54s^2$;
(*b*) $1s^22s^22p^63s^23p^63d^{10}4s^24p^4$;
(*c*) $1s^22s^22p^63s^23p^63d^{10}4s^24p^64d^15s^2$.

65. (*a*) 2.5×10^{74};
(*b*) 5.1×10^{74}.

67. $5.24r_0$.

69. (*a*) 45°, 90°, 135°;
(*b*) 35.3°, 65.9°, 90°, 114.1°, 144.7°;
(*c*) 30°, 54.7°, 73.2°, 90°, 106.8°, 125.3°, 150°;
(*d*) 5.71°, 0.0573°, yes.

71. (*b*) $\overline{K} = -\frac{1}{2}\overline{U}$.

73. (*a*) Forbidden; (*b*) allowed;
(*c*) forbidden; (*d*) forbidden;
(*e*) allowed.

75. 4, beryllium.

77. (*a*) 3×10^{-171}, 1×10^{-202};
(*b*) 1×10^{-8}, 6×10^{-10};
(*c*) 7×10^{15}, 4×10^{14};
(*d*) 4×10^{22} photons/s,
7×10^{23} photons/s.

CHAPTER 40

1. 5.1 eV.

3. 4.7 eV.

5. 1.28 eV.

9. (*a*) 18.59 u;
(*b*) 8.00 u;
(*c*) 0.9801 u.

11. 1.10×10^{-10} m.

13. (*a*) 1.5×10^{-2} eV, 0.082 mm;
(*b*) 3.0×10^{-2} eV, 0.041 mm;
(*c*) 4.6×10^{-2} eV, 0.027 mm.

15. (*a*) 6.86 u;
(*b*) 1850 N/m, $k_{CO}/k_{H_2} = 3.4$.

17. 2.36×10^{-10} m.

19. $m_1 x_1 = m_2 x_2$.

21. 0.2826 nm.

23. 0.34 nm.

25. (*b*) −6.9 eV;
(*c*) −11 eV;
(*d*) −2.8%.

27. 9.0×10^{20}.

29. (*a*) 6.96 eV;
(*b*) 6.89 eV.

31. 1.6%.

33. 3.2 eV, 1.1×10^6 m/s.

39. (*a*) $\dfrac{h^2N^2}{32m\ell^2}$;
(*b*) $\dfrac{h^2(N+1)}{8m\ell^2}$;
(*c*) $\dfrac{4}{N}$.

43. 1.09 μm.

45. (*a*) 2*N*;
(*b*) 6*N*;
(*c*) 6*N*;
(*d*) $2N(2\ell + 1)$.

47. 4×10^6.

49. 1.8 eV.

51. 8.6 mA.

53. (*a*) 1.7 mA; (*b*) 3.4 mA.

55. (*a*) 35 mA; (*b*) 70 mA.

57. 3700 Ω.

59. 0.21 mA.

61. $I_B + I_C = I_E$.

63. (*a*) 3.1×10^4 K;
(*b*) 930 K.

65.

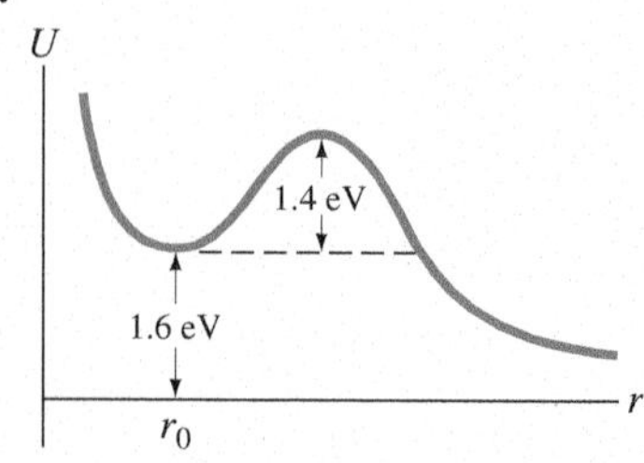

67. (*a*) 0.9801 u;
(*b*) 482 N/m, $k_{HCl}/k_{H_2} = 0.88$.

71. Yes, 1.09 μm.

73. 1100 J/mol.

75. 5.50 eV.

77. 3×10^{25}.

79. 6.47×10^{-4} eV.

81. 1.1 eV.

83. (*a*) 0.094 eV; (*b*) 0.63 nm.

85. (*a*) 150 V $\leq V \leq$ 486 V;
(*b*) 3.16 k$\Omega \leq R_{load} < \infty$.

87. (*a*)

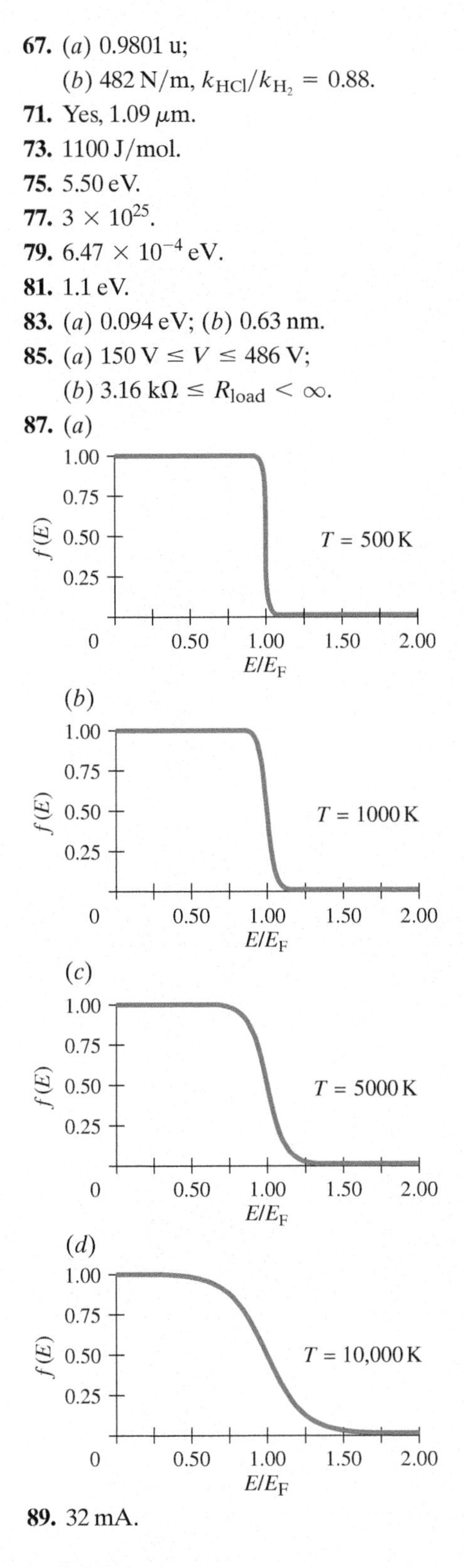

89. 32 mA.

CHAPTER 41

1. 0.149 u.

3. 0.85%.

5. 3727 MeV/c^2.

7. (*b*) 180 m; (*c*) 2.58×10^{-10} m.

9. 30 MeV.

11. 6×10^{26} nucleons, no, mass of all nucleons is approximately the same.

13. 550 MeV.

15. 7.94 MeV.

17. ${}^{23}_{11}$Na: 8.113 MeV/nucleon;
${}^{24}_{11}$Na: 8.063 MeV/nucleon.

19. (*b*) Yes, binding energy is positive.

21. 0.782 MeV.

23. 2.6×10^{-12} m.

25. (*a*) β^-;
(*b*) ${}^{24}_{11}\text{Na} \rightarrow {}^{24}_{12}\text{Mg} + \beta^- + \bar{\nu}$, 5.52 MeV.

27. (*a*) ${}^{234}_{90}$Th; (*b*) 234.04368 u.

29. 0.079 MeV.

31. (*a*) ${}^{32}_{16}$S;
(*b*) 31.972071 u.

33. 0.862 MeV.

35. 0.9612 MeV, 0.9612 MeV, 0, 0.

37. 5.31 MeV.

39. (*a*) $1.5 \times 10^{-10}\,\text{yr}^{-1}$;
(*b*) 6.0 h.

41. 0.16.

43. 0.015625.

45. 6.9×10^{19} nuclei.

47. (*a*) 3.60×10^{12} decays/s;
(*b*) 3.58×10^{12} decays/s;
(*c*) 9.72×10^{7} decays/s.

49. 0.76 g.

51. 2.30×10^{-11} g.

53. 4.3 min.

55. 2.98×10^{-2} g.

57. 35.4 d.

59. ${}^{228}_{88}$Ra, ${}^{228}_{89}$Ac, ${}^{228}_{90}$Th, ${}^{224}_{88}$Ra, ${}^{220}_{86}$Rn;
${}^{231}_{90}$Th, ${}^{231}_{91}$Pa, ${}^{227}_{89}$Ac, ${}^{227}_{90}$Th, ${}^{223}_{88}$Ra.

61. $N_D = N_0(1 - e^{-\lambda t})$.

63. 2.3×10^{4} yr.

65. 41 yr.

69. $6.64T_{1/2}$.

71. (*b*) 98.2%.

73. 1 MeV.

75. (*a*) ${}^{191}_{77}$Ir;
(*b*)

${}^{191}_{76}$Os
β^- (0.14 MeV)
${}^{191}_{77}$Ir*
γ (0.042 MeV)
${}^{191}_{77}$Ir*
γ (0.129 MeV)
${}^{191}_{77}$Ir

(*c*) The higher excited state.

77. 550 MeV, 2.5×10^{12} J.

79. 2.243 MeV.

81. (*a*) 2.4×10^{5} yr;
(*b*) no significant change, maximum age is on the order of 10^5 yr.

83. 5.48×10^{-4}.

85. (*a*) 1.63%;
(*b*) 0.663%.

87. 1.3×10^{21} yr.

89. 8.33×10^{16} nuclei,

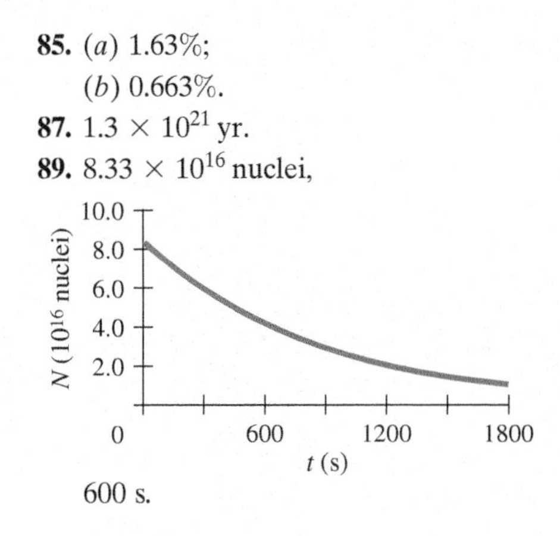

600 s.

CHAPTER 42

1. ${}^{28}_{13}$Al, β^-, ${}^{28}_{14}$Si.

3. Yes, because $Q = 4.807$ MeV.

5. 5.701 MeV released.

7. (*a*) Yes;
(*b*) 20.8 MeV.

9. 4.730 MeV.

11. $\text{n} + {}^{14}_{7}\text{N} \rightarrow {}^{14}_{6}\text{C} + \text{p}$, 0.626 MeV.

13. (*a*) The He has picked up a neutron from the C;
(*b*) ${}^{11}_{6}$C;
(*c*) 1.856 MeV, exothermic.

15. 18.000938 u.

17. 0.671 MeV.

19. $\pi(R_1 + R_2)^2$.

21. 10 cm.

23. 173.3 MeV.

25. 6×10^{18} fissions/s.

27. 0.34 g.

29. 5×10^{-5} kg.

31. 25 collisions.

33. 0.11.

35. 3000 eV.

39. (*a*) 5.98×10^{23} MeV/g,
4.83×10^{23} MeV/g,
2.10×10^{24} MeV/g;
(*b*) 5.13×10^{23} MeV/g; Eq. 42–9a gives about 17% more energy per gram, 42–9b gives about 6% less, and 42–9c gives about 4× more.

41. 0.35 g.

43. 6100 kg/h.

45. 2.46×10^{9} J, 50 times more than gasoline.

47. (*b*) 26.73 MeV;
(*c*) 1.943 MeV, 2.218 MeV, 7.551 MeV, 7.296 MeV, 2.752 MeV, 4.966 MeV;
(*d*) larger Coulomb repulsion to overcome.

49. 4.0 Gy.

51. 220 rad.

53. 280 counts/s.

55. 1.6 days.

57. (*a*) $^{131}_{53}\text{I} \rightarrow {}^{131}_{54}\text{Xe} + \beta^- + \bar{\nu}$;
(*b*) 31 d;
(*c*) 8×10^{-12} kg.

59. 8.3×10^{-7} Gy/d.

61. (*a*) $^{218}_{84}\text{Po}$;
(*b*) radioactive, alpha and beta decay, 3.1 min;
(*c*) chemically reactive;
(*d*) 9.1×10^6 Bq, 4.0×10^4 Bq.

63. 7.041 m, radio wave.

65. (*a*) $^{12}_{6}\text{C}$;
(*b*) 5.701 MeV.

67. 1.0043 : 1.

69. 6.5×10^{-2} rem/yr.

71. 4.4 m.

73. (*a*) 920 kg;
(*b*) 3×10^6 Ci.

75. (*a*) 3.7×10^{26} W;
(*b*) 3.5×10^{38} protons/s;
(*c*) 1.1×10^{11} yr.

77. 8×10^{12} J.

79. (*a*) 3700 decays/s;
(*b*) 4.8×10^{-4} Sv/yr, yes (13% of the background rate).

81. 7.274 MeV.

83. 79 yr.

85. 2 mCi.

CHAPTER 43

1. 5.59 GeV.

3. 2.0 T.

5. 13 MHz.

7. Alpha particles,
$\lambda_\alpha \approx d_{\text{nucleon}}, \lambda_p \approx 2d_{\text{nucleon}}$.

9. 5.5 T.

11. 1.8×10^{-19} m.

15. 33.9 MeV.

17. 1879.2 MeV.

19. 67.5 MeV.

21. (*a*) 178.5 MeV;
(*b*) 128.6 MeV.

23. (*a*) Charge, strangeness;
(*b*) energy;
(*c*) baryon number, strangeness, spin.

25. (*b*) The photon exists for such a short time that the uncertainty principle allows energy to not be conserved during the exchange.

27. 69.3 MeV.

29. $K_{\Lambda^0} = 8.6$ MeV, $K_{\pi^-} = 57.4$ MeV.

31. 52.3 MeV.

33. 9 keV.

35. 7.5×10^{-21} s.

37. (*a*) 700 eV;
(*b*) 70 MeV.

39. (*a*) uss;
(*b*) dss.

41. (*a*) Proton;
(*b*) $\overline{\Sigma}^-$;
(*c*) K^-;
(*d*) π^-;
(*e*) D_S^-.

43. $c\bar{s}$.

45.

p: u d u; μ^-; W^-; u d d: n; ν_μ

47. (*a*) 0.38 A;
(*b*) 1.0×10^2 m/s.

49. 2.1×10^9 m, 7.1 s.

51. (*a*) Possible, strong interaction;
(*b*) possible, strong interaction;
(*c*) possible, strong interaction;
(*d*) not possible, charge is not conserved;
(*e*) possible, weak interaction.

55. 64.

57. (*b*) 10^{29} K.

59. 798.7 MeV, 798.7 MeV.

61. 16 GeV, 7.8×10^{-17} m.

63. Some possibilities:

π^- ($\bar{u}$ d), π^0 ($\bar{u}$ u), π^0 ($\bar{u}$ u), π^- ($\bar{u}$ d)
$\bar{u}\,\bar{u}\,\bar{d}$ ($\bar{p}$), d d u (n), $\bar{u}\,\bar{d}\,\bar{u}$ ($\bar{p}$), u d d (n)

or [see Example 43–9b]

π^- ($\bar{u}$ d), π^0 ($\bar{d}$ d), π^0 ($\bar{d}$ d), π^- ($\bar{u}$ d)
$\bar{u}\,\bar{d}\,\bar{u}$ ($\bar{p}$), u d d (n), $\bar{d}\,\bar{u}\,\bar{u}$ ($\bar{p}$), u d d (n)

65. $v/c = 1 - (9.0 \times 10^9)$.

67.

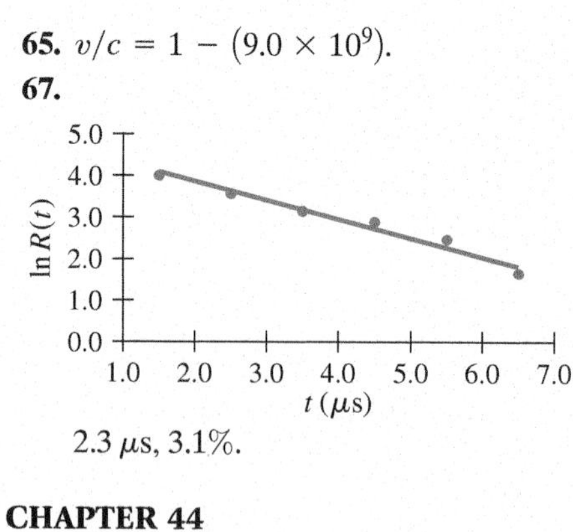

2.3 μs, 3.1%.

CHAPTER 44

1. 3.1 ly.

3. 0.050″, 20 pc.

5. Less than, a factor of 2.

7. 0.037.

9. 2×10^{-3} kg/m³.

11. −0.092 MeV, 7.366 MeV.

13. 1.83×10^9 kg/m³, 3.33×10^5 times.

15. $D_1/D_2 = 0.13$.

19. 540°.

21. 1.3×10^{-16} m.

23. 200 Mly.

25. (*a*) 656 nm;
(*b*) 659 nm.

27. 0.0589 *c*.

31. 1.1×10^{-3} m.

33. 6 nucleons/m³.

35. (*a*) 10^{-5} s;
(*b*) 10^{-7} s;
(*c*) 10^{-4} s.

37. (*a*) 6380 km, 20 km, 8.85 km;
(*b*) 700 : 2 : 1.

39. 8×10^9.

41. A: Temperature increases, luminosity stays the same, and size decreases;
B: Temperature stays the same, and luminosity and size decrease;
C: Temperature decreases, and luminosity and size increase.

43. 2×10^{28} N.

45. $d_{480}/d_{660} = 1.7$.

47. 2×10^{16} K, hadron era.

49. (*a*) 13.93 MeV;
(*b*) 4.7 MeV;
(*c*) 5.5×10^{10} K.

51. $R_{\text{min}} = GM/c^2$.

53. $\approx 10^{-15}$ s.

55. Venus, $b_{\text{Venus}}/b_{\text{Sirius}} = 16$.

57. $\dfrac{h^2}{4m_n^{8/3}\,GM^{1/3}}\left(\dfrac{9}{4\pi^2}\right)^{2/3}$.

Index

Note: The abbreviation *defn* means the page cited gives the definition of the term; *fn* means the reference is in a footnote; *pr* means it is found in a Problem or Question; *ff* means "also the following pages."

Photo Credits

Cover photos top left clockwise NASA/John F. Kennedy Space Center; Mahaux Photography/Getty Images, Inc.–Image Bank; The Microwave Sky: NASA/WMAP Science Team; Giuseppe Molesini, Istituto Nazionale di Ottica Florence

CO-36 Cambridge University Press; "The City Blocks Became Still Shorter" photo from page 4 of the book "Mr Tompkins in Paperback" by George Gamow. Reprinted with the permission of Cambridge University Press **36-1** Albert Einstein and related rights TM/© of The Hebrew University of Jerusalem, used under license. Represented exclusively by Corbis Corporation **36-15** Cambridge University Press; "Unbelievably Shortened" photo from page 3 of the book "Mr Tompkins in Paperback" by George Gamow. Reprinted with the permission of Cambridge University Press **CO-37** P. M. Motta & F. M. Magliocca/Science Photo Library/Photo Researchers, Inc. **37-10** Photo by Samuel Goudsmit, courtesy AIP Emilio Segrè Visual Archives, Goudsmit Collection **37-11** Education Development Center, Inc. **37-15a** Lee D. Simon/Science Source/Photo Researchers, Inc. **37-15b** Oliver Meckes/ Max Planck Institut Tubingen/Photo Researchers, Inc. **37-19b** Richard Megna/Fundamental Photographs, NYC **37-20** Wabash Instrument Corp./Fundamental Photographs, NYC. **CO-38** Institut International de Physique/American Institute of Physics/Emilio Segrè Visual Archives **38-1** Niels Bohr Archive, courtesy AIP Emilio Segrè Visual Archives **38-2** Photograph by F. D. Rasetti, courtesy AIP Emilio Segrè Visual Archives, Segrè Collection **38-4** Advanced Research Laboratory/Hitachi, Ltd. **CO-39** © Richard Cummins/Corbis **39-16** Paul Silverman/Fundamental Photographs, NYC **39-23** NIH/Photo Researchers, Inc. **39-24b** Philippe Plaily/ Photo Researchers, Inc. **CO-40** Intel Corporation Pressroom Photo Archives **40-41** © Alan Schein Photography/CORBIS All Rights Reserved **CO-41** Reuters Newmedia Inc./Corbis/Bettmann **41-3** French Government Tourist Office **41-8** Enrico Fermi Stamp Design © 2001 United States Postal Service. All Rights Reserved. Used with Permission from the U.S. Postal Service and Rachel Fermi **41-16** Fermilab Visual Media Services **CO-42** ITER International Fusion Energy Organization (IIFEO) **42-7** Archival Photofiles, Special Collections Research Center, University of Chicago Library **42-10** Igor Kostin/Corbis/Sygma **42-11** Novosti/ZUMA Press–Gamma **42-12** Corbis/Bettmann **42-19a** Robert Turgeon, Cornell University **42-19b** Courtesy of Brookhaven National Laboratory **42-20b** Custom Medical Stock Photo **42-24a** Martin M. Rotker **42-24b** Scott Camazine/Alamy Images **42-27** Colin Studholme **42-31b** Southern Illinois University/Peter Arnold, Inc. **42-33** Scott Camazine/Photo Researchers, Inc. **CO-43** Fermilab/ Science Photo Library/Photo Researchers, Inc. **43-1** Smithsonian Institution, Science Service Collection, photograph by Watson Davis/ Ernest Orlando Lawrence Berkeley National Laboratory, University of California, Berkeley, courtesy AIP Emilio Segrè Visual Archives, Fermi Film **43-3a/b** Fermilab Visual Media Services **43-5** CERN/ Science Photo Library/Photo Researchers, Inc. **43-6** ATLAS Experiment/CERN–European Organization for Nuclear Research **43-10a/b** Science Photo Library/Photo Researchers **43-12a** Brookhaven National Laboratory **43-13** Lawrence Berkeley National Laboratory **CO-44** WMAP Science Team/NASA Headquarters **44-1a** Space Telescope Science Institute **44-1b** Allan Morton/ Dennis Milon/Science Photo Library/Photo Researchers, Inc. **44-2c** NASA/Johnson Space Center **44-3** U.S. Naval Observatory Photo/NASA Headquarters **44-4** National Optical Astronomy Observatories **44-5a** Reginald J. Dufour, Rice University **44-5b** U.S. Naval Observatory **44-5c** National Optical Astronomy Observatories **44-9a/b** © Anglo-Australian Observatory **44-9c** The Hubble Heritage Team (AURA/STScI/ NASA) **44-9c (inset)** STScI/NASA/ Science Source/Photo Researchers, Inc. **44-15a** NASA Headquarters **44-22** NASA, ESA, S. Beckwith (STScI) and the HUDF Team **44-22 (inset)** NASA, ESA, R. Bouwens and G. Illingworth (University of California, Santa Cruz) **44-24** Alcatel-Lucent **44-26** Fredrik Persson/AP Wide World Photos **44-27** NASA/ WMAP Science Team

Table of Contents Photos p. iii left © Reuters/Corbis; **right** Agence Zoom/Getty Images **p. iv left** Ben Margot/AP Wide World Photos; **right** Kai Pfaffenbach/Reuters Limited **p. v** Jerry Driendl/Taxi/Getty Images **p. vi left** Richard Price/Photographer's Choice/Getty Images; **right** Frank Herholdt/Stone/Getty Images **p. viii** Richard Megna/Fundamental Photographs, NYC **p. ix left** Richard Megna/Fundamental Photographs, NYC; **right** Giuseppe Molesini, Istituto Nazionale di Ottica Florence **p. x** © Richard Cummins/Corbis **p. xi left** Fermilab/Science Photo Library/Photo Researchers, Inc.; **right** The Microwave Sky: NASA/WMAP Science Team **p. xvii** Christian Botting